5th International Conference on Electrical Engineering, Control, and Robotics (EECR 2019)

AA006039

IOP Conference Series: Materials Science and Engineering
Volume 533

Guangzhou, China
12 – 14 January 2019

ISBN: 978-1-5108-8737-4
ISSN: 1757-8981

Printed from e-media with permission by:

Curran Associates, Inc.
57 Morehouse Lane
Red Hook, NY 12571

Some format issues inherent in the e-media version may also appear in this print version.

This work is licensed under a Creative Commons Attribution 3.0 International Licence. Licence details: http://creativecommons.org/licenses/by/3.0/.

No changes have been made to the content of these proceedings. There may be changes to pagination and minor adjustments for aesthetics.

Printed by Curran Associates, Inc. (2019)

For permission requests, please contact the Institute of Physics at the address below.

Institute of Physics
Dirac House, Temple Back
Bristol BS1 6BE UK

Phone: 44 1 17 929 7481
Fax: 44 1 17 920 0979

techtracking@iop.org

Additional copies of this publication are available from:

Curran Associates, Inc.
57 Morehouse Lane
Red Hook, NY 12571 USA
Phone: 845-758-0400
Fax: 845-758-2633
Email: curran@proceedings.com
Web: www.proceedings.com

Table of contents

Volume 533

2019 The 5th International Conference on Electrical Engineering, Control and Robotics (EECR 2019)

12–14 January 2019, Guangzhou, China

Accepted papers received: 16 April 2019
Published online: Pl Online DATE

Preface

Power Electronics and Power Transmission Technology

Mechanical and Vehicle Engineering

Modern Electronic Technology and Application

Image Processing

IOP Conf. Series: Materials Science and Engineering **533** (2019) 011001 doi:10.1088/1757-899X/533/1/011001

PREFACE

The organizing committee of the 5th International Conference on Electrical Engineering, Control and Robotics (EECR 2019) aimed to facilitate interaction among participants for the latest progress and development in the fields of electrical engineering, control and robotics. Through this conference, the committee intends to enhance the sharing of individual experiences and expertise in electrical engineering, control and robotics with particular emphasis on the technical challenges associated with varied applications in these fields.

The conference was held in Xihu Hotel of SCUT, Guangzhou, China, January 12-14, 2019, organized by Sichuan Institute of Electronics with the support of South China University of Technology, University of Electronic Science and Technology of China and Southwest Minzu University. Despite the high quality of most of the submissions, the final proceedings of EECR 2019 includes 62 papers, which were presented at the conference and that were selected after a thorough reviewing process. The authors of these papers come from different countries from Europe, Australia and Asia etc..

The contributions of the technical program committee and the referees are deeply appreciated. Most of all, we would like to express our sincere thanks to the authors for submitting their most recent works and the organizing committee for their enormous efforts to turn this event into a smoothly running meeting.

We sincerely hope that this publication will prove to be an important resource for the scientific community.

Prof. Chun-Yi Su
Concordia University, Canada
Jan. 18, 2019

Content from this work may be used under the terms of the Creative Commons Attribution 3.0 licence. Any further distribution of this work must maintain attribution to the author(s) and the title of the work, journal citation and DOI.
Published under licence by IOP Publishing Ltd

IOP Conf. Series: Materials Science and Engineering **533** (2019) 011002 doi:10.1088/1757-899X/533/1/011002

Conference Committee

Conference Chairs

Chun-Yi Su, Concordia University, Canada
May Huang, International Technological University, USA

Program Chairs

Hailong Pei, South China University of Technology, China
Jinlong Lin, Peking University, China
Quang Ha, University of Technology Sydney, Australia
Alvin Y. Chua, De La Salle University, Philippines
Xuelin Lei, East China University of Science and Technology, China

Local Organizing Chairs

Shilu Dai, South China University of Technology, China
Chenguang Yang, South China University of Technology, China

Technical Committees

Gheorghe-Daniel Andreescu, Politehnica University of Timisoara, Romania
Hasan Çimen, Afyon Kocatepe University, Turkey
Long Jin, Lanzhou University, China
Dong Huang, Chinese Academy of Sciences, China
Samuel P. Kozaitis, Florida Institute of Technology, USA
Chih-Jer Lin, National Taipei University of Technology, Taiwan
Thelma D. Palaoag, University of the Cordilleras, Philippines
Mahir Dursun, Gazi University, Turkey
Konstantinos Parsopoulos, University of Ioannina, Greece
Ray-Hwa Wong, Hwa-Hsia University of Technology, Taiwan
Zahid Halim, Ghulam Ishaq Khan Institute of Engineering Sciences and Technology, Pakistan
Najmi Ghani Haider, NED University of Engineering and Technology, Pakistan
Kardi Teknomo, Ateneo de Manila University, Philippines
Ning Sun, Nankai University, China
Yongqin Fan, Xi'an University of Posts and Telecommunications, China
Cheung-Chieh Ku, National Taiwan Ocean University, Taiwan
Svetlana Vasileva-Boyadzhieva, International College - Dobrich, Bulgaria
Fadzilah Siraj, Universiti Utara Malaysia, Malaysia
Sheila Abaya, University of East-Caloocan, Philippines
B. B. Biswal, National Institute of Technology, India
Jan Muhammad, BUITEMS, Pakistan
KWOK Lam For, City University of Hong Kong
Cengiz ÖRENCİK, Beykent University, Turkey
Eric Chen, Uber, USA
Jennifer L. Santos, Centro Escolar University, Philippines
Emese Gincsainé Szádeczky-Kardoss, Budapest University of Technology and Economics, Hungary
Wen-Jyi Hwang, National Taiwan Normal University, Taiwan
Abdel-Hameed Badawy, New Mexico State University, USA
Jukkrit Kluabwang, Rajamangala University of Technology Lanna Tak, Thailand
Simon K. S. Cheung, The Open University of Hong Kong, Hong Kong
Muhammad Murtaza Yousaf, Punjab University, Pakistan
Waqar ul Qounain, Punjab University, Pakistan
Faisal Bukhari, Punjab University, Pakistan

Content from this work may be used under the terms of the Creative Commons Attribution 3.0 licence. Any further distribution of this work must maintain attribution to the author(s) and the title of the work, journal citation and DOI.
Published under licence by IOP Publishing Ltd

IOP Conf. Series: Materials Science and Engineering **533** (2019) 011003 doi:10.1088/1757-899X/533/1/011003

Peer review statement

All papers published in this volume of *IOP Conference Series: Materials Science and Engineering* have been peer reviewed through processes administered by the proceedings Editors. Reviews were conducted by expert referees to the professional and scientific standards expected of a proceedings journal published by IOP Publishing.

Content from this work may be used under the terms of the Creative Commons Attribution 3.0 licence. Any further distribution of this work must maintain attribution to the author(s) and the title of the work, journal citation and DOI.
Published under licence by IOP Publishing Ltd

Disaster Risk Assessment of Transmission Lines Based on TOPSIS

Taowei Chen[1], Ling Zhu[1] , Qiao Xia[2], Honglei Deng[2] and Chen Zhou[2]

[1]Huizhou Power Supply Bureau of Guangdong Power Grid Corporation , 516001 Huizhou, China
[2]Shcool of Electric Power, South China University of Technology, 510640 Guangzhou, China

xq734956725@163.com

Abstract. It is of great significance to ensure the safe and reliable operation of the power grid by effectively evaluating the disaster risk of the power grid lines and arranging the sequence of line guard reasonably. A disaster risk assessment method of transmission lines based on TOPSIS is proposed in this paper where six typical natural disasters of the line are considered, the comprehensive risk values of different lines under different disasters combined with the hazards of line accidents is determined, the risk assessment of multiple transmission lines are realized, and its feasibility is verified by applying it to multiple transmission lines in a certain area of China Southern power grid.

1. Introduction

Extreme natural disasters are the most important external threats to the power grid, especially overhead transmission lines. According to relevant documents, the proportion of power grid accidents caused by natural disasters such as lightning strikes, typhoons, ice storms, bird pests, mountain fires, mudslides, and mountain collapses is increasing year by year in China. These disasters not only bring great damage to power facilities, but also bring great challenges to the meteorological disaster prevention and mitigation work of power system[1-2].

In order to cope with the damage caused by extreme natural disasters to overhead transmission lines and strengthen the construction and protection of power systems, many research institutions and provincial grid power companies at home and abroad have carried out research on the operational risks of transmission lines under natural disasters. To overcome the shortcomings of Lightning distribution map and quantitative calculation method of lightning stroke, a differential and multi-level lightning risk assessment model of "tower - tower section - line" is brought forward[3]. Song Xiaozhe et al. proposed a risk assessment method for power grid transient stability under typhoon weather conditions[4]. Tom B et al. build a power system icing risk assessment model by using the fault tree analysis method, which can evaluate the disconnection of the transmission line and the collapse of the tower effectively[5]. Liu Mingjun et al. established a risk assessment model for the fault trip of the transmission line by computing the probability of line failure under the condition of mountain fire and the occurrence probability of mountain fire on the line corridor[6]. Xiong Xiaofu et al. put forward a transmission line failure rate model that taking into account different meteorological grades and meteorological factors, and proposed a risk analysis method of transmission line based on grey fuzzy

Content from this work may be used under the terms of the Creative Commons Attribution 3.0 licence. Any further distribution of this work must maintain attribution to the author(s) and the title of the work, journal citation and DOI.
Published under licence by IOP Publishing Ltd

theory[7]. Aiming at the shortcomings of current study of mutual impact on varied meteorological disasters on the overhead transmission line, Deng Honglei et al. introduced a comprehensive risk assessment method of transmission line based on AHP and entropy weight method by considering the risk weights, risk probability and risk consequences of various meteorological disasters[8].

In view of the above assessment methods, some studies considered the effects of a natural disaster alone and while the simultaneous effects of various disasters is neglected. Some studies only focused on the failure probability of transmission lines and ignored the consequences of failure so that the assessment results are not comprehensive.

Therefore, in order to remedy the deficiency of the above methods, a disaster risk assessment method of transmission lines based on TOPSIS is proposed in this paper where not only the impact of six natural disasters on transmission lines over the years, but also the hazards of line accidents determined by the load reduction ratio of power grid are considered, a disaster risk assessment model for multiple transmission lines is established, the rationality and feasibility of the proposed method are verified by comparing different methods and analysing the actual situation.

2. Risk assessment framework based on TOPSIS

In this paper, six typical natural disasters are selected including lightning strikes, typhoons, ice storms, bird pests, mountain fires, and external force destruction. Risk assessment framework is shown in Figure 1. The combination of disaster trip-out rate and accident levels of transmission lines is used as assessment system. Combined with the degree of hazard under different disasters, multiple transmission lines are evaluated according to the TOPSIS method.

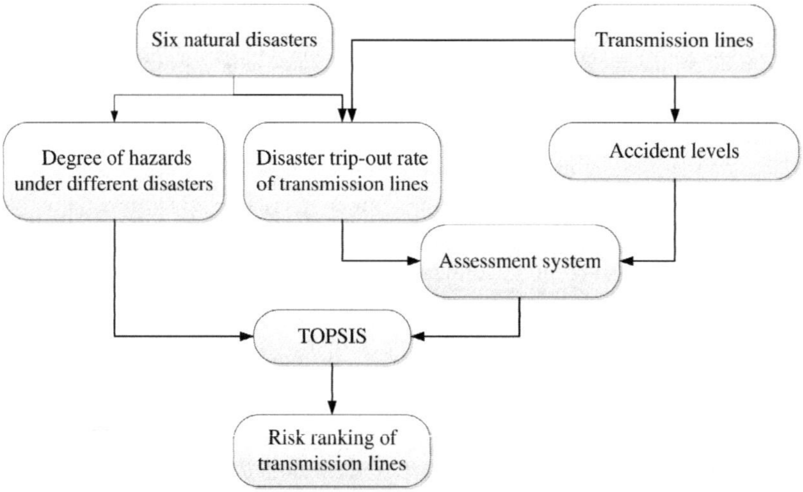

Figure 1. Risk assessment framework based on TOPSIS.

According to the regulations on emergency disposal, electric power safety accidents are classified into special major accidents, serious accidents, major accidents and general accidents based on the extent to which they affect the safe and stable operation of the power system or affect the normal supply of electric power (thermal power). Therefore, the accident of transmission lines is divided into the above four levels in this paper.

For the convenience of calculations, according to *the Regulations on Power Grid Operation Safety Risk Management of China Southern Power Grid Corporation*, the accident level is quantified and the quantified score table is shown in Table 1.

Table 1. Quantification of line accident levels.

S/N	Accident levels	Symbol	Quantized score
1	special major accidents	I	4000~8000
2	serious accidents	II	2000~2400
3	major accidents	III	400~600
4	general accidents	IV	200~250

2.1. Risk assessment method based on TOPSIS

The TOPSIS method which is presented by Hwang and Yoon is a MCDA (multi criteria decision analysis) method that is used to identity solutions from a finite set of alternatives. This method attempts to choose alternatives that simultaneously have the shortest distance from the positive ideal solution (the solution that maximizes benefit criteria and minimizes cost criteria) and the farthest distance from the negative ideal solution (the solution that maximizes the cost criteria and minimizes the benefit criteria). TOPSIS has been identified as a MCDA method which is suitable for problems with large number of criteria and alternatives which are provided with numerical or quantitative data. The method consists of the following steps:

1) Form decision matrix.

For multi-objective decision making system, m as the number of targets to be assessed, n as the number of assessment indicators, then the decision matrix is $R^* = (r_{ij}^*)_{m\times n}$.

2) Establish standardized decision matrix.

A standardized decision matrix $V = (v_{ij})_{m\times n}$ is obtained by dimensionless processing of decision matrix.

$$v_{ij} = (r_{ij}^* - r_{\min(j)}^*) / (r_{\max(j)}^* - r_{\min(j)}^*) \tag{1}$$

Where v_{ij} is the element of standardized matrix, and $r^*_{max(j)}$ and $r^*_{min(j)}$ are the maximum and minimum of the j th indicator respectively.

3) Establish weighted decision matrix.

The weighted decision matrix $X = (x_{ij})_{m\times n}$ is obtained by multiplying the weights of each indicator with the element of standardized decision matrix.

$$x_{ij} = w_j \cdot v_{ij} \tag{2}$$

4) Calculate the distances between the positive ideal solution and the negative ideal solution.

$$\begin{cases} S_j^+ = \max\limits_{1\le i\le m}\{x_{ij}\} \\ S_j^- = \min\limits_{1\le i\le m}\{x_{ij}\} \end{cases} \tag{3}$$

$$\begin{cases} S_{d_i}^+ = \sqrt{\sum_{j=1}^{n}(S_j^+ - x_{ij})^2} \\ S_{d_i}^- = \sqrt{\sum_{j=1}^{n}(S_j^- - x_{ij})^2} \end{cases} \tag{4}$$

5) Calculate close-degree.

$$\beta_i = S_{d_i}^- / (S_{d_i}^+ + S_{d_i}^-) \tag{5}$$

IOP Conf. Series: Materials Science and Engineering **533** (2019) 012001 doi:10.1088/1757-899X/533/1/012001

Where β_i is close-degree of the i th target. We can judge the order of the decision target according to close-degree, The larger the value of β_i, the closer to the positive ideal solution, that is, the better the target.

2.2 . Indicator weights of multi-objective systems

2.2.1 Subjective weight of indicators. Subjective weight calculation is a method that reflects the experience or subjective preference of decision makers. The commonly used subjective weight calculation methods include analytic hierarchy process(AHP) , improved analytic hierarchy process (IAHP), G1 and Delphi method. In this paper, G1 and IAHP are used to determine the subjective weight.

G1 is a method to determine the weights of indicators by calculating the relative importance between indicators. its value is determined by experiences of experts[9]. The method consists of the following steps:

1) Determine the relative order of indicators.

The assessment system consist of n indicators, that is, $\{a_1, a_2, \cdots, a_n\}$. The order of importance of n indicators is determined according to the assessment rules , such as $\{ a_1^* \succ a_2^* \succ \cdots \succ a_n^* \}$.

2) Calculate the relative importance between adjacent indicators.

$$r_j = a_{j-1}^* / a_j^*, \quad j = 2, 3, \cdots, n \tag{6}$$

Where r_j is determined by the experts according to the parameter table of G1 method.

3) Calculate the subjective weight.

$$w_n^* = (1 + \sum_{k=2}^{n} \prod_{j=k}^{n} r_j)^{-1} \tag{7}$$

$$w_{j-1}^* = r_j w_j^* \tag{8}$$

The AHP method is a systemic analysis method proposed by *Saaty* in the mid-1970s. It is an approach to determine the relative importance of a set of activities in a multi-criteria decision problem and is possible to incorporate judgments on intangible qualitative criteria alongside tangible quantitative criteria. The improved analytic hierarchy process (IAHP) is established on the basis of analytic hierarchy process. This method makes full use of the subjective experience of the decision maker, divides the various factors into different levels to quantify and ranks the advantages and disadvantages of the decision scheme. At the same time, the consistency checking error caused by the unreasonable value of the judgment matrix element is avoided[10].

2.2.2. Objective weight of indicators. Objective weight calculation method is a quantitative analysis method, which calculates the indicator weight according to the sample data by a certain mathematical method. The commonly used objective weight calculation methods include entropy method, grey correlation analysis method and the coefficient of variation (CV). In this paper, entropy method and CV are used to determine the objective weight.

In order to facilitate the calculation of objective weights, a primitive data matrix $R = (r_{ij})_{m \times n}$ is first established, which assumes that there are m targets to be assessed and n indicators in the assessment system. $S = (s_{ij})_{m \times n}$ is a dimensionless matrix of the primitive data matrix and obtained by extreme value method.

Entropy method was first introduced by *Claude Elwood Shannon* into information theory. At present, it has been widely applied in engineering technology, social economy and other fields. The basic idea of entropy method is to determine objective weights according to the variability of indicators.

Generally speaking, the smaller the information entropy of an indicator, indicating that the greater the degree of variation of the indicator value, the more information it provides, the greater the role it plays in the comprehensive evaluation, the greater its weight [11].

The coefficient of variation (CV) is a widely used scaleless measure of variability in many disciplines. CV is estimated by the ratio of the sample standard deviation over the sample mean[12]. The method consists of the following steps:

1) Calculate the mean and standard deviation of each assessment indicator.

$$\mu_j = \sum_{i=1}^{m} s_{ij} / m \tag{9}$$

$$\sigma_j = \sqrt{\frac{1}{m}\sum_{i=1}^{m}(s_{ij} - \mu_j)^2} \tag{10}$$

2) Calculate the coefficient of variation.

$$v_j = \sigma_j / \mu_j \tag{11}$$

3) Calculate objective weight.

$$w_j = v_j / \sum_{j=1}^{n} v_j \tag{12}$$

2.2.3. Comprehensive weight of indicators. In order to make the indicator weight not only reflect the subjective experience of the decision maker but also conform to the objective facts, this paper integrates the above subjective and objective weights and obtains a comprehensive optimization indicator weight[13]. The relative importance coefficients of subjective weight and objective weight are γ and 1-γ. The optimization model is established as follows.

$$\min L = \sum_{j=1}^{n}\left[\gamma\sum_{a=1}^{2}(w_j - w_{aj})^2 + (1-\gamma)\sum_{b=3}^{4}(w_j - w_{bj})^2\right] \tag{13}$$

Where w_j , w_{aj} and w_{bj} is the j th weight coefficient of the optimization indicator weight vector W, subjective weight vector W_a and objective weight vector W_b respectively.

The constraints of eq. (13) are as follows.

$$\sum_{j=1}^{n} w_j = 1,\ 0 \le w_j \le 1 \tag{14}$$

The subjective and objective expectations of each indicator weight are calculated based on the moment estimation theory.

$$\begin{cases} E(w_j) = (w_{1j} + w_{2j}) / 2 \\ E(w'_j) = (w_{3j} + w_{4j}) / 2 \end{cases} \tag{15}$$

Calculate the subjective weight coefficient of each indicator.

$$\gamma_j = E(w_j) / (E(w_j) + E(w_j')) \tag{16}$$

Calculate the relative importance coefficient of subjective weight.

$$\gamma = \sum_{j=1}^{n} \gamma_j / n \tag{17}$$

Calculate the comprehensive weight of indicators according to eq. (13) ~ (17).

3. An application of proposed model

The model is applied to the disaster risk assessment of 8 transmission lines in a certain area of China Southern Power Grid. The voltage level of the transmission line to be evaluated is 220kV. In this paper, lightning strikes, typhoons, ice storms, bird pests, mountain fires, and external force destruction are used as indicators to be evaluated, set up C={C_1, C_2, C_3, C_4, C_5, C_6} as indicators set to be evaluated. Taking 8 overhead transmission lines as targets to be evaluated, set up M = {M_1, M_2, M_3, M_4, M_5, M_6, M_7, M_8} as target set to be evaluated.

3.1. Calculation of comprehensive indicator weights

3.1.1. Calculation of subjective weights. According to the degree of risk and harm caused by six kinds of disasters to transmission lines, the relative order of indicators is established, where lightning strikes > ice storms > mountain fires > external force destruction > typhoons > bird pests. The relative importance of indicators is determined through expert experience, where r_2= 1.6, r_3 = 1.4, r_4 = 1.2, r_5= 1.2, r_6= 1.4. Reference eq.(6) ~ (8) calculates the weights of indicators in relative order, where w_1^* = 0.3361, w_2^* = 0.2101, w_3^* = 0.1501, w_4^* = 0.1251, w_5^* =0.1042, w_6^* = 0.0744. Therefore, the subjective weight vector W_1 determined by G1 method is as follows, W_1= (0.3361, 0.1042, 0.2101, 0.0744, 0.1501, 0.1251).
According to the relative order of indicators mentioned above, we can calculate the subjective weight based on IAHP, and its weight vector W_2 is as follows. W_2 = (0.4679, 0.0384, 0.2613, 0.0214, 0.1385, 0.0724).

3.1.2 Calculation of objective weights. The trip distributions of all 220kV transmission lines in a certain area of the South China Power Grid in recent 8 years is shown in Table 2.

Table 2. A regional trip distributions of transmission lines in recent eight years.

	C_1	C_2	C_3	C_4	C_5	C_6
2009	33	3	12	1	2	6
2010	27	6	22	1	4	3
2011	36	1	8	2	1	1
2012	18	4	5	1	1	2
2013	29	3	15	3	7	2
2014	22	2	28	1	3	11
2015	13	2	23	2	3	1
2016	34	7	7	2	1	9

We can establish the primitive data matrix according to table 2. In order to facilitate calculation of objective weights, the maximum value method is applied to dimensionless processing of the primitive data matrix, and the matrix S is obtained.

$$S = \begin{bmatrix} 0.9167 & 0.4286 & 0.4286 & 0.3333 & 0.2857 & 0.5455 \\ 0.7500 & 0.8571 & 0.7857 & 0.3333 & 0.5714 & 0.2727 \\ 1.0000 & 0.1429 & 0.2857 & 0.6667 & 0.1429 & 0.0909 \\ 0.5000 & 0.5714 & 0.1786 & 0.3333 & 0.1429 & 0.1818 \\ 0.8056 & 0.4286 & 0.5357 & 1.0000 & 1.0000 & 0.1818 \\ 0.6111 & 0.2857 & 1.0000 & 0.3333 & 0.4286 & 1.0000 \\ 0.3611 & 0.2857 & 0.8214 & 0.6667 & 0.4286 & 0.0909 \\ 0.9444 & 1.0000 & 0.2500 & 0.6667 & 0.1429 & 0.8182 \end{bmatrix}$$

We can calculate the objective weight vector based on entropy weight method, and its weight vector W_3 is as follows, W_3 = (0.0463, 0.1545, 0.1475, 0.0905, 0.2287, 0.3326).
According to eq.(9) ~(12), the objective weight vector W_4 based on CV is calculated, where W_4 = (0.0871, 0.1666, 0.1593, 0.1289, 0.2102, 0.2479).

3.1.3. Calculation of comprehensive weights. The subjective weight vectors W_1, W_2 and objective weight vectors W_3, W_4 obtained above are brought into eq.(13) ~ (17) to calculate the comprehensive weight vector W. That is, W = (0.2190, 0.1200, 0.1908, 0.0816, 0.1853, 0.2033).

3.2. Disaster risk assessment of transmission lines based on TOPSIS

The multi-year trip distribution of 8 transmission lines to be evaluated is shown in Table 3. Length of transmission line to be evaluated and accident levels determined according to the load reduction ratio of the power grid are shown in Table 4.

Table 3. Trip distributions of overhead transmission lines to be evaluated in recent years.

	C_1	C_2	C_3	C_4	C_5	C_6
M_1	8	0	2	0	0	1
M_2	7	1	6	0	1	0
M_3	10	0	4	1	0	1
M_4	6	1	0	0	0	1
M_5	5	0	3	1	2	0
M_6	5	0	6	0	1	2
M_7	3	1	4	0	0	0
M_8	9	1	1	0	0	1

Table 4. Length and accident levels of transmission lines.

	Length（km）	Accident levels	Symbol
M_1	30.815	general accidents	IV
M_2	13.511	serious accidents	II
M_3	47.685	general accidents	IV
M_4	16.255	major accidents	III
M_5	24.146	serious accidents	II
M_6	44.860	general accidents	IV
M_7	25.788	general accidents	IV
M_8	36.482	major accidents	III

We can calculate the disaster trip-out rate of 8 transmission lines by using the data of Tables 3 and 4, and the decision matrix of the assessment system R^* is established by multiplying the quantized score in Table 1.

$$R^{*}=\begin{bmatrix}0.0519 & 0.0000 & 0.0130 & 0.0000 & 0.0000 & 0.0065\\ 1.0362 & 0.1480 & 0.8882 & 0.0000 & 0.1480 & 0.0000\\ 0.0419 & 0.0000 & 0.0168 & 0.0042 & 0.0000 & 0.0042\\ 0.1476 & 0.0246 & 0.0000 & 0.0000 & 0.0000 & 0.0246\\ 0.4141 & 0.0000 & 0.2485 & 0.0828 & 0.1657 & 0.0000\\ 0.0223 & 0.0000 & 0.0267 & 0.0000 & 0.0045 & 0.0089\\ 0.0233 & 0.0078 & 0.0310 & 0.0000 & 0.0000 & 0.0000\\ 0.0987 & 0.0110 & 0.0110 & 0.0000 & 0.0000 & 0.0110\end{bmatrix}$$

A standardized decision matrix V is established according to eq. (1).

$$V=\begin{bmatrix}0.0292 & 0.0000 & 0.0146 & 0.0000 & 0.0000 & 0.2642\\ 1.0000 & 1.0000 & 1.0000 & 0.0000 & 0.8932 & 0.0000\\ 0.0193 & 0.0000 & 0.0189 & 0.0507 & 0.0000 & 0.1707\\ 0.1236 & 0.1662 & 0.0000 & 0.0000 & 0.0000 & 1.0000\\ 0.3864 & 0.0000 & 0.2798 & 1.0000 & 1.0000 & 0.0000\\ 0.0000 & 0.0000 & 0.0301 & 0.0000 & 0.0272 & 0.3618\\ 0.0010 & 0.0527 & 0.0349 & 0.0000 & 0.0000 & 0.0000\\ 0.0754 & 0.0743 & 0.0124 & 0.0000 & 0.0000 & 0.4472\end{bmatrix}$$

To combine with the degree of hazard under different disasters, we need consider the comprehensive weight of indicators mentioned above. A weighted decision matrix X is established based on eq. (2).

$$X=\begin{bmatrix}0.0064 & 0.0000 & 0.0028 & 0.0000 & 0.0000 & 0.0537\\ 0.2190 & 0.1200 & 0.1908 & 0.0000 & 0.1655 & 0.0000\\ 0.0042 & 0.0000 & 0.0036 & 0.0041 & 0.0000 & 0.0347\\ 0.0271 & 0.0199 & 0.0000 & 0.0000 & 0.0000 & 0.2033\\ 0.0846 & 0.0000 & 0.0534 & 0.0816 & 0.1853 & 0.0000\\ 0.0000 & 0.0000 & 0.0057 & 0.0000 & 0.0050 & 0.0736\\ 0.0002 & 0.0063 & 0.0067 & 0.0000 & 0.0000 & 0.0000\\ 0.0165 & 0.0089 & 0.0024 & 0.0000 & 0.0000 & 0.0909\end{bmatrix}$$

According to eq. (3) ~ (4), we calculate the distance between the evaluated target and the positive ideal solution and the negative ideal solution, and get S_d^+ = (0.3979, 0.2200, 0.4054, 0.3525, 0.3044, 0.3906, 0.4207, 0.3775), S_d^- = (0.0542, 0.3552, 0.0354, 0.2061, 0.2258, 0.0739, 0.0092, 0.0929).

Finally, according to eq.(5), we calculate close-degree of each target to get β = (0.1198, 0.6176, 0.0803, 0.3689, 0.4259, 0.1592, 0.0214, 0.1974)

The value of β indicates that the order of importance of the 8 assessment targets is $M_2 > M_5 > M_4 > M_8 > M_6 > M_1 > M_3 > M_7$. Therefore, the relative magnitude of disaster risk is ranked as follows, line 2 > line 5 > line 4 > line 8 > line 6 > line 1 > line 3 > line 7.

According to the evaluation results above, the power grid operation management departments need to focus on strengthening the risk prevention of line 2, line 5 and line 4. For the convenience of analysis and explanation, the disaster risk ranking results and comparison charts of transmission line under two assessment standards are shown in Table 5 and Figure 2 respectively.

Table 5. Ranking results of disaster risk for transmission lines under two assessment standards.

Line	Close-degree	Total trip-out rate	Risk ranking	
			TOPSIS	Traditional model
Line 1	0.119 8	0.000 357	6	4
line 2	0.617 6	0.001 110	1	1
line 3	0.080 3	0.000 336	7	5

line 4	0.368 9	0.000 492	3	2
line 5	0.425 9	0.000 456	2	3
line 6	0.159 2	0.000 312	5	7
line 7	0.021 4	0.000 310	8	8
line 8	0.197 4	0.000 329	4	6

As can be seen from Table 5 and Figure 2, the results of risk assessment based on total trip-out rate and the model proposed in this paper are inconsistent, and the risk ranking of transmission lines with the highest total trip-out rate may be improved or reduced after considering accident levels and the degree of hazard under different disasters. Considering the distribution of different disasters, the total trip-out rate of line 1 is higher than that of line 6, but line 6 suffers more times from ice storms and external force destruction. Compared with other disasters, the damage degree of ice storms and external force destruction is higher, so the risk of line 6 is higher. This is more in line with the actual situation. Considering accident levels of transmission lines, the total trip-out rate of line 4 is higher than line 5, but the load reduction caused by line 5 accident is larger, resulting in a wider range of blackouts, so the result of line 5 with higher risk is more reasonable. Based on the above analysis, compared with the traditional model, the model proposed in this paper are more realistic and feasible.

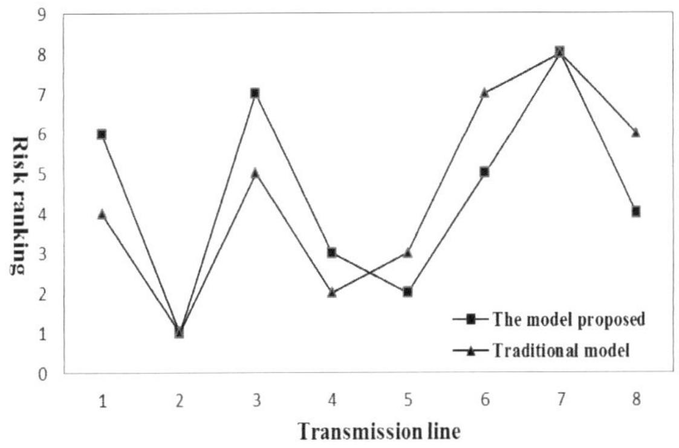

Figure 2. Comparison of disaster risk for transmission lines under two assessment standards

4. Conclusions

Developing comprehensive risk assessment of transmission lines is an important work to ensure the safe and reliable operation of power systems. In view of the influence of various natural disasters on transmission lines, a disaster risk assessment method based on TOPSIS for transmission lines is proposed in this paper. This method considers six typical natural disasters, accident levels of transmission lines and the degree of hazard under different disasters. Compared with the traditional method, the method proposed in this paper is more reasonable and scientific, more consistent with the actual comprehensive risk ranking of transmission lines. Application of this model provide an effective reference for the early prediction of line accidents and the formulation of safety plans.

References

[1] Zhou Ning, Xiong Xiaofu. Electrical and Meteoro-logical Technology and Its Application[M]. BeiJing: China Electric Power Press [M], 2015： 3.
[2] Duan Tao, Luo Yi, Shi Lin, et al. A real-time fault probability assessment model of transmission line considering meteorological factors[J]. Power System Protection and Control, 2013, 41(15): 59-67.

[3] Deng Honglei, Li Shuwen, Dai Dong, Lightning risk assessment of overhead transmission line based on hierarchical and differentiation[J]. Power System Protection and Control, 2016, 44(04): 69-75.
[4] Song Xiaozhe, Wang Zhen, Gan Deqiang, et al. Transient stability risk assessment of power grid under typhoon weather[J]. Power System Protection and Control, 2012, 40(24): 1-8
[5] Tom B, Gutwin P. Application of BCTC standardized risk estimation model to assess risk due to ice storms[C] //8th International Conference on Probabilistic Methods Applied to Power Systems. Ames: IowaStateUniversity, 2004: 970-974.
[6] Zhao Mingjun, Shao Zhouce, Shang Guantie, et al. Study on models and methods of risk assessment of transmission line fault caused by fire[J]. Power System Protection and Control, 2016, 44 (06):82-89.
[7] Xiong Xiaofu, Wang Weijun,Yu Yan, et al. Risk Analysis Method for Transmission Line Combining of Various Meterorological Factors[J]. Proceedings of the Chinese Society of Universities for Electric Power System and Automation, 2011, 23 (06): 11- 15+28.
[8] Deng Honglei, Dai Dong, Li Shuwen. Compreh-ensive operation risk assessment of overhead transmission line based on hierarchical analysis-entropy weight method[J]. Power System Protection and Control , 2017, 45(01): 28-34.
[9] Li Jun, Li Jiguang, Yao Jiangang, et al. Application of Attribute Recognition and G1-Entropy Method in Assessment of Power Quality[J]. Power System Technology, 2009, 33 (14): 56-61.
[10] Zhang Xiaoming,Wu Chaojun,Gan Yan, et al. Risk Assessment Model of Lightning Disturbance for Transmission Lines based on Improved AHP[J]. Electric Power Construction, 2012, 33 (08): 35- 39.
[11] Fei Zhicong. Research on entropy weight-analytic hierarchy process and grey-analytic hierarchy process[D]. Tianjin: TianjinUniversity, 2009.
[12] Guo Haiyang, Chen Haozhong, Ma Zeliang, et al. Comprehensive Assessment of Power Quality Based on Probability Theory and Combination Weighting Method [J]. Smart Grid, 2017, 5(06): 524-528.
[13] Shen Yangwu, Peng Xiaotao, Shi Tongqing, et al. A Grey Comprehensive Assessment Method of Power Quality Based on Optimal Combination Weight[J]. Automation of Electric Power Systems, 2012, 36 (10): 67-73.

Research on Distributed Protection in Switch Stations Based on GOOSE

Xin Jiang[1], Baoji Yun[1]

[1]Electrical Engineering, Electrical Engineering Department, 710054 Xi'an University of Science and Technology, CHINA

E-mail: 912819538@qq.com

Abstract. The application of IEC61850 standard and the development of smart terminal unit (STU) technology provide a powerful platform for secondary protection research of distribution network. Aiming at the problems of override trip in the protection of the incoming line and feeders on traditional 10kV switching stations, a scheme combining distributed voltage loss protection and overcurrent protection is proposed. The smart terminal unit collects and processes the operation data, realizes data exchange through GOOSE network communication, and cooperates to realize the protection function of the switch stations. In this paper, a protection logic diagram based on the GOOSE mechanism is constructed, which realizes the protection function of feeder lines fault lockout incoming line breakers and the bus line fault lockout busbar automatic transfer switch.

1. Preface

The traditional protection technology for switch stations can be divided into local control methods (such as current protection) and centralized control methods according to the protection device layout structure. The local control method only uses the information of the installation location of the protection device, is easy to implement, and moving fast, but the information used is limited and the protection performance is not perfect. The centralized control mode is mainly based on the primary station, which can integrate global information, optimize control, and improve protection performance. However, there are many links involved, low reliability, and slow protection response. On the other hand, from the perspective of the protection principle, the traditional three-stage current protection is mainly used for the switch stations. Since the three-stage protection is achieved according to the current setting grading and the action time grading, it has its inherent defects, and it is prone to problems such as override trip. In the literature 1, the ultra-short incoming lines and feeder lines current protection of the switch stations in the urban power grid is improved, but the information used for the protection implementation is limited to the local information of the protection installation, and the protection performance is not perfect[1]. In the literature 2, the incoming lines and feeder lines protection of the switch stations based on the GOOSE mechanism is studied, and the GOOSE action logic diagram is constructed. However, the protection implementation principle is still based on the traditional three-stage current protection[2]. Under different operating modes and different fault types of the distribution network, the protection current value has a problem of setting difficulty.

Content from this work may be used under the terms of the Creative Commons Attribution 3.0 licence. Any further distribution of this work must maintain attribution to the author(s) and the title of the work, journal citation and DOI.
Published under licence by IOP Publishing Ltd

With the promulgation and application of the IEC61850 standard, it laid the foundation for the application of distributed current protection technology. The IEC61850 standard enables intelligent terminals to implement interworking between devices according to a unified data model, and also enables interoperability between intelligent terminals of different manufacturers. The automation system based on IEC61850 standard can realize GOOSE function. The intelligent terminal can exchange real-time data through GOOSE network communication quickly. The real-time data transmission time is generally less than 10ms, which can guarantee the response speed of distributed protection technology[3,4].

In addition, the rapid development of smart terminal unit (STU) technology provides a realistic path for the application of distributed protection technology. STU uses high-performance digital signal processor (DSP), microprocessor (MCU), large-scale field programmable logic array (FPGA), etc., with powerful data storage and data processing capabilities[5]. Its main functions include: collecting and processing operational data, exchanging data with relevant STUs in distributed protection, and then obtaining fault information, acting on circuit breaker tripping or issuing warning information to realize protection control function. STU's interface circuit design meets modularity and standardization and can be flexibly selected according to engineering demands.

2. Structure and protection of the switch station

In order to save the 10kV substation outlet corridor and solve the problem of insufficient spacing of the power distribution equipment, the switch station is widely used at the exit of the 10kV substation. It generally adopts the main bus form of a single bus bar or a single bus bar segment, and generally two power sources are incoming.

When two power supplies are used, the power supply comes from different substations or different busbars of the same substation. The number of feeders is from 3 to 5. Figure 1 shows the main wiring diagram of a typical single busbar section switch station. When the bus coupling switch is closed, the switch station is in the parallel running state, and when the bus coupling switch is disconnected, switch station is in the divided operating state. When the switch station is in normal operation, the bus-coupled switch is in the open state, and the bus-coupled switch is configured with the busbar automatic transfer switch (BATS) device.

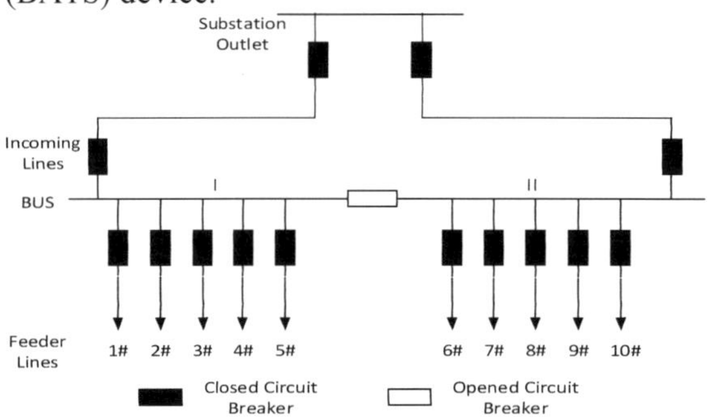

Figure 1. The Structure of Switch Station

The functions to be implemented by the protection of switch station are:

a) When the incoming line fault occurs, the upper substation outlet circuit breaker trips. After the reclosing, if it is a permanent fault, it is put into the BATS device and is powered by the standby incoming line.

b) When the busbar fault occurs in the switch station, it will be tripped by the incoming circuit breaker, and the BATS device will be blocked to realize the isolation of the fault area and ensure the power supply in the non-faulty area.

c) When the feeders fault occurs, the incoming line protection is blocked to prevent the override trip, and the faulty circuit is tripped by the action of the circuit breaker on the feeder line branch.

3. Distributed protection principle of switch station

The distributed protection system consists of a substation outlet circuit breaker, an incoming line circuit breaker, a feeder line branch circuit breaker, a smart terminal unit (STU), and a GOOSE communication network for exchanging fault detection information between STUs. For the protection system of the switch station system, the traditional protection system focuses on the combination of current setting difference and time level difference[3], while the distributed protection system focuses on the cooperation between STUs. STU obtains local measurement data for local processing. At the same time, based on GOOSE networked communication, the measurement and control data of other relevant STU transmissions are received, thereby synthesizing the information of each site, judging the fault-related situation and making a decision to realize the protection function. As shown in Figure 2, the distributed protection system of switch station, in which the substation outlet circuit breaker protection devices STU1 and STU2 are equipped with overcurrent protection, and the incoming line circuit breaker protection devices STU3 and STU4 are equipped with voltage loss protection and overcurrent protection, and the line protection device for feeder lines branches STU5~STU10 are equipped with overcurrent protection, and the busbar protection device STU0 is equipped with overcurrent protection and BATS function.

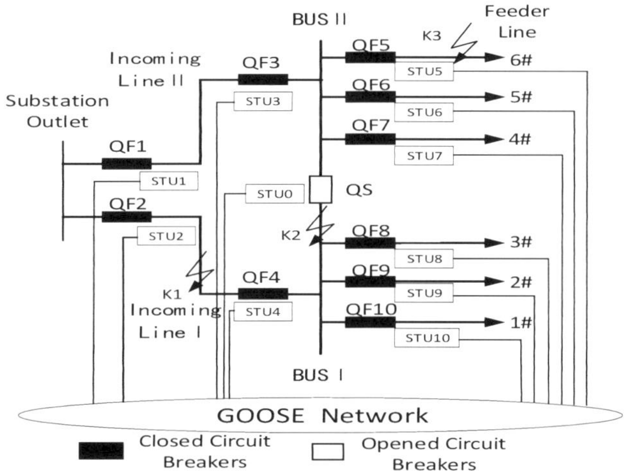

Figure 2. The Distributed Protection System of Switch Station

According to the installation position, the distributed current protection of switch stations can be divided into incoming line protection and feeder line protection. When a fault occurs at different locations on the line, the protection action is as follows:

1) When the incoming line K1 fault, the protection device STU2 detects that the fault current is started, and other protections are not activated. After a short delay, the STU2 can receive no other protection start information through the GOOSE network, then it is judged as the incoming line fault, QF2 disconnect. After the QF2 is disconnected, the reclosing is performed. If it is a transient fault, the STU2 detects that the fault current disappears. In the case of a permanent failure, STU2 immediately jumps off. At this time, STU4 detects that the incoming line 1 is no voltage, the voltage loss protection is activated, and simultaneously sends a GOOSE message to STU0. If STU3 detects that there is voltage in the incoming line 2 , STU0 cannot receive the STU3 loss-of-voltage protection start information, and puts in preparation BATS function, QS switch is closed. If STU3 detects that there is no voltage in incoming line 2, the voltage loss protection is started, and STU0 receives the STU3 voltage loss protection start information, then the STU4 and STU3 information is combined to make a judgment, and the BATS function is blocked.

2) When the bus K2 fails, the protection device STU2 detects the fault current start, STU4 detects that the fault current is also started, and STU8, STU9, and STU10 do not start, it is judged as the bus fault. STU2 receives the STU4 overcurrent start message within a short delay, and then judges to be an out-of-zone fault and lock-up protection. STU4 acts directly on the trip and the QF4 switch is off. At

the same time, the GOOSE message is sent to STU0, and the BATS is locked.

3) When the fault occurs at the feeder line K3, the protection devices STU1, STU3, and STU5 detect that the fault current is started, and it is determined that the feeder branch 6# is faulty. After a short delay, STU1 receives the overcurrent blocking signal of STU3, STU3 receives the overcurrent blocking signal of STU5, judges that an out-of-zone fault occurs, and the blocking protection, STU5 acts on the trip, and QF5 switch is off.

4. Incoming and feeder line protection setting in switch stations

4.1. Incoming and feeder line protection setting

The traditional switch station local protection and micro-computer protection adopt three-stage current protection, and the current setting and action time are matched in the limit of the upper and lower protection devices. The intelligent level is not high, and the protection delay near the power supply is too long. The number of protection stages increases, and the protection setting is complicated. The distributed protection developed by the protection platform based on STU, GOOSE network communication and distributed interaction greatly simplifies the protection setting of switch stations.

The distributed protection of the switch station adopts the principle of combining the voltage loss protection and the overcurrent protection to improve the protection performance. The setting of the voltage loss protection is 70% of the rated voltage, and the action time limit is set according to the time of avoiding the reclosing of the substation outlet circuit breaker. When the substation outlet circuit breaker is unsuccessful and jumps off, the incoming line voltage is rapidly reduced and the voltage loss protection is started.

For the overcurrent protection of the distributed structure, the networked communication and logic judgment are used to judge the fault information, and the current setting can reflect the fault information. Therefore, the current III protection is used as the primary protection and backup protection, no need to install current quick-break or time-limited current quick-break protection. The current setting of the current III protection is set according to the maximum load current. When the current III segment protection is used as the main protection of the incoming line, the current at the end of the two-phase short circuit at the end of the line is verified by the minimum operation mode, and the sensitivity coefficient K_{sen} is taken as 1.5. If the sensitivity factor is not met, it can be adjusted by extending the line protection range.

4.2 Protection time performance calculation

For the implementation process of distributed protection, the delays generated by each link include: fault current detection time T0, generally less than 20ms; GOOSE message command transmission time T1, is about 10ms; circuit breaker disconnect fault current time T2, Generally less than 70ms; the time T3 of the incoming circuit breaker reclosing is 300ms; the delay of the current III protection T4 should avoid the transmission time of the GOOSE message and the logical judgment time of the STU, is about 30ms; The protection time under different line fault conditions is:

a) When the incoming line fails, the substation outlet circuit breaker completes a reclosing by T0+T4+T2+T3=20+30+70+300=420ms.

b) When the busbar line fails, the incoming line circuit breaker cuts off the fault by T0+T4+T2=20+30+70=120ms.

c) When the outgoing branch line fails, the outgoing branch circuit breaker cuts off the fault by T0+T4+T2=20+30+70=120ms.

5. Protection action logic based on GOOSE mechanism

5.1 The incoming line fault protection logic in switch stations

As shown in Figure 3, when the incoming line I of switch station fails, STU2 detects that the overcurrent information is activated, and STU4, STU8, STU9, and STU10 are not started. After a delay of 10ms, STU2 does not receive the GOOSE message initiated by other STUs, and it is determined that the incoming I has a fault, so it directly acts on the substation exit breaker QF2 to trip. After a reclosing of QF2, if it is a permanent fault, it will jump again to achieve isolation of the fault incoming line. At this point, the incoming line I loses power, the incoming line voltage drops to 70% of the rated voltage, the STU4 loss of voltage protection starts, and sends a GOOSE message to STU0. after 10ms delay, if STU0 does not receive the loss from STU3 When the voltage protection start signal is sent, the bus-connected circuit breaker QS closing message is sent to the STU4 to lock the voltage loss protection, and at the same time, the circuit breaker QS is closed and put into the BATS device. If STU0 receives the STU3's voltage loss protection start signal within 10ms delay, it will lock the BATS device.

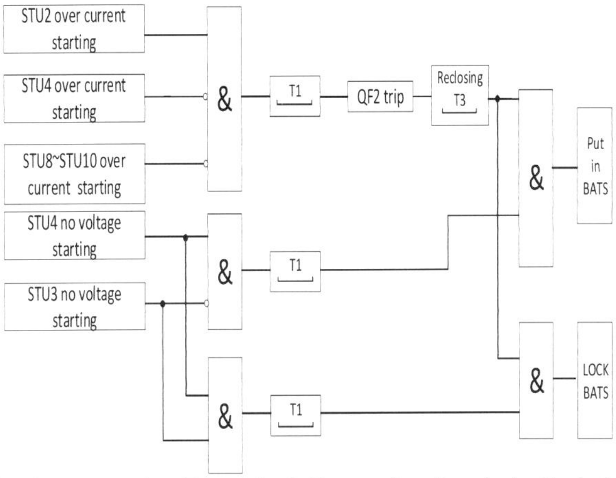

Figure 3. The Incoming Line Fault Protection Logic in Switch Stations

5.2 Bus fault protection logic in switch stations

When the bus I fails, STU2 and STU4 detect that the overcurrent information is activated, and STU8, STU9, and STU10 are not started. After 10ms delay, STU2 receives the GOOSE message from the STU4 overcurrent start, and blocks QF2 protection. When STU4 does not receive the startup information of the STU8~STU10, it is determined that the bus I has a fault, directly acts on the QF4 trip, isolates the faulty bus, and blocks the BATS device. When QF4 is opened, STU4 sends a GOOSE message to STU2 to cancel the blocking. As shown in Figure 4:

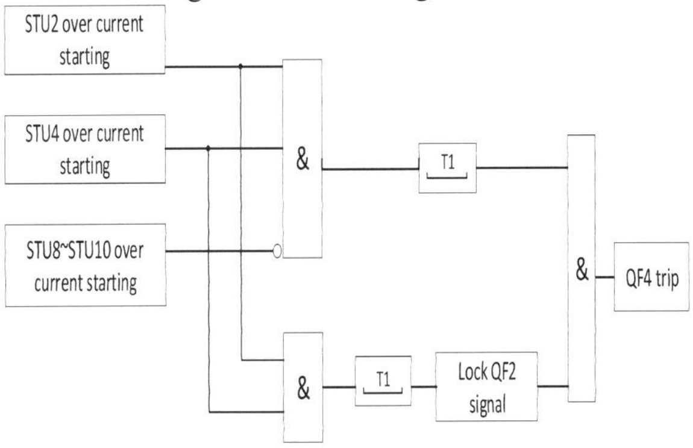

Figure 4. Bus Fault Protection Logic In Switch Stations

5.3 Feeder branch fault protection logic in switch station

When the feeder branch of the switch station is faulty, the protection logic is as shown in Figure 5. STU2 and STU4 detect that the overcurrent information is started, and the branch line 3# (such as

STU8) detects that the fault current is also started. After 10ms GOOSE message transmission delay, STU2 receives STU4 startup information and locks protection. At the same time, STU4 receives the STU8 start information and locks the protection. It determines that the feeder branch 3# has failed, and directly acts on the feeder branch breaker QF8 to trip and cut off the faulty line. When QF8 trips, STU8 sends a GOOSE message to STU2 and STU4 to cancel the lock.

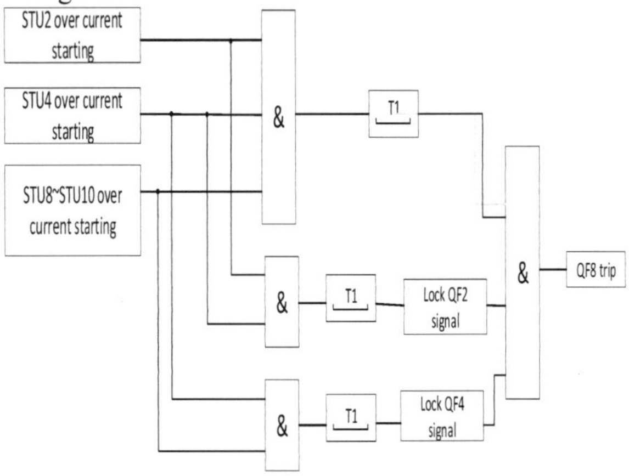

Figure 5. Feeder Branch Fault Protection Logic in Switch Station

6.Setting Examples

The structure of the switch station with distributed protection is shown in Figure 6. The 10kV incoming line includes the incoming line 1 and the incoming line 2, both of which are introduced by the 110kV low-voltage side busbar, which is used for hybrid power supply of overhead line and cables. The length of the overhead line is 1.3km. The length of cable is 0.7km. Single bus segmentation structure is adopted. The number of feeder line for the busbar I segment and the busbar segment II are four. Among them, the 4# feeder and the 8# feeder are connected to the transformer, and the load capacity is 30kVA.

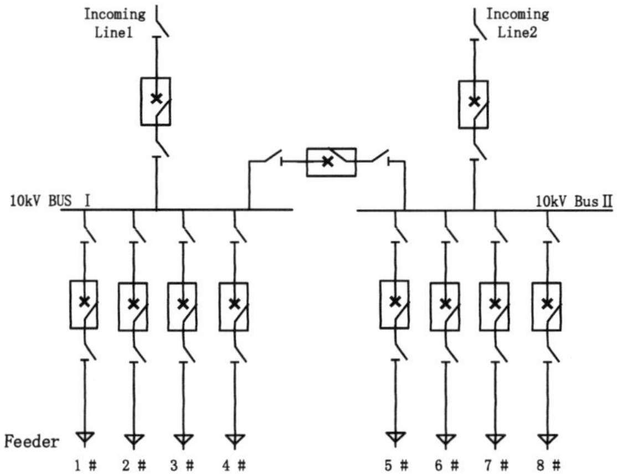

Figure 6. The Structure of The Switch Station

Feeder lines data are shown in Table 1:

IOP Conf. Series: Materials Science and Engineering **533** (2019) 012002 doi:10.1088/1757-899X/533/1/012002

Table 1. Feeder Lines Data

Feeder Number	1#	2#	3#	4#	5#	6#	7#	8#
Load Capacity (kVA)	500	630	1000	30	630	630	315	30
Rated Current(A)	28.9	36.4	57.7	—	36.4	36.4	18.2	—
Cable Length(km)	0.23	0.7	0.3	—	0.6	0.53	0.48	—

Taking the reference capacity S_b=100MVA and the reference voltage U_d=10.5kV, the reference current I_d=5.5kA, and the maximum incoming line load current is 207 A. The 10kV bus short-circuit impedance standard valueX_{max}=0.4193,X_{min}=0.8581. When in the maximum operating mode, the short-circuit current of the II section busbar is: three-phase short circuit 6.3kA , two-phase short circuit 5.46kA , and the short-circuit current for feeder line (for example, 1#): three-phase short circuit 0.62kA, two-phase short circuit 0.54kA . When in the minimum operation mode, the short-circuit current of the II section busbar is: three-phase short circuit 4.11kA and two-phase short circuit 3.63kA; the short-circuit current at the end of 1# feeder is: three-phase 0.59kA, two-phase 0.51kA.

a) Incoming line overcurrent protection setting:

$$I_{dz} = \frac{k_k}{k_{fh}} I_{fm_{ax}} = 1.3\times 207/0.95 = 283A \quad (1)$$

Among them, K_k is the reliability coefficient, which is 1.3; $K_{fh}$is the protection return coefficient, which is 0.95.

When the line overcurrent protection is used as the main protection of the line, the current value of the two-phase short circuit at the end of the line is checked in the minimum operation mode, and the protection sensitivity is:

$$K_{sen} = \frac{I^{(2)}_{min}}{I_{Dz}} = 3630/283 = 12.8 > 1.5 \quad (2)$$

Therefore, the requirements are met.

When the incoming line overcurrent protection is used as the backup protection for the feeder line, the current value of the two-phase short circuit at the end of the feeder line is checked in the minimum operating mode. Taking the 1# feeder line as an example, the protection sensitivity is:

$$K_{sen} = \frac{I^{(2)}_{L1\,min}}{I_{Dz}} = 510/283 = 1.8 > 1.2 \quad (3)$$

Therefore, the requirements are met.

b) Feeder line overcurrent protection setting (take 1# as an example)

$$I_{dz} = \frac{k_k}{k_{fh}} I_{fm_{ax}} = 1.3\times 28.9/0.95 = 40A \quad (4)$$

Among them, K_k is the reliability coefficient, which is 1.3; $K_{fh}$is the protection return coefficient, which is 0.95.

When the feeder line overcurrent protection is used as the main protection of the line, the current value of the two-phase short circuit at the end of the line is checked in the minimum operation mode, and the protection sensitivity is:

$$K_{sen} = \frac{I^{(2)}_{min}}{I_{Dz}} = 510/40 = 12.7 > 1.5 \quad (5)$$

Therefore, the requirements are met.

7. Conclusion

The distributed protection method based on GOOSE networked communication and logic judgment can greatly simplify the protection setting of the switch station, and can effectively solve the problem of multi-level coordination setting difficulty in the protection of traditional switch station, especially in the case of the cascading of the switch station, it is possible to effectively solve the problem of the override trip in the protection.

References

[1] Dong Zhangzhuo, Duan Xin. Improvement of current protection for ultra-short ingress and egress lines in open and close [J]. Power System Protection and Control, 2010, 38(6): 80-83

[2] Chen Haidong, Yan Baoji, Acting article, et al. Research on protection of opening and closing based on GOOSE mechanism [J]. Power System Protection and Control, 2013, 41(13): 134-138

[3] Liu Jian, Pei Baoji, Cui Qi, et al. A Rapid Self-healing Distributed Intelligent Feeder Automation System[J]. Automation of Electric Power Systems, 2010,34(10): 62-66

[4] Chen Jieming. Research and Application of 10 kV Simple Busbar Protection Based on GOOSE[J]. Automation of Electric Power Systems, 2011, 35(4): 96-99.

[5] Fu xuhua, Xiao ming, Wang Song, and other digital GOOSE substation engineering technology [J] Electric Power Automation Equipment, 2011, 31 (11): 112-116.

[6] Zhi qiang, Xu jiayu, & T. substation protection applications such as research and implementation of GOOSE [J] China Electric Power, 2011 (6) based on: 41-46.

IOP Conf. Series: Materials Science and Engineering **533** (2019) 012003 doi:10.1088/1757-899X/533/1/012003

Study on stability evaluation in multi-send hybrid AC/DC transmission system

Rong Li

Kunming Power Supply Bureau of Yunnan Power Grid Co. Ltd, Kunming City, Yunnan 650011, China

84371185@qq.com

Abstract. Our country has shown the significant features of high voltage direct current (HVDC) including multi-infeed and multi-send. Compared to multi-infeed system, there are a lot of weakness in the multi-send system, such as the grid structure is single and weak, which makes the interaction between AC and DC to a greater extent affect the stability of the multi-send hybrid AC/DC transmission system. In order to evaluating the system strength in multi-send hybrid AC/DC transmission system, for the first time, this paper proposed the general formula and simplified formula for multi-send short-circuit ratio (MSSCR), which accounts for the coupling between multi-dc projects. Considering the control characteristics of rectifier in sending-side and combining the maximum power curve method, the criticality of MSSCR is proposed. Multi-send critical short-circuit ratio (MSCSCR) is deduced by the quasi steady-state equation of rectifier, to determine the multi-send system is strength enough or not. The proposal of MSSCR is of considerable significance for effectively evaluating the stability performance of the multi-send AC/DC hybrid transmission system, completing the short-circuit ratio theoretical system, and consolidating the short-circuit ratio index status.

1. Introduction

At present, China's high voltage direct current (HVDC) transmission is mostly concentrated in East China and South China power grids is the most typical, with 7 HVDC in East China Power Grid: 4 HVDC project: Yihua, Genan, Linfeng, Longzheng, , and 3 UHVDC project: Fengfu, Xizhe, Sujin, 8 HVDC in Guangdong Power Grid: 6 high-voltage DC project: Tianguang, Sanguang, Xiluodu to Guangdong double-back, Guiguang double-back, and 2 UHVDC project: Yunguang, Nuozhadu. The drop points of HVDC transmission are concentrated in energy-intensive areas such as northwest and southwest. several power grids of China have shown the important characteristics of DC multi-infeed and multi-send, and their scale is rare in the world.

The sending end of the multi-send hybrid AC/DC transmission system most chose at an energy enrichment place, the DC falling points are dense, the DCs are closely coupled, and the power grid structure is simple and relatively weak. Compared with the receiving end system, interactions between DC and DC, DC and AC have a greater impact on the stability of the sending end system. When the sending end system of the hybrid AC/DC transmission system is relatively weak, the slight power disturbance will also cause the converter bus voltage to fluctuate drastically, causing the rectifier to trigger phase shift, which affects the DC system operation. In severe cases, it may cause inverter commutation failed. When the commutation failure occurs on the inverter side, causing DC blocking, a large power surplus occurs at the sending end, and a huge power shortage occurs at the receiving end.

Content from this work may be used under the terms of the Creative Commons Attribution 3.0 licence. Any further distribution of this work must maintain attribution to the author(s) and the title of the work, journal citation and DOI.
Published under licence by IOP Publishing Ltd

At this time, due to the closed DC channel, the reactive power compensation device such as the sending end system filter has a slower removing action, and excess reactive power is sent to the sending end system, resulting in overvoltage of the transmitting terminal commutation bus. And DC is equivalent to the load on the sending end system, and the load shedding caused by DC blocking will also cause temporary overvoltage and frequency problems. These stability problems threaten the safe and stable operation of multi-send hybrid AC/DC transmission system. Therefore, it is urgent to propose indicators to measure the strength of multi-send hybrid AC/DC transmission system, and it is of great significance to study the stability of multi-send hybrid AC/DC transmission system.

This paper starts from the static stability evaluation index of power system, short-circuit ratio (SCR) and multi-infeed short-circuit ratio (MISCR), and affirms its index significance. Applying the concept of SCR and MISCR in multi-send hybrid AC/DC transmission system, the multi-send short-circuit ratio(MSSCR) is proposed. The maximum power curve method is used to study the criticality of MSSCR, and the value of the multi-send critical short-circuit ratio (MSCSCR) is given. The MSSCR has enriched and improved the concept of the SCR index, and effectively evaluated the stability performance of the multi-send hybrid AC/DC transmission system.

2. SCR and MISCR

2.1. SCR

The mutual coupling between the AC system and the DC system depends to a large extent on the relative power of the HVDC transmission rated capacity of the AC system. For the equivalent single-infeed DC system shown in Fig.1, the International Council on Large Electric Systems (CIGRE) proposed short-circuit ratio (SCR) as a quantitative indicator of the strength of the AC system [1]. It refers to the ratio of the three-phase short-circuit capacity of the AC system S_{ac} to the rated DC power P_{dN} of the HVDC transmission connected to it, as shown in Equation 1. This indicator can better measure the voltage support capability of the AC system, and provides an evaluation index for the static stability of the AC system, which has been widely recognized. In Fig.1, the virtual frame part is the receiving end AC system; E is the AC system equivalent electromotive force; P_d, Q_d are the DC active power and reactive power of the converter station; U is the commutating bus voltage; Q_C is the converter station reactive power compensation capacity; Z_S is the system equivalent impedance.

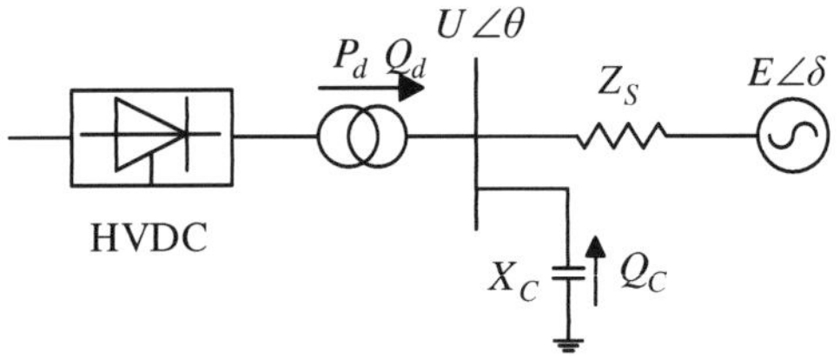

Figure 1. Diagram of single infeed system

$$SCR = \frac{S_{ac}}{P_{dN}} \tag{1}$$

Based on the short-circuit ratio indicator, taking into account the influence of the reactive power compensation of the converter station, the effective short-circuit ratio (ESCR) index can be obtained.

$$ESCR = \frac{S_{ac} - Q_c}{P_{dN}} \tag{2}$$

For the short-circuit ratio indicator, it is usually used for design planning, and the denominator often takes the DC rated active power P_{dN}. For operation, such as DC power reduction operation or system operation mode change, the actual operation P_d is less than the rated active power P_{dN}, so the short-circuit ratio indicator obtained always has

$$SCR = \frac{S_{ac}}{P_{dN}} \leq \frac{S_{ac}}{P_d} \tag{3}$$

That is, the SCR under rated operating conditions is more stringent. If the rated operation can meet the requirements of static and stable operation, the operating requirements can be met after the power reduction or operation mode is changed, and the stability margin is greater. This further determines the index meaning of the SCR, and the SCR indicators are discussed below in terms of rated operating conditions.

2.2. MISCR

With the development of high-voltage direct current transmission, there are many cases where the DC drop point is close and mutual influence. In the research report, CIGRE defined the multi-infeed interaction factor (MIIF) to account for the coupling between DCs[2].

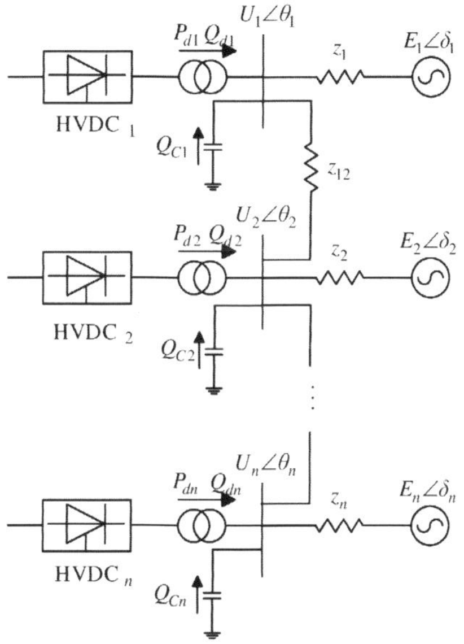

Figure 2. Diagram of hybrid AC/DC transmission system

The meaning is: multi-infeed as shown in Fig.2, in the simplified model of the hybrid AC/DC transmission system, the symmetrical three-phase reactor is input to the commutating bus bar i, so that the voltage drop on the bus bar is exactly 1%. At this time, the ratio of the voltage variation of the commutating bus bar of the DC subsystem j ΔU_j and the voltage changes of the i-th commutation bus ΔU_i, is the MIIF, which is $MIIF_{j,i} = \frac{\Delta U_j}{\Delta U_i}$.

With the introduction of MIIF, the SCR indicator has also been developed. Following the SCR index in single-infeed DC system, the multi-infeed interaction factor MIIF and the multi-DC coupling relationship are evaluated, and the other DC coupling effects are evaluated according to the coupling degree and superimposed on the current DC power. The SCR forms a multi-infeed short-circuit ratio (MISCR) and a multi-infeed effective short-circuit ratio (MIESCR)[3]. Its definition is as follows

$$MISCR_i = \frac{S_{aci}}{\tilde{P}_{diN}} = \frac{S_{aci}}{P_{diN} + \sum\limits_{j=1, j\neq i}^{n} MIIF_{j,i} \cdot P_{djN}} \tag{4}$$

$$MIESCR_i = \frac{S_{aci} - Q_{ci}}{\tilde{P}_{diN}} = \frac{S_{aci} - Q_{ci}}{P_{diN} + \sum_{j=1, j\neq i}^{n} MIIF_{j,i} \cdot P_{djN}} \tag{5}$$

In fact, in the definition of SCR, MIIF and MISCR, it can be seen that although the target is the receiving end of hybrid AC/DC transmission system, the concept is equally applicable to the sending end, can carry out concept migration to evaluate the stability performance of multi-send hybrid AC/DC transmission system.

3. MSSCR

For multi-send hybrid AC/DC transmission system, similar multi-infeed interaction factor indicators can also be used to measure the mutual coupling between multiple-send DCs, which is defined as multi-send interaction factor (MSIF).

$$MSIF_{j,i} = \frac{\Delta U_j}{\Delta U_i} \tag{6}$$

Analogous to the SCR and MISCR, a multi-send short-circuit ratio (MSSCR) is proposed, which takes into account the interaction between multiple DCs. The expression is as shown in equation (7), where S_{aci} is the three-phase short-circuit capacity, $\tilde{P}_{diN} = P_{diN} + \sum_{j=1, j\neq i}^{n} MSIF_{j,i} \cdot P_{djN}$ is the equivalent DC rated power, in which P_{diN} is the rated power of i-th DC, and P_{djN} is the rated power of other n-1 DC.

$$MSSCR_i = \frac{S_{aci}}{\tilde{P}_{diN}} = \frac{S_{aci}}{P_{diN} + \sum_{j=1, j\neq i}^{n} MSIF_{j,i} \cdot P_{djN}} \tag{7}$$

In the multi-send hybrid AC/DC transmission system shown in Fig.3, the virtual frame part is the sending end power grid, and the DC is sent out for two of them. Considering the i-th and *j*-th DC, according to the node voltage equation $\dot{U}_{ac} = Z_{BUS} \cdot \dot{I}_{ac}$, the current $\dot{I}_j$ is injected at the node j, and the voltage change caused by this current at the node i is as equation (8).

$$\Delta\dot{U}_{ij} = Z_{ij} \cdot \dot{I}_j \tag{8}$$

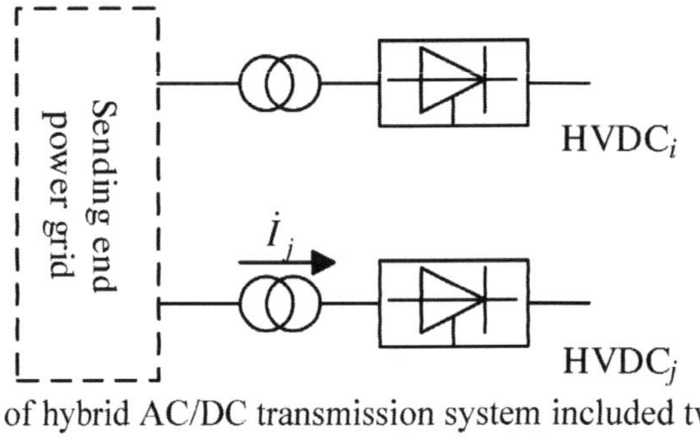

Figure 3. Diagram of hybrid AC/DC transmission system included two DC sending ends

At this point, suppose there is an equivalent current $\dot{I}_{ij}$, and the current is injected into the i point, which can cause the same voltage change at the i point, that is

$$\Delta\dot{U}_{ij} = Z_{ii} \cdot \dot{I}_{ij} \tag{9}$$

Therefore, it can be considered that the effect of the j point injection current $\dot{I}_j$ at the i point can be equivalent to directly injecting the equivalent current $\dot{I}_{ij}$ at the node-i. Where Z_{ij} is the mutual impedance between nodes i and j, Z_{ii} is the self-impedance of node *i*. We can obtain equation (10) by equations (8) and (9),

$$\dot{I}_{ij} = \frac{Z_{ij}}{Z_{ii}} \cdot \dot{I}_j \tag{10}$$

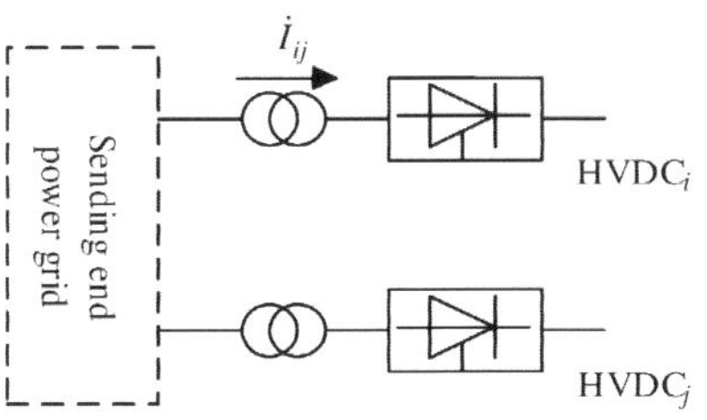

Figure 4. Node j injection current is equivalent to the node i

At this point, the point j injection current $\dot{I}_j$ in the system generates the AC bus power at point i can expressed as

$$S_{ij} = \dot{U}_i \cdot \dot{I}_{ij}^* = \dot{U}_i \cdot \frac{Z_{ij}}{Z_{ii}} \cdot \dot{I}_j^* \tag{11}$$

Among them, $\dot{U}_i = |\dot{U}_i| \angle\delta_i = U_i\angle\delta_i$ and $\dot{U}_j = |\dot{U}_j| \angle\delta_j = U_j\angle\delta_j$ are the system bus voltage, the voltage level in the transmission system is so high that can ignore the resistance effect, so that the system impedance is mainly reactance; assuming that the self-impedance and mutual impedance in the system are purely inductive, that is, the impedance phase angle is 90°, according to equation (11) can be reduced to

$$S_{ij} = \left|\frac{Z_{ij}}{Z_{ii}}\right| \cdot \frac{\dot{U}_i}{\dot{U}_j} \cdot \dot{U}_j \cdot I^*{}_j = \left|\frac{Z_{ij}}{Z_{ii}}\right| \cdot \frac{U_i}{U_j} \cdot S_j\angle\delta_{ij} \tag{12}$$

Where $S_j = P_{dj} - jQ_{dj} = |S_j|\angle\varphi_j$, $\delta_{ij} = \delta_i - \delta_j$, after simplification we can get

$$\begin{aligned} S_{ij} = P_{dij} - jQ_{dij} &= \left|\frac{Z_{ij}}{Z_{ii}}\right| \cdot \left|\frac{U_i}{U_j}\right| \cdot |S_j| \angle(\varphi_j + \delta_{ij}) \\ &= \left|\frac{Z_{ij}}{Z_{ii}}\right| \cdot \frac{U_i}{U_j} \cdot \left[|S_j| \cdot \cos(\varphi_j + \delta_{ij}) + j|S_j| \cdot \sin(\varphi_j + \delta_{ij})\right] \end{aligned} \tag{13}$$

Equation (13) corresponds to the left and right sides, you can get the next two

$$P_{dij} = \left|\frac{Z_{ij}}{Z_{ii}}\right| \cdot \frac{U_i}{U_j} \cdot \left(P_{dj} \cos\delta_{ij} + Q_{dj} \sin\delta_{ij}\right) \tag{14}$$

$$Q_{dij} = \left|\frac{Z_{ij}}{Z_{ii}}\right| \cdot \frac{U_i}{U_j} \cdot \left(Q_{dj} \cos\delta_{ij} - P_{dj} \sin\delta_{ij}\right) \tag{15}$$

Equation (14) and (15) means in two DCs, the influence of active power and reactive power of one DC on the other when the mutual coupling between DCs is considered.

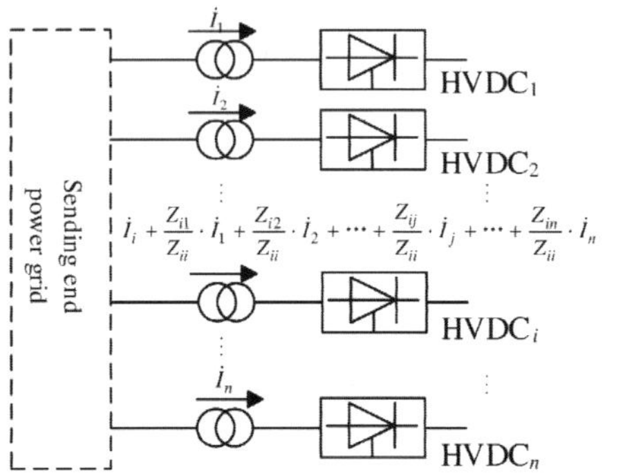

Figure 5. Node n-1 injection current equivalent to the node i

The situation of two DCs can be generalized. When the system contains n DCs, consider the equivalent power of the other n-1 DCs superimposed on the i-th DC converter bus, we can get

$$\tilde{P}_{dij} = \sum_{j=1, j\neq i}^{n} \left|\frac{Z_{ij}}{Z_{ii}}\right| \cdot \frac{U_i}{U_j} \cdot \left(P_{dj} \cos\delta_{ij} + Q_{dj} \sin\delta_{ij}\right) \tag{16}$$

$$\tilde{Q}_{dij} = \sum_{j=1, j\neq i}^{n} \left|\frac{Z_{ij}}{Z_{ii}}\right| \cdot \frac{U_i}{U_j} \cdot \left(Q_{dj} \cos\delta_{ij} - P_{dj} \sin\delta_{ij}\right) \tag{17}$$

From the above derivation, we can know that the coupling effect of other n-1 DCs on the i-th DC can be expressed as the equivalent superposition of its power, thus the total equivalent DC's active power in i-th DC can be expressed as the sum of active power of other n-1 DC and i-th DC self-active power.

$$\begin{aligned} MSSCR_i &= \frac{S_{aci}}{\tilde{P}_{di}} = \frac{S_{aci}}{P_{di} + P_{dij}} \\ &= \frac{S_{aci}}{P_{di} + \sum_{j=1, j\neq i}^{n} \left|\frac{Z_{ij}}{Z_{ii}}\right| \cdot \frac{U_i}{U_j} \cdot \left(P_{dj} \cos\delta_{ij} + Q_{dj} \sin\delta_{ij}\right)} \end{aligned} \tag{18}$$

Equation (18) is the general formula for multi-send short-circuit ratio. It can be seen that the influence of other DC on the active power of the i-th DC is not only included in the influence of DC's active power, but also includes other DC's reactive power.

Under normal circumstances, the phase angles of the commutated bus voltages of the DCs in the system are not much different, which can be considered $\delta_{ij} = 0$. For further simplification, assuming that the voltage amplitude and phase angle of all commutated bus bars are rated and equal, $U_i = U_j$, equations (16) and (17) can be reduced to

$$\tilde{P}_{dij} = \sum_{j=1, j\neq i}^{n} \left|\frac{Z_{ij}}{Z_{ii}}\right| \cdot P_{dj} \tag{19}$$

$$\tilde{Q}_{dij} = \sum_{j=1, j\neq i}^{n} \left|\frac{Z_{ij}}{Z_{ii}}\right| \cdot Q_{dj} \tag{20}$$

At this point, the multi-send interaction factor and multi-send short-circuit ratio can also be expressed as

$$MSIF_{j,i} = \frac{\Delta U_j}{\Delta U_i} \approx \left|\frac{Z_{ij}}{Z_{ii}}\right| \tag{21}$$

$$MSSCR_i = \frac{S_{aci}}{P_{di} + \sum\limits_{j=1, j \neq i}^{n} \left| \frac{Z_{ij}}{Z_{ii}} \right| \cdot P_{dj}} \tag{22}$$

So far, the multi-send short-circuit ratio, MSSCR, under simplified conditions has been derived.
In equation (22), the first term of the denominator of the expression P_{di} represents the influence of the disturbance at the node i of the converter station causing the change of the i-th DC active power to affect the AC system; and the second term of the denominator $\sum\limits_{j=1, j \neq i}^{n} \left| \frac{Z_{ij}}{Z_{ii}} \right| \cdot P_{dj}$ indicates when disturbance occurs at the converter station i causes the voltage change of the j-th DC, which leads to the changes of power in j-th DC to affect the AC system. The superposition of each item fully considers the degree of influence of each DC, indicating that the disturbance occurs at i, taking into account the DC-coupling effect on the AC system.
In the concept of MSSCR, the influence of HVDC transmission reactive power compensation is not taken into account, so that the evaluation result obtained is biased optimistic. When the voltage of the commutating bus bar is reduced, the reactive power compensation should increase the reactive power to provide voltage support. However, since the reactive power compensation provides the reactive power and voltage satisfaction relationship $Q_{fc} = \omega_1 C U^2$, the reactive power is further reduced at the voltage squared rate when the voltage is lowered, leading to weaker voltage support. So it takes into account this vicious cycle of reactive power compensation devices such as AC filters, and defines it as multi-send effective short-circuit ratio (MSESCR), which has the following expression

$$MSESCR_i = \frac{S_{aci} - Q_{ci}}{\tilde{P}_{di}} = \frac{S_{aci} - Q_{ci}}{P_{di} + \sum\limits_{j=1, j \neq i}^{n} \left| \frac{Z_{ij}}{Z_{ii}} \right| \cdot P_{dj}} \tag{23}$$

It can be seen from the above derivation process that the influence of the DC system on the AC system is mainly reflected in the change of the DC power level, and the influence of the AC system on the DC system mainly causes the voltage fluctuation of the converter bus through the disturbance. The coupling effect between multiple DCs is also caused by the electrical coupling effect, which has a fluctuation effect on the voltage of the commutation bus, and thus affects the active power of the DC.

4. MSCSCR

When studying the static stability problem of the AC system, it is considered that the power stability margin of the DC system is closely related to the static stability of the AC system. The greater the static reserve in the system, the better the static stability of the system. In the single-infeed DC system shown in Fig.1, when the inverter adopts the γ control, it is assumed that the turn-off angle is constant, and the system filter, reactive power compensation, and transformer tap action are not considered. It is considered that the electromotive force of the AC system is constant. Under the ideal premise of these assumptions, the curve of DC power with DC current is the maximum power curve (MPC) of the DC system. The extreme point on the maximum power curve, which satisfied $\frac{dP_d}{dI_d} = 0$, is the maximum point, that is, maximum available power (MAP), the limit of the active power that the HVDC transmission system can deliver. The larger the value, the better the rated runtime stability. The greater the stability margin, the stronger the system's overload capacity.
When the system is running at the left side of the MAP, it is at the rising edge of the power curve, that is $\frac{dP_d}{dI_d} > 0$, the system can keep the power stable. When the system is running at the right side of the

MAP, it is at the falling edge of the power curve, that is $\frac{dP_d}{dI_d} < 0$, the system will run in the low-voltage and high-current working conditions, causing static and stable destruction of the AC system, which may cause the power system to oscillate asynchronously, and it is difficult to determine the power stable operation.

Usually in a stronger system, the system equivalent reactance is much smaller than the converter transformer leakage reactance, so the system equivalent reactance can be ignored. At this time, the maximum stable running DC current of the inverter is always around 5.5 times the DC current rating, and the system transmission power is always within the transmission limit, leaving a large margin. However, when the system is weak, its stable operating range will be limited. Therefore, it is important to send out the criticality of the stable operation of the hybrid AC/DC transmission system, that is, the critical value of the multi-send short-circuit ratio.

On the MPC, when the rated operating point (I_{dN} , P_{dN}) of the system coincides with the MAP (I_{dMAP} , P_{dmax}), it is said that the multi-send critical short-circuit ratio (MSCSCR) is sent at this time. For the rectifier of the sending end system, the control mode is constant current control. In the initial stage, as the current increases, the angle α decreases, and when the current increases to a certain extent, the angle α decreases to a minimum firing angle of 5°. The rectifier control mode is switched to a fixed minimum firing angle control. Therefore, when the current value is small, it is still in constant current control, the operating point of MPC in the rectifier side changes on the curve under different α angles. When it is switched to the fixed minimum firing angle control, its operating point is move on α=5° curve. However, the maximum power operating extreme point is always obtained after switching the control mode, that is, always at the maximum power point where α is 5°.

Analogous to the SCR, the commutation bus voltage is selected as the system voltage reference value, and the DC system rated active power is the system power reference value, which can be used to standardize the MSSCR.

$$MSSCR_i = \frac{1}{|Z_{ii}| \cdot P_{di} + \sum_{j=1, j\neq i}^{n} |Z_{ij}| \cdot P_{dj}} \tag{24}$$

Considering the influence of multiple DC coupling, the system is equivalent to the single sending DC system shown in Fig.6. The effect of other DC on the i-th DC is reflected in the system impedance, and the equivalent impedance can be defined as $Z_{eqi} = |Z_{ii}| \cdot P_{di} + \sum_{j=1, j\neq i}^{n} |Z_{ij}| \cdot P_{dj}$. At this time, similar to the SCR, the MSSCR can also be expressed as the reciprocal of the equivalent impedance $\frac{1}{Z_{eqi}}$, that is, the critical short-circuit ratio can be equivalent to the critical value of $\frac{1}{Z_{eqi}}$.

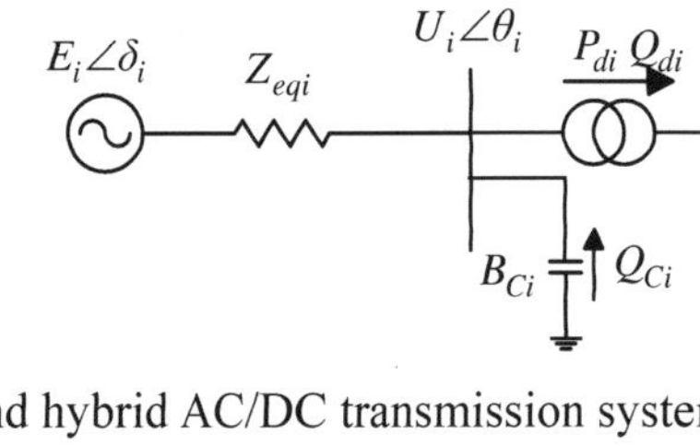

Figure 6. Multi-send hybrid AC/DC transmission system equivalent to single-send HVDC system

For the simplified single-send HVDC transmission system of the multi-send AC/DC system, the rectifier side quasi-steady-state equation of the single HVDC transmission system can be used to describe the operation characteristics of the sending end system. The AC system is in the form of

Thevenin equivalent circuit, and the DC system uses a steady-state average model. Its analytical expression is as shown in equations (25)~(33).

$$P_{di} = C_i U_i^2 \left[\cos 2\alpha_i - \cos\left(2\alpha_i + 2\mu_i\right) \right] \tag{25}$$

$$Q_{di} = C_i U_i^2 \left[2\mu_i + \sin 2\alpha_i - \sin\left(2\alpha_i + 2\mu_i\right) \right] \tag{26}$$

$$I_{di} = K_i U_i \left[\cos \alpha_i - \cos\left(\alpha_i + \mu_i\right) \right] \tag{27}$$

$$U_{di} = \frac{P_{di}}{I_{di}} \tag{28}$$

$$P_{aci} = \frac{1}{\left|Z_{eqi}\right|} \left[U_i^2 \cos\theta_i - E_i U_i \cos\left(\delta_i + \theta_i\right) \right] \tag{29}$$

$$Q_{aci} = \frac{1}{\left|Z_{eqi}\right|} \left[U_i^2 \sin\theta_i - E_i U_i \sin\left(\delta_i + \theta_i\right) \right] \tag{30}$$

$$Q_{ci} = B_{ci} \cdot U_i^2 \tag{31}$$

$$P_{di} - P_{aci} = 0 \tag{32}$$

$$Q_{di} + Q_{aci} - Q_{ci} = 0 \tag{33}$$

Where Ci and Ki are converter station equipment parameters, the expression is $C_i = \frac{3}{4\pi} \cdot \frac{S_T}{P_{dN}} \cdot \frac{1}{u_k\%} \cdot \frac{1}{\tau^2}$ and $K_i = \frac{\sqrt{2}}{2} \cdot \frac{S_T}{P_{dN}} \cdot \frac{1}{u_k\%} \cdot \frac{1}{\tau}$, where S_T is the converter transformer capacity, generally taking 1.1~1.2 times DC rated power P_{dN}, here assumed $S_T = 1.15P_{dN}$; u_k is commutation transformer short-circuit impedance, generally take 15° ~20°, this paper takes u_k=18°; τ is the converter transformer ratio, take 1 under rated operating conditions; P_{dN} is the rated active power of DC transmission. Typical operating parameters of the sending end system when rated operation and maximum power limit are: α_i=5°, μ_i=25°. Solving the equations, it is possible to obtain a multi-send critical short-circuit ratio of about 1.7 in a multi-send hybrid AC/DC transmission system. When the MSSCR considering other DC-DC effects is less than the critical value of 1.7, the DC power delivered by the system will exceed the limit value, and the system cannot maintain stable operation.

For a single-input DC system and a multi-infeed DC system, the critical short-circuit ratio is 2, which is higher than the multi-infeed critical short-circuit ratio. It can be seen that since the rectifier side generally does not cause commutation failure, the stability threshold of the system on the rectifier side is lower than that on the inverter side, and the strength requirement of the AC system on the transmission side is also reduced. However, there is still a criticality. When the sending end system is too weak, small disturbances may also cause the DC power to exceed the limit, causing system instability.

5. Conclusion

In this paper, the static stability evaluation index—SCR and MISCR are studied. On the basis of this, the applicability of the SCR index to the multi-send hybrid AC/DC transmission system is studied, and the MSSCR index is proposed to evaluate the stability of the multi-send hybrid AC/DC transmission system. Through research, the following conclusions can be drawn:

(1) The SCR index has a certain index status. The SCR index applicable to the planned design under rated conditions can characterize the operating characteristics under other conditions.

(2) The general formula of multi-send short-circuit ratio MSSCR index with multi-send interaction factor and multi-send DC coupling is proposed, and a simplified formula considering certain conditions is used for evaluation the strength of the sending end system. In the general formula of MSSCR, it can be seen that other DC's reactive power have an effect on the active power transmission of DC. Enriched the meaning of the SCR.

(3) The definition of the MSCSCR is given. The MSCSCR reference value is obtained under the typical rated operating parameters of the rectifier. According to the research, for the multi-send hybrid AC/DC system, the MSCSCR is about 1.7. When the MSSCR is less than 1.7, the system will not be able to maintain stable operation. In the planning and design, calculation and verification of MSSCR should be carried out to avoid sending more DC access to weaker AC system.

References

[1] CIGRE Working Group 14.07 (AC/DC System Interaction) IEEE Working Group 15.05.05 (HVDC Interaction with low SCR AC Systems. Guide for planning DC links terminating low short-cicuit capacities[C]. 1992.
[2] Denis Lee Hau Aik.Voltage and Power Interactions in Multi-infeed HVDC Systems[R].2007.
[3] CIGRE WG B4.41. Systems with Multiple DC Infeed[R].2008.

The Complex Torque Coefficient Applied in Subsynchronous Oscillation Caused by HVDC

Hanhua Zhang[1], Jiaqi Li[2,a], Jun Wen[2], Shuang Zhang[1] and Bei Tian[1]

[1] State Grid, Ningxia Electric Power Research Institute, 750002 Ningxia, China

[2] North China Electric Power University, School of Electric and Electronics Engineering, 102206 Beijing, China

[a]corresponding author:lijiaq7@163.com

Abstract. Subsynchronous oscillation is an unstable state in power system, which may damage rotors of turbine generators and break the steady operation of the power system. In China, some accidents of subsynchronous oscillations had happened in the power stations where there are HVDC projects nearby. So that it is necessary to study the Subsynchronous oscillation caused by HVDC. After analysis of the principle of subsynchronous oscillation, the mathematical calculation model of the system containing HVDC is built and analysed in method of complex torque coefficient. And the simulation model is established in PSCAD/EMTDC to prove the validation of the method of complex torque coefficient used for the system containing HVDC, which combines calculation and simulation of the method for the system.

1. Introduction

According to the IEEE Committee Report by the Subsynchronous Resonance Working Group of the System Dynamic Performance Subcommittee[1], subsynchronous resonance (SSR) is an abnormal state of power system, where the power system exchanges energy with turbine generators at one or more frequencies below the synchronous frequency. Besides the interaction of the turbine generators and the series capacitor compensated power system network, subsynchronous oscillation (SSO) also results from the interactions with other sources, including high voltage direct current (HVDC), static var compensator (SVC), power system stabilizer (PSS).

In China, HVDC meets the large demand of transmission capacity and distance, thus becoming the powerful complement to AC power transmission. By far, China has the largest number of HVDC projects and the highest voltage level around the world[2,3]. Generally, SSO occurred the turbine generators near the HVDC projects, which may cause the torsional vibration, out of steady state and even shaft breakage threatening the safety of power system operation. In 1977, The first SSO caused by HVDC happened in Square Butte project of the United Sates[4]. The research finally indicated that the torsional vibration was caused by the interaction of power oscillation of the turbine generator with the HVDC. The similar accidents also happened in China several times. In October of 2008, the Gaoling back-to-back HVDC caused SSO of the turbine generators in Suizhong power plant, which connects the northeast grid and the north grid of China[5]. In September of 2010, the HVDC project

Content from this work may be used under the terms of the Creative Commons Attribution 3.0 licence. Any further distribution of this work must maintain attribution to the author(s) and the title of the work, journal citation and DOI.
Published under licence by IOP Publishing Ltd

IOP Conf. Series: Materials Science and Engineering **533** (2019) 012004 doi:10.1088/1757-899X/533/1/012004

from Hulun Buir to Liaoning caused SSO of the turbine generators in Yimin power plant[6]. Therefore, it is urgent to research the SSO characteristic between the HVDC and turbine generators.

The analysis methods fall into two categories: qualitative methods and quantitative methods. Qualitative methods are applied to roughly analyse the possibility of SSO in power system and find out the risky generators. The qualitative methods need a small amount of original data and is easy to calculate, but the error of calculation results is considerable. The qualitative methods include UIF and frequency scanning. The quantitative methods are applied to concretely analyze the SSO characteristic, with detailed original data for more precise research. The qualitative methods include eigenvalue analysis, complex torque coefficient and the time domain simulation. eigenvalue analysis has a clear theory and high analysis accuracy. But it is difficult to apply in big power system networks, because the factors matrix of the model is too complicated to be calculated. Although the complex torque coefficient also needs detailed turbine generators' rotors parameters as eigenvalue analysis, its calculation is easier. It is feasible for applying in big power system networks. The time domain simulation can evaluate the power system stability by the laws of changes of variables with time in electromagnetic transient simulation application. It has high speed and accuracy, however the mechanism of SSO cannot be presented.

In conclusion, the complex torque coefficient is suitable to study the subsynchronous oscillation caused by HVDC, which can both ensure the accuracy of big power system network and the feasible calculation.

Firstly, the mechanism of SSO caused by HVDC is studied. Then the mathematical models of turbine generators and HVDC are established for analysis by the complex torque coefficient. After that simulation model is built in PSCAD/EMTDC to verify the validation of the method combining theory calculation and simulation, which is applied to SSO caused by HVDC.

2. The mechanism of SSO caused by HVDC

The common reason that HVDC caused SSO is the fast control of HVDC, in which the regulating effect provide feedback to the turbine generator to torsional vibration, causing unstable oscillation.

The simplified network configuration of the effect of HVDC on SSO is shown in Figure 1., in which the left part is the grid and the right part is HVDC. In the left part, the aim turbine generators are separated from the grid to focus on the SSO happening. In the right part, u and i are voltage and current of the AC voltage of commutation bus, x_T is the commutation reactance, and the U_d and I_d are the voltage and current of the DC.

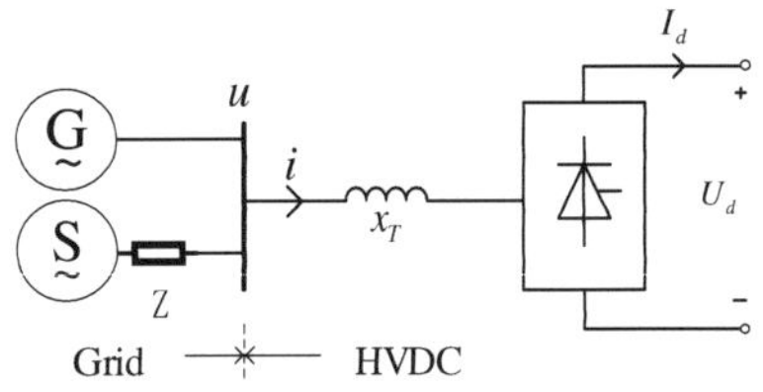

Figure 1. The simplified network

In commutation theory, the DC voltage in steady state is

$$U_d = \frac{3\sqrt{2}}{\pi} U \cos\alpha - \frac{3}{\pi} x_T I_d \tag{1}$$

When the voltage of the turbine generator rotor changes, the voltage of stator accordingly changes leading to the change of the amplitude and the phase of the voltage of the AC voltage of commutation bus. According to Equation (1), the changes of the voltage amplitude of the AC voltage of commutation bus (U) and the firing angle (α) can result in the change of DC voltage (U_d), thus the DC current and power changing.

The change of DC current activates the constant-current control, thus changing the firing angle (α). Then the active power of generators to HVDC, changing the electrical torque of the generator rotor.

When the phase lag of the rotor shaft speed and the final change in electrical torque on the generator rotor exceed 90°, the unstable torsion oscillation may occur.

3. The complex torque coefficient analysis of SSO caused by HVDC

3.1. The principle of the complex torque coefficient

3.1.1. The turbine generator rotor. If the amplitude of the disturbance on turbine generators rotor is A, the displacement increment on the rotor is

$$\Delta\delta = A\sin\omega_m t$$

in which ω_m is the natural frequency.
This disturbance makes the rotor speed increase

$$\Delta\omega = \frac{d\Delta\delta}{dt} = A\omega_m\cos\omega_m t$$

The speed increasement causes the according changes on voltage and flux linkage of the generator stator. The voltage increasement of phase a in dq-coordinates is

$$\begin{aligned}\Delta u_{a1} &= \Delta u_d\cos\omega t - \Delta u_q\sin\omega t + \\ &\Delta\delta(\Delta u_d\sin\omega t - \Delta u_q\cos\omega t) - \Delta\delta(u_d\sin\omega t - u_q\cos\omega t)\end{aligned} \tag{2}$$

After calculation with higher order factors ignored, the voltage increasement of phase a is

$$\begin{aligned}\Delta u_{a2} &= \sqrt{\psi_d^2+\psi_q^2}\{\frac{A}{2}(\omega-\omega_m)\sin[(\omega-\omega_m)t+\rho] - \\ &\frac{A}{2}(\omega+\omega_m)\sin[(\omega+\omega_m)t+\rho]\end{aligned} \tag{3}$$

in which ψ_d and ψ_q are the flux linkages of d-axis and q-axis, and $\rho = \arctan(\psi_q/\psi_d)$.
Equation (3) shows that the voltage increasement contains subsynchronous (ω-ω_m) and supersynchronous (ω+ω_m) components.
(1) The subsynchronous component of the voltage increasement causes current increasement with a frequency of (ω-ω_m), and the current has the same phase as the voltage increasement, which can be expressed as

$$\begin{cases}\Delta i'_{a2} = \Delta I\sin[(\omega-\omega_m)t+\rho] \\ \Delta i'_{b2} = \Delta I\sin[(\omega-\omega_m)t+\rho-2\pi/3] \\ \Delta i'_{c2} = \Delta I\sin[(\omega-\omega_m)t+\rho+2\pi/3]\end{cases}$$

in which,

$$\Delta I = \sqrt{\psi_d^2+\psi_q^2}\,\frac{A}{2}\frac{(\omega-\omega_m)}{R}$$

and the R is the resistance of the stator circuit.
The subsychronous electromagnetic torque of the rotor calculated by current the increasement in dq-coordinates is

$$\Delta T'_e = \psi_d\Delta i'_q - \psi_q\Delta i'_d = -\frac{A}{2}\cdot\frac{\omega-\omega_m}{R}(\psi_d^2+\psi_q^2)\cos\omega_m t \tag{4}$$

(2) The current increasement with a frequency of (ω+ωm) caused by supersynchronous component of the voltage increasement is

$$\begin{cases} \Delta i_{a2}^{"}=\Delta I\sin[(\omega+\omega_m)t-\pi/2+\rho] \\ \Delta i_{b2}^{"}=\Delta I\sin[(\omega+\omega_m)t-\pi/2+\rho-2\pi/3] \\ \Delta i_{c2}^{"}=\Delta I\sin[(\omega+\omega_m)t-\pi/2+\rho+2\pi/3] \end{cases}$$

in which

$$\Delta I=\sqrt{\psi_d^2+\psi_q^2}\frac{A}{2}\frac{(\omega+\omega_m)}{X}$$

and X is the reactance of the stator circuit.
In the similar way, the supersynchronous electromagnetic torque of the rotor is

$$\Delta T_e^{"}=\psi_d\Delta i_q^{"}-\psi_q\Delta i_d^{"}=\frac{A}{2}\cdot\frac{\omega+\omega_m}{R}(\psi_d^2+\psi_q^2)\sin\omega_m t \tag{5}$$

To sum up the equation (4) and (5), the total electromagnetic torque caused by disturbance on the rotor is

$$\Delta T_e=\Delta T_e^{'}+\Delta T_e^{"}=\frac{\omega+\omega_m}{2\omega_m X}(\psi_d^2+\psi_q^2)\Delta\delta-\frac{\omega-\omega_m}{2\omega_m R}(\psi_d^2+\psi_q^2)\Delta\omega \tag{6}$$

3.1.2. The criteria of complex torque coefficient. The concept and judgement of the complex torque coefficient are firstly proposed by Canay in 1982[7], which uses two complex torque coefficients to present the electrical and mechanical torques. Then the possibility of occurrence of SSO in the system is evaluated by analyse the torque coefficients.
Equation (6) shows that the total electromagnetic torque can be expressed as $\Delta T_e=K_e\Delta\delta+D_e\Delta\omega$. The equal mechanic coefficient is defined as $K_M=\Delta T_M/\Delta\delta$, and the equal electric coefficient is defined as $K_E=\Delta T_E/\Delta\delta$. K_M and K_E are complex, which are

$$K_M(j\eta)=K_m(\eta)+j\eta D_m(\eta)$$

$$K_E(j\eta)=K_e(\eta)+j\eta D_e(\eta)$$

in which $\eta=2\pi f$; K_m and D_m are the equal mechanic coefficients of elastic and damping, K_e and D_e are the equal electric coefficients of elastic and damping, and they are all functions of frequency (f).
The damping characteristic of the system can be analyzed though the laws of D_m and D_e by f ranging from 0~50Hz. The criteria of SSO occurrence is

$$(D_m+D_e)\big|_{K_m+K_e=0}<0 \tag{7}$$

Below it is about how to get the equal mechanic coefficients of elastic (K_m) and damping (D_m) and the equal electric coefficients of elastic (K_e) and damping (D_e), Hence it can be evaluated whether the SSO will occur the system by Equation (7), and when $D_m+D_e<0$, unstable oscillation may happen in the system.

3.2. Method to get the complex torque coefficient

3.2.1. The mechanic coefficient. The mechanic torque increasement equals to the electric torque after the equations of the rotor linearized and the armature reaction torque and the function of turbine generator governor are ignored, therefore which is

$$K_M(j\eta) = K_m(\eta) + j\eta D_m(\eta) \tag{8}$$

in which $K_m(\eta)$ is proportional to rotor phase, $D_m(\eta)$ is proportional to the rotor speed, which are defined as equal mechanic coefficient of elastic and damping.

3.2.2. The electric coefficient. When HVDC is in steady state of Figure 1., there are other equations besides Equation (1)

$$\begin{cases} I = \dfrac{\sqrt{6}}{\pi} I_d \\ \cos\varphi = \dfrac{U_d}{\dfrac{3\sqrt{2}}{\pi} U} \\ \Delta\alpha = \Delta\alpha_o + \Delta\delta_U \\ \Delta\delta_I = \Delta\delta_U - \Delta\varphi \end{cases} \tag{9}$$

in which δ_U and δ_I are phases of the voltage and current of the AC voltage of commutation bus, $\Delta\delta_U$ and $\Delta\delta_I$ are their phase errors, φ is the power factor, $\Delta\alpha_o$ is the output increasement of the instant-current control.

The turbine generators are set to operate in island, which means u_S=U and $Z\rightarrow+\infty$. The voltage of the AC voltage of commutation bus is set to be $\dot{U}=U\angle 0^\circ$, and the output apparent power is set to be $S=P+jQ$. Refer to the calculations in [8], there are these equations of the armatures of direct-axis and quadrature-axis and exicitation system by Park mode.

$$X_G = j\eta\Delta\delta(j\eta I - \begin{bmatrix} H_d^{-1}N_d \\ H_q^{-1}N_q \\ 0 \end{bmatrix} - \begin{bmatrix} H_d^{-1}M_d \\ H_q^{-1}M_q \\ 0 \end{bmatrix} \cdot P_U \cdot N_I N_H^{-1} N_0 I_\beta \cdot P_I^{-1} \cdot N_2)^{-1} \begin{bmatrix} H_d^{-1}P_d \\ H_q^{-1}P_q \\ 0 \end{bmatrix} \tag{10}$$

in which N_H, P_U, P_I, I_β are parameters of HVDC, H_d, N_d, M_d, P_d are parameters of armature of d-axis, H_q, N_q, M_d, P_d are parameters of armature of q-axis and N_0, N_1, N_2 are constants matrixes.
Therefore, the increasement of electromagnetic torque is presented as

$$\begin{aligned}\Delta P_e &= (\psi_{d0}\Delta i_q - \psi_{q0}\Delta i_d) + (i_{q0}\Delta\psi_d - i_{d0}\Delta\psi_q) \\ &= P \cdot X_G\end{aligned} \tag{11}$$

in which $P=\begin{bmatrix} -x_d i_{q0} - \psi_{q0} & x_{ad} i_{q0} & x_{ad} i_{q0} & x_q i_{d0} + \psi_{d0} & -x_{aq} i_{d0} & -x_{aq} i_{d0} & 0 & 0 \end{bmatrix}$, the reactance, current and flux linkage are variable of stator, and the subscript '0' presents the original state.

Hence the equal electric complex torque coefficient is $K_E = \dfrac{\Delta P_e}{\Delta\delta}$.
As

$$K_E(j\eta) = K_e(\eta) + j\eta D_e(\eta) \tag{12}$$

therefore the equal electric complex torque coefficient is

$$K_e(\eta) = \mathrm{Re}\{K_E(j\eta)\} \tag{13}$$

and the equal mechanic complex torque coefficient is

IOP Conf. Series: Materials Science and Engineering **533** (2019) 012004 doi:10.1088/1757-899X/533/1/012004

$$D_e(\eta) = \frac{\mathrm{Im}\{K_E(j\eta)\}}{\eta} \tag{14}$$

Above all, by calculating the complex torque coefficients, the equal elastic and damping complex torque coefficients of electric and mechanic are gained to evaluate the possibility of SSO occurrence in the system through Equation (7).

4. Simulation verification

In this chapter, a simulation model in PSCAD/EMTDC is established to verify the validation of equations of complex torque coefficients.

4.1. Caculation of complex torque coefficients

In the model, the HVDC section is the same as the CIGRE Benchmark model for HVDC controls and the generator is the same as in the IEEE first Benchmark case for subsynchronous resonance studies. The capacity of the generator is 892.4MVA, the voltage of the output of the generator is 26kV, and the mechanic complex torque is 0. According to the parameters of the model, complex torque coefficients can be calculated by Equation (9) ~ (14), of which the laws show the SSO characteristic.

Set a disturbance at time of 5s with the whole simulation time of 10s. The commutation failure occurs the HVDC at the disturbance and then the system goes back to steady operation. The steady equal electric complex torque coefficient (D_e) changes by frequency (f) as Figure 2.

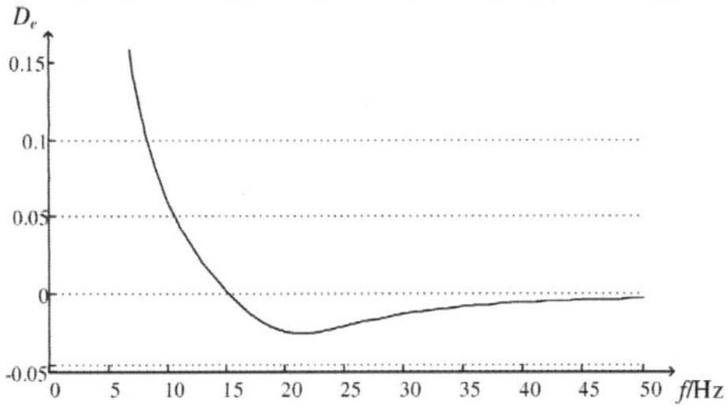

Figure 2. D_e (in calculation)

Figure 2. shows that there is obvious electrical negative damping in the subsynchronous frequencies range. As the mechanical damping of complex torque is 0, the negative damping of electric equals the system unstable, thus verifying to be able to analyse the SSO characteristic of the model system by damping coefficients (D_m and D_e).

4.2. Simulation analysis

In the model, equal electric damping coefficient can be scanning by frequency ranging from 0~50Hz, which is shown in Figure 3. When equal mechanic complex coefficient is 0, the equal elastic and damping coefficients of mechanic are 0.

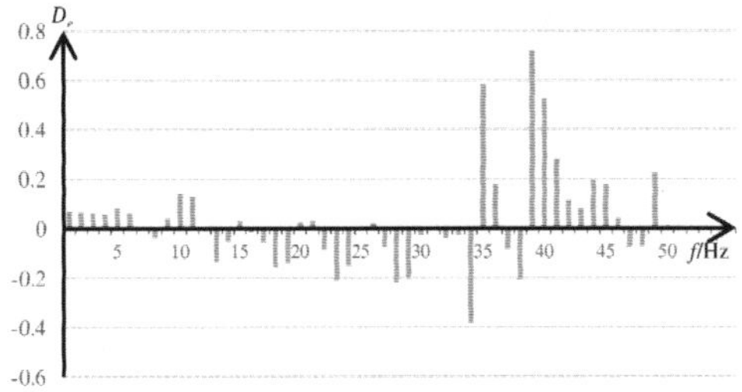

Figure 3. D_e (in frequency scanning)

Comparing Figure 2 and 3, the SSO frequency range is wider in the calculation result than in the simulation. Meanwhile the damping coefficient is less negative in the calculation result than in the simulation. Nonetheless the calculation and simulation results both demonstrate the complex torque coefficient can estimate the occurrence of SSO with the error making little effect. In other words, when the equal electric damping is negative in the model, unstable oscillation on the generator will happen.

The waveforms of rotor speed and electromagnetic torque in total simulation time are shown in Figure 4., in which SSO can be seen.

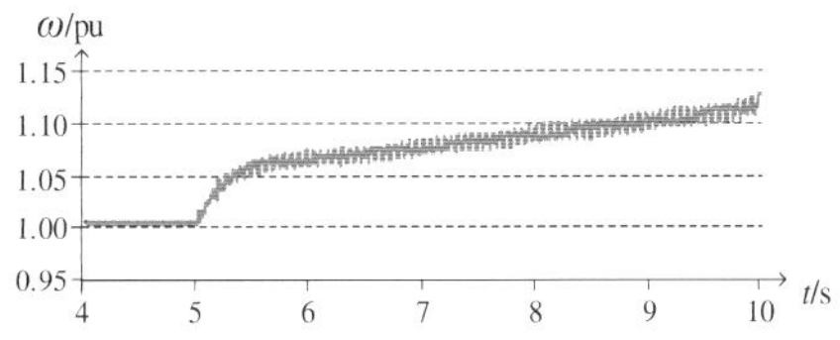

(a) the rotor speed

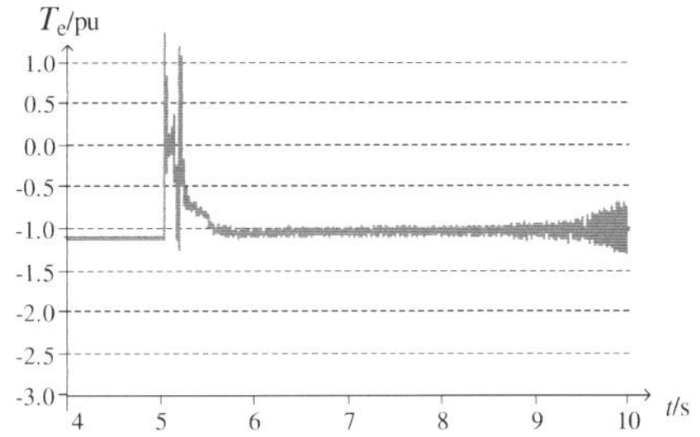

(b) the electromagnetic torque

Figure 4. the SSO occurred

The waveforms in simulation are coincident with the complex torque coefficient results, which the SSO occurred the model system. From the above consequence, the complex torque coefficient can effectively evaluate the possibility of a system containing HVDC. The calculation and simulation both verify that when K_m, D_m, K_e and D_e meet the condition of Figure (7), unstable oscillation may happen in the system containing HVDC.

5. Conclusion

Through the complex torque coefficient, the subsynchronous oscillation caused by HVDC is analysed. The control of HVDC caused the electrical negative damping on the turbine generators rotors, bringing about torsional vibration and SSO. Not only can the complex torque coefficient analyse the big power system in detail, but also it ensures the feasible calculation amount. Calculation and simulation results together prove that the complex torque coefficient can evaluate whether unstable oscillation will occur the system containing HVDC. When the the equal elastic and damping complex torque coefficients of electric and mechanic (K_m, D_m, K_e and D_e) meet the condition of $(D_m + D_e)\big|_{K_m+K_e=0} < 0$, it can be judged that SSO may happen in the system containing HVDC.

6. References

[1] IEEE Committee Report.Reader's guide to subsynchronous resonance. *IEEE Transactions on Power Systems*, **7(1)**,150-157(1992).

[2] X.Q He. The status of UHVDC technology and the application prospect. *Telecom World*, **02**, 216-217(2018).

[3] G.X Zhao. The development research of UHVDC technology. *Electronic Technology & Software Engineering*, **04**, 233-234(2017).

[4] G.F Tang,H P,Z.Y He.R&D and Application of Advanced Power Transmission Technology in China.Proceedings of the CSEE,**07**,1760-1771(2016).

[5] M. Bahrman,E.V.Larsen,R.J. Piwko,et al.Experience with HVDC-Turbine-Generator Torsional Interaction at Square Butte.IEEE Transactions on Power Apparatus and Systems, **99(3)**,966-

975(1980).

[6] P. Zhang, T.S Bi, Q.X Yang,et al.Research on Subsynchronous Oscillation of Parallel Connected Generators With Close Torsional Frequencies. Proceedings of the CSEE,**35(20)**,5181-5187(2015).

[7] I. M. Canay.A novel approach to the torsional interaction and electrical damping of the synchronous machine. *IEEE Transactions on Power Apparatus and Systems*,**101(10)**,3630-3647(1982).

[8] Sh.J Cheng, Y.J Cao, Q.Y Jiang.*The Theory and Methods of subsynchronous oscillation in Power System*. Beijing:Science Press,(2009).

IOP Conf. Series: Materials Science and Engineering **533** (2019) 012005 doi:10.1088/1757-899X/533/1/012005

Research on predictive current control method of PMSM for continuous switched boost inverter

Liyong Yang, Shuo Han, Shuo Liu and Xiaojuan He

College of Electrical and Control Engineering in North China University of Technology, Beijing 100144, China

hanshuo930204@163.com

Abstract. In this paper, the continuous switched boost inverter (CSBI) is used to improve permanent magnet synchronous motor (PMSM) control performance through predictive control, aiming at the drawback of the traditional PMSM drive system. Between the inverted circuit and the input DC power, a switched boost circuit is added to form a switched boost inverter circuit to achieve higher voltage gain and eliminate the dead band. Meanwhile, the drive system has higher reliability. Moreover, the paper adopts predictive current control (PCC) to control PMSM, comparing with the traditional field oriented control (FOC), PCC has faster dynamic performance and reduces current ripple. Finally, the simulation and experimental results verify the feasibility of the based on the continuous switched boost inverter of predictive current control for PMSM.

1. Introduction

Permanent magnet synchronous motor (PMSM) has high efficiency and power factor comparing with direct current machines and induction machines. It possesses characteristics such as less heat, small, simple structure, large overload current and high reliability. It has been widely used in industrial and agricultural production, industrial automation and information technology field [1]. The traditional drive system modulation usually adopts the space vector pulse width modulation (SVPWM) [2], which can reduce harmonic distortion. Meanwhile, SVPWM technology with the continuous switched boost inverter (CSBI) has good ability to resist electromagnetic interference (EMI) and avoids the distortion of the output waveform [3].

CSBI has high boost capability and shoot-through immunity, it combines the high gain of Z-source inverter
(ZSI) [4] and switched boost inverter (SBI) [5] characteristics using a small number of passive components. The CSBI has only one LC filter from the inverter structure. Meanwhile, based on continuous switched boost inverter, establishing the relationship of DC input and alternating current (AC) output analyzes steady state of the inverter.

PMSM control methods are proposed to have good dynamic response performance, such as adaptive PID controller [6], neuro-fuzzy controller [7], direct predictive control (DPC) [8], dead-beat PC [3]. These control schemes have different defects which include fast response, torque ripple, system delay, uncertain parameters and so on. However, predictive current control (PCC) combines advantages of DPC and deadbeat PC, and has been applied successfully in PMSM control.

Content from this work may be used under the terms of the Creative Commons Attribution 3.0 licence. Any further distribution of this work must maintain attribution to the author(s) and the title of the work, journal citation and DOI.
Published under licence by IOP Publishing Ltd

The cost function is chosen appropriately to force the control process to reach the desired standard for PCC method. It contains the current control squared error, and also adds additional conditions to the controller to satisfy the implementation of several additional criteria. The PCC optimizes the voltage vector and the duty cycle to the cost function which guarantees the global optimum of the selected voltage vector, and obviously improves the steady-state performance of the system.

2. Continuous switched boost inverter

2.1. CSBI topology

CSBI gets high gain and can improve EMI immunity in the boost mode. The inverter combines the advantages of the traditional voltage source inverter (VSI) [5], the SBI and the ZSI. It has the same gain as the ZSI, and possesses the same number of components as the SBI. It doesn't need the dead band of the switch signals, so the output waveform distortion could be avoided [5]. The topology of CSBI with the PMSM as shown in Fig. 1.

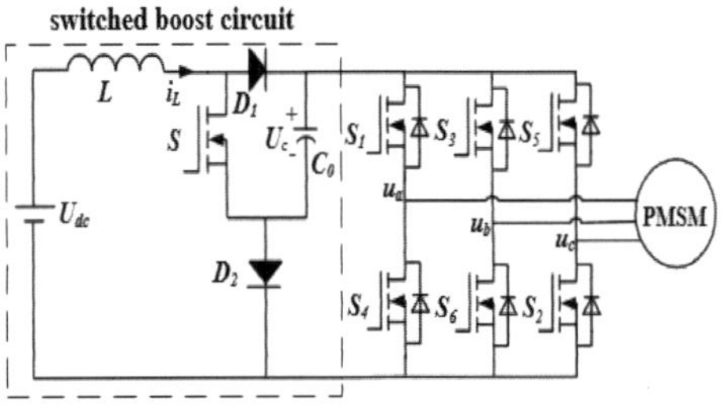

Figure 1. Topology of CSBI

2.2. CSBI operation

The CSBI can provide high voltage gain and can work in two states: shoot-through state and non-shoot-through state. In the shoot-through state, the inductor L is charged by capacitor C_0 and input power U_{dc}. In the non-shoot-through state, the capacitor C_0 is charged by the inductor L. The equivalent circuits of the CSBI in the shoot-through and non-shoot-through state are shown in Fig. 2.

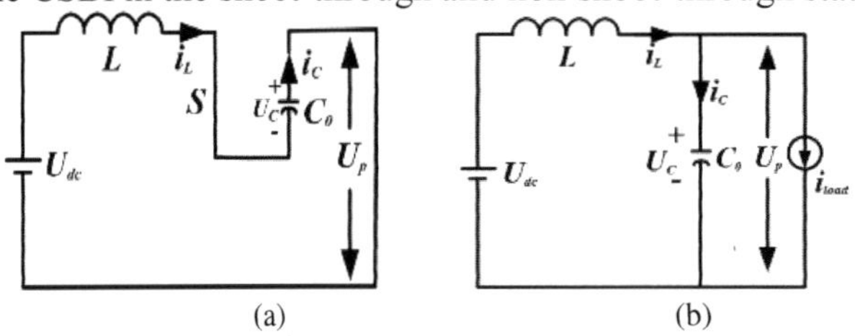

Figure 2. Equivalent circuit of the CSBI: (a) shoot-through state and (b) non-shoot-through state

When the inverter operates in the shoot-through state, the voltage of inductor L and the current of capacitor C_0 are given by Fig.2(a).

$$\begin{cases} U_L = U_{dc} + U_C \\ i_C = -i_L \end{cases} \tag{1}$$

When the inverter operates in the non-shoot-through state, the voltage of inductor L and the current of capacitor C_0 are given by Fig.2(b).

$$\begin{cases} U_L = U_{dc} - U_C \\ i_C = i_L - i_{load} \end{cases} \tag{2}$$

where U_L is the voltage of the inductor L; U_{dc} is the input source; U_C is the voltage of the capacitor C_0; i_L is the current of the inductor L; i_C is the current of the capacitor C_0, and i_{load} is the output current of the inverter.

When the steady state equilibrium, using the volt-second balance of inductance and the amp-second balance of capacitor can get the voltage conversion ratio B of the CSBI [9],

$$B=\frac{U_C}{U_{dc}}=\frac{1}{1-2D} \tag{3}$$

Where D is the shoot-through duty cycle, and B is the voltage conversion ratio of CSBI. The relationship between voltage conversion ratio and shoot-through duty cycle is shown in Figure 3. Fig. 3 can be seen the B of the system increases with the increase of D.

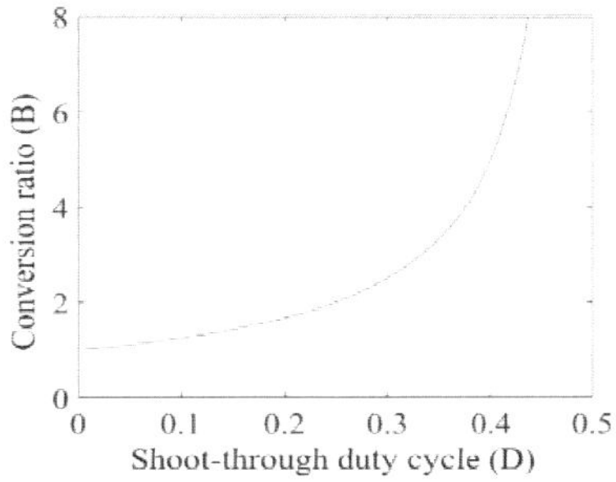

Figure 3. Relationship between voltage conversion ratio and shoot-through duty ratio

3. PMSM model and control strategy

PCC [6] aims to improve the response speed and steady-state error of PMSM, and the PCC replaces the PI controller to use current control in the FOC. Similar to other predictive control methods, the PCC based on the motor model in the switching period to get the desired current, the feedback of the desired currents are i_{dk} and i_{qk}, the value of rotor speed ω_s from coder and the value of the d-q currents. Fig.4 shows the diagram of the PCC algorithm using a PI controller as a speed control loop. This paper uses i^*_{dk}=0 control method.

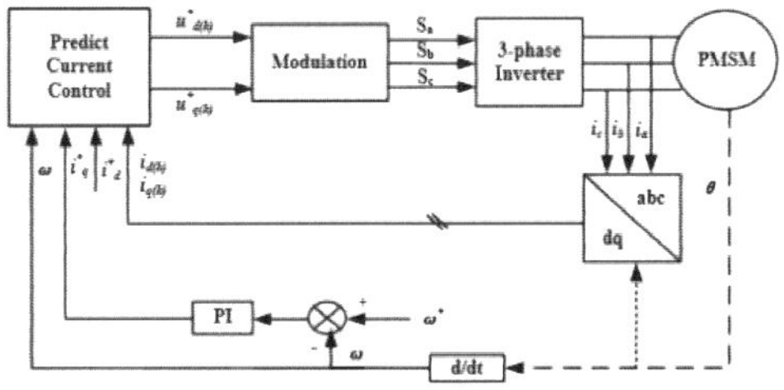

Figure 4. Block diagram of the PCC algorithm

3.1 Discrete model of PMSM

In the PCC algorithm [6], the control variables are stator currents. In order to predict the stator currents of given switching states, the differential equation model needs to transform to discrete time model based on the dynamic model of the motor. In this way, the next state of d-q currents is predicted by the current value of the d-q currents, the measured electrical angle and the input d-q voltages. When the sampling period is very short, the surface mounted PMSM (L=L_d=L_q) is created a discrete time model by Taylor series expansion [6].

$$\begin{cases} i_{dk+1}=(1-\frac{RT_s}{L_d}i_{dk}+\omega_{sk}T_s i_{qk})+\frac{T_s}{L_d}u_{dk} \\ i_{qk+1}=(1-\frac{RT_s}{L_q})i_{qk}-\omega_{sk}(T_s i_{dk}+\frac{T_s\psi_f}{L_q})+\frac{T_s}{L_q}u_{qk} \end{cases} \tag{4}$$

3.2 Predictive Current Control Method

The design of the controller involves the calculation of the reference voltage that should be implemented in the current time step, so as to ensure that the current will follow the reference voltage in the next step. This goal is to get the minimum value of the difference between the predicted currents i_{dk+1}, i_{qk+1} and the reference values i^*_{dk}, i^*_{qk}.

The cost function selects the voltage vector that minimizes between the predicted current and given current. It contains the current control squared error, and adds additional conditions to the controller to satisfy the implementation of several additional criteria. The value of predictive current can be given by the discrete model (4) of PMSM, a cost function [7] is calculated as follows:

$$g(u^*_{dk}, u^*_{qk}) = i_{dk+1}{}^2 + (i_{qk+1} - i^*_{qk})^2 \tag{5}$$

The cost function includes u^*_{dk} and u^*_{qk} , it must ensure that achieves the minimum value in each step. Through (4), the predictive currents i_{dk+1}, i_{qk+1} can be expressed the reference voltages u^*_{dk}, u^*_{qk}, and measured currents i_{dk}, i_{qk}, and speed value ω_s, respectively.

The cost function is rewritten as,

$$g(\mathbf{u}_k) = \mathbf{u}_k^T \lambda^2 \mathbf{u}_k + \mu_k \mathbf{u}_k \tag{6}$$

The parameters of this cost function are defined as:

$$\mathbf{u}_k = \begin{bmatrix} u_{dk} \\ u_{qk} \end{bmatrix}$$

$$\lambda = \frac{T_s}{L_{dq}}$$

$$\mu_{1k} = 2\lambda\left(i_{dk} - \frac{RT_s}{L_{dq}} i_{dk} + \omega_{sk} T_s i_{qk+1}\right)$$

$$\mu_{2k} = 2\lambda\left(-\omega_{sk} T_s i_{dk} + i_{qk} + \frac{RT_s}{L_{dq}} i_{qk} - i^*_{qk} - \omega_{sk} \frac{T_s}{L_{dq}} \psi_f\right)$$

$$\mu_k = [\mu_{1k} \quad \mu_{2k}]$$

The diagonal-constant matrix $\boldsymbol{\mu}_k$ of ĝ is time variable, because they depend on the feedback currents i_{dk}, i_{qk} and rotor speed ω_{sk}. Moreover, $\boldsymbol{\mu}_{2k}$ depends on the output of the speed control loop orthogonal reference current i^*_{qk}. Therefore, the optimization process takes place at each time step and produces a variable voltage reference signal.

Because of the safety and reliability of the motor, an additional constraint is required that the current amplitude does not exceed a certain maximum value. The technical specifications of the motor select the maximum current as the maximum transient permissible current.

The maximum current constraint needs a certain limit; that is,

$$\begin{cases} i_{dk+1} < i_{MAX} \\ i_{qk+1} < i_{MAX} \end{cases} \tag{7}$$

Substitute (4) into (7) to obtain the optimal voltage value for the controller,

$$\begin{cases} u^*_{dk} < \dfrac{1}{\lambda}\left(i_{MAX} - i_{dk} + \dfrac{RT_s}{L_{dq}} i_{dk} - \omega_{sk} T_s i_{qk}\right) \\ u^*_{qk} < \dfrac{1}{\lambda}\left(i_{MAX} + \omega_{sk} T_s i_{dk} - i_{qk} + \dfrac{RT_s}{L_{dq}} i_{qk}\right) + \omega_{sk} \psi_f \end{cases} \tag{8}$$

By using the Newton's iterative optimal algorithm, the system can run the optimization process, and the r+1 iteration function is expressed by the function of the rth iteration result [6], that is,

$$\mathbf{u}_{kr+1} = \mathbf{u}_{kr} - \gamma\left(\mathbf{u}_{kr} + \frac{1}{2\lambda^2} \mu_K^T\right) \tag{9}$$

The coefficient γ ($0<\gamma<1$) of the Newton's method is introduced to modify. It can reduce the step size to prevent potential numerical instability, divergence, and overshooting.

Table 1. Parameter of the used PMSM

Parameter	Value
Rated Voltage U_n	160V
Rated Current I_n	5.5A
Rated Speed ω_n	1200r/min
Inductance L	17mH
Per Phase Winding Resistance R	1.3Ω
Torque Constant k_m	1.23Nm/A
Number of Pole Pairs n_p	3
Inertia J	0.00087kg·m^2

Table 2. Parameters of the control system

Parameter	Value
Switching frequency f_s	5kHz
Sample time T_c	0.2μs
Boost circuit inductance L	2mH
DC bus capacitor bank C_0	560μF
Optimization step coefficient γ	0.7

4. Simulation results

In this part, Fig. 5 shows the relationship between the input voltage U_{dc} and the bus voltage U_c. Then, the PCC algorithm in SVPWM was carried out using a PMSM. Nominal parameters are given in Table 1, and the performance of the PCC scheme has been analyzed by simulation. Table 2 shows the parameters of the Control System, and the shoot-through duty cycle is D=0.2 in this simulation.

4.1. The input voltage and the bus voltage simulation results of CSBI

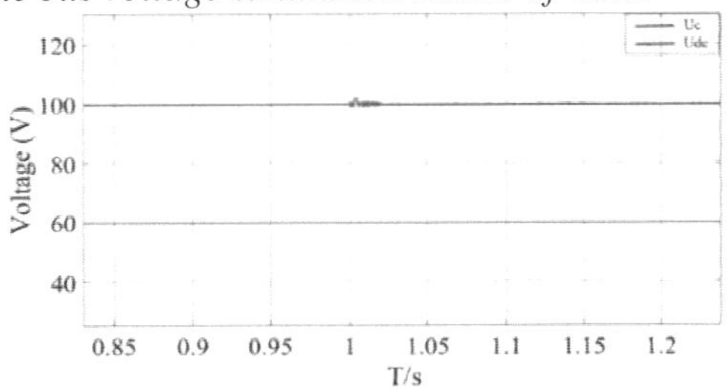

Figure 5. The input voltage and the bus voltage

In Fig.5, the input voltage U_{dc} is 60V, according to (3) gets the bus voltage U_c after boost circuit. The steady state after external load, the bus voltage is 100V satisfying the equation (3).

IOP Conf. Series: Materials Science and Engineering **533** (2019) 012005 doi:10.1088/1757-899X/533/1/012005

4.2. PMSM simulation results

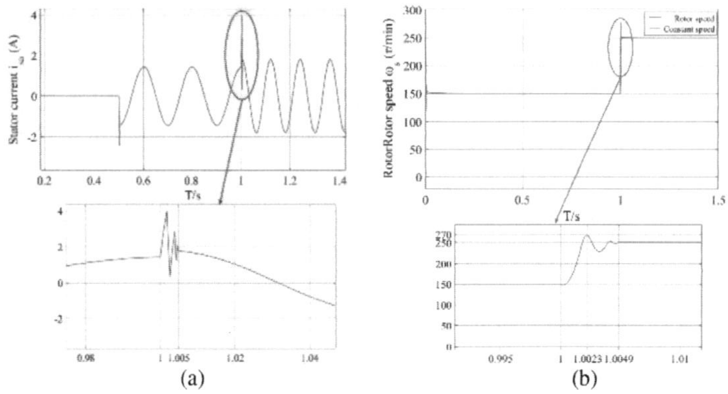

Figure 6. Simulation results: (a) stator current waveform and (b) rotor speed waveform

The motor starts at no-load. Then, a step of external load is 0.2 N· m and a constant speed is about 150r/min that are applied at about t=0.5s and then removed at t=1s. Finally, a step of external load is 0.5 N· m and a constant speed is about 250r/min that are applied at about t=1s and then removed at t=1.5s.

In Fig.6(a), after external load, the waveform of the stator current could recover quickly. In Fig.6(b), under the SVPWM, the PCC peak time is 0.0023s, the maximum speed overshoot is 8%, and the transient time is 0.0049s.

The performance of the PCC has been analyzed by simulation results as shown in the Table 3.

Table 3. Simulation Results for PCC.

	PCC
Speed peak time	0.0023s
Speed transient time	0.0049s
Speed overshoot	8%
Speed steady-state error	0

5. Experimental results

The experimental setup basically consisted of a PMSM with a control board implemented in stm32f407ZGT6 and EMP570T100C5. Then, it verifies the relationship between the input voltage U_{dc} and the bus voltage U_c. The PMSM and the control system parameters are the same as Table 1 and Table 2, and the shoot-through duty cycle is D=0.3 in this experiment.

5.1. The input voltage and the bus voltage experimental results of CSBI

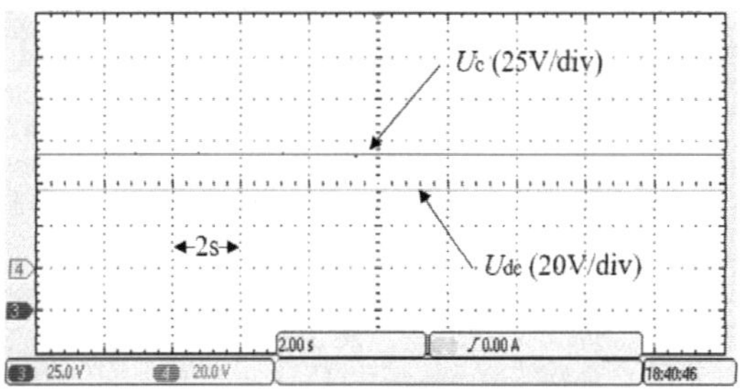

Figure 7. The input voltage and the bus voltage (2s/div)

In Fig.7, the input voltage U_{dc} is 36V, according to the equation (3) gets the bus voltage U_c after boost circuit. The bus voltage is 92.5V satisfying the equation (3) after the steady state.

5.2. PCC experimental result

In this part, the motor starts at no-load. Then, a constant speed is 50r/min, and a constant speed changes 200r/min a few minutes later.

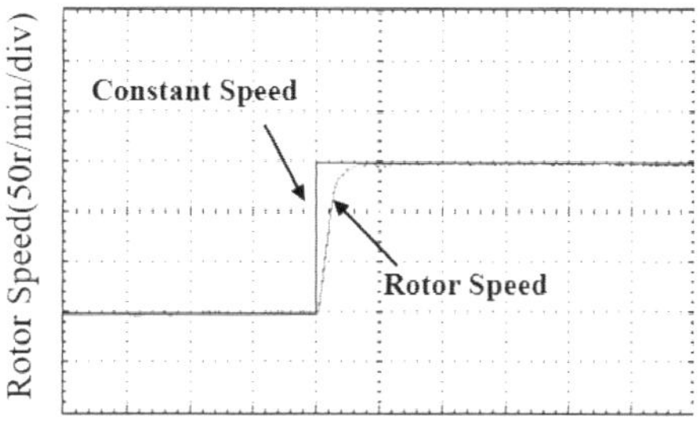

Figure 8. Rotor speed experimental results (1s/div)

Fig.8 shows the rotor speed experimental results of a speed controller with the PCC method. The speed rising time is 0.8s. Meanwhile, the PCC method has not overshoot and steady-state error. The PCC algorithm possesses preferable response speed.

6. Conclusion

This paper designs a kind of PCC of PMSM for the CSBI, it combines and improves the CSBI, SVPWM method and PCC algorithm. The PMSM drive system can obtain the continuous input current from DC power source, and avoid the output waveform distortion. When the input voltage fluctuation, the voltage of the capacitor remains the same. It can be seen from the simulation and experimental results that the same results can be obtained by using the PCC method based on the CSBI. Meanwhile, the control system also acquired excellently steady control effect.

References

[1] Alacoque J C. Direct Eigen Control for Induction Machines and Synchronous Motors[M]. Wiley & Sons, 2013.
[2] Singh S, Tiwari A N. Analysis and Simulation of Vector Controlled PMSM Drive using SVPWM Inverter[C]. International Conference for Convergence in Technology. 2017.
[3] Zhang X, Hou B, Mei Y. Deadbeat Predictive Current Control of Permanent-Magnet Synchronous Motors with Stator Current and Disturbance Observer[J]. IEEE Transactions on Power Electronics, 2017, 32(5):3818-3834.
[4] Yu K, Xiong B, Zhao J. A comprehensive study of space vector pulse-width modulation technique for three-phase Z-source inverters[J]. International Journal of Circuit Theory & Applications, 2016, 44(2):364-381.
[5] Meinagh F A A, Babaei E, Tarzamni H. Modified high voltage gain switched boost inverter[J]. Iet Power Electronics, 2017, 10(13):1655-1664.
[6] Alexandrou A D, Adamopoulos N K, Kladas A G. Development of a Constant Switching Frequency Deadbeat Predictive Control Technique for Field-Oriented Synchronous Permanent-Magnet Motor Drive[J]. IEEE Transactions on Industrial Electronics, 2016, 63(8):5167-5175.
[7] Bozorgi A M, Farasat M, Jafarishiadeh S. Model predictive current control of surface-mounted permanent magnet synchronous motor with low torque and current ripple[J]. Iet Power Electronics, 2017, 10(10):1120-1128.
[8] Antonello R, Carraro M, Peretti L, et al. Hierarchical Scaled-States Direct Predictive Control of Synchronous Reluctance Motor Drives[J]. IEEE Transactions on Industrial Electronics, 2016, 63(8):5176-5185.
[9] Nag S S, Mishra S. Current-Fed Switched Inverter[J]. IEEE Transactions on Industrial Electronics, 2014, 61(9):4680-4690.

IOP Conf. Series: Materials Science and Engineering **533** (2019) 012006 doi:10.1088/1757-899X/533/1/012006

Study on the Reactive Power Control in Hybrid DC Asynchronous Interconnection System

Yan Han, Yang Jia, Che Keshan, Chen Hui, Yang Wenli, Liu Ke, Bao Jixiu

Power Research Institute of Qinghai Power Grid Company, Xining City, China

yh19940103@163.com

Abstract. Reactive power control of hybrid DC asynchronous interconnection system has a vital impact on the safety, reliability and high quality of AC/DC power transmission system. The LCC-HVDC unit and MMC-HVDC unit share the same station control system in the hybrid DC asynchronous interconnected system. Firstly, this paper studies the reactive power switching control of LCC-HVDC system, including five subfunctions and control strategies of the reactive power switching control. Secondly, the reactive power auxiliary control of LCC-HVDC system including low load reactive power regulation function and Gamma-Kick function is studied. And the shortcomings of using reactive power auxiliary control are pointed out. Finally, the reactive power control of MMC-HVDC system is deeply studied. The mathematical model of MMC-HVDC control system is established. Reactive power control of MMC-HVDC can be used to replace reactive power auxiliary control of LCC-HVDC to improve the converter utilization and service life

1. Introduction

The imbalance of energy storage and load distribution determines that long-distance and large-capacity transmission becomes the inevitable choice for developing the power grid and meeting the demand of electricity in China. High Voltage Direct Current (HVDC)transmission technology has obvious advantages in long-distance and large-scale transmission. In the recent years, with the vigorous development of HVDC transmission projects, the grid structures with parallel AC-DC and multi-infeed direct current have gradually formed. The safe and stable operation of power grid is facing many problems and challenges. In 2016, China southern power grid realized the asynchronous interconnection between Yunnan power grid and southern China power grid by Guanyin Yan DC and Luxi Back-to-Back DC projects. It optimized network structure and eliminated the risk of large-scale blackouts. After comprehensively comparing the economy, land occupation, technology and engineering demonstration significance the parallel connection structure of LCC-HVDC and MMC-HVDC are used to form the hybrid DC asynchronous interconnection system in the Luxi back-to-back HVDC project .The superior control performance of the flexible transmission technology is utilized to

Content from this work may be used under the terms of the Creative Commons Attribution 3.0 licence. Any further distribution of this work must maintain attribution to the author(s) and the title of the work, journal citation and DOI.
Published under licence by IOP Publishing Ltd

further improve the operation of the hybrid DC transmission system. The voltage stability of LCC-HVDC transmission system is improved.

Hybrid DC asynchronous interconnection system integrates the advantages of LCC-HVDC and MMC-HVDC transmission system. LCC-HVDC is connected by two 12 pulse converters through a smoothing reactor (L_d). Each 12-pulse converter is composed of two 6-pulse converters. The DC grounding electrode is located between the two 6-pulse converters on the inverter side. MMC-HVDC consists of two back-to-back modular multilevel converters(MMC). Because the probability of DC faults in back-to-back DC system is very low, the cascaded MMC composed of half-bridge sub-modules without DC side fault self-cleaning ability will not seriously reduce the power supply reliability and availability of MMC-HVDC. The schematic diagram of hybrid DC asynchronous interconnection system is shown in Figure 1.

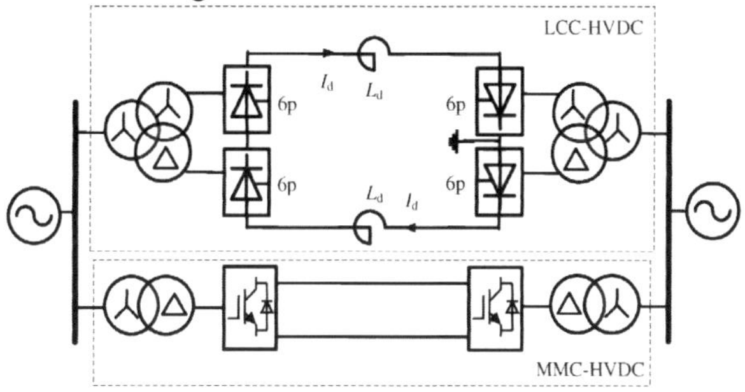

Figure 1. schematic diagram of hybrid DC asynchronous interconnection system

LCC-HVDC and MMC-HVDC share the same station level control system in the hybrid DC asynchronous interconnection system. Reactive power control belongs to the DC station control system. In this paper, reactive power control of LCC-HVDC and MMC-HVDC are studied respectively. Firstly, the reactive power switching control strategy and function of LCC-HVDC are analyzed. And then the reactive power auxiliary control of LCC-HVDC: low load reactive power regulation function and Gamma-Kick function are studied. Finally, the mathematic model of MMC-HVDC control system is established and the principle of reactive power control is deeply studied. It is analyzed that reactive power control of MMC-HVDC replaces reactive power auxiliary control of LCC-HVDC to improve the utilization rate and service life of converter.

2. Reactive power switching control of LCC-HVDC

LCC-HVDC system needs absorbing a lot of reactive power. At the same time the required reactive power changes with circumstance. Besides, LCC-HVDC system is a huge harmonic source for the AC system. If it is not controlled, it will seriously damage the reactive power balance and power quality of the grid. Therefore, the reactive power compensation configuration of LCC-HVDC system has a crucial impact on the designing and planning of HVDC project. The reactive power switching control of LCC-HVDC can effectively control the switching of various reactive power compensation devices in the converter station in real time so that the reactive power exchange between the AC system and DC system or voltage of converter bus is within the prescribed range. The reactive power compensation equipments in the converter station include AC filter, shunt capacitor and shunt reactor. All kinds of reactive power compensation equipment have the characteristics of "step-type" compensation.

2.1. Function of reactive power switching control

Reactive power switching control of LCC-HVDC system includes reactive power control, AC voltage control and filter control. According to the preset priority and criteria, the reactive power control sub-function realizes its specific control function without conflict. The coordination among the sub-functions is realized through the scheduled reactive power control strategy. And the optimal control effect is achieved. According to the order of the highest priority 1 to the lowest priority 5, the subfunctions of reactive power control are shown in Table 1.

Table1. Subfunction of reactive power control.

Priority	Subfunction of reactive power control	Content of subfunction
1	Abs Min Filter Control	Ensure that the filter is not overloaded.
2	$U_{max/min}$ Control	Ensure that the AC system voltage connected to the converter station during the operation of the DC system does not exceed its maximum and minimum values
3	Q_{max} Control	Ensure that the maximum capacitive reactive power delivered to the AC system at the converter station does not exceed the maximum value
4	Min Filter Control	In order to satisfy the filtering performance of the system, the filter banks are put into the filter group according to the required filter type and number.
5	Q_{con}/U_{con} Control	Q_{con} takes reactive power switching as the criterion to control the reactive power exchange within the allowable deviation $\pm Q$ by switching filters. U_{con} takes reactive power switching as the criterion to control the converter station bus voltage variation within the allowable deviation $\pm Q$ by switching filters.

2.2. Strategy of reactive power switching control

According to the logic of reactive power control priority , the corresponding reactive power control strategy is formulated to achieve the coordination of each sub-function and meet the requirements of reactive power control. The specific control strategy is as follows:

Step 1:Abs Min Filter control has the highest priority. According to different working conditions and different power levels, the number and type of filter banks are determined by the filter capacity limitation. Then the absolute minimum filter banks are put into operation, and the control conditions of Abs Min Filter control are monitored in the full power range of DC operation. When other sub-functions request the removal of reactive power units, it determines whether the removal meets the Abs Min Filter control conditions. If it is not satisfied, the Abs Min Filter control will prohibit the removal operation.

Step 2: If the Umax/min control detects that the AC bus voltage is higher than the maximum $U_{MAX\text{-}LIMIT}$ and lasts for a period of time after the Abs Min Filter control condition is satisfied, the reactive units are removed sequentially until the AC bus voltage returns to normal;and vice versa.

Step 3: After the Umax/min control condition is satisfied, the Q_{max} control starts to monitor the reactive power of the converter station feeding into the AC system. When the detection result exceeds its predetermined maximum Q_{acmax} ,the instruction to remove the reactive power unit is executed. Q_{max} control can not two-way control reactive power unit to input or cut but can only control the removal of reactive power unit.

Step 4: After the Q_{max} control condition is satisfied, the Min Filter control first calculates the type of filter banks and the minimum capacity required to satisfy the filtering requirements of the converter station according to the operating state of the converter station and the DC transmission power. When the filter condition is not satisfied, the filter bank is input according to the result of precomputation.

Step 5: *Q*con/Ucon controls the reactive power exchange Q_{exc} and the commutation bus voltage U_{pcc} in the AC/DC system to be within the prescribed range by the reactive power unit switching on or off. *Q*con/*U*con controls are respectively used to suppress reactive power balance and AC bus voltage fluctuation in AC/DC system. When the DC system is fed into the weak grid, reactive power will significantly affect the stability of AC voltage. Therefore, U_{con} control is usually used when the AC system is weak to suppress the voltage fluctuation of the converter bus more effectively.

3. Reactive power auxiliary control of LCC-HVDC

3.1. Low load reactive power regulation function

Low-load reactive power regulation function is to increase the trigger angle or turn-off angle of the converter to absorb excess reactive power when HVDC project is running at low power, which plays a good role in back-to-back HVDC project. When the DC system operates at low load and can not continue to remove reactive power units, the converter absorbs less reactive power than reactive power compensation. The converter will inject a large number of excess reactive power into the AC system. It is easy to cause the voltage of the converter station and nearby stations to rise. Even the voltage may overrun. It poses a threat to the stable operation of the grid. Increasing the trigger angle to increase the reactive power absorbed by the converter is a natural characteristic of the converter, but the operation performance of the HVDC transmission project will deteriorate after increasing the trigger angle, which also has a very adverse impact on the life of the equipment. Therefore, the HVDC transmission project is not suitable for long-term operation in the transmission project.

3.2. Low load reactive power regulation function

LCC-HVDC reactive power switching control can cause voltage fluctuation of converter bus. The Gamma-kick function can reduce AC voltage fluctuation by controlling turn-off angle quickly and reversely. When the reactive power control of converter station detects the requirement of switching in the reactive power unit, the turn-off angle is suddenly increased within a relatively short period of time after the switching signal of switching in the filter is sent out, which makes the converter absorb reactive power increasing suddenly to counteract the voltage of the converter bus increasing suddenly. Then the turn-off angle will gradually decrease to normal value at a relatively long time. Similarly, when the reactive power control of converter station detects the requirement of removing reactive power unit, the turn-off angle increases gradually in a relatively long time. And the turn-off angle decreases abruptly to the normal value within a relatively short time after the signal of removing reactive power unit is sent out, which makes the reactive power absorbed by converter reducing suddenly to counteract the sudden drop of the voltage of converter bus bar.

As is shown in Figure 2, when the reactive power unit is switched on the output of the Gamma-Kick function increases to γmax in T-short time and then decreases from γmax to normal in T-long time. Besides, when the reactive power unit is removed the output of Gamma-Kick function increases to γmax in T-long time and then decreases from γmax to normal in T-short time. The value of γmax, T-long, T-short on the rectifier and inverter should be determined according to the actual system. T-long and T-short should match the response time of the voltage control system. The exact time constants can be tested in the factory system test and the final commissioning process. In engineering applications, the increment of turn-off angle is usually between 2 and 7 degrees. T-short is more than ten milliseconds. T-long can be between hundreds of milliseconds and several seconds. Similar to the low-load reactive power regulation function, the Gamma-Kick function suppresses AC voltage fluctuations by increasing the turn-off angle at the expense of converter utilization, which has greater drawbacks.

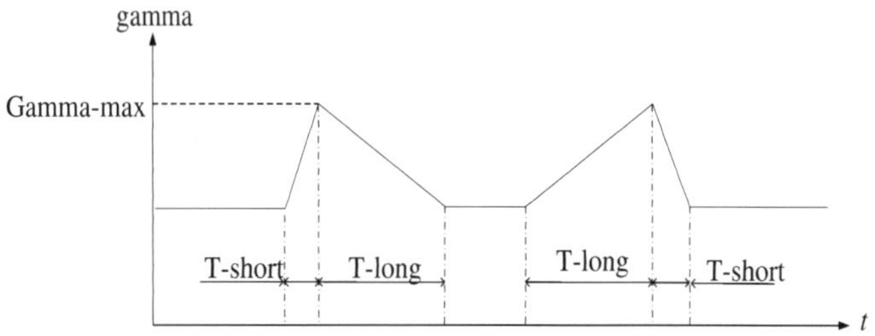

Figure 2. Coordination between Gamma-Kick function and reactive power switching control

4. Reactive power control of MMC-HVDC

Comparing with LCC-HVDC, MMC-HVDC not only does not need to provide reactive power compensation, but also can play the role of static synchronous compensator (STATCOM) when the capacity permits. MMC-HVDC provides dynamic reactive power to the AC system and improves the voltage stability of the connected system. MMC-HVDC control system includes internal loop control and external loop control. Reactive power control is implemented in external loop control. The block diagram of MMC-HVDC control system is shown in Figure 3.

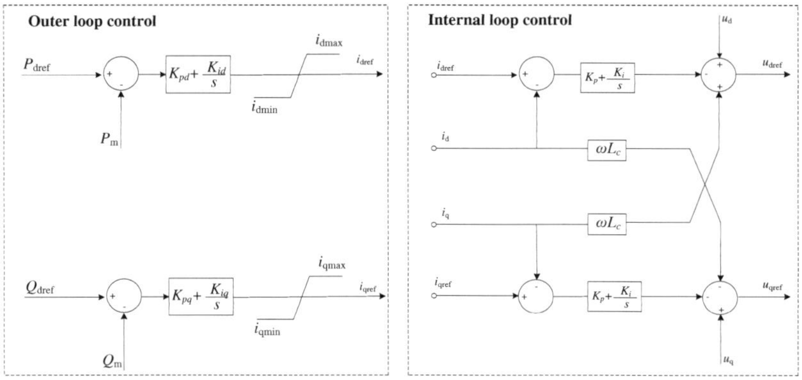

Figure 3. Controlcontrol block diagram of MMC-HVDC

4.1. *Internal loop control of MMC-HVDC*

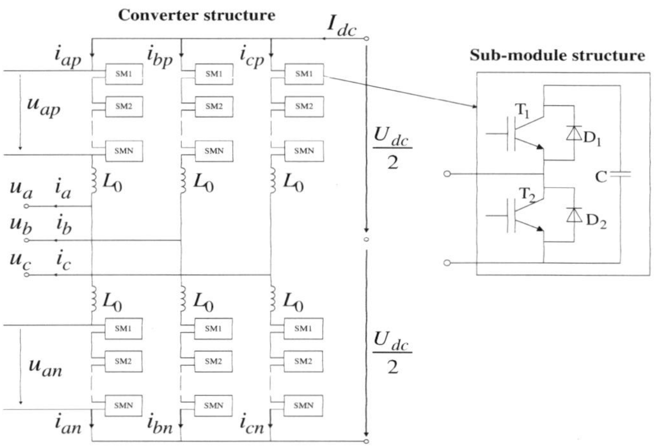

Figure 4. Topology structure diagram of MMC-HVDC

According to Figure 4, the KVL law can be used for the upper and lower arm of k(a,b,c) phase units

$$\begin{cases} u_k = \dfrac{U_{dc}}{2} - e_{kp} - R_0 i_{kp} - L_0 \dfrac{di_{kp}}{dt} \\ u_k = -\dfrac{U_{dc}}{2} + e_{kn} + R_0 i_{kn} + L_0 \dfrac{di_{kn}}{dt} \end{cases} \tag{1}$$

R_0 represents the equivalent resistance of the bridge arm. e_{kp} and e_{kn} respectively represent the equivalent electromotive force of the upper and lower bridge arms. The voltage u_{kp} and u_{kn} are formed under the action of the bridge arm current. Due to the symmetry of the phase units, the DC current I_{dc} is evenly distributed among the three phase units, and the DC current in each phase unit is $I_{dc/3}$. Because the reactance L_0 of the upper and lower arm are equal, the alternating current i_k is equally distributed between the upper and lower arm. The current of the upper and lower arm are respectively equal.

$$\begin{cases} i_{kp} = \dfrac{I_{dc}}{3} + \dfrac{i_k}{2} \\ i_{kn} = \dfrac{I_{dc}}{3} - \dfrac{i_k}{2} \end{cases} \tag{2}$$

Join the formula(1) and formula(2) and make $e_k = \dfrac{e_{kn} - e_{kp}}{2}$

$$e_k = u_k + \frac{1}{2} R_0 i_k + \frac{1}{2} L_0 \frac{di_k}{dt} \tag{3}$$

e_k is the AC electromotive force of $k(a, b, c)$ phase. formula(3) describes the relationship between the controllable electromotive force e_k of the phase unit and the voltage and current of the AC side. For the formula(3), the park transform and the Laplasse transform are used.

$$\begin{cases} e_d = u_d + \dfrac{1}{2}(R_0 + sL_0) i_d + \dfrac{1}{2}\omega L_0 i_q \\ e_q = u_q + \dfrac{1}{2}(R_0 + sL_0) i_q - \dfrac{1}{2}\omega L_0 i_d \end{cases} \tag{4}$$

i_d and i_q are controlle$_d$ variables. e_d and e_q are controlling variables. u_d and u_q are disturbance variables.

4.2. External loop control of MMC-HVDC

The reference values of active and reactive control variables provide current reference values for the internal loop current control in the external loop control receiving system. Commonly external loop control methods include active power control(constant active power control, constant DC voltage control, constant frequency control) and reactive power control (constant reactive power control and constant AC voltage control). In order to ensure the stability of DC voltage and the balance of active power, one of converter station must adopt constant DC voltage control, the other must adopt constant active power control or constant frequency control. In order to obtain the relationship between active and reactive power and AC current ik, the following deduction is carried out. Decompose the formula(4) into two subsystems: positive sequence and negative sequence component.

$$\begin{cases} e_d^{\ +} = u_d^{\ +} + \dfrac{1}{2}(R_0 + sL_0) i_d^{\ +} + \dfrac{1}{2}\omega L_0 i_q^{\ +} \\ e_q^{\ +} = u_q^{\ +} + \dfrac{1}{2}(R_0 + sL_0) i_q^{\ +} - \dfrac{1}{2}\omega L_0 i_d^{\ +} \end{cases} \tag{5}$$

$$\begin{cases} e_d^- = u_d^- + \frac{1}{2}(R_0 + sL_0)i_d^- + \frac{1}{2}\omega L_0 i_q^- \\ e_q^- = u_q^- + \frac{1}{2}(R_0 + sL_0)i_q^- - \frac{1}{2}\omega L_0 i_d^- \end{cases} \tag{6}$$

The asymmetrical three-phase voltage uk without zero sequence components can be represented by two orthogonal positive sequence and negative sequence components.

$$u_k = e^{j\omega t}u_{dq}^+ + e^{-j\omega t}u_{dq}^- \tag{7}$$

$$\begin{cases} u_{dq}^+ = u_d^+ + ju_q^+ \\ u_{dq}^- = u_d^- + ju_q^- \end{cases} \tag{8}$$

Similarly,

$$i_{abc} = e^{j\omega t}i_{dq}^+ + e^{-j\omega t}i_{dq}^- \tag{9}$$

$$\begin{cases} i_{dq}^+ = i_d^+ + ji_q^+ \\ i_{dq}^- = i_d^- + ji_q^- \end{cases} \tag{10}$$

$$S = P + jQ = (e^{j\omega t}u_{dq}^+ + e^{-j\omega t}u_{dq}^-)(e^{j\omega t}i_{dq}^+ + e^{-j\omega t}i_{dq}^-)^* \tag{11}$$

$$\begin{cases} P(\mathrm{t}) = P_0 + P_{c2}\cos 2\omega t + P_{s2}\sin 2\omega t \\ Q(\mathrm{t}) = Q_0 + Q_{c2}\cos 2\omega t + Q_{s2}\sin 2\omega t \end{cases} \tag{12}$$

It can be seen from formula(12) that the double frequency fluctuation component will appear in the active power and reactive power due to the unbalanced voltage. When the steady state (uq=0) and the negative sequence current component is suppressed, the DC component of the active and reactive power sent into the MMC by the AC system can be expressed as

$$\begin{cases} P_0 = 1.5u_d i_d \\ Q_0 = -1.5u_d i_q \end{cases} \tag{13}$$

The reactive power control of MMC-HVDC makes the AC bus voltage of converter station or the reactive power exchanged between converter station and system maintain at the set value by fast and accurate reactive power regulation. Finally, according to the voltage deviation of the converter bus, the reference current Iq of the MMC internal loop is obtained by the proportional-integral(PI)link, and the reactive power control is realized by controlling Iq.

Because of the shortcomings of LCC-HVDC reactive power auxiliary control and the superiority of MMC-HVDC reactive power control MMC-HVDC can be used to coordinate with reactive power control of LCC-HVDC to improve system operation economy and equipment utilization. When the MMC-HVDC has reached its output capacity limit, LCC-HVDC can activate its reactive power auxiliary control function. When the hybrid DC asynchronous interconnection system works stably, the reactive output of the MMC-HVDC system is kept at zero. The reactive power produced by MMC-HVDC can be rapidly changed when the filter is switched on and off, and then the reactive power produced by MMC-HVDC can be reduced to zero at a gentle rate. This can reduce the disturbance to AC system when the filter is switched on and off.

As is shown in Figure 5, when the reactive power compensation equipment is input, the output of reactive power control of MMC-HVDC reaches QMMC_ref in T-short time and then slowly drops to 0 in T-long time. Besides, when the reactive power compensation equipment is removed, the output of the reactive power control of MMC-HVDC reaches QMMC_ref in T-short time and then slow recover zero in T-long time.

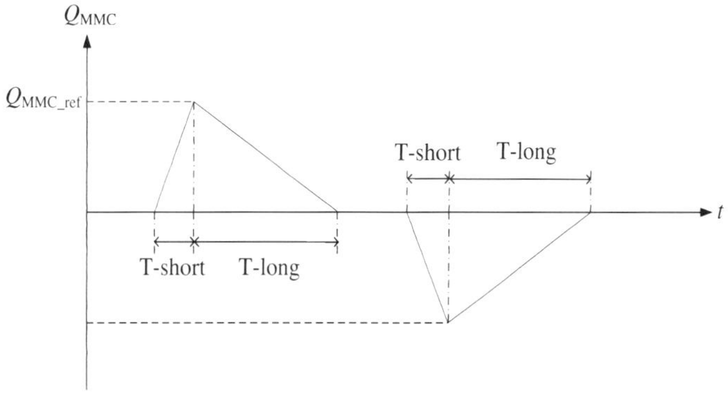

Figure 5. Coordination between reactive power control of LCC-HVDC and MMC-HVDC

5. Conclusion

Based on the existing research results of reactive power control in HVDC transmission project, this paper introduces the reactive power control functions of hybrid DC asynchronous interconnection system in detail. The coordination between the subfunctions of reactive power switching control of LCC-HVDC and the main work of reactive power auxiliary control functions of MMC-HVDC are analyzed. The reactive power control of MMC-HVDC is analyzed in detail, and the mathematical model of reactive power control of MMC-HVDC is established. The reactive power coordination control between LCC-HVDC and MMC-HVDC is studied.

(1)The reactive power switching control of LCC-HVDC needs to complete the principle of reactive power grouping according to the actual project requirements. The reactive power switching control subfunctions include Abs Min Filter control, Umax/min control, Qmax control, Min Filter control and Qcon/Ucon control. The reactive power switching control is formulated according to the corresponding priority logic. The reactive power control strategy realizes the coordination of all subfunctions to achieve reactive power control requirements.

(2)The reactive power auxiliary control of LCC-HVDC includes low load reactive power regulation function and Gamma-Kick function. By adjusting the trigger angle or turn-off angle of the converter, the reactive power absorbed or emitted by the converter can be changed to achieve reactive power control. However, this kind of control has a negative impact on the operational characteristics of HVDC transmission at the expense of converter utilization. Although the control of reactive power balance and commutation bus voltage is realized, there are great defects.

(3) The reactive power control of MMC-HVDC can track the AC bus voltage of the converter station or the reactive power exchanged between the converter station and the system through fast and accurate reactive power regulation. In view of the shortcomings of LCC-HVDC reactive power auxiliary control and the superiority of MMC-HVDC reactive power control, MMC-HVDC can be used to coordinate with LCC-HVDC reactive power control to improve system operation economy and equipment utilization.

References

[1] Tang G, Xu Z, Xue Y L, et al. A LCC-MMC Hybrid HVDC Transmission System. J. Transactions of China Electrotechnical Society, (2013)

[2] Lee Y, Cui S, Kim S, et al. Control of hybrid HVDC transmission system with LCC and FB-MMC. C. // Energy Conversion Congress and Exposition. IEEE, (2014)

[3] Kjørholt Å M H. HVDC Transmission Using a Bipolar Configuration Composed of an LCC and MMC: Operating Characteristics of Skagerrak 3 and Skagerrak 4. J. Department of Electrical Power Engineering, (2014)

[4] Li K, Li X, Chen S, et al. Investigation of hybrid pseudo bipolar HVDC performances supply power to passive AC network. J. Sensors & Transducers, (2014)

[5] Guo C, Zhao C, Montanari A, et al. Investigation of Hybrid Bipolar HVDC System Performances. J. Proceedings of the Csee, (2012)
[6] Lei W, Xingyuan L I, Kuan L I, et al. Research of pseudo bipolar LCC-VSC hybrid HVDC system supplying passive network. J. Power System Protection & Control, (2015)
[7] Murray N J, Arrillaga J, Liu Y H, et al. Flexible Reactive Power Control in Multigroup Current-Sourced HVDC Interconnections. J. IEEE Transactions on Power Delivery, (2008)
[8] Zhao C, Guo C. Complete-Independent Control Strategy of Active and Reactive Power for VSC Based HVDC System. J. American Journal of Clinical Nutrition, (2009)
[9] Kanngiesser K W, Lips H P. Control Methods for Improving the Reactive Power Characteristic of HVDC Links. J. IEEE Transactions on Power Apparatus & Systems, (1970)
[10] Zhang J, Yao W, Yang M, et al. Power Stability Control Strategy of Hybrid-HVDC Transmission System Based on Triggering Angle Monitoring. J. Electric Power Construction, (2016)

Influence of transmission distance on magnetically coupled resonant radio energy transmission system

TONG JUN, Yang Xingchen, LI Facheng and Li Xiang

College of Electrical and Control Engineering,Xi'an University of Science And Technology,xi'an 710054,China

E-mail :1148212063@qq.com

Abstract. Transmission distance is a key factor in designing a wireless energy transmission system. The change in coil spacing will have a huge impact on the output characteristics of the system.Therefore, this paper theoretically analyzes the mutual inductance model of two coils from the perspective of circuit theory.Studies have shown that coil spacing, impedance, self-inductance, operating frequency, load resistance and many other factors can affect the output characteristics of the system.So, this thesis uses the control variable method to study the relationship between transmission distance and output characteristics, and proposes the optimal output power point.Finally, an experimental prototype was made to verify the feasibility of the method.

1. Introduction

The traditional transmission mode can no longer meet the demand for power supply of today's high-tech products. Getting rid of the constraints of transmission lines has become the only way for many electronic products to go further. As early as 1889, the famous American scientist Tesla began to study radio energy transmission technology. However, limited by the theoretical level, experimental conditions and research funding at the time, Tesla did not make significant scientific research results in the study of radio energy transmission theory. Furthermore, the research on the theory of radio energy transmission has entered a trough. Although many scholars have carried out many projects, the contradiction between transmission distance and transmission efficiency has always been like a mountain that hinders the development of this technology[1-3].

Until 2007, the research team of the Massachusetts Institute of Technology proposed a new theory of radio energy transfer, the theory of magnetically coupled resonant radio energy transmission, and made the research of radio energy transmission theory a hot topic of the era technology[4]. The theory of magnetically coupled resonant radio energy transmission is to use a set of energy transmission devices with the same resonant frequency to realize the transmission of electrical energy over long distances with medium and high frequency magnetic fields as the medium. The magnetic coupling resonance type uses strong coupling technology to improve the output power and transmission efficiency of the system, and has good penetrability. The non-radiation technology adopted by the system also effectively ensures personal safety[5-8].

In this thesis, the system modeling and analysis are carried out by using circuit theory, and the relationship between transmission distance, transmission efficiency and output power is studied. Focusing on a new thinking mode, the specific algorithm of maximum output power point is proposed, and the influence of various factors on transmission efficiency and output power is analyzed. Finally,

Content from this work may be used under the terms of the Creative Commons Attribution 3.0 licence. Any further distribution of this work must maintain attribution to the author(s) and the title of the work, journal citation and DOI.
Published under licence by IOP Publishing Ltd

an improved method is proposed. This provides a useful reference for optimizing the design of magnetically coupled resonant radio energy transmission.

2. Transmission model selection and theoretical analysis

The magnetic coupling resonant radio energy transmission system is mainly composed of three parts: high frequency power system, energy transmission system and load. The energy transmission system is the key research part of the radio energy transmission system, and its model is shown in Fig. 1.

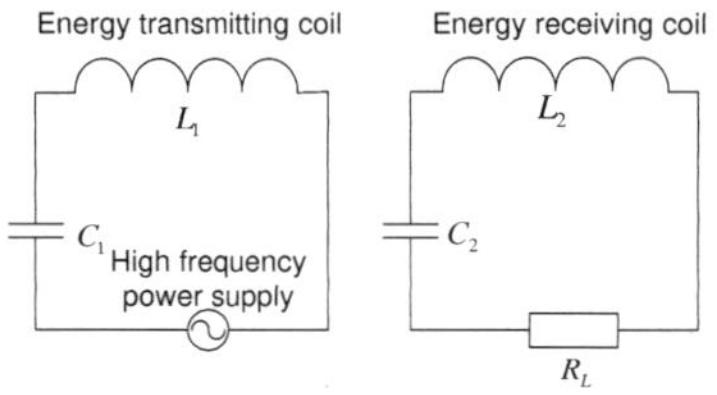

Figure 1. Transmitter and receiver model of wireless power transfer

The energy transmission system is composed of a transmitting module and a receiving module, and the energy is transmitted from the transmitting coil to the receiving coil through the magnetic magnetic field by means of magnetic coupling resonance, thereby realizing wireless transmission of electric energy.

System transmission model selection

The transmission structure of the two coils is different according to the connection mode of the inductor and the capacitor, and can be divided into four types: string type, string type, parallel type, and parallel type. When the system is fully resonant, the resonant capacitor value can be obtained according to the topology of the circuit. It can be seen from the analysis that the resonant capacitance of the SS-type transmission structure is only related to the system frequency and the self-inductance of the coil, and has nothing to do with the parasitic resistance and the mutual inductance of the coil, which is convenient for system design and analysis, and has high transmission efficiency. Therefore, the SS-type transmission structure is selected as the research model of this paper.

Circuit Analysis

As shown in Figure 2, it is an SS type transmission model. Among them, R1 and R2 are the equivalent resistance of the coil, C1 and C2 are compensation capacitors, L1 and L2 are coil self-inductance, M is coil mutual inductance, US is high frequency power supply, and RL is load.

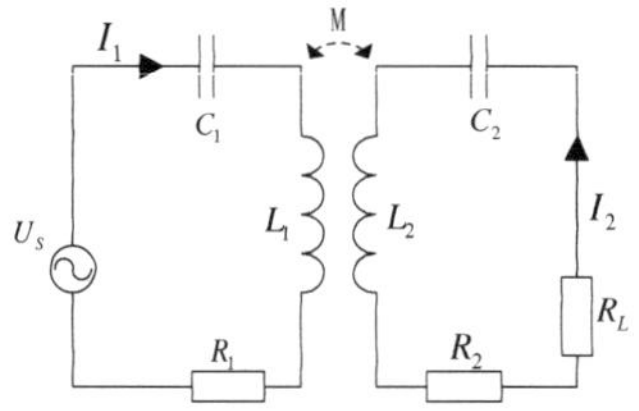

Figure 2. SS type transmission model

According to Kirchhoff's voltage law, the circuit analysis of Figure 2 can be obtained by equation (1)(2)

$$U_s = (R_1 + \frac{1}{j\omega C_1} + j\omega L_1)I_1 + j\omega M I_2 \quad (1)$$

$$0 = (R_1 + R_L + \frac{1}{j\omega C_2} + j\omega L_2)I_2 + j\omega M I_1 \quad (2)$$

When the radio energy transmission system is operating in a resonant state, the loop equation can be reduced to:

$$U_s = R_1 I_1 + j\omega M I_2 \quad (3)$$

$$0 = (R_2 + R_L) I_2 + j\omega M I_1 \quad (4)$$

The simultaneous equation (3) (4) can be used to find the primary and secondary currents of the system:

$$I_1 = \frac{(R_2 + R_L) U_s}{R_1 R_2 + R_1 R_L + \omega^2 M^2} \quad (5)$$

$$I_2 = -\frac{j\omega M U_s}{R_1 R_2 + R_1 R_L + \omega^2 M^2} \quad (6)$$

When the system resonates, its input impedance is:

$$Z_{in} = R_1 + \frac{\omega^2 M^2}{R_2 + R_L} \quad (7)$$

Then, the input power and output power of the system are:

$$P_1 = P_{in} = I_1^2 Z_{in} = \frac{U_S^2 (R_2 + R_L)(\omega^2 M^2 + R_1 R_2 + R_1 R_L)}{(\omega^2 M^2 + R_1 R_2 + R_1 R_L)^2} \quad (8)$$

$$P_2 = P_{out} = I_2^2 R_L = \frac{\omega^2 M^2 U_S^2 R_L}{(\omega^2 M^2 + R_1 R_2 + R_1 R_L)^2} \quad (9)$$

In turn, the transmission efficiency of the system can be solved:

$$\eta = \frac{P_{out}}{P_{in}} = \frac{\omega^2 M^2 R_L}{(R_L + R_2)(R_1 R_2 + R_1 R_L + \omega^2 M^2)} \quad (10)$$

Deriving the expression of the output power:

$$\frac{dP_2}{dM} = 0$$

Got

$$M_1 = 0 \quad M_2 = -\frac{\sqrt{R_1 (R_L + R_2)}}{\omega} \quad M_3 = \frac{\sqrt{R_1 (R_L + R_2)}}{\omega}$$

Because M1 and M2 do not meet the actual situation, they are discarded.

According to the derivative relationship of the output power, it can be obtained that when 0<M<M3, the output power monotonically increases; when M3<M<1, the output power monotonically decreases. Therefore, when the system output power is at the M3 point, the maximum value is taken. The mutual inductance of the coil is related to the impedance, self-inductance, load resistance and angular frequency of the secondary side coil. By changing one of the parameters, the mutual inductance of the coil of the system can be changed, thereby affecting the output power.

When the secondary side is a spiral coil and the central axis is on the same straight line, the distance and mutual inductance of the magnetically coupled resonant radio energy transmission system are:

$$M = M_{12} = M_{21} = \frac{\pi \mu_0 (n_1 n_2)^{0.5} (r_1 r_2)^2}{2D^3} \quad (11)$$

Where μ_0 is the vacuum permeability, D is the horizontal distance of the coil center point, and n1, n2, r1, and r2 are the turns and radii of the primary side and the secondary side coil, respectively. When n1=n2=n, r1=r2=r, the above formula can be reduced to:

$$M = \frac{\pi \mu_0 n r^4}{2D^3} \quad (12)$$

The mutual inductance of the maximum output power point is:

$$M_3 = \frac{\sqrt{R_1(R_L + R_2)}}{\omega} \tag{13}$$

The simultaneous output formula (12) (13) can be used to derive the maximum output power point:

$$D = \sqrt[3]{\frac{\pi\mu_0 n r^4 \omega}{2\sqrt{R_1(R_L + R_2)}}} \tag{14}$$

Similarly, for the system transmission efficiency, the derivative function about the mutual inductance M is obtained:

$$\frac{d\eta}{dM} = 0$$

According to the system transmission efficiency derivation relationship, when M=0, x takes the minimum value. When $0 < M < 1$, the transmission efficiency monotonically increases. The mutual inductance M of the coil is inversely proportional to the transmission distance D. When the transmission distance increases, the mutual inductance of the coil will decrease and the transmission power of the system will also decrease.

3. EXPERIMENT

In order to verify the correctness of the theoretical analysis, based on the SS-type transmission model, combined with the circuit theory, an experimental prototype of the radio energy transmission system is designed, as shown in Figure 3.

Figure 3. Experiment platform

The prototype of this experiment is mainly composed of signal generation controller, DC high-frequency inverter and electromagnetic emission and receiving system. Among them, in order to avoid the influence of the skin effect on the experimental data, the litz wire is used instead of the solid copper wire to wind the high frequency coil. The main experimental parameters of this experiment are shown in Table 1:

Table 1. Experimental parameters

Element	Numerical value
f	201KHz
L_1	28.3uH
L_2	28.7uH
C_1	0.022uF
C_2	0.022uF
R_1	0.394 Ω
R_2	0.408 Ω

R_L	25 Ω
r	4.5e-2m
n	14.5

The experimental results are shown in Figures 4 and 5:

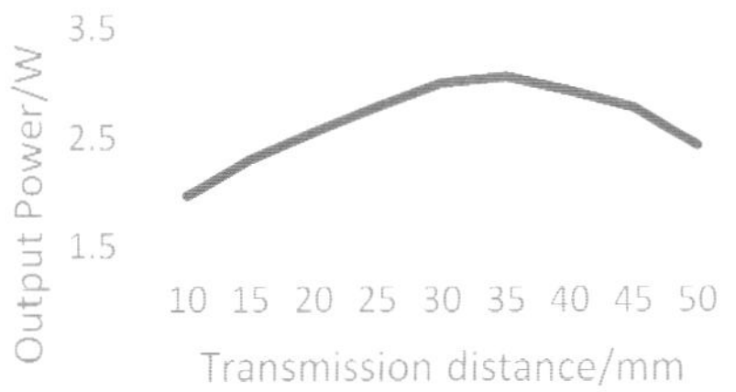

Figure 4. Influence of transmission distance on the output power

Figure 5. Influence of transmission distance onthe system efficiency

It can be seen from Fig. 4 and Fig. 5 that as the distance increases, the transmission efficiency decreases continuously, while the output power shows a trend of increasing first and then decreasing, and the output power of the system reaches the maximum value when the transmission distance is 35 mm, and the formula (14) The calculated values are basically consistent. This shows that the variation law of the output power of the magnetically coupled resonant radio energy transmission system increases first and then decreases, and the maximum value can be reached at a certain point, but the output efficiency will gradually decrease as the transmission distance increases. Further analysis of Figures 4 and 5 shows that when the transmission distance is increased, the transmission efficiency is significantly reduced, but the output power is not significantly improved. Therefore, when designing the wireless energy transmission system, the pursuit of the maximum output power can be appropriately abandoned. The distance between the system transmission efficiency and the output power can be considered.

4. Optimization Analysis

4.1. Improve the working frequency

When the system works under high frequency conditions, the internal resistance of the coil is mainly composed of two parts: ohmic loss resistance and radiation loss resistance. Generally, in the frequency range of 1-50MHz, the radiation loss resistance is negligible, then the internal resistance of the coil is mainly ohmic loss resistance.

$$R_0 = \sqrt{\frac{\omega\mu_0}{2\delta}}\frac{l}{4\pi a} = \sqrt{\frac{\pi f\mu_0}{\delta}}\frac{l}{4\pi a} \tag{15}$$

Where a is the wire radius, L is the wire length, μ_0 is the vacuum permeability, δ is the conductivity, and f is the system operating frequency.

As can be seen from equation (15), when the system frequency is increased, the internal resistance of the coil increases at the square root speed. Increasing the internal resistance of the coil will reduce the transmission efficiency of the system, but its growth rate is far less than the growth rate of the angular frequency. Therefore, the distance of the maximum output power point can be increased by increasing the operating frequency of the system.

4.2. Reduce the internal resistance of the coil

According to the formula analysis, when the internal resistance of the coil is reduced, the maximum transmission distance point can be effectively increased. Therefore, when designing the coil, a plurality of twisted silver-plated copper thick wires should be selected as much as possible to reduce the skin effect of the coil, increase the conductivity of the wire, and reduce the self-resistance of the coil.

5. Conclusion

In this paper, the distance characteristics of the magnetically coupled resonant radio energy transmission system are analyzed, and the experimental demonstration is carried out based on the derivation results. Finally, an optimization analysis was carried out. The specific conclusions are as follows:

（1）As the transmission distance increases, the output power of the system increases first and then decreases, while the transmission efficiency of the system is always reduced. Also, near the maximum output power point, as the transmission distance continues to increase, the transmission efficiency will be significantly reduced, but the output power is not significantly improved.

（2）When the transmission distance is (14), the output power of the system will reach the maximum value.

（3）Increasing the operating frequency of the system and reducing the coil self-resistance can effectively increase the distance of the maximum output power point.

References

1. Cheng shijie , Chen Xiaoliang , Wang Junhua , et al. Key technologies and appliantions of wireless power transmission [J] . Transactions of China Electrotechnical Society , 2015 , 30 (19) :68-84
2. Fu Wenzhen , Zhang Bo , Qiu Dongyuan , et al . Maximum efficiency analysis and design of self-resonance coupling coils for wireless power transmission system [J] . Proceedings of the CSEE , 2009, 29 (18): 21-26
3. Zhang Xian , Yang Qingxin , Chen Haiyan , et al. Research on characteristics of frequency splitting in electromagnetic coupling resonant power transmission systems [J] . Proceedings of the CSEE, 2012, 39 (9):167-173 .
4. Kurs A, Karalis A, Moffatt R, et al. Wireless power transfer via strongly coupled magnetic resonances[J]. science, 2007, 317(5834): 83-86.
5. Zhao Zhengming , Zhang yiming , Chen Kainan , New progress of magnetically-coupled resonant wireless power transfer technology [J] . Proceedings of the CESS , 2013 , 33 (3):1-13.
6. Li Zongqi,Huang Shoudao,Yi Jiliang,Li Junjun.Frequency split suppression method for magnetically coupled resonant radio energy transmission system [J] .Automation of Electric Power Systems,2017,41(02):21-27.
7. Li Yang,Yang Qingxin,Yan Zhuo,et al. Analysis on effective range of wireless power transfer and its impact factors[J]. Transactions of China Electrotechnical Society, 2013, 28(1): 106-112.
8. Li Zongfeng,Xu Lei . Research on detuning control of general contractless power supply platfrom with multiple loads [J] . Modern Electronics Technique , 2009 , 32 (3) :133-136

Performance comparison of series and parallel DC power supply system

Shengli Kong [1,a], Xudong Leng [2], Weidong Ma[1], Lei Liu [1] and Feng Zhang [3]

[1]State Grid Henan Electric Power Research Institute, China
[2]Shenzhen Tieon Energy Technology Co.,Ltd., China
[3]Henan EPRI Hitech Group Co.,Ltd., China

kongshengli@139.com

Abstract. Since the structure of series DC power supply system, if any battery cell breaks down, the group of battery cells can't play its due role. The parallel DC power supply system has been successfully developed and applied in recent years. We want to know whether the parallel DC power system is worth vigorously popularizing and applying in the substation in the future? So we made a few special tests, and this paper will compare the series and parallel DC power supply systems with some data in some ways, such as performance, safety, reliability, maintenance quantity, convenience, economic benefits , etc.

1. Introduction

The traditional DC power supply system used in substation is composed of charger and series-connected batteries. The parallel DC power system developed in recent years consists of several parallel battery components. Because the batteries of the former are connected in series, the DC power supply system is greatly affected by such a structure. If the circuit is open, or single battery breaks down, or old and new batteries are mixed, the group of battery cells won't play its due role [1]. The battery of the latter is a parallel structure, so the latter is called "parallel DC power system". Therefore in this paper, we call the traditional DC power system "series DC power system".

At present, series DC power supply system is widely used in substation and has been running for decades. Parallel DC power supply system is an emerging product, which has obvious advantages on structure, but we want to know whether it also has advantages over performance, safety and other aspects. This paper intends to carry out detailed comparison of series and parallel DC power system through tests as comprehensive as possible.

2. Parallel DC power system technology

2.1. Parallel battery component

In order to solve the problem caused by the structure of series DC power system, "parallel battery component" was innovatively designed by combining one 12V battery with matching AC/DC and DC/DC charge-discharge circuits [2]. The module designed with AC/DC and DC/DC circuits is called "parallel battery module" [3].

Content from this work may be used under the terms of the Creative Commons Attribution 3.0 licence. Any further distribution of this work must maintain attribution to the author(s) and the title of the work, journal citation and DOI.
Published under licence by IOP Publishing Ltd

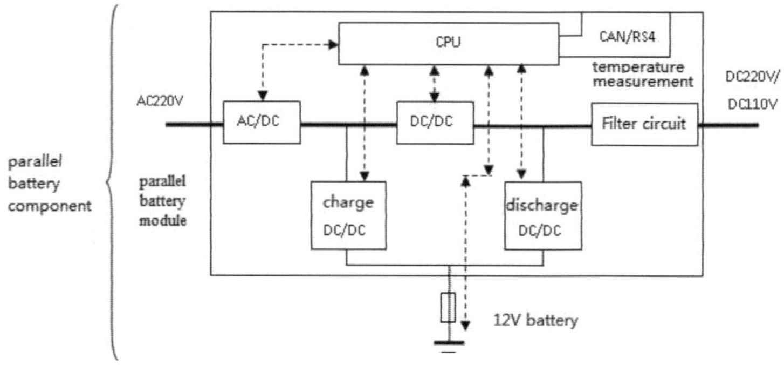

Figure 1. Schematic diagram of parallel battery component.

2.2. *Technical performance and parameters of parallel battery modules*

2.2.1. *Input characteristic.*

Table 1. Input characteristic.

	Min. value	Rated value	Max. value	Remarks
AC input voltage	95Vac	220Vac	285Vac	Normally starting and running at entire voltage range
AC input frequency	45Hz	50Hz	65Hz	Normally running at entire frequency range
AC input current			8.5A	Input:90~108Vac; Output:440W
AC starting impulse current			30A	Input:265Vac; 25°C; Duration>1mS
Battery input voltage	10.8Vdc	13.4Vdc	14.1Vdc	Normally starting and running at entire voltage range

2.2.2. *Output characteristic.*

Table 2. Output characteristic.

	Value	Remarks
Main rated output voltage	220/110Vdc	
Main rated output current	2/4A	
Rated power of output	440W	
Efficiency	≥87%	AC power supply; Main output: 2A; Charging output: 20A, environment ; 25° C
Efficiency	≥90%	Battery discharge; Nominal voltage input; Rated output voltage current; 25° C
Module softly starting time	3~8S	Resistance load; AC power supply
Current stability accuracy (charging output)	≤±1%	Test method following GB/T 19826-2014 6.3.2
Voltage stability accuracy (charging output)	≤±0.5%	Test method following GB/T 19826-2014 6.3.3
Voltage stability accuracy (main output)	≤±0.5%	Test method following GB/T 19826-2014 6.3.3
Ripple factor (charging output)	≤0.5%	Test method following GB/T 19826-2014 6.3.4
Ripple factor (main output)	≤0.5%	Test method following GB/T 19826-2014 6.3.4
Flow unbalance of parallel module	≤±5%	Test method following GB/T 19826-2014 6.7

2.3. Parallel DC power supply system

The parallel DC power supply system is made up of several parallel battery components, and the corresponding monitoring equipment and feeders [4-5].

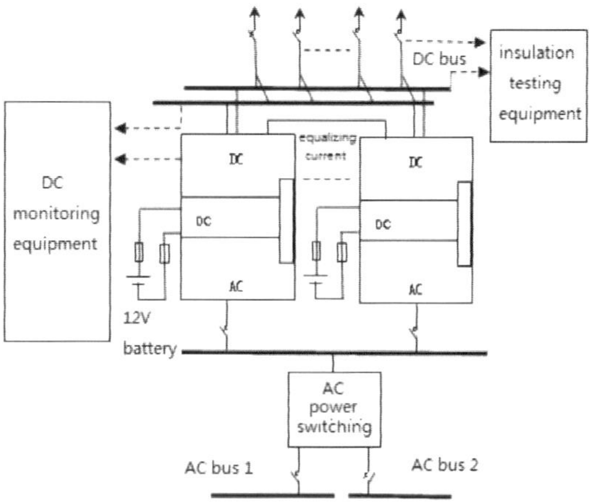

Figure 2. The structure of single - bar parallel DC power supply system.

3. Performance comparison of series and parallel DC power supply system

3.1. Comparison of system structure

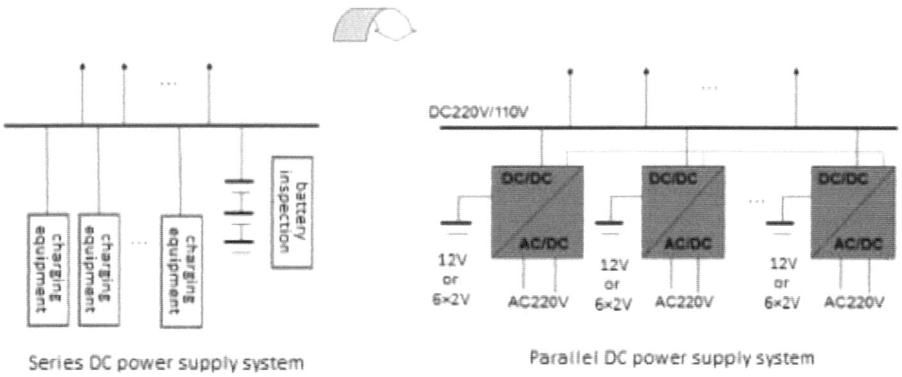

Figure 3. The structure comparison of two different principle system.

3.2. Performance comparison

Table 3. Performance comparison.

	Parallel DC power supply system	Series DC power supply system
Voltage stability accuracy	≤±0.5%	≤±0.5%
Current stability accuracy	≤±1%	≤±1%
Efficiency	Rectifier efficiency≥90%; Battery discharging efficiency≥95%	Rectifier efficiency≥90%; Battery directly discharging without loss
Noise level（dB）	Self-cooling <45	Air cooling <55; Self-cooling <45
Charge and discharge	One-to-one management	One-to-many management

management		
Battery connection mode	Indirectly parallel(can be mixed with old and new)	In series(High consistency requirement) [6]

3.3. Safety comparison

Table 4. Safety comparison.

	Parallel DC power supply system	Series DC power supply system
Anti-explosion	Anti-explosion and flame-proof treatment for 12V cell	Special battery room for series-connected batteries with capacity≥300AH

3.4. Reliability comparison

Table 5. Reliability comparison.

	Parallel DC power supply system	Series DC power supply system
Effects of the breakdown of cell battery	Will have no effects on the system [7]	Will affect the output of whole battery group
Effects of the breakdown of several cell batteries	Batteries with fault can be replaced independently without affecting the system reliability	Several batteries with fault will cause the whole battery group to be scrapped

In reliability mathematics, the reliability rate of the system is expressed as K, and the failure rate is expressed as G, so K+G=1. DC power supply system in series would reduce the reliability of the system. If the two same reliability rates are K1 and K2, the reliability rate K of series DC power supply system is K1K2, and the failure rate G is 1-K1K2. Two parallel systems of the same function are called redundant systems, which can improve the system reliability. If the two same reliability rates are K1 and K2, and the failure rates G1 is 1- K1and G2 is 1- K2[8].

On the basis of experience, we can assume that the reliability rate of a 2V VRLA(Valve Regulated Lead Battery) is 99.95%, the reliability rate of a 12V VRLA is 99.9%, and the reliability rate of a parallel battery module is 96%. Since a parallel battery component is connected with a parallel battery module and a 12V VRLA in series, its reliability rate is 96%×99.9%=95.9%.

The theoretical reliability calculation of the two DC power supply systems belonged to 110kV substations is stated as follows (according to the reliability formula of series and parallel system):

● Series: For 108 VRLA (2V, 200AH) in series, the reliability rate of the battery group is $(99.95\%)^{108}$=94.7%, and the failure rate is 5.3%. The reliability rate of two series DC power supply systems is $1\text{-}(5.3\%)^{2}$=99.72%.

●Parallel: According to the bidding technical data of a 110kV substation in Guizhou，there are 5 10A /220V charging modules in the original series DC power supply system. If we remove the stand-by module and the battery charging current (20A), we can get that the actual fault current is 20A. The parallel DC power supply system is equipped with 24 parallel battery components, each of which is rated at DC220V/2A. We can calculate that there are 14 stand-by components. While considering discharge capacity, there are actually 3 stand-by components. Since the failure rate of each component is (1-95.9%) =4.1%, the reliability rate of the parallel DC power supply system is $1\text{-}(4.1\%)^{3}$=99.99%..

3.5. Convenience comparison

Table 6. Convenience comparison.

	Parallel DC power supply system	Series DC power supply system
Installation	High(No battery inspection device； 12V battery； the number of cell battery is reduced)	Middle
running	High(Can check 100% capacity online with	Low

	0.1C discharging current)	
Maintenance	High(Can replace cell battery online)	Middle
Technical improvement	High(Old and new battery can be used in the same system)	Middle

3.6. Maintenance workload comparison

Table 7. Maintenance workload comparison.

	Parallel DC power supply system	Series DC power supply system
Workload of checking capacity and cost	Low(Can check 100% capacity online with 0.1C discharging current; 12V battery; the number of cell battery is reduced)	High(Can check capacity offline; external backup battery group will be needed; heavy workload)
Repairing workload and cost	Low(Can repair every cell battery)	High(Several batteries with fault will cause the whole battery group to be scrapped)

4. Economic comparison of series and parallel DC power supply system

4.1. Configuration comparison

Suppose there is a 110kV substation with series DC power supply system which is composed with double chargers and double battery groups. This paper is going to compare such a series DC power supply system with parallel DC power supply system.

Table 8. Comparison table of main components.

	Parallel type	Series type
AC switching circuit	1 set	2 sets
Charging module	24 parallel battery module. Parallel battery module isn't just a rectifier, but also a management of battery charging and discharging.	2 sets of 5 charging modules (DC220V/10A)
Battery	24 lithium iron phosphate batteries (12V/200Ah). The cell batteries are independent and have no direct electrical connection.	2 sets of 108 VRLA (2V/200Ah). The cell batteries are connected in series.
Battery inspection equipment	No. Parallel battery module is a one-to-one battery management.	2 sets
Stepdown silicon chain	No. DC bus-bar voltage is stable since the battery is not connected to the DC bus-bar.	2 sets. DC bus-bar voltage is sometimes a little high while charging the battery group on the bus-bar.
DC monitoring equipment	1 DC monitoring equipment +1 parallel battery module.The monitoring equipment of parallel battery module mainly controls checking battery capacity.	2 sets
External backup battery group	No. Parallel DC power supply system can check 100% capacity online.	1 set
Insulation testing equipment	2 sets	2 sets
DC feeder switch	As needed	As needed

4.2. Full life cycle (12 years) benefit comparison

●Cost of equipment purchase: The purchase cost of Parallel DC power supply system and series DC power supply system is similar.

●Cost of checking battery capacity: For such a series DC power supply system, there will be 10 times to check the battery capacity. If once it will cost $8,000, the total cost of checking battery capacity is $80,000. While parallel DC power supply system can check battery capacity online, this cost will be saved.

●Cost of replacing battery:

For such a series DC power supply system, we have to shut off the electric power supply to replace the battery group. On the basis of replacing one set every 6 years, one new set should be purchased in the 7th year which costs $120,000. As so on, another new set should be purchased in the 12th year which also costs $120,000. The remaining 90% value of the batteries is about $108,000, so the actual cost of replacing the battery during the 12-year life cycle is $132,000.

For parallel DC power supply system, in the first six years, 20% of the batteries have to be replaced which costs $28,000. The rest of the batteries should be replaced in the 10th year which costs $112,000. The remaining 30% value of the batteries 6 years before and the remaining 90% value of the batteries 2 years before are $109,200. So the cost of replacing the battery during the 12-year life cycle is $30,800.

Through the above comparison we can know that parallel DC power supply system costs 30% less than Series DC power supply system during the 12-year life cycle.

5. Comparison of overload capacity based on improvement

5.1. Overload output characteristic curve of parallel battery module

Overload output characteristic curve of parallel battery module is shown in Figure 4.

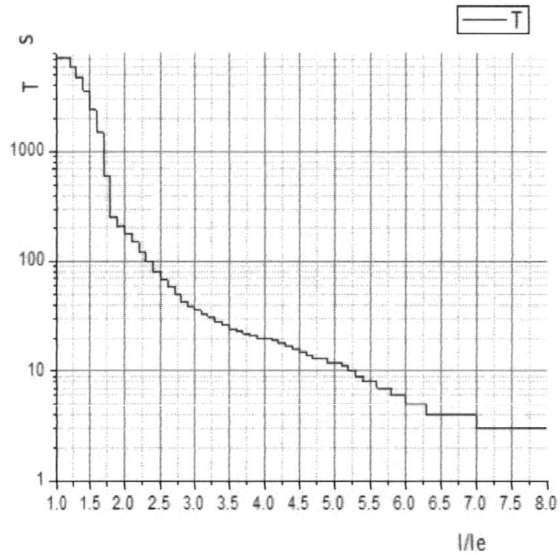

Figure 4. Overload output characteristic curve of parallel battery module.

5.2. Continuous current circuit of series battery based on parallel DC Power Supply System

On account of every battery branch independent from with AC bus-bar, DC bus-bar, and other battery branch, we connect 16 or 8 2V cell batteries to the DC bus-bar with discharge diodes and fuses, whose voltage is lower than the voltage of DC bus-bar. The batteries in series only have the discharge path to the DC bus-bar. The schematic diagram of continuous current circuit of series battery based on parallel DC Power Supply System is shown as Figure 5.

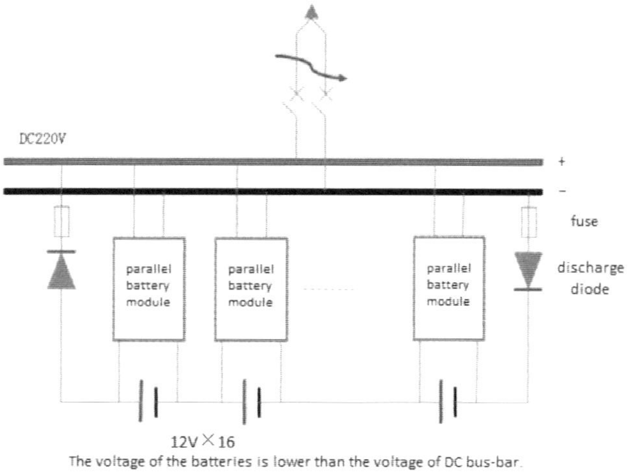

Figure 5. The schematic diagram of continuous current circuit of series battery based on parallel DC Power Supply System.

Normally, it is carried by the parallel battery module with stable voltage. When the system is overloaded or short-circuit fault occurs, the current is still provided by the parallel battery module. If the DC bus-bar voltage is pulled down to the series battery voltage, the batteries in series will provide continuous current at the same time [9-10].

The short circuit test waveform of feeder lines for 16 batteries in series is shown as Figure 6. If 16 12V batteries in series and 16 parallel battery modules run simultaneously, the short circuit test waveform of feeder lines is shown as Figure 7.

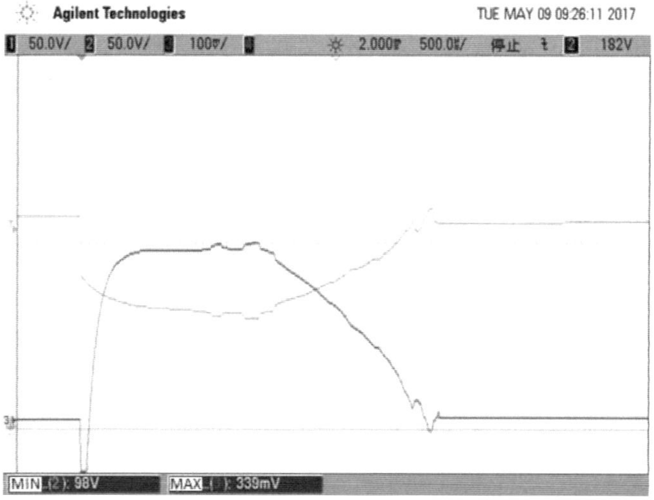

Figure 6. Short circuit test waveform of feeder lines for 16 batteries in series (1mV is corresponding 4A).

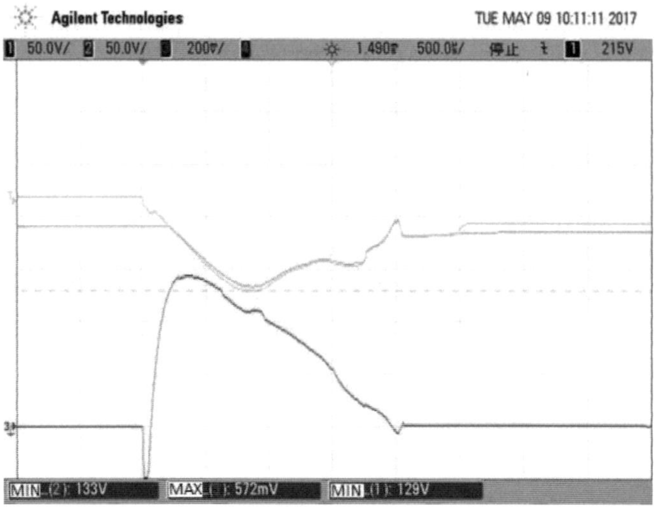

Figure 7. Short circuit test waveform of feeder lines for 16 12V batteries in series and 16 parallel battery modules running simultaneously (1mV is corresponding 4A).

The test results show that the overload effect of the parallel DC power supply system is higher than that of the traditional series DC power supply system. When the short-circuit fault of feeder occurs, the bus-bar voltage will drop less, and the bus-bar voltage recovery time is shorter after fault removal.

6. Conclusion

The parallel DC power supply system has been applied in 110kV substation, 35kV substation, 20kV switching station and 10kV switching station since 2012. The advantages of the parallel DC power supply system are obvious: (1) It solves the problem of single cell battery's fault affecting the output of whole battery; (2) It solves the problem of several batteries with fault causing the whole battery group to be scrapped; (3)Mixing old and new battery in one system is allowed, and it improves battery utilization; (4) It makes check 100% capacity online and repair every cell battery come true; (5) The application is flexible, which can be centralized and distributed, and promote the integration development of primary and secondary equipments of substation. On the other hand, at present, the power of parallel battery modules is relatively small, which is limited in high-power applications.

References

[1] H. Wang, X. Lin, G. Zhang, S. Yang, X. Cui, Chn. Journ. Pow. Sour., 37, 492 (2013)
[2] J. Yin, F. Peng, HB. Elec. Pow., 38, 36 (2014)
[3] T. Zhao, D. Yan, GX. Elec. Pow., 36, 76 (2013)
[4] Q. Luo, Appl. Volt. Technol., 15, 74 (2012)
[5] L. Shi, D. Liu, M. Lee, Z. Zong, Chn. Journ. Pow. Sour., 41, 612 (2017)
[6] Y. Cao, H. Jiang, K. Hu, G. Cai, S. Yang, Chn. Journ. Pow. Sour., 40, 358 (2016)
[7] B. Han, F. Zhang, Z. Lin, Elec. Engi. Technol., (4), 74 (2016)
[8] Y. Guo, Reliability Modeling and Analvsis on k/n:F(G) System (Xi'an university of electronic science and technology, 2013)
[9] 9.X. Leng, J. Wang, Pow. Sys. Prot. Contr., 37, 114 (2009)
[10] 10.J. Yin, B. Yang, S. Yang, Worl. Pow. Suppl., (09), 26 (2016)

IOP Conf. Series: Materials Science and Engineering **533** (2019) 012009 doi:10.1088/1757-899X/533/1/012009

Research on Push-pull-type Inductively Coupled Power Transfer System for ZVS

Liu Shulin, Yan Jizhi, Tao Shitong, Hu chuanyi and Wu hao

College of Electrical and Control Engineering, Xi'an University of Science and Technology, Xi'an, China

E-mail: yanjizhi111@163.com

Abstract. According to the poor stability of the output constant current and the control complexity of the existing Inductive Coupled Wireless Power Transmission System, a series-parallel compensation network is applied to push-pull converter. And the soft switch is realized by configuring the parameters of compensation network, and the constant-current output of the wireless power is realized without additional complex control system. Firstly, analysis of the push-pull converter to achieve the conditions of soft-switching. Secondly, based on the leakage inductance model of the transformer, combined with the Thevenin equivalent theorem, the parameters of the string-and-compensation are designed. Lastly, the push-pull soft-switching inductive coupled power transfer converter is designed and built.The feasibility of theoretical analysis and parameter design method are verified by the experimental results.

1. Introduction

Inductively coupled power transfer (ICPT) is a safe, reliable and flexible power access technology. The electrical energy is transmitted to the powered device through electromagnetic induction between the two coils. It has been widely used in recent years[1-3].

In order to improve the transmission efficiency and reduce the loss of the switching tube, soft switching technology is generally adopted. In [4], the push-pull transformer is introduced on the basis of the traditional ICPT converter, The voltage-current zero-crossing detection circuit existing in the conventional ICPT soft-switching converter simplifies the design of the converter; however, the article cannot realize constant current output under the condition of variable load. In [5], the BUCK variator is added to the secondary side to adjust the charging current, and the PI control algorithm is used to realize the constant-current charging of the variable load. However, this document is mainly applied to high power.

Based on the above problems, this paper uses the serial-to-parallel (SP) topology to compensate the leakage inductance of the loosely coupled transformer based on the push-pull converter. The system can realize soft switching, achieve constant current within a certain load range and improve system stability[6-7].

2. Circuit structure and soft switch working principle

2.1 Push-pull converter circuit

Content from this work may be used under the terms of the Creative Commons Attribution 3.0 licence. Any further distribution of this work must maintain attribution to the author(s) and the title of the work, journal citation and DOI.
Published under licence by IOP Publishing Ltd

Figure 1 is a block diagram of the push-pull converter ICPT used in this paper[8], which mainly includes: switching transistors Q1 and Q2, push-pull transformers T, C_p and C_s which are the compensation capacitors of the primary and secondary sides, the transmitting coil L_p , and the receiving coil L_s . L_p and C_p are series structures, while L_s and C_s are parallel structures. D1~D4 are rectifier diodes, followed by filter capacitor C_o and load RL. C_{s1} and C_{s2} are drain-source parallel capacitors for switching transistors Q1 and Q2.

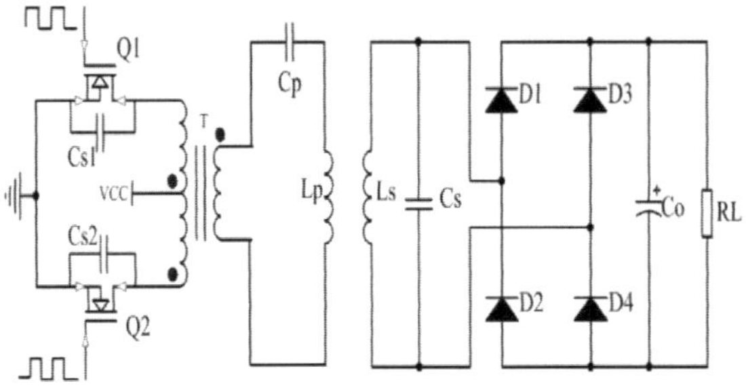

Figure 1. Push-Pull Converter ICPT.

During normal operation, the two power switches Q1 and Q2 are alternately turned on and off by PWM complementary pulse driving; during the half cycle of each conduction, the input DC voltage *Vcc* is converted into a high-frequency square wave pulse voltage, which is alternately applied to each other.

2.2 Zero voltage working conditions

The switch tube Q1 and Q2 of the push-pull converter are respectively connected with a large capacitance at both ends of the drain and the source. Since the voltage of the capacitor cannot be abruptly changed, the circuit can realize zero voltage shutdown. To realize the zero voltage of the circuit, the compensation network needs to be inductive; The primary compensation network of the loosely coupled transformer is L and C series. To make it inductive, the traditional method is to take the circuit operating frequency greater than the L and C natural resonant frequency. In this paper, the operating frequency of the circuit is the primary leakage inductance of the loosely coupled transformer and the frequency of the compensation capacitor resonance. At this time, the compensation network is inductive.

The working principle of the ZVS push-pull resonant converter is analyzed below. When it reaches steady state, the working principle waveform is shown in Figure 2.

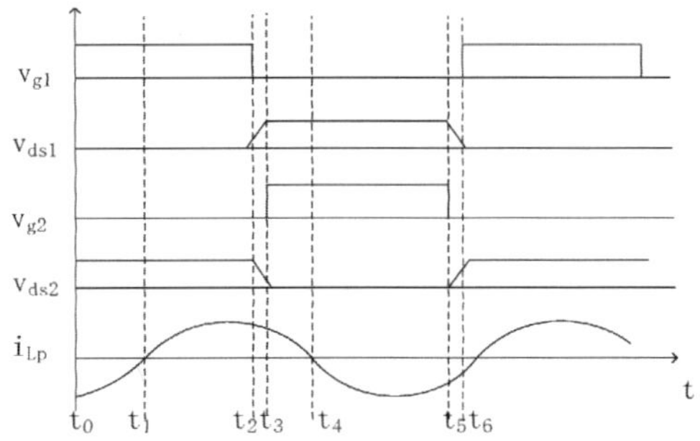

Figure 2. Working waveform of Converter Circuit.

Make the following assumptions before the analysis:

(1) All switching tubes and diodes are ideal components;

(2) Inductance and capacitance are ideal components, C_{s1}=C_{s2};

(3) The circuit has entered steady state.

In one switching cycle, there are six switching modes, and the operation of each mode is described as follows:

Mode 1: when it is at t_0, Q1 zero voltage is turned on. Due to the clamping action of the body diode, the voltage across Q1 is approximately 0. So Q1 is zero voltage open. At this time, the resonant network composed of C_p and L_{l1} has resonated, the current flowing through the primary side of the transformer gradually decreases, and the resonant current begins to prepare for zero crossing. At time t_1, i_p rises to zero and switch mode 1 ends.

Modal 2: when it is at t1,the primary current is zero. Due to the forward voltage, the forward current flows and the primary compensation network continues to resonate.

Mode 3: Turn off Q1 at time t_2. At this time, the current flowing through the power tube is relatively large. Since the power tube has a large capacitance in parallel, the voltage across the capacitor cannot be abruptly increased from zero, so Q1 is zero voltage off. C_{s1} starts to charge while C_{s2} starts to discharge.

In the next three modes, Q2 is turned on, and the working condition of the circuit is similar to the previous three modes, which will not be described here.

3. SP leakage compensation principle and its characteristics analysis

3.1. Transformer leakage inductance model

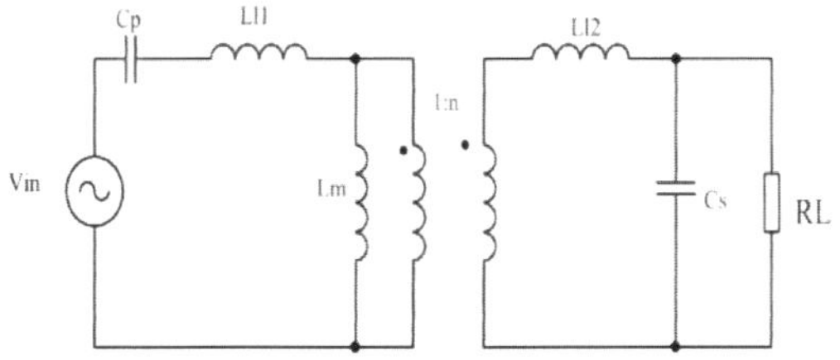

Figure 3. Leakage sensing model of transformer

Like the T-type equivalent model of the ordinary transformer, the secondary side of the loosely coupled transformer can also be converted to the primary side. Regardless of the loss of iron among others, the leakage inductance model of the coupling transformer is shown in Figure 3.In order to simplify the analysis, the parasitic resistance of the primary and secondary coils of the loosely coupled transformer is neglected. L_1 and L_2 respectively represent the primary and secondary side self-inductance of the transformer, and L_{l1} and L_{l2} respectively represent the leakage inductance on the equivalent primary side and secondary side of the loosely coupled transformer. n is the equivalent voltage ratio. L_m is the equivalent magnetizing inductance, and M is the mutual inductance between the transmitting and receiving coils[9]. According to the two-port equivalent characteristics:

$$L_m = \frac{M}{n} \tag{1}$$

$$L_{l1} = L_1 - \frac{M}{n} \tag{2}$$

$$L_{l2} = L_2 - nM \tag{3}$$

3.2. SP leakage compensation principle analysis

Traditional ICPT compensation is divided into two types: unilateral compensation and bilateral compensation. The unilateral compensation is to add the resonant compensation capacitor only on the side of the loosely coupled transformer, and the bilateral compensation is to add the compensation capacitor on both sides of the loosely coupled transformer; the resonant mode of the resonant network

composed of the loosely coupled transformer coil and the compensation capacitor may have two modes: series resonance and parallel resonance. In [10], the voltage gain, current gain and power factor characteristics of various modes of the two compensation types are analyzed in detail, and the effect of the bilateral compensation network is far superior to that of unilateral compensation.
There are four basic compensation networks (SS, SP, PS, and PP) for bilateral compensation. According to the principle, they can be divided into two types: constant voltage output and constant current output.In the actual circuit, to achieve constant voltage and constant current output within a certain load range, the compensation network must be operated in a resonant state. Besides the equivalent impedance of the coil and the loss of the MOS tube should be as small as possible. For example, SS can theoretically achieve constant current output. In actual circuits, it is necessary to reduce the switching and coil losses as much as possible in a certain load range. The coil loss can be made by using multiple strands and winding the Litz wire to make a loosely coupled transformer coil. To reduce switching losses, soft-switching techniques must be used to achieve ZVS by adding a voltage zero-crossing detection circuit and corresponding control circuitry[11-12].
In this paper, SP can compensate for leakage inductance, which can realize soft switching and realize output constant current. It does not need additional control circuit, and the constant current characteristics are better. The principle of SP leakage compensation is as follows:
The SP compensation method is analyzed by the leakage inductance model. If the primary and secondary compensation capacitors C_p and C_s resonate with the leakage inductance respectively, the equation is satisfied:

$$f_{sw} = \frac{1}{2\pi\sqrt{L_{l1}C_p}} = \frac{1}{2\pi\sqrt{L_{l2}C_s}} \tag{4}$$

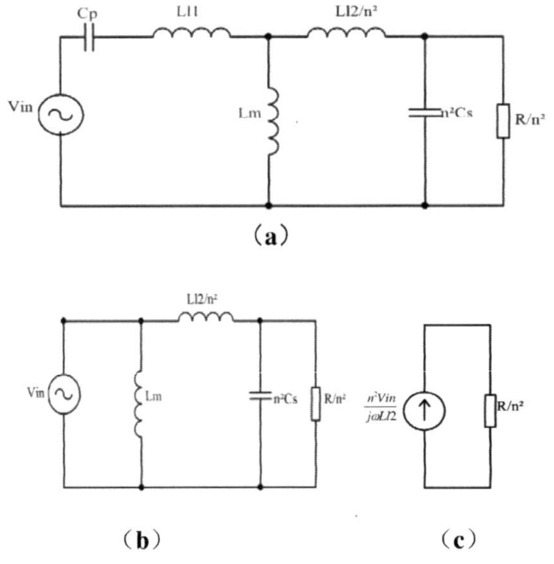

Figure 4. Compensating leakage of SP topology
Figure. 4(a) is an equivalent circuit diagram of the secondary side converted from Figure 3 to the primary side. The series resonance of C_p and L_{l1} is equivalent to a short circuit, so that Figure 4(b) can be obtained. Figure 4(b) L_{l2}/n^2 and n^2C_s resonance, after the equivalent conversion of the power supply can be obtained Figure 4(c), the output is the current source. When the system parameters are determined, the secondary output current is only related to the compensation network port voltage V_{IN}. The current expression is:

$$i_o = \frac{nv_{IN}}{j\omega L_{l2}} \tag{5}$$

The input impedance ZIN obtained from Figure 4(a) is:

$$\begin{aligned} Z_{IN} &= j\omega L_M // (\frac{j\omega L_{l2}}{n^2} + \frac{1}{j\omega n^2 C_S} // \frac{R}{n^2}) \\ &= j\omega L_M // (\frac{1}{n^2} \frac{1}{\omega^2 C_S^2 R - j\omega C_S}) \end{aligned} \tag{6}$$

From equation (6), we can obtain equation (7):

$$Z_{IN} = \mathrm{j}\frac{\omega R L_M n^2}{\omega^2 C_S{}^2 R^2 + n^4} + \omega^2 (\frac{R^2 L_M C_S}{\omega^2 C_S{}^2 + n^4} - \frac{L_M L_{L2}}{n^2}) \tag{7}$$

From the formula (7), Z_{IN} can be obtained as an inductive property, which satisfies the condition for realizing ZVS.

3.3. Analysis of the influence of coil parasitic resistance

Considering the parasitic resistance of the coil,the SP leakage compensation circuit model is shown in Figure 5(a). C_p and L_{l1} are in series resonance. In the case of system parameter determination, the ICPT system can be equivalent to the actual current source model according to the Norton equivalent transformation. The dotted line portion in Figure 5(b) is equivalent to the impedance Z, thereby analyzing the influence of the coil parasitic parameters on the output constant current.

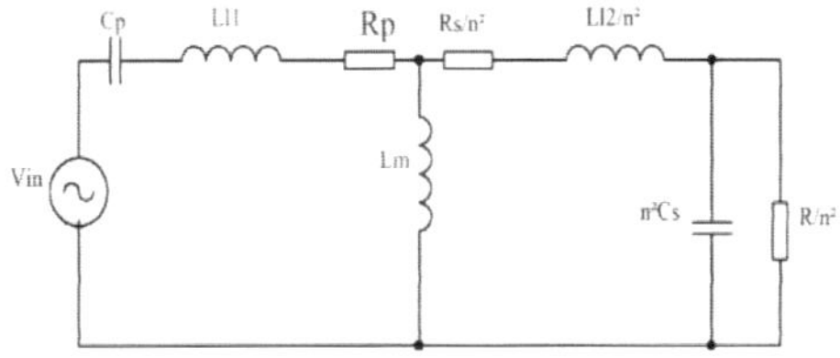

(a)

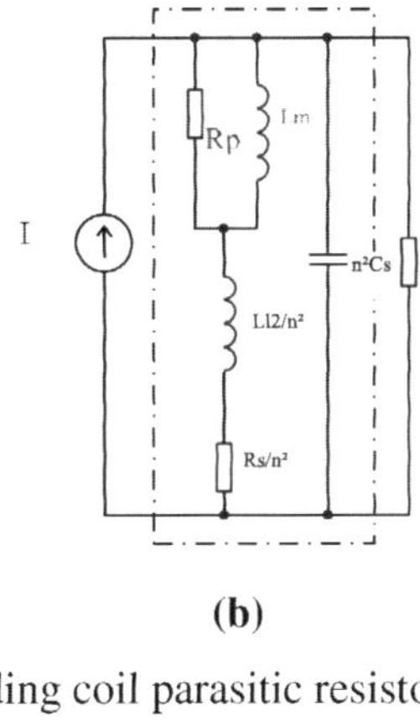

(b)

Figure 5. Including coil parasitic resistor equivalent circuit

The original secondary side ratio of the loosely coupled transformer is approximately 1, and the equivalent current and equivalent impedance are obtained. The effective value I_o of the output current can be obtained by simplification:

$$I_o = \frac{C_S L_M U_{in} \omega}{\sqrt{A^2 + B^2}} \tag{8}$$

where:

$$A = (C_S + L_{l2} + L_M) R R_P + C_S R_P R_S$$
$$B = \omega(L_{l2} L_M R + C_S (L_{l2} R_P + L_M (R + R_P + R_S - L_{l2}\omega)))$$

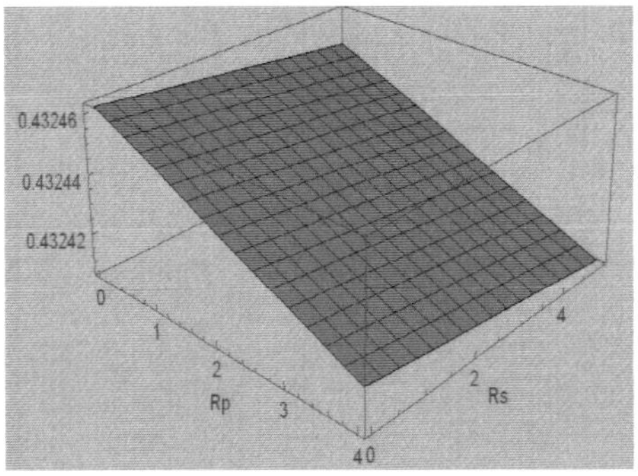

Figure 6. Effect of R_p and R_s on output current I_o

Assuming that the system parameters are fixed, the influence of R_p and R_s on Io is studied according to equation (8). As shown in Figure 6, the output current I_0 decreases slightly with the increase of Rp and Rs. Among them, Rp has a greater influence on the output current Io. When the coil is actually designed, the coil should be optimized to reduce the influence of the parasitic resistance of the coil on the constant current characteristics.

4. Experimental verification

In order to verify that the SP can compensate for the leakage inductance under the push-pull soft switch, the constant current output can be realized. According to Figure 1, the push-pull soft-switch ICPT experimental prototype is built.

In this experiment, the self-inductance of the primary coil of the loose coupling transformer is 142.4μH at a distance of 3.5cm. The leakage inductance of the primary side is 132.5μH.The self-inductance of the secondary coil is 123μH.The leakage inductance of the secondary coil is 115.4μH, and the mutual inductance is 32.6μH.The primary compensation capacitor C_p is 28nF, and the secondary compensation capacitor is 32nF. The input voltage is DC30V, and the load current is tested for 5Ω, 50Ω, 100Ω and 200Ω output currents.

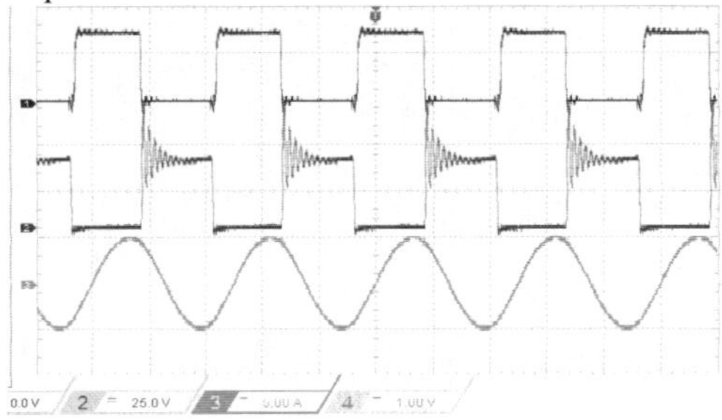

Figure 7. Converter operating wave forms

Figure 7 is a working waveform diagram of the bit converter. From top to bottom, they are the waveform of the switch tube driven, the waveform of the drain and source of the switch tube and the current waveform of the resonant network. It can be seen from the figure that the resonant current lags the voltage.

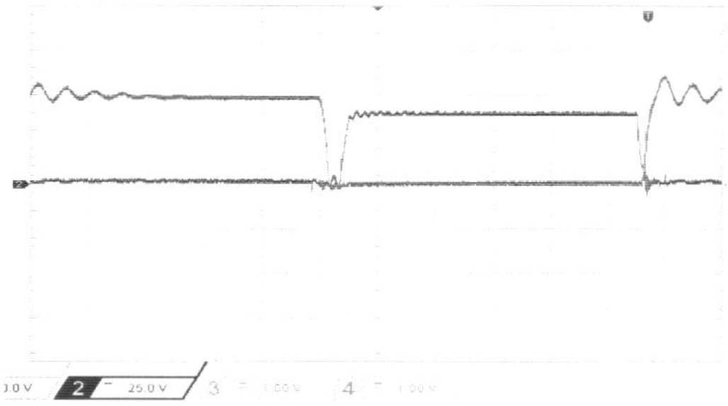

Figure 8. Switching tube voltage waveform

Figure 8 is a voltage waveform diagram of the driving and draining sources when the switching transistor is turned on and off. When the switch tube starts to work, the voltage across the DS has dropped to zero, achieving zero voltage turn-on. When the driving voltage becomes zero, the voltage across the drain source rises from zero, achieving zero voltage turn-off.

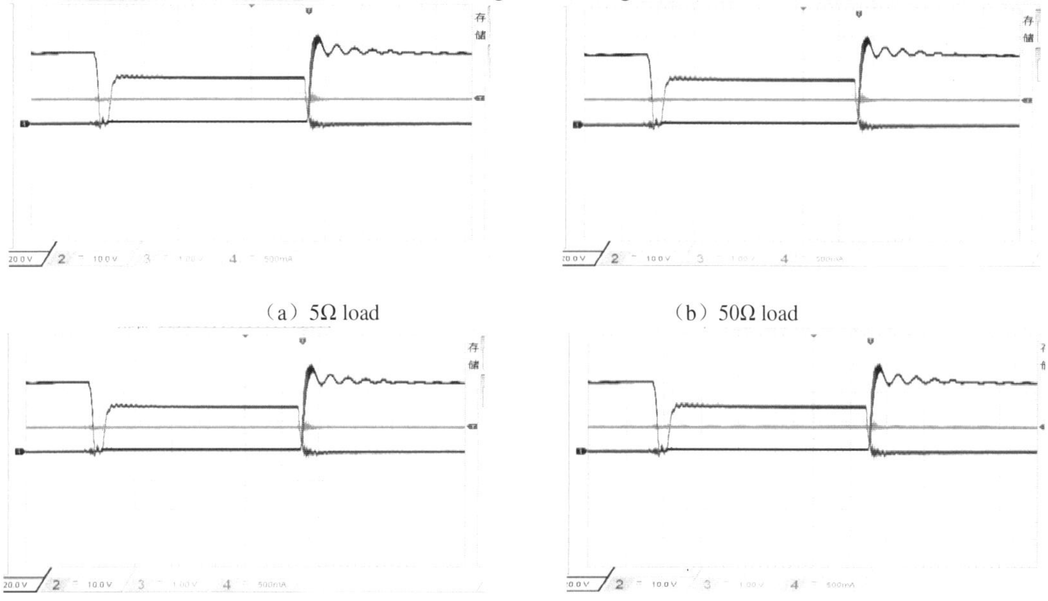

(a) 5Ω load (b) 50Ω load

(c) 100Ω load (d) 200Ω load

Figure 9. Switch-tube drive, drain source and output current waveform under each load

Figure 9 shows the switching transistor drive, the drain and source ends, and the output current waveform for each load value. It can be seen that the soft switch can be realized in the 5-200 Ω load range, and the output current is basically maintained at 0.43A with the load increase. As can be seen from the experimental waveforms, the simulation and experiment are in very good agreement.

5. Conclusion

Based on the push-pull converter, the SP compensation compensation method is used to realize the soft switching of the converter without increasing the voltage zero-crossing detection circuit. At the same time, the ICPT is realized in the 5-200 Ω load range. The constant current output of the system is analyzed. The influence of switching loss on the constant current characteristics of ICPT is analyzed. The soft switching circuit is designed based on the SP leakage compensation characteristic. The theoretical verification is carried out by using Matlab simulation software. The simulation and experimental results show that the SP leakage is adopted. The sense compensation combined with the soft switching characteristics of the push-pull converter can achieve a constant current output characteristic.

Acknowledment
This work is supported by the National Natural Science Foundation of China(51777167,51604217); Shaanxi province technical innovation guidance special plan project(2017CGZH-XNGJ-09)Supported.

References

[1] Xiong Wei, Li An-ming, Ren Qiao-lin, etal. Review on Application of Wireless Charging Device on Electric Vehicles[J]. Telecom Power Technology. 33(3):26-28 (2016)
[2] Shin J, Shin S, Kim Y, et al. Design and implementation of shaped magnetic resonance based wireless power transfer system for roadway powered moving electric vehicles[J]. IEEE Transactions on Power Electronics, 61(3):1179-1192 (2014)
[3] Zhang Wenjing, Liu Bin. Research and application of radio transmission technology[J], Electronic Test, 11:98-99 (2016)
[4] Rahimi Baharom, Mohammad Nawawi Seroji, and Ahmad Ihsan Mohd Yassin,Verification of Lossless ZVS Condition for Three-Phase AC-DC CIHRC IJEETC[J]. 7-20(2018)
[5] Song Kai, Li Zhenjie Du Zhijiang, et al. Constant Current Charging Technology for Variable Load Wireless Charging System[J]. Transactiongs Of China Electrotechnical Society, 13:130-136 (2017)
[6] Jing Yanyan, Qu Xiaohui, Han Hongdou, et al. The Magnetic Coupled Wireless Power Transfer Driver Based on Adjustable Gain Constant-Current Compensation Network[J]. Transactiongs Of China Electrotechnical Society, S1:1-8(2016)
[7] Mai Ruikun,Liu Yern.Chen Yang.Studies of Efficiency Optimization Methods Baesd on Optimal Equivalent Load Control in IPT Systems[J].Proceedings of the CSEE, 36(23): 6468-6475 (2016)
[8] Han Feng. Research on Push-Pull Resonant Converters[D]. Nanjing University of Aeronautics and Astronautics, (2005)
[9] Chen Qingbin, hang Weihao, uyang yixin,et al.A Design Method for Compensation Network Parameters with Variable Constant Current Output and its Characteristics Analysis Based on Leakage Inductance Compensation[J]. Transactiongs Of China Electrotechnical Society, 32(22) (2017)
[10] Zhou Wenqi, Ma Hao, He Xiangning. Investigation on Different Compensation Topologies in Inductively Coupled Power Transfer System[J]. Transactiongs Of China Electrotechnical Society, 24（1）:133-139 (2009,)
[11] Kurs A, Karalis A, Moffatt R. Wireless power transfer via strongly coupled magnetic resonances[J].Science, 317 2017)
[12] EleniGati, Georgios Kampitsis, Stefanos Manias. Variable Frequency Controller for Inductive Power Transfer in Dynamic Conditions[J]. IEEE Transactions on Power Electronics, 32(2), 1684-1696 (2017)

IOP Conf. Series: Materials Science and Engineering **533** (2019) 012010 doi:10.1088/1757-899X/533/1/012010

Study on three-phase unbalanced control of low voltage distribution network

Chu Zhang[1,a], Zhouxing Fu[1], Huanfei Ye[1] and Dan Niu[1]

[1]College of electrical and control engineering of Xi'an University of Science And Technology, Xi'an, China

Corresponding author: JulyDonut@163.com

Abstract. Three-phase unbalance is one of the power quality problems. For the power quality problem caused by the three-phase unbalance phenomenon on the load side in the low-voltage distribution network, combined with the characteristics of the three-phase four-wire system of the distribution network in China, the three-phase unbalance adjustment is studied. The device control strategy is selected, and the three-phase four-bridge arm topology and the three-phase current real-time detection method based on instantaneous reactive power theory are selected. The experimental model is built by MATLAB/Simulink to verify the feasibility of the method.

1 Introduction

In recent years, China's economy has flourished, and the demand for electricity consumption has also increased sharply[1]. Among them, the impact and non-linear load devices have also increased, resulting in voltage waveform distortion, voltage fluctuation, flicker and three-phase imbalance in the power grid[2]. Power quality problems sometimes occur, so three-phase unbalanced control is an important measure to improve power quality[3].

The treatment of three-phase unbalance in the low-voltage distribution network can improve the power factor of the components in the system, reduce the loss of equipment, lines and transformers, improve the power supply quality of the power grid, extend the service life of equipment in the power grid, and improve the stability of the system to avoid[4]. When the system encounters a large disturbance, a stable breakdown accident such as a voltage collapse occurs, which is of great significance for the safe, economical and high-quality operation of the power system[5].

In this paper, the power quality problems caused by the three-phase unbalance of the load side in the low-voltage distribution network are combined with the characteristics of the three-phase four-wire system of the distribution network in China[6]. Select the main topology. The hysteresis comparison method in the PWM tracking control method is used to control the compensation current. The theory of instantaneous reactive power is studied, and the method of accurately extracting positive sequence, negative sequence and zero sequence current in real time is obtained[7]. The experimental model is built based on MATLAB/Simulink for simulation analysis. The system simulation model is built according to the main system, detection system, control system and adjustment system. The selected method is verified and the simulation results are analyzed[8].

2 Three-phase unbalanced principle

Content from this work may be used under the terms of the Creative Commons Attribution 3.0 licence. Any further distribution of this work must maintain attribution to the author(s) and the title of the work, journal citation and DOI.
Published under licence by IOP Publishing Ltd

2.1 Unbalance measure

In practical engineering applications, the three-phase four-wire system with zero sequence components is defined as follows[9]:

$$\varepsilon_U^- = \frac{U_2}{U_1} \times 100\% \tag{1}$$

$$\varepsilon_U^0 = \frac{U_0}{U_1} \times 100\% \tag{2}$$

Where ε_U^- represents the three-phase voltage negative sequence imbalance; ε_U^0 represents the three-phase voltage zero sequence imbalance; U_1 represents the three-phase voltage fundamental positive sequence RMS; U_2 represents the three-phase voltage fundamental negative sequence RMS; U_0 represents the zero-order rms value of the three-phase voltage fundamental.

Similarly, by replacing the voltage in the above equation with a current, the negative sequence and zero sequence imbalance of the three-phase current can be obtained[10].

2.2 Three-phase load imbalance management principle

The simple power system model is shown in Figure 1. The purpose of using the adjustment device to manage the three-phase load imbalance is to transfer the unbalanced power between the power supply and the load to the load and the adjustment device, so that the system network side three Phase power balance, three-phase current symmetry, negative sequence, zero sequence current is zero. In Figure 1, e_s represents the grid voltage, i_s is the grid side current, i_L is the load current, and i_c is the output compensation current of the adjustment device. According to the symmetrical component method, the three-phase current is decomposed into a positive sequence, a negative sequence, and a zero sequence. If the amount of negative sequence and zero sequence components in the load current is injected into the grid by the device, so that only the positive sequence component is left in the current on the system network side, the imbalance treatment of the system is realized[11].

The adjustment device is mainly composed of two parts, namely a power electronic converter and an auxiliary circuit. The converter completes the function of injecting unbalanced current into the system, and the auxiliary circuit realizes the function of detecting, controlling, and driving the IGBT. The working principle of the adjusting device can be described as: detecting the negative sequence and the zero sequence component of the three-phase current through the detecting module, forming a command signal for compensating the current, forming a modulation signal of the IGBT through the control module, and controlling the IGBT after being amplified by the driving circuit, so that The device injects and directs a current with the same amplitude of the signal in the system, so that the current on the network side of the system is three-phase symmetrical.

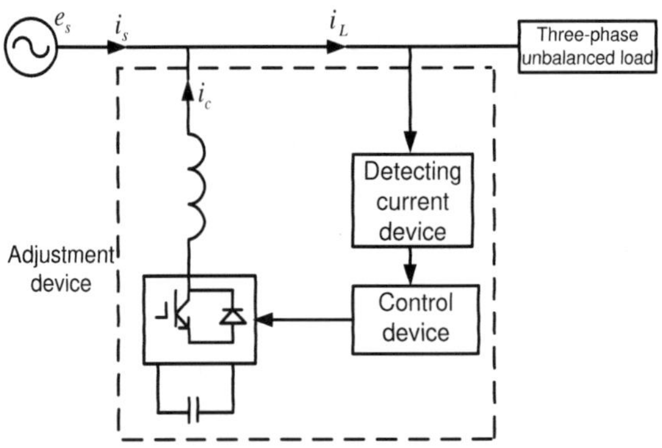

Figure 1. governance schematic

This paper is to deal with the three-phase unbalanced load of the low-voltage distribution network. Considering the cost and design difficulty, the two-level structure can meet the requirements. Because the system is a three-phase four-wire system, the system needs to output a large number of zero-sequence currents. Therefore, the single-capacitor three-phase four-bridge arm single-inductor grid-connected interface structure as shown in Fig. 2 is selected as the research object.

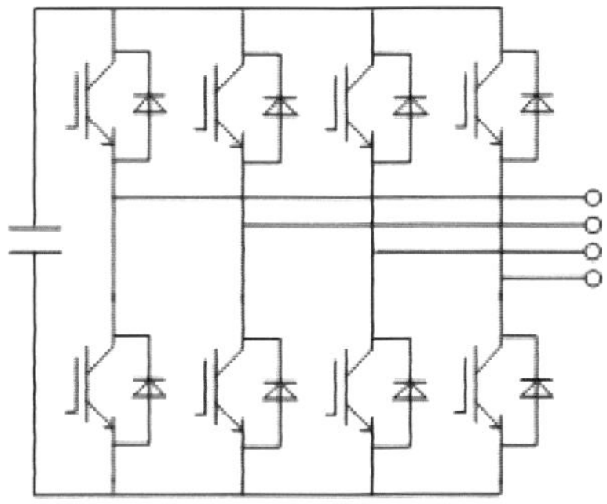

Figure 2. Three-phase four-wire topology main circuit topology

3 MATLAB Simulation

3.1 Establishment of simulation model for three-phase load imbalance management

The parameters for establishing the simulation model system are shown in Table 1. The model is mainly divided into main system, current detection module and control module. The overall simulation timing control of the system is as follows: the capacitor is pre-charged to 540V, the 0.05s adjustment device is put into operation, and the positive sequence active current detection module is input, the DC capacitor is charged and the DC voltage control is controlled by the ramp signal, and the slope is 6500V/ s, 0.09s began to control the capacitor with constant voltage, the reference voltage is 800V. The 0.1s system manages the unbalanced current and puts into the system unbalanced current detection module. The 0.7s load is abrupt, which is twice the original load.

Table 1. Simulation model system parameters

element	parameter
System power	380V, 50HZ
Phase A load	1mH, 30Ω
Phase B load	1mH, 50Ω
Phase C load	1mH, 20Ω
DC side capacitor	2000uF, 540V
Step size	10us
On-off level	18kHz

3.2 Signal detection module

For the detection of the three-sequence current, the i_p-i_q detection method based on the instantaneous reactive power theory can reduce the influence of harmonics in the current on the control current. At the same time, in order to be more realistic and cooperate with the working sequence of the main system, the step signal is used to control the input time of the detection module.

The simulation results of the three-phase load current can be obtained through simulation operation as shown in Figure 3.

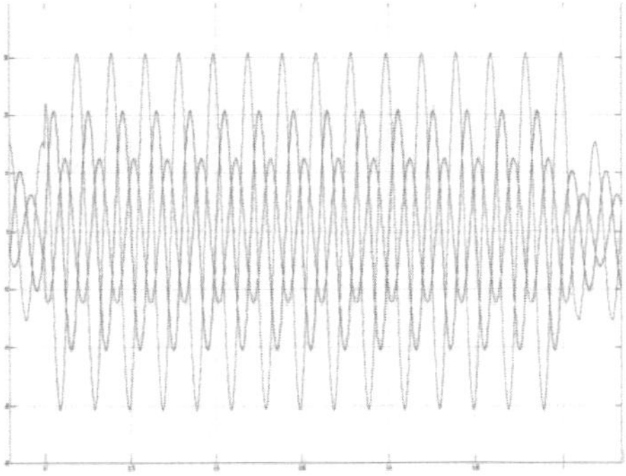

Figure 3. Three-phase load current test results

Therefore, the i_p-i_q detection algorithm based on the instantaneous reactive power theory can detect the positive sequence and negative sequence current components of the three-phase current. The zero sequence current component can be measured directly from the center line.

3.3 DC voltage control module

DC capacitor voltage control is a constant value control system, as shown in Figure 4. In the simulation, the initial voltage of the capacitor is set to 540V, and the adjustment device starts running at 0.05s. In order to avoid the inrush current caused by the switching device, the slope of the DC side voltage is ramped up to 6500V/s, so that the DC voltage of the system can reach 800V after 0.04s. After 0.09s, the DC voltage is controlled by constant voltage. The control result is shown in Figure 4-8.

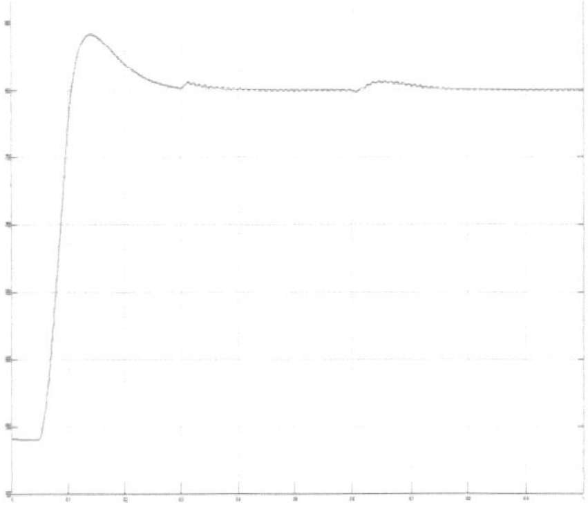

Figure 4. DC voltage control simulation results

3.4 Governance outcome analysis

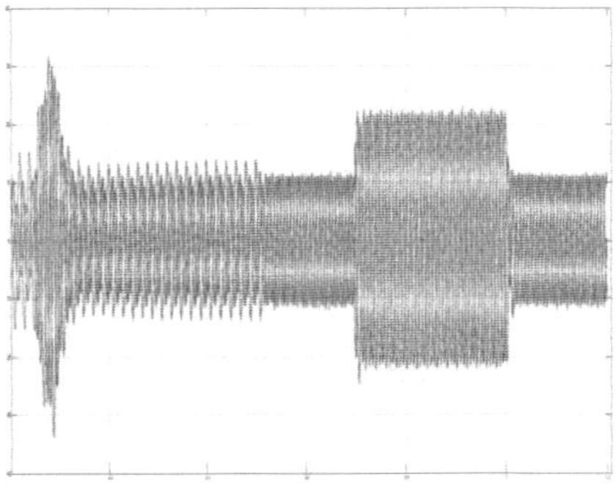

Figure 5. Three-phase network side current change process

IOP Conf. Series: Materials Science and Engineering **533** (2019) 012010 doi:10.1088/1757-899X/533/1/012010

Figure 5. shows the change process of the three-phase grid side current in the process of system three-phase load imbalance. The stages are now analyzed as follows:

3.4.1 Steady state analysis. Fig 6. is a three-phase current waveform after the switching device is turned on, and Fig. 7 is a three-phase current waveform after the load is abrupt.

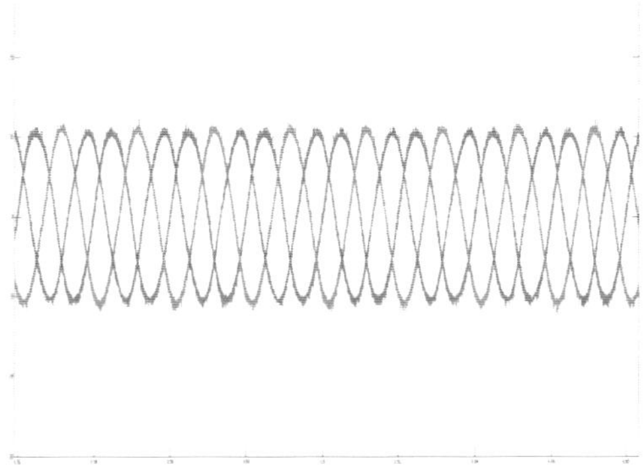

Figure 6. Steady-state waveform after three-phase current imbalance treatment

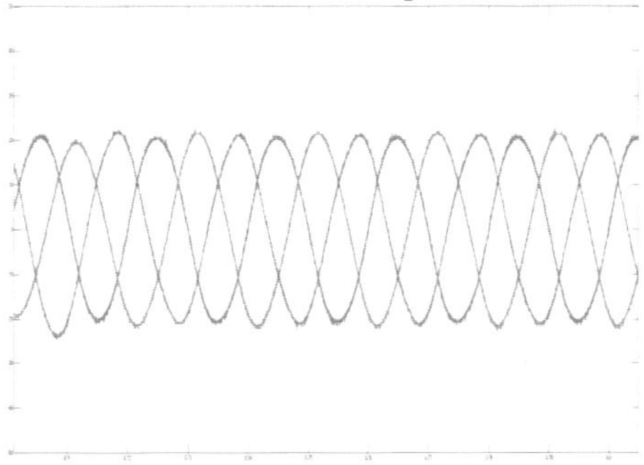

Figure 7. Three-phase current waveform after sudden load change

From the three-phase unbalance formula, the comparison of the three-phase unbalance before and after the treatment can be obtained, as shown in Table 2.

It can be seen that although the imbalance is not possible after the treatment, the effect of the treatment is obvious, and the three-phase negative sequence imbalance and the three-phase zero sequence imbalance are obviously dynamically stabilized in a small range. Due to the small fluctuation of the DC side voltage, there is still a certain negative sequence and zero sequence current on the system network side.

Table 2. Comparison of three-phase unbalance

Governance status	negative sequence imbalance	zero sequence imbalance
Before governance	106%	287%
Before the load is mutated	15%	9%
After the load is mutated	1.5%	3.7%

3.4.2 Dynamic Analysis. As shown in Fig. 8, (a) is a waveform of a change in current when the adjustment device is put into operation, and (b) is a three-phase grid-side current waveform when the load is abrupt. It can be seen from the figure that when the adjustment device is put into operation to start the three-phase unbalanced current, the inrush current is not large; when the load is abrupt, the transition time is appropriate and the next steady state can be quickly entered. The control system performs well.

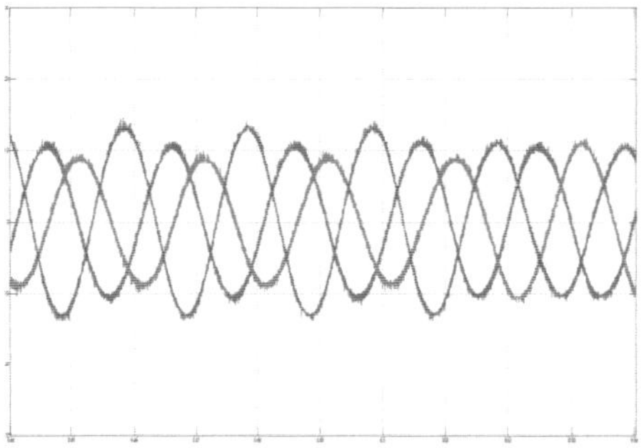

(a)The waveform of the current change when the adjustment device is put into treatment

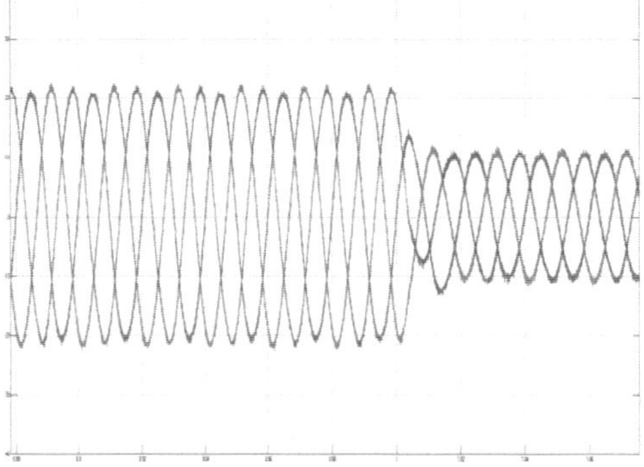

(b)Three-phase grid side current waveform when load is abrupt

Figure 8. Current management results

The simulation results show that the three-phase unbalance phenomenon of the system has a good effect in the case of grid voltage symmetry.

4 Conclusion

According to the characteristics of the three-phase four-wire system, this paper selects the single-capacitor four-bridge single-inductor grid-connected interface as the main topology of the compensation device. The i_p-i_q current detection algorithm based on instantaneous reactive power theory is used to detect the unbalanced current. Filtering is performed using a dual filter approach combining a notch filter and a low pass filter. The simulation results show that the detection algorithm has fast detection speed and high precision, which can meet the system requirements. The hysteresis comparison method in PWM tracking control theory is used to control the compensation current, and various parameters are set. The control performance is obtained by simulation. In the Simulink environment, a simulation model was built. The simulation results show that the three-phase unbalanced control effect of the system is good when the grid voltage is symmetrical.

However,there are still some shortcomings in this paper that need to continue to study in future work:This paper only analyzes the three-phase unbalance phenomenon when the grid voltage is symmetrical.And the control strategy selected in this paper only compensates for the negative sequence and zero sequence component currents in the fundamental current, and the influence of harmonics in the system on the treatment effect needs further discussion.

References

[1] Wang Zhaoan. Harmonic suppression and reactive power compensation [M]. Beijing, Mechanical Industry Press. 1998
[2] Lin Haixue. Three-phase unbalance of power system [M]. Beijing: China Electric Power Press. 1998.
[3] Sun Guoping, Sun Haiyan. Discussion on the harm of low voltage three-phase load imbalance [J]. China Electric Power Education. 2010(S1): 485-4 89.
[4] Qiu Guanyuan. Modern Circuit Theory [M]. Beijing: Higher Education Press. 2001
[5] He Yangzan, Wen Zengyin. Power System Analysis [M]. Wuhan, Huazhong University of Science and Technology Press. 2002.
[6] Ding Renjie, Liu Jian, Zhao Yuwei, et al. Instantaneous power analysis and asymmetric load compensation method for unbalanced circuits[J].Transactions of China Electrotechnical Society. 2007，22(1)：120-124.
[7] Zhang Bo.Generalized Instantaneous Reactive Power Theory for Three-Phase Circuits[J].Journal of Electric Power Systems and Automation. 2005
[8] Ye Jiaming.Research on Harmonic and Reactive Current Detection Methods Based on Instantaneous Reactive Power Theory[D].Chongqing University. 2012.
[9] GB/T 15543-2008. "Power Quality Three-Phase Voltage Imbalance" [S]. Beijing: China Standard Press. 2009
[10] Wang Zhaoan, Li Min. Zhuo Fang. Instantaneous reactive power theory and power active filter [C]. Proceedings of the Academic Annual Conference on Electrical Engineering Progress and Development in the 1990s, Beijing. 1991.
[11] Yu Qun, Cao Na. Modeling and Simulation of MATLAB/Simulation Power System [M]. Beijing, Mechanical Press. 2011.

IOP Conf. Series: Materials Science and Engineering **533** (2019) 012011 doi:10.1088/1757-899X/533/1/012011

Linear regression model for screening SiC MOSFETs for paralleling to minimize transient current imbalance

James Abuogo[1], Zhibin Zao[1] and Junji Ke[1]

[1]State Key Laboratory of Alternate Electrical Power Systems with Renewable Energy Sources, North China Electric Power University, 102206 Changping District, Beijing, China

E-mail: zhibinzhao@126.com

Abstract. This paper describes the development of a machine learning model which can be used to screen SiC MOSFETs for paralleling with minimized transient current imbalance. The spread of device parameters is determined to isolate the device parameters which are likely to significantly influence transient current distribution. A linear regression model is then developed and trained using device parameters' data measured from forty devices of the same production lot. The resulting model is an expression for current imbalance as a function of device parameters, with weight of each parameter determined. The performance of the trained model on a set of twenty testing devices is then determined and verified via a double pulse test experiment incorporating paralleled devices. The model is found to perform satisfactorily and can effectively be deployed in screening devices for paralleling.

1. Introduction

Over the recent years, Silicon Carbide (SiC) Metal Oxide Semiconductor Field Effect Transistor (MOSFET) has received a lot of attention due to its fast switching speed, high breakdown voltage, good thermal conductivity and low switching loss [1-3]. Despite having superior material properties over conventional Silicon Insulate Gate Bipolar Transistor (Si IGBT) and Silicon MOSFET, one significant drawback of SiC MOSFET is its low current carrying capacity [1-3]. This is a hindrance to deployment of SiC MOSFETs in high power and high current applications. To navigate this problem, parallel connection of discrete SiC MOSFET devices or building a multichip module have been the most common solutions [1-4]. However, studies of paralleled SiC MOSFET devices or chips reveal that mismatches in device and circuit parameters lead to an imbalanced current distribution among the paralleled units. The mismatch in device parameters is significant in SiC MOSFETs due to the relatively immature fabrication technology of SiC MOSFETs compared to other conventional power electronic devices like Si IGBTs and Si MOSFETs [5]. This leads to overcurrent in some units which further causes excessive heating with the final result being device degradation or eventually failure.

According to [5, 6], device parameters which affect current sharing among paralleled SiC MOSFET devices include the on-resistance, pinch-off voltage, reverse breakdown voltage of gate, and static transfer characteristics. In [6-8], distribution of transient current is attributed to gate threshold voltage (V_{th}) while that of steady state current is dependent on on-resistance. In some works like [9, 10], only V_{th} is considered for balancing transient current distribution while in other works like [5, 6] other parameters are found to contribute to the current imbalance as well. Therefore, in this paper, balancing of transient current for paralleled devices by considering multiple parameters simultaneously is sought.

Content from this work may be used under the terms of the Creative Commons Attribution 3.0 licence. Any further distribution of this work must maintain attribution to the author(s) and the title of the work, journal citation and DOI.
Published under licence by IOP Publishing Ltd

Some parameters which can potentially affect transient current distribution are considered. First, the variation of each of these parameters among a group of devices picked from a single production batch is investigated. The parameters which show significant variation are then subjected to screening using a machine learning approach. Linear regression method of machine learning is used to determine the extent to which each of these parameters contribute to current sharing and express this contribution in terms of weights. The resulting model is then used to screen twenty testing devices to assess the model's performance.

The organization of the paper is as follows. Section I introduces the paper. Section II describes supervised machine learning and outlines how it has been applied for screening SiC MOSFETs in this paper. Section III presents device parameters' test results and shows the statistical description of the batch of devices used in developing the screening model. Section IV describes the training of the linear regression model while Section V assesses the model's performance through experiment. Section VI concludes the paper.

2. Supervised Learning Algorithm

The screening approach adopted in this paper takes advantage of the supervised machine learning algorithms to screen SiC MOSFET devices. A number of different static and dynamic device parameters affect the current distribution of SiC MOSFETs. However, the relationship between the individual parameters with the current distribution is quite complex and one single function that can relate the current distribution with all the parameters has not been derived. Even though many researchers have identified the individual parameters that affect the current distribution, the extent to which each parameter affects current distribution has not been established. Screening has therefore been limited to consideration of a single parameter at a time.

Machine learning is the process of giving computers the ability to learn from data without being explicitly programmed [11, 12]. Application of machine learning to SiC MOSFETs screening involves training a model with data of device parameters and corresponding current imbalances to get a trained model. This is a supervised learning problem since device parameters are given as well as the label of corresponding current imbalances. From the trained model, the weights for individual parameters are obtained. These represent the contribution of each parameter to the current imbalance. The resulting model can be used to screen a new set of devices provided their parameters are measured and fed into the model. A linear regression method using gradient descent is deployed for this exercise. The flow of tasks is as shown in Figure 1.

3. Spread of SiC MOSFET Parameters

Due to relatively immature manufacturing technology of SiC MOSFET chips, the parameters of the devices vary significantly. The important device parameters include threshold gate voltage (V_{th}), transconductance (g_{fs}), MOSFET's on-resistance (R_{ds}), gate resistance (R_g), gate-drain capacitance (C_{gd}), gate-source capacitance (C_{gs}) and drain-source capacitance (C_{ds}). The spreads of these parameters among sixty devices drawn from a single manufacturing batch were determined using Agilent Curve Tracer. These devices were produced by one of the global leaders in semiconductor device manufacturing. Figure 2 to Figure 8 show the spread of each parameter among the sixty devices. C_{gs} is ignored in this paper since its influence on transient current imbalance is dependent on R_g [13]. The two parameters determine the switching delay time ($\tau=R_gC_{gs}$), where R_g is more dominant.

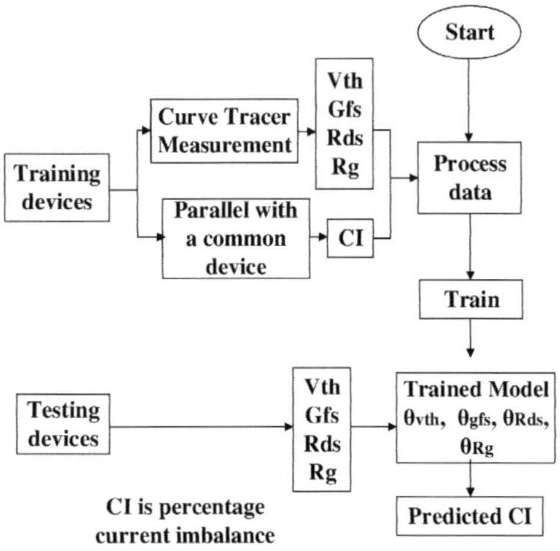

Figure 1. Overview of supervised learning screening approach

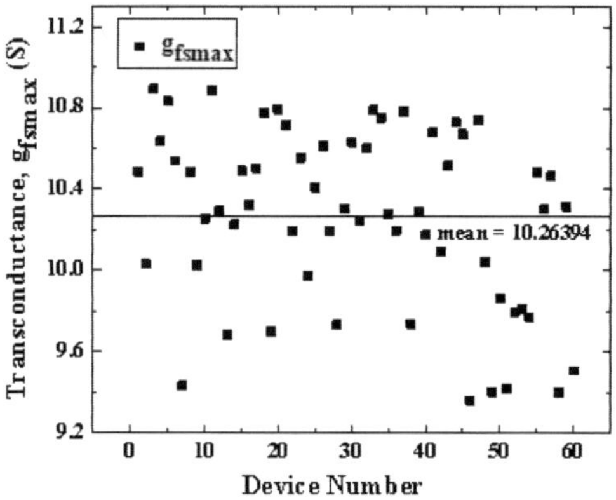

Figure 4. Spread of maximum transconductance of the devices

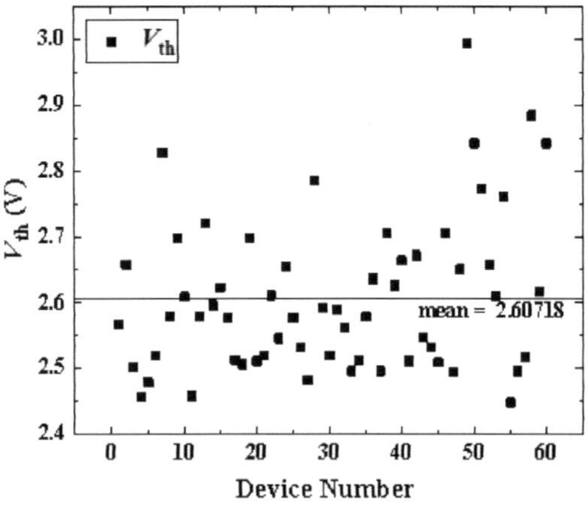

Figure 2. Spread of gate threshold voltage

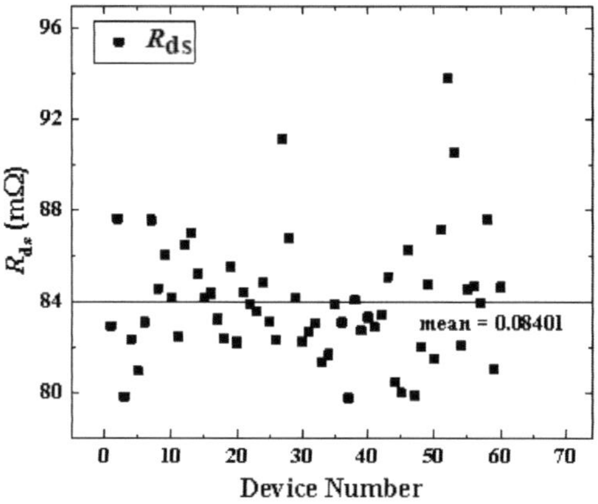

Figure 5. Spread of on-resistances of the devices

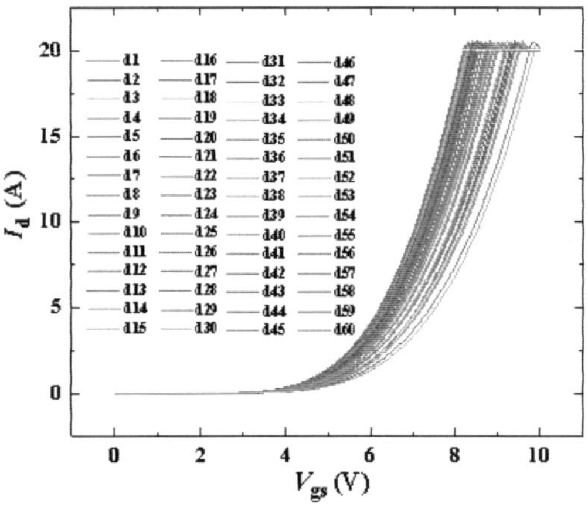

Figure 3. Spread of device transfer curves

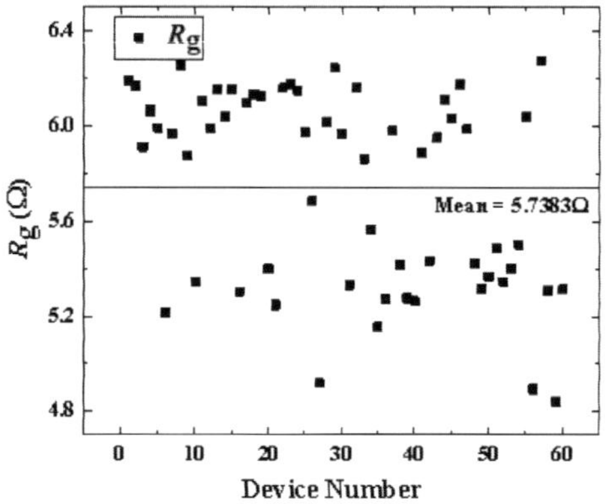

Figure 6. Spread of internal gate resistances of the devices

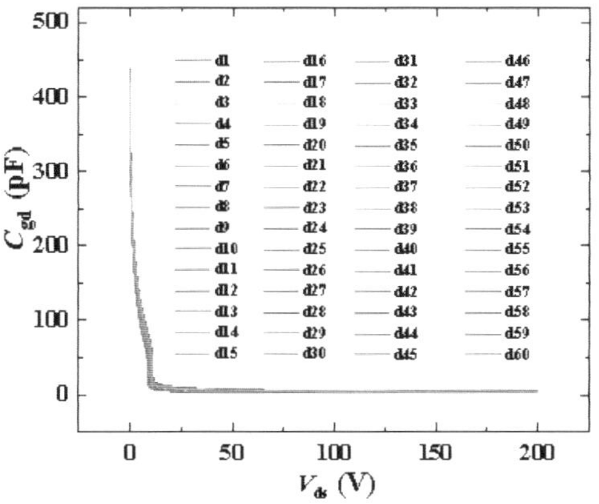

Figure 7. Spread of gate-drain capacitances

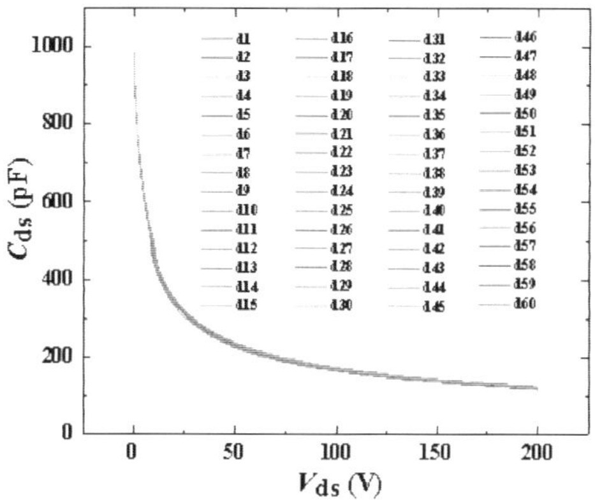

Figure 8. Spread of drain-source capacitances

The output capacitance of SiC MOSFET package (C_{gd} and C_{ds}) is dependent on the drain-source voltage. The Agilent Curve Tracer measures these capacitances as functions of V_{ds}. C_{gd} curves for all the sixty devices are plotted on the same axes to display the variation in the parameter as shown in Figure 7. Likewise, C_{ds} curves are plotted together in Figure 8.

From the spread of parameters plotted above, V_{th}, g_{fs}, R_{ds} and R_g are the ones which show large variations. C_{gd} and C_{ds} are approximately the same for the whole production batch. Hence current imbalance is mainly as a result of the former four parameters. The statistical description of the four parameters for the batch of sixty devices is summarized in Table 1 below where s.d. abbreviates standard deviation.

Table 1. Statistics of the four selected device parameters

	Mean	s.d.	Min	Max	Range	Median
V_{th}	2.607	0.117	2.448	2.994	0.546	2.5795
g_{fsmax}	10.26	0.434	9.361	10.89	1.529	10.304
R_{ds}	0.084	0.0027	0.080	0.094	0.0141	0.0837
R_g	5.738	0.408	4.842	6.273	5.930	1.4315

4. Training Algorithm

Machine learning algorithm was developed in MATLAB and devices' data, already presented in the previous section, used to train a linear regression model. The test devices' data was applied to the resulting model for prediction of their current imbalances. The remainder of this section describes the components of the learning algorithm used in this paper for developing a model for screening SiC MOSFETs for paralleling.

4.1. Data Processing

The training data for any supervised machine learning project consists of the input data (presented as an m by n matrix, X) and label (which is an m dimension vector, y). Here m represents the number of training examples while n is the number of features. In this exercise, sixty devices are used. Out of the sixty devices, forty are used to train the model while twenty are used to verify the performance of the trained model. For the forty training devices, their parameters are measured to get the X matrix and later each of them is paralleled with a common device (device 0) which is not a part of this batch. The current imbalance (CI) is determined for parallel connection of each device with device 0. CI is the ratio of difference between Id values to their average expressed as a percentage. This gives the label vector y. Since the input data has features with considerably different numerical values, for instance, the average value of R_{ds} is 0.084Ω while that of g_{fsmax} is 10.26S, it is important to carry out feature normalization (or feature scaling) on the data before using it in training. This will make the gradient descent algorithm converge faster. Z-score feature normalization, with mean of zero and standard deviation of one, is used in this exercise. Equation (1) is the formula used for this normalization method.

$$x^{i}_{j\_norm} = \frac{x^{i}_{j} - \mu_j}{\sigma_j} \tag{1}$$

Here, $x^{i}_{j\text{-norm}}$ is the normalized value of the ith training example of feature j. μ_j and σ_j are mean and standard deviation for feature j respectively.

4.2. Gradient Descent Algorithm

In linear regression method, the objective is to minimize the cost function expressed as (2) [11, 14].

$$J(\theta) = \frac{1}{2m}\sum_{i=1}^{m}\left(h_\theta\left(x^{(i)}\right) - y^{(i)}\right)^2 \tag{2}$$

Where m is the total number of training examples and hθ is a linear hypothesis given as (3), where n is the number of training features.

$$h_\theta(x) = \theta^T x = \theta_0 x_0 + \theta_1 x_1 + \theta_2 x_2 + \ldots + \theta_n x_n \tag{3}$$

Batch gradient descent algorithm adjusts the values of θ_j in order to minimize the cost. Each iteration of batch gradient descent simultaneously updates the weight parameter θ_j for all features j. The update function can be expressed as (4) where α is the learning rate.

$$\theta_j := \theta_j - \alpha\frac{\partial J(\theta)}{\partial \theta_j} = \theta_j - \alpha\frac{1}{m}\sum_{i=1}^{m}\left(h_\theta\left(x^{(i)}\right) - y^{(i)}\right)x_j^{(i)} \tag{4}$$

To vectorize the implementation of this algorithm, a feature x_0 is introduced whose value is a vector of ones to multiply the intercept term θ_0 of (3).

4.3. Visualizing the Cost Function

In order know whether the algorithm is working well or not, it is important to plot values of the cost after every iteration. The cost should monotonically reduce as the number of iterations increase. Plotting cost values (training, validation and test errors) enables debugging of the code, determining if the learning rate is too large or too small, and identifying presence of problems of over-fitting and under-fitting.

4.4. Normal Equation Method

Besides the gradient descent algorithm, normal equation method can also be used to solve the linear regression machine learning problem. This method uses the closed form solution of linear regression which is represented as (5). Normal equation method does not require feature scaling. It is also implemented without writing loops as the problem is completely vectorized and is achieved conveniently through matrix multiplication. This method gives exactly the same answer as gradient descent algorithm and is therefore used in this work to verify the results of gradient descent training algorithm.

$$\theta = \left(X^T X\right)^{-1} X^T y \tag{5}$$

Here, X is the matrix of input training values, y is the vector of input training labels and θ is the vector of trained model parameters (weights).

5. Testing the Trained Model

5.1. Prediction using the Trained Model

After training with the forty devices, a model is obtained with parameters (weights) corresponding to the contribution of each feature (device parameter) on current imbalance. This model is used to predict the current imbalance for each of the twenty remaining (testing) devices. This is similar to obtaining the current imbalance which could have been measured had each of them been paralleled with device 0. Since device 0 which these test devices are tested against is common, the devices with close current imbalance values must have minimum current imbalance between them. Table II below shows the predicted current imbalances for the twenty testing devices.

Table 2. Current Imbalances of test devices as obtained from the model's prediction

Device	CI (%)	Device	CI (%)
41	2.4885	51	16.4045
42	10.7199	52	11.8968
43	4.1265	53	10.2524
44	2.3723	54	14.4785
45	2.0956	55	1.1138
46	13.0290	56	5.9505
47	1.4588	57	2.6892
48	10.3182	58	20.2235
49	23.4038	59	9.5586
50	16.8281	60	18.4721

5.2. Experimental Verification

From the predictions shown in Table II, selection for paralleling of devices whose CI values are close results in minimized CI. To test this, a double pulse test (DPT) experiment was conducted with paralleled devices. The test bench for DPT was optimized to minimize circuit parameter mismatches. This ensures that CI is mainly as a result of device parameter mismatches. The test bench for DPT comprised gate driver, dc bus capacitors, two paralleled SiC MOSFET devices (DUT), SiC Schottky Barrier Diode (SBD) as a freewheeling diode, load inductance and discharge resistor. Figure 9 below shows the DPT set-up.

From the Table 2, device 48 and device 53 have close CI values. Paralleling them is expected result in a very low CI. Another set of devices with close CI values are device 50 and device 51. These two sets of devices were tested to verify this assertion. Figure 10 and Figure 11 below are the waveforms of drain current for both pairs of devices during turn-on transient. The transient drain currents for device 48 and device 53 are very close. This implies that when the two devices are paralleled, a good transient current distribution is achieved. Likewise, device 50 and device 51 also have a good transient current distribution when paralleled. The model developed through machine learning technique can therefore aid in selecting devices for paralleling such that transient current distribution is nearly uniform.

Still, according to the model's prediction, device 47 and device 58 have CI values which are quite far apart. Also, device 49 and device 55 have CI values which are significantly different. Paralleling these sets of devices is expected to result in large transient current imbalances. A DPT experiment was conducted to ascertain this and the results tally with the assertion. Figure 12 and Figure 13 show the turn-on transient current waveforms obtained from paralleling each pair. Again, the transient drain currents are significantly different in this case as expected, from the model prediction. This further validates the effectiveness of the trained model in screening SiC MOSFETs for paralleling.

In order to get a quantitative measure of accuracy of the model, the actual current imbalances for these paralleled pairs of devices were determined. The values were compared with the difference in CI for each device as obtained from the model. It is expected that when two devices are paralleled then the current imbalance should be equal to the difference in their CI values predicted by the model. This is because the model assumes paralleling each of the devices with a common device. Since the model represents a linear relationship between device parameters and current imbalance, with the device parameters remaining constant, this additive nature of current imbalances is feasible. Model's accuracy is then calculated from test CI and model's CI as shown in Table 3. In this paper, CI is calculated using (6).

IOP Conf. Series: Materials Science and Engineering **533** (2019) 012011 doi:10.1088/1757-899X/533/1/012011

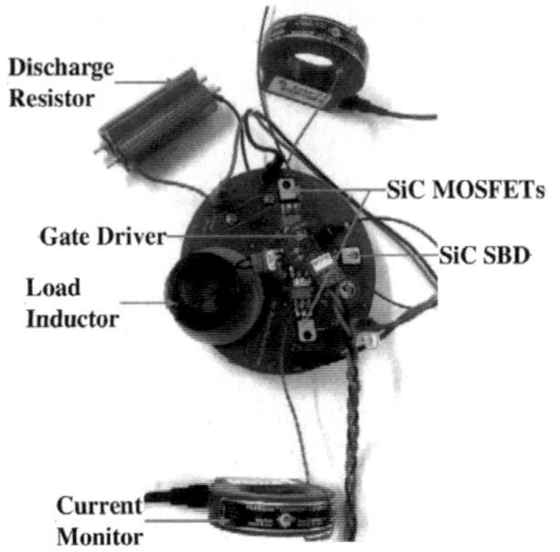

Figure 9. DPT experiment test bench for paralleled devices

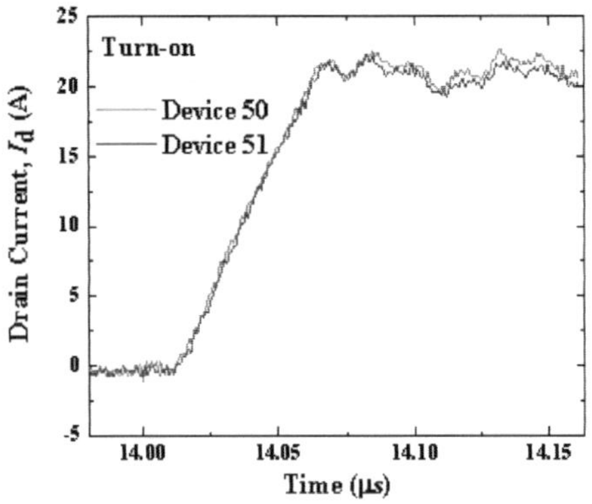

Figure 10. Drain currents when device 50 and device 51 are paralleled

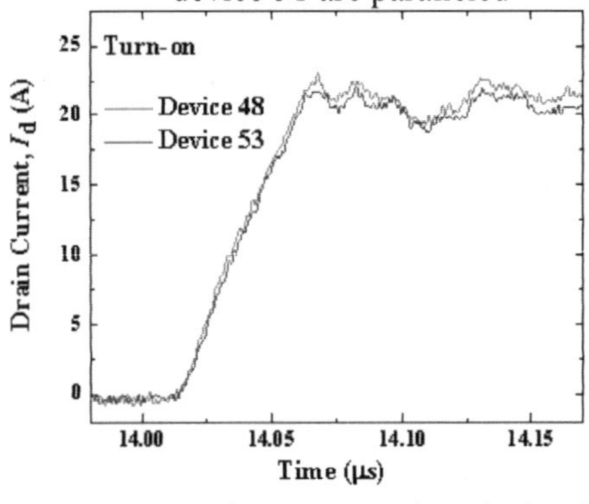

Figure 11. Drain currents when device 48 and device 53 are paralleled

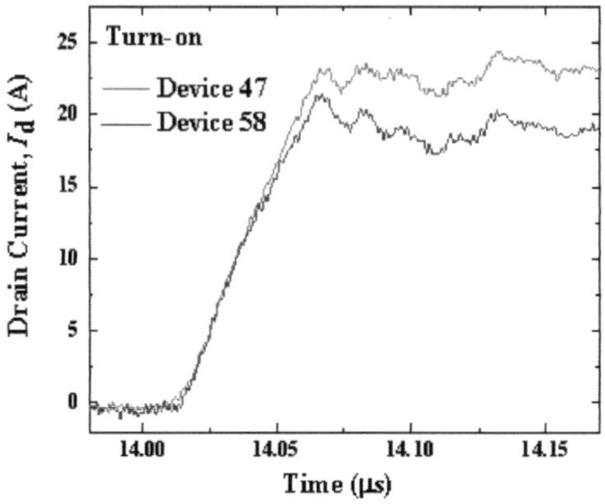

Figure 12. Drain currents when device 47 and device 58 are paralleled

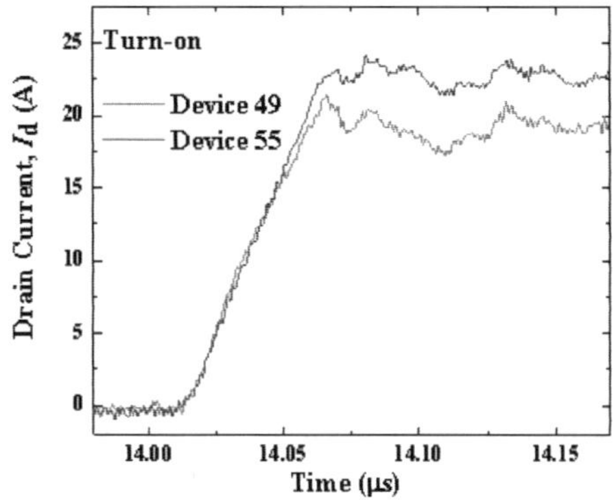

Figure 13. Drain currents when device 49 and device 55 are paralleled

$$CI = \frac{2(I_{d2} - I_{d1})}{I_{d1} + I_{d2}} \times 100 \qquad (6)$$

Table 3. Comparison of model with experimental results

1st	2nd	Exp. I_{d1} (A)	Exp. I_{d2} (A)	Exp. CI (%)	ΔCI model (%)	Accuracy
47	58	18.94	22.538	17.36	18.76	92%
49	55	18.95	23.419	21.10	22.29	95%
48	53	21.32	21.331	0.056	0.066	83%
50	51	20.75	20.663	0.420	0.424	99%
*Exp. represents experimental						

From the experimental results, the developed model can effectively screen devices for paralleling. The predicted CI values can be used to determine which devices to parallel to achieve low transient current imbalance. Thedevices with small difference in predicted CI values have small actual current imbalance when paralleled. Those with large predicted CI values on the other hand have large actual CI when paralleled. The accuracy of prediction of the model is also satisfactorily high.

6. Conclusion

Using machine learning algorithms to learn the level of influence of device parameters on current imbalance can help screen SiC MOSFET devices for paralleling. This paper has demonstrated implementation of such an approach by using linear regression method to train a model which is later used to screen devices for paralleling. The screened devices are tested using DPT experiment and the performance of the model is found to be satisfactory. Further work will be done to develop a model which can screen devices to achieve both minimum transient and steady state current imbalance when paralleled.

Acknowledgement

The authors of this paper wish to send their gratitude to National R&D Program of China for supporting this research through the project numbered 2018YFB0905703.

References

[1] H. Li et al 2016 Influences of device and circuit mismatches on paralleling silicon carbide MOSFETs, in *IEEE Transactions on Power Electronics*, **vol. 31**, no. 1, pp. 621-634.
[2] H. Li, S. Munk-Nielsen, C. Pham and S. Bęczkowski, 2014 Circuit mismatch influence on performance of paralleling silicon carbide MOSFETs, *2014 16th European Conference on Power Electronics and Applications*, Lappeenranta, pp. 1-8.
[3] J. Fabre and P. Ladoux, 2016 Parallel connection of 1200-V/100-A SiC-MOSFET half-bridge modules, in *IEEE Transactions on Industry Applications*, **vol. 52**, no. 2, pp. 1669-1676.
[4] D. Peftitsis et al 2013 Challenges regarding parallel connection of SiC JFETs, in *IEEE Transactions on Power Electronics*, **vol. 28**, no. 3, pp. 1449-1463.
[5] J. K. Lim et al 2014, Analysis and experimental verification of the influence of fabrication process tolerances and circuit parasitics on transient current sharing of parallel-connected SiC JFETs, in *IEEE Transactions on Power Electronics*, **vol. 29**, no. 5, pp. 2180-2191.
[6] G. Wang, J. Mookken, J. Rice and M. Schupbach, 2014 Dynamic and static behavior of packaged silicon carbide MOSFETs in paralleled applications, *2014 IEEE Applied Power Electronics Conference and Exposition - APEC 2014*, Fort Worth, TX, pp. 1478-1483.
[7] J. Fabre and P. Ladoux 2015 Parallel connection of SiC MOSFET modules for future use in traction converters, *2015 International Conference on Electrical Systems for Aircraft, Railway, Ship Propulsion and Road Vehicles (ESARS)*, Aachen, pp. 1-6.
[8] Y. Mao et al 2017 Balancing of peak currents between paralleled SiC MOSFETs by source impedances, *2017 IEEE Applied Power Electronics Conference and Exposition (APEC)*, Tampa, FL, pp. 800-803.
[9] Y. Xue et al Active current balancing for parallel-connected silicon carbide MOSFETs, *2013 IEEE Energy Conversion Congress and Exposition*, Denver, CO, 2013, pp. 1563-1569 (2013).
[10] S. Bęczkowski et al 2017 Switching current imbalance mitigation in power modules with parallel connected SiC MOSFETs, *2017 19th European Conference on Power Electronics and Applications (EPE'17 ECCE Europe)*, Warsaw, 2017, pp. P.1-P.8.
[11] I. Goodfellow and Y. Bengio 2016 Deep learning, MIT Press.
[12] N. M. Nasrabadi 2007 Pattern recognition and machine learning, *Journal of Electronic Imaging.*
[13] J. Ke et al 2018 Influence of parasitic capacitances on transient current distribution of paralleled SiC MOSFETs, *IEEE Workshop on Wide Bandgap Power Devices and Applications in Asia - WiPDA Asia 2018*, Xi'an, Shaanxi, China.
[14] R. A. Farrugia 2012 Improving motion vector prediction using linear regression, *2012 5th International Symposium on Communications, Control and Signal Processing*, Rome, pp. 1-4 .

IOP Conf. Series: Materials Science and Engineering **533** (2019) 012012 doi:10.1088/1757-899X/533/1/012012

Research on saturation model of ferromagnetic resonance voltage transformer

Qinghao Zheng[1], Yunge Li[2], Nana Zhang[1], Fan Gan[1] and Zhouxing Fu[1]

[1]College of electrical and control engineering of Xi'an University of Science And Technology, Xi'an, China
[2]State Grid Shaanxi Electric Power Research Institute, Xi'an, China

Corresponding author:18330215072@163.com

Abstract. The voltage transformer is a common nonlinear inductor in the power system. The core nonlinearity caused by saturation, hysteresis and eddy current makes it difficult to describe the dynamic magnetization of the core. At present, there are many different saturation models for nonlinear inductors. Different models are used to simulate ferromagnetic resonance, which will result in different results. In this paper, based on ATP-EMTP electromagnetic transient simulation software, the simulation of different voltage transformer saturation model is carried out, the calculation error is analyzed, and the results of different models are compared.

1.Introduction

In the power system, the voltage transformer is a core inductor component. If there is some kind of large disturbance or operation, the nonlinear core of the PT may be saturated, thus forming a resonance circuit. Ferromagnetic resonance occurs in both neutral grounded and ungrounded systems. Abnormal overvoltages during resonance cause insulation flashover and lightning tube explosion which seriously threaten the safe operation of the power grid [1]. However, the core nonlinearity makes it difficult to describe the dynamic magnetization characteristics of the core. Therefore, it is necessary to study the model of the voltage transformer.

The literature [1] uses a single-valued magnetization curve to describe the nonlinear characteristics of the voltage transformer, ignoring the loss, and there is no parallel resistance on the nonlinear inductance of the PT equivalent. In [2], starting from the magnetic flux waveform, based on the relationship between the magnetic flux and the excitation branch voltage and current, a method for calculating the magnetization curve is proposed. In [3], the piecewise linearization method is adopted. The processing of the saturated segment is approximated as a straight line segment. The loss is simulated by a linear resistor and paralleled with the nonlinear inductor. In [4], the curve fitting method is used to establish the saturation model of the voltage transformer by establishing an analytical expression to approximate the discrete data. It is divided into parameters and non-parametric fitting. The calculation of ferromagnetic resonance using different models will produce different results, so it is necessary to study the exact saturation model of the voltage transformer.

In this paper, based on the ferromagnetic resonance of a 110kV substation and ATP-EMTP simulation software, different voltage transformer saturation models are used for calculation. The calculation error is analyzed. The results of different models are compared.

Content from this work may be used under the terms of the Creative Commons Attribution 3.0 licence. Any further distribution of this work must maintain attribution to the author(s) and the title of the work, journal citation and DOI.
Published under licence by IOP Publishing Ltd

2.Basic circuit

Figure 1 is a resonant circuit of a substation in which ferromagnetic resonance occurs. The isolating blades D0-1, D0-2, D1-1, D1-2 are in the closed position, the circuit breaker B0 or B1 is already open, and when the other is turned off, ferromagnetic resonance overvoltage that causes a sustained higher amplitude.

Applying the Thevenin's theorem, Figure 1 can be simplified to the basic resonant circuit shown in Figure 2, where E is the equivalent power supply potential, C is the equivalent capacitance, L is the nonlinear inductance of PT, and R is the PT core equivalent resistance. The parameter is:: E=47951V, C=2.38x10^{-9}F, R=80MΩ, The PT ratio is 1100.

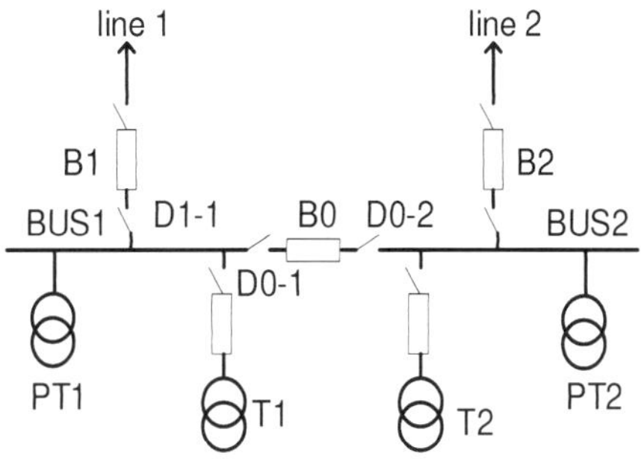

Figure 1. Typical configutation of substations prone to Ferroresonance

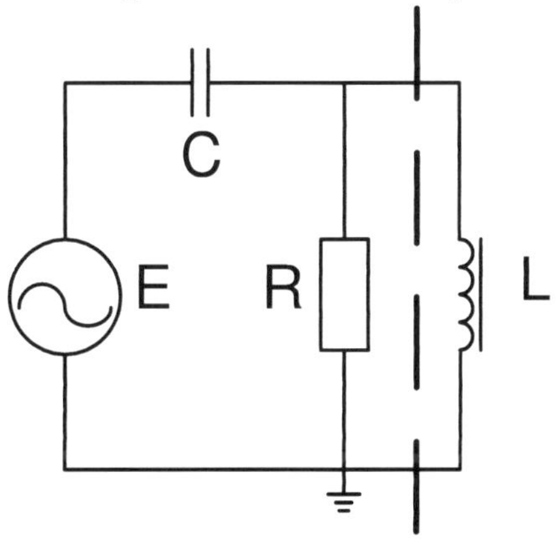

Figure 2. Simplification of ferroresonant circuit

The current-flux characteristic reflects the excitation characteristics of the nonlinear inductor. However, the manufacturer does not provide an i-φ curve with the product and can be converted by a U-I curve when needed. The subroutine supporting this function in ATP-EMTP is SATURA The experimental data obtained from a set of secondary U-I curves on the U-I curve are shown in Table 1. The sub-program SATURA is used to obtain the nonlinear inductive excitation characteristic curve required for the simulation calculation

Table 1. Voltage-current pairs on U-I curve

Voltage /V	Current /A
55.0	1.22
70.7	3.95
85.6	9.83
111.0	37.10
152.0	89.80
212.0	177.00

IOP Conf. Series: Materials Science and Engineering **533** (2019) 012012 doi:10.1088/1757-899X/533/1/012012

This article will simulate the single-phase circuit shown in Figure 2, and simulate the case of a single open circuit breaker. Set the circuit breaker to open at 0.02s, the phase angle of the opening is set to 0°.

3.Matlab numerical simulation

3.1. Numerical solution of differential equation

In the neutral point direct grounding system, if the load of the line is not counted, the three-phase circuit can be simplified into three single-phase circuits due to the three-phase symmetry, as shown in Figure 2. Using the basic principles of circuit analysis, a set of equations of state (1) with φ and u as state variables is obtained.

$$\frac{d\varphi}{dt} = u$$
$$\frac{du}{dt} = -\frac{1}{C}\left[\frac{u}{R} + f(\varphi)\right] + \omega E_{\mathrm{m}} \cos \omega t$$

(1)

Where φ is the magnetic flux in the nonlinear inductor; u is the voltage across the nonlinear inductor; E is the equivalent power supply potential; C is the equivalent capacitance; R is the PT core equivalent resistance.

In the analytical calculation, the excitation characteristics of the core inductor are as shown in equation (2).

$$i_{\mathrm{L}} = f(\varphi) = a\varphi + b\varphi^{n}$$

(2)

Where i_L is the current flowing through the PT, φ is the magnetic flux of the PT core inductance, n=3, 5, 7 or 9, in the calculation of this example, n is 3. The excitation characteristics of the core inductor are obtained by polynomial fitting from the i-φ curve in Table 1, wherein a=-3.04x10^{-5} , b=4.51x10^{-10}。

The simulation program is compiled by Matlab, mainly using the function ode23 (using the trapezoidal rule algorithm), which is suitable for solving moderately rigid differential equations. The results of the numerical analysis are shown in Figure 3. The analysis waveform can be found that the system fails to transition to the normal steady state, and a continuous, higher amplitude overvoltage occurs, and the steady-state voltage amplitude is 420 kV.

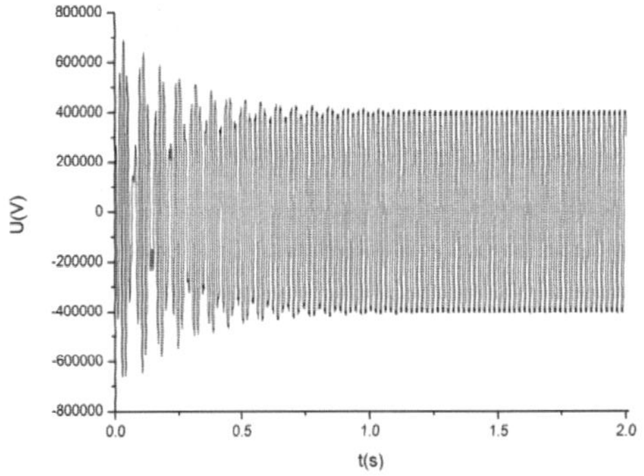

(a)

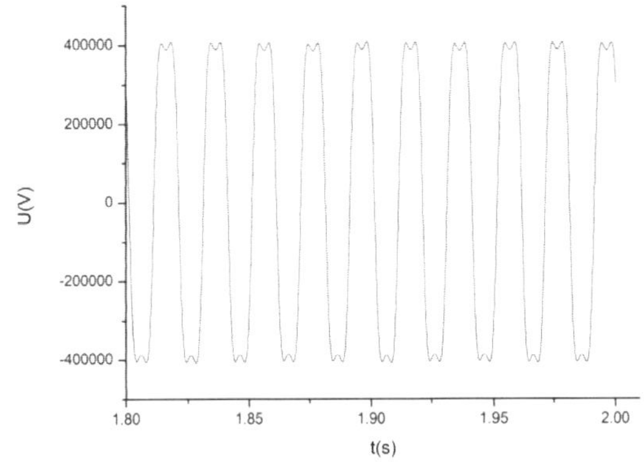

(b)

Figure 3. Waveform of ferroresonance voltage

3.2. SIMULINK modeling simulation

The case of opening the circuit breaker was simulated, and the waveform results of the simulation are shown in Figure 4.

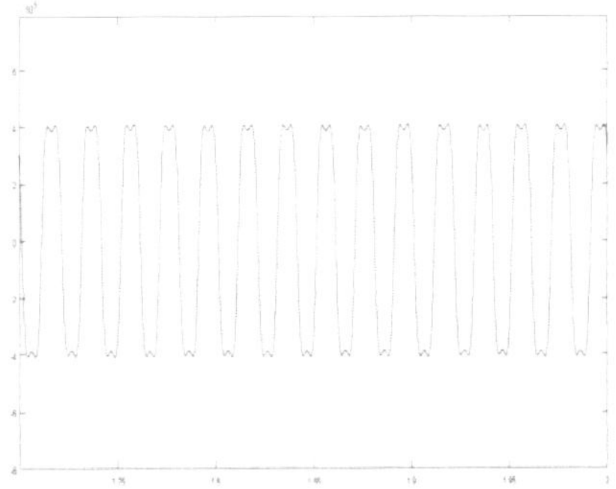

Figure 4. Waveform of ferroresonance voltage with the circuit breaker open(MATLAB)

After opening the circuit breaker, the simulation results using Matlab are close to the numerical analysis. The system fails to transition to the normal steady state, and the transient process excites the resonance. The steady-state voltage amplitude is 471kV, which is about 12.14% higher than the numerical analysis. It can be seen that the ferromagnetic resonance simulation using Matlab's saturated transformer model produces errors.

4.ATP-EMTP simulation

4.1. Type98 type component nonlinear inductor

In the ATP, the modeling method of the literature [5] is used to establish a simulation model for the resonant circuit. The model of the nonlinear inductor Type98 type component shunt resistor is used instead of the PT. The Type98 inductor component is represented by the piecewise linearization method.

The case of opening the circuit breaker was simulated, and the waveform results of the simulation are shown in Figure 5. After opening the circuit breaker, the simulation results using ATP-EMTP are close to the numerical analysis. The system fails to transition to normal steady state, and a continuous, higher amplitude overvoltage occurs. The steady state voltage amplitude is 465kV. The results of the analysis increased by approximately 10.71%. Since the Type98 inductor component adopts the piecewise linearization method, the simulation results also produce errors compared with the numerical analysis, but the accuracy of the Matlab simulation results is improved.

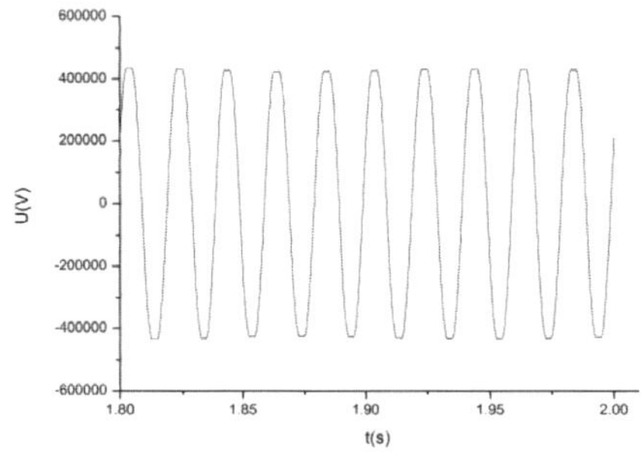

Figure 5. Waveform of ferroresonance voltage with the circuit breaker open(ATP)

4.2. Model using hysteresis eddy current loss

*4.2.1. Type98 type component nonlinear inductor.*In this paper, the EMTP simulation method is used to obtain the hysteresis loop of the nonlinear inductor. The transient process of the simple circuit shown in Figure 6 is simulated by EMTP, and the accurate hysteresis curve can be obtained. The direct result of the simulation is the discrete voltage and current values.). In the figure, E is the AC power supply potential, which varies over a wide range. In the ATP, L is a nonlinear inductance expressed by a 98-type element. Given the E value, the circuit response after the analog switch is closed, the simulation time should be long enough for the loop to reach steady state. The voltage is integrated by using the TACS integral component in ATP, and the appropriate initial value is selected to obtain the hysteresis curve. The specific data is shown in Table 2.

Table 2. Flux-current pairs on i-φ curve

Current/A	Flux/Wb
0.0086	-272.35
0.0216	0
0.0434	490.22
0.0906	623.92
0.2117	938.88
0.6047	1912.16
0.8611	2747.72

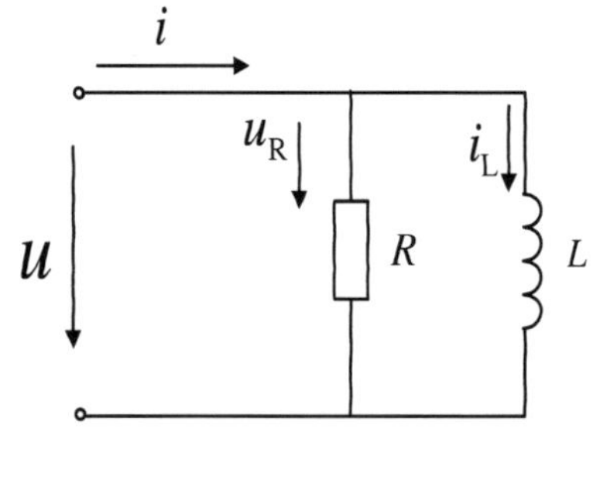

Figure 6. Equivalent circuit with magnetic hysteresis and eddy current

The nonlinear inductor model is used to replace the PT. The Type 96 inductor is a nonlinear inductor model represented by the method of directly using the hysteresis loop. The excitation characteristic is obtained by the EMTP simulation method described above. Arbitrarily set a group of moments to open the circuit breaker, perform a simulation, and obtain the resonance waveform as shown in Figure 7. It can be seen from the figure that the transient process excites the resonance, and as time goes by, the amplitude gradually decreases until it is stable. The steady-state voltage amplitude is 433kV, which is about 3.1% higher than that of the numerical analysis. It is very close to the numerical analysis. It can be seen that the model considering the hysteresis eddy current loss can improve the accuracy of the simulation. Therefore, when performing ferromagnetic resonance simulation calculations, a saturation model that accounts for hysteresis eddy current losses should be used.

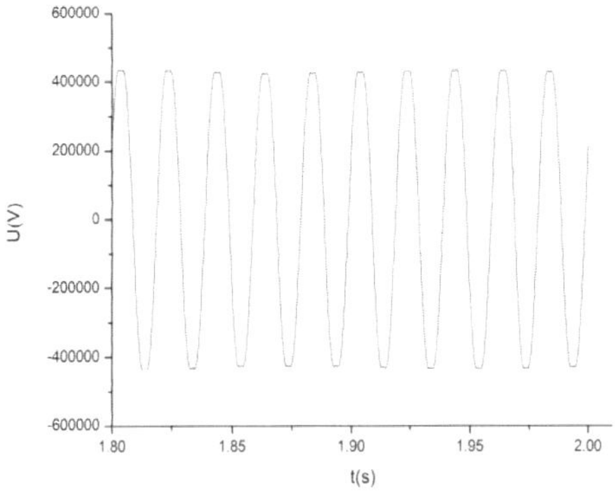

Figure 7. Waveform of ferroresonance voltage with the circuit breaker open（Type96）

*4.2.2. Using Hevia 98->96 type component nonlinear inductance.*There is also a method in the ATP that directly converts the hysteresis loop into the hysteresis loop. The Hevia 98->96 model of the hysteresis loop nonlinear inductor is obtained. The excitation characteristic curve adopts the i-φ curve of the Type98 component in 3.1. Arbitrarily set a group of moments to open the circuit breaker, perform a simulation, and obtain the resonance waveform as shown in Figure 8. It can be seen from the figure that the transient process is numerically unstable. This may be due to the nonlinear inductive magnetization characteristics obtained by this conversion method, and a negative hysteresis loop occurs, resulting in amplification of the nonlinear inductor oscillation amplitude. And the result is an error. It shows that this model cannot be used to calculate the ferromagnetic resonance calculation model of the neutral point direct grounding system.

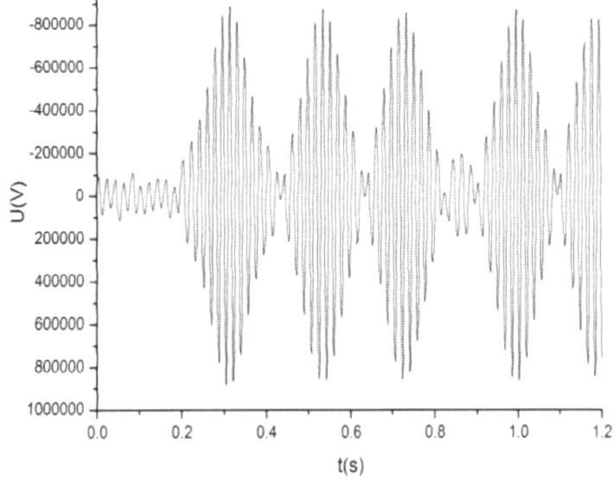

(a)

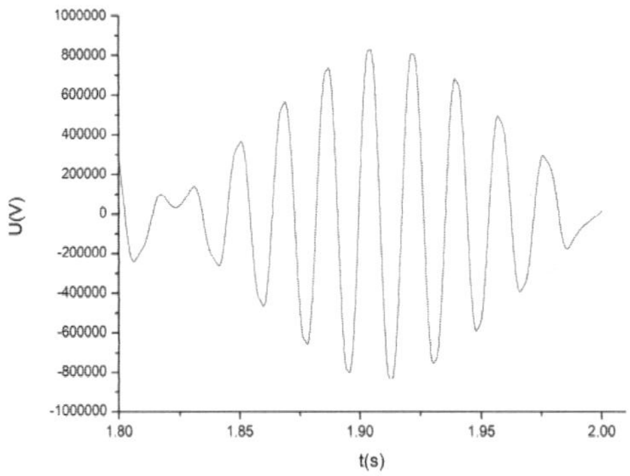

(b)

Figure 8. Waveform of ferroresonance voltage with the circuit breaker open（Hevia 98->96）

5.Model comparison

The polynomial fitting method is used for numerical calculation. When the circuit breaker is opened and the system does not reach the steady state, the maximum voltage appearing on the busbar is greater than the amplitude of the bus voltage at steady state, and the ferromagnetic resonance overvoltage occurs.

Using the saturation transformer model in Matlab and the Type98 model in ATP-EMTP, the system fails to transition to normal steady state, and the transient process excites resonance, but the amplitude of the resonant overvoltage produces a larger error than the numerical solution of the differential equation.

Using the Type96 model in ATP-EMTP, the transient process excites the resonance, which is very close to the numerical solution of the differential equation.。

Using the Hevia 98->96 model in ATP-EMTP, the numerical process is unstable in the transient process. This model cannot be used to calculate the ferromagnetic resonance calculation model of the neutral point direct grounding system.

6.Conclusion

Aiming at the simulation calculation of ferromagnetic resonance in power system, the nonlinear inductance saturation model for ATP-EMTP simulation calculation is compared, and the influence of different models on ferromagnetic resonance is analyzed. When the fundamental resonance occurs in the system, by comparing the calculation results of different models, it can be known that the Type 96 saturation model considering the hysteresis eddy current loss, the fundamental frequency resonance amplitude is closer to the numerical calculation result, and is more accurate than the piecewise linearization model. Flexible and convenient to use, it helps to conduct a more comprehensive and in-depth study of ferromagnetic resonance from a system perspective.

References

[1] ZENG Xiangjun, YANG Xiangui, WANG Wen, et al. Based on the fiexible control of zero sequence voltage distribution network ferroresonance suppression method[J].Proceeding of the CSEE, 2015, 51(1): 80-85.

[2] SIMA Wenxia, ZHENG Zheren, YANG Qing, et al. Suppressing the ferroresonance overvoltage by impulsive synchronization of chaotic systems with parameter mismatched, 2012,27(6):218-226.

[3] LIANG Zhirui, DONG Wei, LIU Wenxuan, et al. Simulation analysis on ferroresonance of potential transformer[J].High Voltage Apparatus, 2012, 48(11): 18-23.

[4] DONG Jimin. Analysis of explosion on 35 kV voltage transformer in 500 kV substation[J]. Power System Protection and Control, 2008, 36(13): 82-85.
[5] LI Yunge, SHI Wei. Study of fundamental ferroresonance on neutral-grounded systems by using analytical method criterion and elimination[J]. Proceedings of the CSEE, 2003(9):142-146.
[6] LI Yunge, SHI Wei. Study of fundamental ferroresonance on neutral-grounded systems by using analytical method solution to the power frequency excitation characteristic of nonlinear inductors[J]. Proceedings of the CSEE, 2003, 23(10): 94-98.
[7] MEI Chenglin, ZHANG Chaoshu. Analysis of voltage transformer ferroresonance[J]. Power System Technology, 2008, 32(s2): 311-313.
[8] ZHANG Bo, LU Tiecheng, DU Xiaolei. Nonlinear dynamic analysis of ferroresonance in neutral grounded power system[J].High voltage engineering, 2007,33(1): 31-35.
[9] WANG Jia. Transient calculation of over-voltage in power system containing ferromagnetic elements[J]. Journal of electric engineering, 2003(6):71-74.

A Line Loss Calculation Method for Arbitrary Partition of Power Grid Based on Measurement Resource and Matpower

Kai Yi[1,a], WeiZhou Wang[2], RunQing Bai[3], Xin He[3] and TianJun Jing[1]

1. College of Information and Electrical Engineering, China Agricultural University,100083 Beijing, China
2. State Grid Gansu Electric Power Company, 730030 Lanzhou, China
3. State Grid Gansu Electric Power Research Institute, 730070 Lanzhou, China
cauyikai@163.com

Abstract. The existing four-division management method for line loss in smart grids has problems such as the inability to verify line loss contemporaneity and low application flexibility. Considering that the rich data resources provided by data acquisition devices such as measurement meters in the smart grid are not fully utilized, an arbitrary partitioning line loss calculation method that fully utilizes measurement data resources is proposed. The method classifies the grid nodes according to the type of measurement data, obtains complete line power data, and then formulates the boundary equivalence principle of grid partition to realize arbitrary region partition of the power network. On the basis of network partition, the power loss of each subnetwork is calculated by using the power flow of matpower, which greatly increases the flexibility of smart grids line loss management. Taking a power grid in a certain area of Gansu as a case study, the effectiveness of the calculation method is verified.

1. Introduction

The Line loss rate is an important economic indicator for power supply companies [1]. At present, the line loss statistical assessment in china mainly adopts the four-division method. The four-division method refers to the four-division line loss management methods used by the power grid within the jurisdiction of the power supply enterprise, that is, the line loss management according to the voltage level, the power supply area, the power supply line and the substation [2,3]. When the grid size is small, the four-division method can accurately calculate the line loss. However, with the expansion of the grid size, the traditional "four-division method" cannot verify the simultaneity of the line loss, and only the line loss analysis can be performed on the designated area. In addition, the statistical error of line loss is large, which is not in line with the flexible development characteristics of today's smart grid. Breaking the limitations of the four-division management and realizing the line loss calculation of any partition is conducive to the precise and flexible management of line loss.

The power stations, substation feeders and nodes equipped with measuring devices in the power grid can obtain real-time data such as voltage, active power, reactive power, and power factor [4]. However, due to different acquisition times and different periods of supply and sale of electricity, the data may not have simultaneity and affect the accuracy of line loss calculation. The traditional line loss calculation method is based on simplified approximation, and the data collected by the measuring device is not fully utilized, so that the line loss calculation result is not ideal [5]. In order to maximize the use of measurement data, it is necessary to overcome the simultaneity of the data, and repair the data missing from the meter.

Content from this work may be used under the terms of the Creative Commons Attribution 3.0 licence. Any further distribution of this work must maintain attribution to the author(s) and the title of the work, journal citation and DOI.
Published under licence by IOP Publishing Ltd

IOP Conf. Series: Materials Science and Engineering **533** (2019) 012013 doi:10.1088/1757-899X/533/1/012013

Matpower is a software written in Matlab to solve power trends and optimize power flow problems. It provides a variety of power flow calculation methods, including the classic Newton-Raphson method, Gauss-Seidel method, and fast decoupling algorithm (XB version) and fast decoupling algorithm (BX version) [6,7]. It is only necessary to input the parameters of the grid and set accordingly, and the power loss in the network can be calculated from the power flow. The network information in Matpower is stored in the form of a matrix, which is easy to implement for reading, writing, modifying and segmenting network parameters, and has strong operability.

In view of the shortcomings of the existing line loss four-division management method and the limitations in the application, this paper has developed a corresponding data processing method with the goal of maximizing the measurement data. Combined with the advantages of Matpower processing multi-node complex system, the calculation method of arbitrary partition line loss of power grid using full measurement data is proposed, and the feasibility of the method is illustrated by actual cases.

2. Processing of measurement data

According to the time effect and real-time, the measurement data of the power grid is divided into two categories: one is real-time measurement data, which can reflect the real-time status of the grid to a certain time, such as voltage, current, line power, etc.; one is non-real-time measurement data, which mainly refers to data such as the quantity of electricity that reflects information within a certain period of time.

2.1. Simultaneous processing of data

Both types of data are subject to interference from simultaneous issues, affecting the use and mining of data.

For real-time measurement data, the problem of different periods is mainly due to the different sampling time of the real-time measurement meter, and there is a clock deviation in the sampling time of each meter. Figure 1 shows the power curve of i-node in one day, and its sampling time is later than the whole point. In this way, when calculating the line loss at the hour, the value used is collected at the previous time, which will bring errors to the line loss. For example, the data for its 12-point line loss calculation is 110kW. Using the method of interpolation [4], the data of different sampling moments of each measuring point is estimated to the data of the same time, in order to avoid the error caused by Non-simultaneity of real-time data. The results of the i-node data processing are shown in Fig. 2. Its 12-point line loss calculation uses real-time data of 100.7 kW, which greatly improves the accuracy.

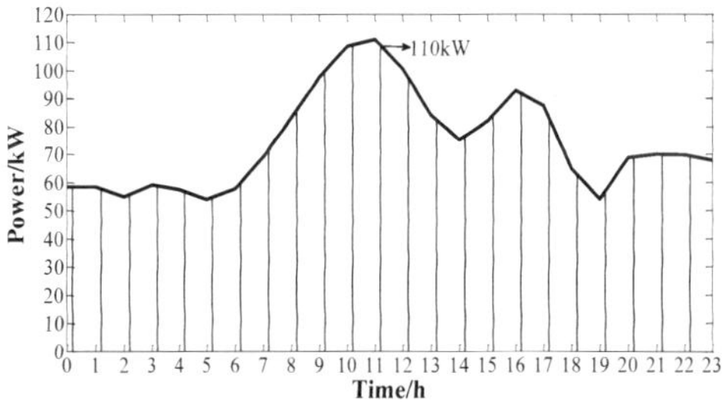

Figure 1. Power load curve of node *i*.

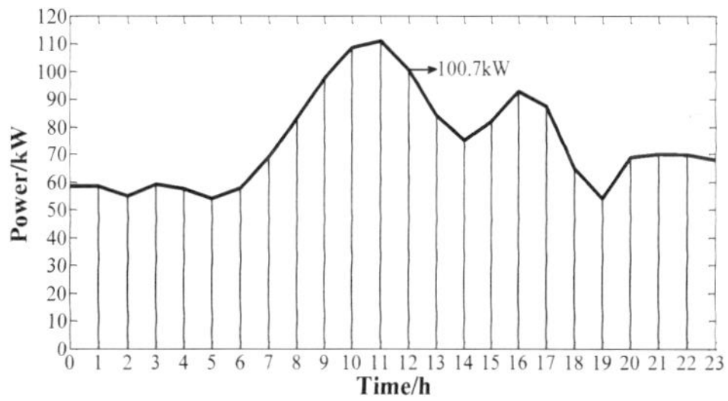

Figure 2. Power load curve of node *i* after processing.

For the quantity of electricity data, the main reason for its non-simultaneity of problems is the different periods of supply and sales statistics. In areas where traditional meter reading methods are used, due to the large number of users who need to go to the household to read the meter, it is difficult to achieve the same cycle statistics for electricity supply and sales. Only the meter reading time set by the power company can be strictly implemented to reduce the phenomenon of different periods as much as possible [8]. For the phenomenon of meter reading at different times that is difficult to avoid, the quantity of electricity obtained by meter reading and the actual meter reading cycle can be used to reduce the error caused by different period statistics [9]. As shown in formula (1):

$$A_0 = A_1 \cdot \frac{T_0}{T_1} \tag{1}$$

In the formula, A_0 represents the actual monthly sales volume of quantity of electricity, A_1 represents the meter reading quantity of electricity, T_0 represents the actual monthly statistical days, and T_1 represents the meter reading statistics days.

2.2. Data processing for different types of nodes

According to the type of sampling data, the nodes are divided into three categories: one is a node that can collect real-time measurement data such as voltage and power, which is called a measurement node; another type of node does not have a collection terminal and can only obtain monthly quantity of electricity information through the electric energy meter ,which is called electricity node; the last one is the node that can neither obtain the real-time measurement data nor the quantity of electricity information, and only knows its distribution capacity, which is called the capacity node [9].

On the basis of realizing the data synchronization processing, in order to make full use of the data collected by the monitoring device and realize the arbitrary partition line loss calculation based on the measurement data, it is necessary to obtain the real-time power values of each node, especially the electricity node and the capacity node. Corresponding data processing methods are proposed for different types of nodes.

The measurement node with the most detailed measurement data, the data processing target is to establish the load reference in each load mode as a reference node, and use the established reference to estimate the real-time injection power value of the remaining nodes. The injection power of the electricity node of the same load type as the measurement node can be estimated for the line loss calculation. Specific steps are as follows:

1) Calculate the daily quantity of electricity of the electricity node:

Calculate the power coefficient of the measured node, as shown in equation (2), which represents the load level on the representative day.

$$\begin{aligned} K_P^m &= \frac{A_P^m}{\left[\sum_{i=1}^{N} A_{P,i}^m\right]/N} \\ K_Q^m &= \frac{A_Q^m}{\left[\sum_{i=1}^{N} A_{Q,i}^m\right]/N} \end{aligned} \tag{2}$$

In the formula, K^m is the power coefficients of the reference node m on the representative day; A^m is the quantity of electricity of the reference node m on the representative day; A_i^m is the quantity of electricity injected into the reference node m on the i-th day of the month; N is the number of days in the month. P,Q in the formula represent active and reactive variables.

Combine the monthly quantity of electricity information and the power coefficient of the reference node, and calculate the power consumption of the electricity node on the representative day by using equation (3), namely:

$$\begin{aligned} A_P^n &= K_P^m \cdot \frac{A_{P,\Sigma}^n}{N} \\ A_Q^n &= K_Q^m \cdot \frac{A_{Q,\Sigma}^n}{N} \end{aligned} \tag{3}$$

In the formula, A_Σ^n is the monthly quantity of electricity of the electricity node n; A^n is the quantity of electricity of the electricity node n on the representative day; N is the number of days in the month. P,Q in the formula represent active and reactive variables.

2) Calculate the average power of the reference node:

$$P_{av}^{m}=\frac{A_P^{m}}{T} \qquad Q_{av}^{m}=\frac{A_Q^{m}}{T} \tag{4}$$

In the formula, P_{av}^{m} and Q_{av}^{m} are the average power injected; A^{m} is the quantity of electricity of the reference node on the representative day; *T* is the measured hours, according to the actual situation, generally take 24 hours.

3) Calculate the ratio of injected power to average power at each time of the reference node:

$$K_{P,i}=\frac{P_i^{m}}{P_{av}^{m}} \qquad K_{Q,i}=\frac{Q_i^{m}}{Q_{av}^{m}} \tag{5}$$

Where P_i^{m} and Q_i^{m} are the power injected at the i-th moment of the reference node.

4) Using the analysis of the above power data, calculate the injection power at each moment of the electricity node.

$$P_i^{n}=\frac{K_{P,i}A_P^{n}}{T} \qquad Q_i^{n}=\frac{K_{Q,i}A_Q^{n}}{T} \tag{6}$$

Where P_i^{n} and Q_i^{n} are the injection power at the i-th moment of the electricity node *n* on the representative day; A^{n} is the quantity of electricity of the electricity node *n* on the representative day.

For the capacity node, there are two practical data processing methods. The first one is to convert capacity nodes to electricity node processing. Calculate the total quantity of electricity of all capacity nodes by using the difference between the quantity of electricity of the line root node and the quantity of electricity of the remaining nodes, and allocate the total quantity of electricity to each capacity node according to the capacity ratio, so as to get the quantity of electricity of each node on the representative day.

$$A_P^{r}=\frac{S_e^{r}}{\Sigma_{t_3} S_e^{t_3}}\left[A_P^{0}-\Sigma_{t_1} A_P^{t_1}-\Sigma_{t_2} A_P^{t_2}\right] \qquad A_Q^{r}=\frac{S_e^{r}}{\Sigma_{t_3} S_e^{t_3}}\left[A_Q^{0}-\Sigma_{t_1} A_Q^{t_1}-\Sigma_{t_2} A_Q^{t_2}\right] \tag{7}$$

Where A^{0} represents the quantity of electricity of the line root node; S_e^{r} is the rated capacity of the distribution transformer at node *r*; t_1,t_2, $t_3$are the collection of measurement nodes, electricity nodes, and capacity nodes, respectively.

The second method is to directly use the estimated power data. Directly use the total power of the root node minus the power of measurement node and the approximate power of the electricity node, and then allocate the node power according to the capacity ratio.

$$P_i^{r}=\frac{S_e^{r}}{\Sigma_{t_3} S_e^{t_3}}\left[P_i^{0}-\Sigma_{t_1} Q_i^{t_1}-\Sigma_{t_2} Q_i^{t_2}\right] \qquad Q_i^{r}=\frac{S_e^{r}}{\Sigma_{t_3} S_e^{t_3}}\left[Q_i^{0}-\Sigma_{t_1} Q_i^{t_1}-\Sigma_{t_2} Q_i^{t_2}\right] \tag{8}$$

Compared with method 1, method 2 has no node conversion process, and the calculation is simple. However, the second method uses the estimated data, which is not as accurate as the method one.

Based on the real-time injection power value of the measurement node, the above method can be used to estimate the injection power of the electricity node and the capacity node, and further provide reliable data support for the line loss calculation of any partition.

3. Line loss calculation for any partition

In the process of power flow calculation, the power grid is purposefully segmented to realize the line loss calculation of the region of interest, and the line loss management is more refined. It is the principle and purpose of calculating the arbitrary partitioning line loss in this paper.

3.1. Arbitrary partition line loss calculation model

On the basis of obtaining the approximate real-time power of each electricity node and capacity node, the measurement information of the node can be fully utilized to realize the calculation of the arbitrary partitioning line loss of the power grid.

Figure 3 is a partial schematic view of a grid partition. In the figure, *m* is the total number of boundary nodes of partition *i* and partition *j*, and x_i and x_j are the internal nodes of the *i*, *j* partition, respectively.

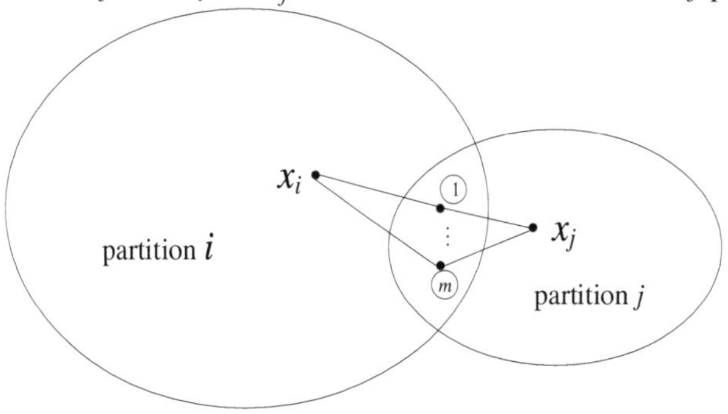

Figure 3. Schematic diagram of grid partition.

On the basis of obtaining the injection power of each non-real-time measurement node, the measurement information of the boundary nodes 1, 2, ..., m is utilized. According to the voltage level and the direction of the power flow, the lines connected to the boundary nodes of the upper power grid are equivalent to the load, and the line connected to the boundary node of the lower power grid is equivalent to the power supply. For the line whose current trend is uncertain, it can be temporarily used as a variable power supply.

For the partition i, as the upper-level power grid, the line connected to its boundary node is equivalent to the load. The load of each boundary node is expressed as:

$$\mathrm{P}_{L,n} = \mathrm{P}^0_{L,n} + \mathrm{P}'_{L,n} \tag{9}$$

Where $\mathrm{P}_{L,n}$ represents the equivalent load of the boundary node *n*; $\mathrm{P}^0_{L,n}$ is the load that supplies power to the user; $\mathrm{P}'_{L,n}$ is the sum of the power delivered to the other partitions.

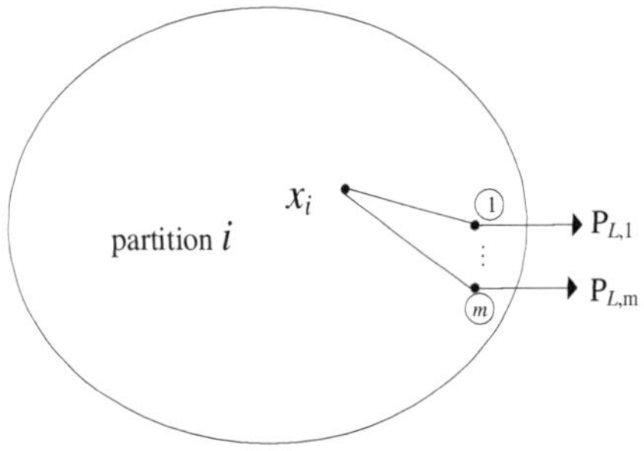

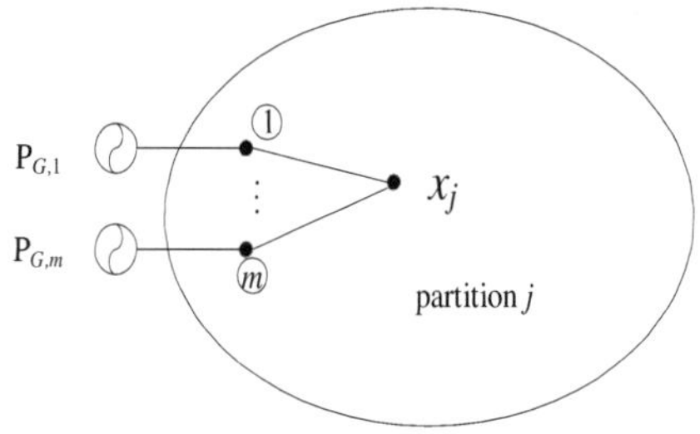

Figure 4. The equivalent result of the *i* partition **Figure 5.** The equivalent result of the *j* partition.

For the partition *j*, as the lower-level network of the power receiving end, the line connected to the boundary node is equivalent to the power supply. The input power of each boundary node is:

$$P_{G,n} = P^{0}_{G,n} + P^{'}_{G,n} \tag{10}$$

Where $P_{G,n}$ represents the equivalent input power of the boundary node n; $P^{0}_{G,n}$ is the power generated by the generator connected to the node and the power generated by the distributed power source; $P^{'}_{G,n}$ is the sum of the power delivered to it by the upper-level grid.

After obtaining the equivalent partitioning network, the theoretical line loss calculation of each partition can be carried out separately. In the partition calculation process, if the difference ΔP between the average power per hour of the node and the actual measured data of the measurement node is too large, the approximate injection power of the electricity node and the capacity node may be corrected and checked whether the capacity of each distribution transformer itself is exceeded. The result is considered accurate until ΔP is within the allowable range.

3.2. Calculation of arbitrary partitioning line loss based on matpower

In this paper, a simple operation software matpower is used to calculate the arbitrary partitioning line loss. Only the power supply, load and branch parameters of the power grid need to be input according to the specified format, and the power flow calculation result and line loss can be obtained [10]. The software is easy to apply and has high calculation accuracy.

The steps to build a matpower network computing model are as follows:

1) Set the network reference capacity and calculation mode. Normally, the system reference capacity is set to 100 MVA.

2) Construct a network node matrix, sequentially number each bus node of the power grid, and input the actual network data into each column of the matrix.

3) Construct a network power matrix. The nodes connecting the generators in the network are regarded as power supply nodes, and the nodes connected to the upper grid at the boundary are also regarded as power supply nodes.

4) Construct a network line matrix, determine the nodes connected at both ends of each route, and input information such as line impedance, admittance, and transformer ratio into the matrix.

5) According to the specified format, save the parameters in a project file under the matpower directory. After inputting the command, the program will automatically run the Newton method power flow calculation, and get the line loss of the network.

6) Based on the partition model described in the previous section, the data parameters are read and written to rewrite and divide the network parameters to realize the calculation of the arbitrary partitioning line loss of the network.

After the power grid is partitioned by manual decision, the boundary of each subnet uses the node measurement information as the power source and load equivalent. The subnets after the boundary is equivalent to the complete network, whether it is the node size, or the power and load parameters will change, you need to modify the parameters of the subnet.

First, the nodes included in the subnet are extracted according to the numbers to form a new network node matrix. Subnet generation can be achieved by deleting redundant nodes of the original network. For the power supply and load information of the subnet, based on the original node information of the subnet, the rewriting of the boundary node is implemented according to the measurement information of the boundary line. In this way, the line loss calculation of any partition is realized by generating a new subnet parameter matrix.

In the process of calculating the line loss of each partition, if the calculation does not converge, the subnet parameter matrix can be checked and a new matrix can be regenerated until the final solution of convergence is obtained.

The overall flow of the partition line loss calculation method proposed in this paper is shown in Figure 6.

4. Case analysis

Taking a power grid in a certain area of Gansu as an example, the method of this paper is used to calculate the line loss of any partition. The geographical connection diagram of the area network is shown in Figure 7. The power network spans two voltage levels of 110kV and 35kV, including 123 computing nodes, 174 transmission lines, and 43 substations. The data collection is 472,340, which is rich in measurement information.

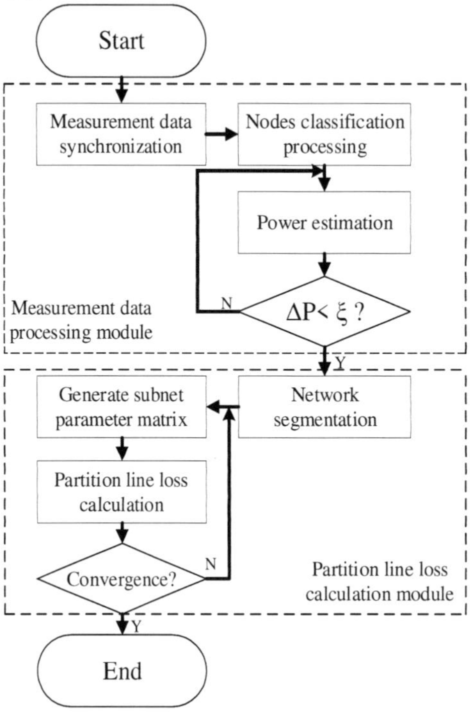

Figure 6. Line loss calculation process for any partition.

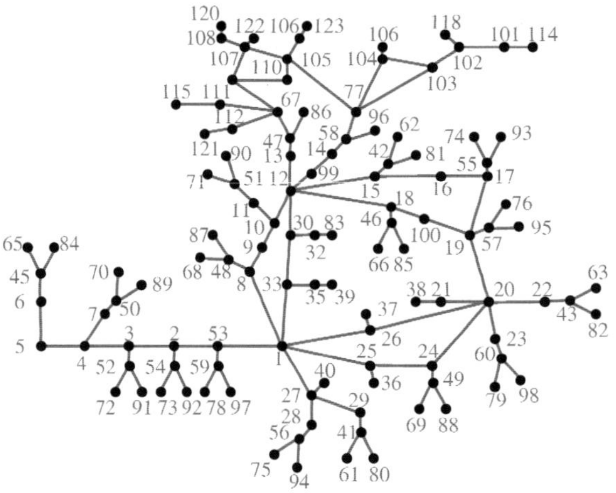

Figure 7. Electrical connection diagram of the area network.

After constructing the matpower 123-nodes network model in the region, the region of interest is selected and the system is divided into a 100-nodes system and a 23-nodes system. There are three boundary nodes in the two subnets, two of which are measured nodes, the node numbers are 66 and 67 respectively; one electricity node, the node number is 77.

IOP Conf. Series: Materials Science and Engineering 533 (2019) 012013 doi:10.1088/1757-899X/533/1/012013

The 66-node incoming and outgoing line power and load can be obtained from the measurement information, which are 3.294+j0.386MVA, 3.293+j0.387MVA and 0MVA respectively. The node is equivalent to a load node in the upper-level power grid, and its equivalent load is the sum of the outgoing power and the user load, which is 3.293+j0.387MVA; in the lower-level power grid, it is equivalent to the power supply node, and its equivalent power is the incoming line power, which is 3.294+j0.386MVA. The same is handled for the 67 node that are the same measurement nodes.

For the 77 electricity node, because of the lack of measurement information of the incoming and outgoing line power, it is necessary to use the measurement node as a reference for power estimation. The resulting incoming and outgoing line powers were estimated to be 12.563 + j3.772 MVA and 12.563 + j3.771 MVA by the method described in Chapter 2. Equivalent to the load node of the upper-level power grid, the equivalent load is 12.563+j3.771MVA; equivalent to the power supply node of the lower-level power grid, the equivalent power supply is 12.563+j3 .772MVA.

After the equivalent processing of the boundary nodes of the two subnets, the node matrix of each subnet can be derived, and the power and load information of each subnet can be corrected. Finally, the power flow of each partition is converged, and the line loss of the partition is calculated. Figures 8 and 9 show the calculated voltage and power for each node. Table 1 shows the calculation result. Table 2 compares the calculation results of the method with the conventional method. The results show that the method is highly accurate when calculating the arbitrary partition line loss of the grid. The method of this paper is very feasible.

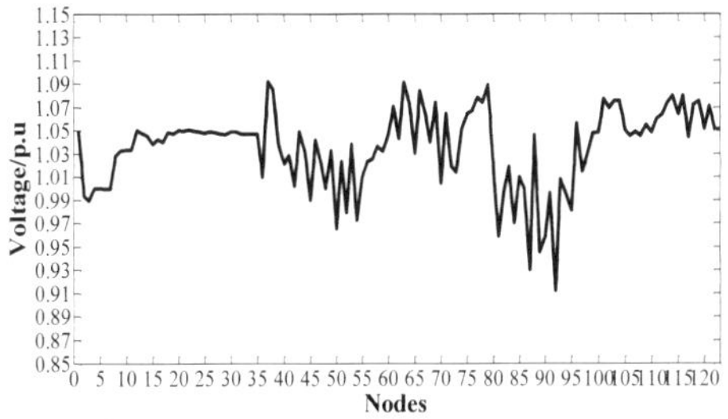

Figure 8. Voltage amplitude of each node of the power grid.

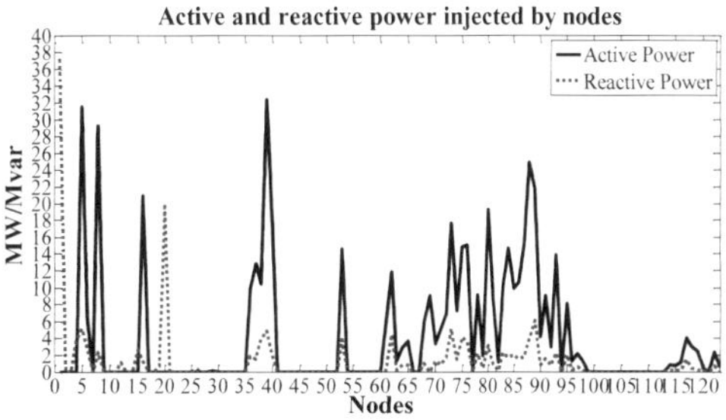

Figure 9. Active power and reactive power of each node.

Table 1. Line loss calculation result

Power network	Active loss(MW)	Reactive loss(Mvar)
100-nodes network	4.094	41.53
23-nodes network	0.864	4.271
Total network	4.958	45.801

Table 2. Comparison of calculated results between proposed method and conventional method

	Active loss(MW)	Reactive loss(Mvar)
Conventional method	4.967	45.54
Proposed method	4.958	45.801
Deviation	0.18%	0.57%

5. Conclusion

In order to adapt to the flexible and changeable development characteristics of the smart grid, it is necessary to break the limitation of the four-division line loss management, realize the transition to arbitrary partition management, and flexibly and effectively select any region for line loss analysis. Based on the data synchronization process, this paper classifies each node to estimate the injected power value of each node. Make full use of measurement information, combined with matpower's convenient power flow calculation and easy-to-modify network parameters, realize line loss calculation in any partition. This method has no redundant calculation, simple and efficient, and provides a new idea for line loss management in any partition of smart grid. The case analysis shows that the method is very feasible.

Acknowledgments

The research was financially supported by the State Grid Corporation Science and Technology Project (contract number: 522722180003).

References

[1] Zhu Zizhao, Ye Faxin. "Improved Method for Theoretical Line Loss Calculation of Low-voltage Distribution Network," *Electrical Measurement &Instrumnenation*, vol. **49**, pp. 6-10 (2012).
[2] Ruan Jianqiu. "Research on Loss Reduction Measures Based on four-division line loss management," *Electromechanical information*, vol. **21**, pp. 23-25 (2014).
[3] R. Lawrence. "Application of path integrals in modeling transmission line loss," *IEEE transactions on components, packaging, and manufacturing technology,* vol. **19**, pp. 775-788 (1996).
[4] Hu Yingjun, Zhang Jianmin, Liu Daye. "Calculation of multi-source distribution network loss based on back/forward sweep method and measurement resource," *Electric Power Automation Equipment*, vol. **29**, pp. 55-59 (2009).
[5] Lin Chen, Ding Xiaoqun, Liu Xiaobo, Zhou Zhihui. "Line loss comprehensive analytical method based on real-time system data and its application," *Electric Power Automation Equipment*, vol. **25**, pp. 47-50 (2005).
[6] Wu Haidong, Ren Xiaoming. "Application of Matpower in Power Flow Calculation of Multi-Node Power System," *Electrical application*, vol. **34**, pp. 34-37 (2015).
[7] Z.R. Daniel, M.S.C. Edmundo, T.R. John. "MATPOWER: Steady-state operations, planning, and analysis tools for power systems research and education," *IEEE Transactions on Power Systems*, vol. **26**, pp. 12-19 (2011).
[8] Hao Li, Wang Wenting. "Discussion of line loss fluctuation caused by power supply and power sale in different stages," *Ningxia Electric Power*, vol. **3**, pp. 14-16 (2010).
[9] Zhang Tiefeng, Lv Xin, Lv Yueying,Yang Wanchun. "Study of incomplete measured synchronous line loss calculation in distribution network," *Electrical Measurement & Instrumnenation*, vol. **52**, pp. 91-96 (2015).
[10] LV Sha. "Study on Successive Backward Method Combined MATPOWER Power Flow Calculation for Regional Power Grid Frame Planning," *Hubei Electric Power*, vol. **36**, pp. 6-8 (2012).

IOP Conf. Series: Materials Science and Engineering **533** (2019) 012014 doi:10.1088/1757-899X/533/1/012014

Simulation of Relay Coordination by Labview and the Effect of Relay Coordination on Cost

Hasan Çimen[1,a], Recep Akkan[1] and İsmail Yabanova[1]

[1]Afyon Kocatepe University Technology Faculty, Electrical-Electronics Department, ANS Kampusu,Afyonkarahisar, TURKEY

hcimen@aku.edu.tr

Abstract. In this study, it is emphasized that simulation of the avoidance of unnecessary energy cuts in a sample electricity distribution network. The electricity cut-off costs can be decreased by ensuring that relay coordination end only at the source of the fault from the point of failure. Over currents are calculated and the sample network is simulated in the LabVIEW environment. The parameters of delayed overcurrent trip setting (I>), instantaneous overcurrent trip setting (I>>), delayed ground overcurrent trip setting (Io>), instantaneous ground current trip setting (Io>>) are set according to a specific logic in relay coordination. As a result, you can get an idea about how much profit can be made by doing the relay coordination correctly and the amount of unnecessarily avoided energy. It is aimed with the application of this simulation in the real field that electrically operated materials of the people, who are unnecessarily energy-consuming, have less damage and that the energy-selling company loses less money.

1. Introduction

One of the most important protection elements in the transmission and distribution of electrical energy is the overcurrent protection relays. Every part required for the integration of transmission and distribution is protected by suitable breakers at the time of failure and damage to all equipment in this integration is prevented. The duty of the overcurrent protection relays (digital, mechanical) is to separate the faulty output from the system in the case of a failure, to send an immediate trip signal to the breaker and to help the personnel inform the type, duration and time of the fault by keeping in its own memory (at least 20 records) [1].

Reliability in distribution systems is very important. This reliability is directly related to the adjustment and operation of the relays. Reports by the North American Reliability Council notes that protection systems often play a role in failures in power distribution systems. For the period from 1984 to 1988, the study of significant failures reported and outlined by the Council states that protection relays play a role in 73.5% of the main faults somehow. The reliability analysis of the protection relays is the measurement of faulty or improper operation of the protection relays [2].

This study was carried out upon the fact that there is non-conforming operation of 15 breakers and relays located in İhsaniye region and that it is consequently seen that settlements, agricultural irrigation transformers etc. are de-energized needlessly. Many studies related to relay coordination have been carried out. In one of these, the article titled "Adaptive protection testbed using real time and hardware -in-the-loop simulation", Montenegro, Montoya and Ramos propose an alternative

Content from this work may be used under the terms of the Creative Commons Attribution 3.0 licence. Any further distribution of this work must maintain attribution to the author(s) and the title of the work, journal citation and DOI.
Published under licence by IOP Publishing Ltd

IOP Conf. Series: Materials Science and Engineering **533** (2019) 012014 doi:10.1088/1757-899X/533/1/012014

method for coordination of overcurrent protection devices with two co-simulation techniques. These are real-time simulation and hardware-loop simulation. These simulations were performed using the IEEE 13-Nodes test system in different software (PSAT® and LABVIEW®) [3]. In this article, we have seen that the over-current protection relays on the simulation have worked in a harmonious way after the required safety distances have been entered into the LabVIEW program and the multipliers that need to be entered into the overcurrent protection relays are calculated considering the field environment rather than the computer environment.

In this study, digital and mechanical relays in the field are set. Settings are calculated on the LabVIEW program and entered into the relays in the field in the most appropriate way. The time multipliers calculated on the LabVIEW program are 0,11-0.13 etc., the values closest to these values have been entered into the mechanical relays in such a way as to minimize the relay coordination because these values couldn't have been entered into mechanical relays.

2. Overcurrent Protection Relay

The overcurrent protection relays must be selective, reliable, easily adjustable and economical. The overcurrent protection relays send a trip signal to the breakers if the current exceeds the current set. They flow current through by current transformers. The current transformer helps the relays to read the nominal current by reducing the current in the normal line. For example; 150-300/5-5 current transformer etc.

2.1 Current-Time Characteristics in Overcurrent Relay

T=(K/((G/G_s)^L-1)+C)×TM(s) (1)

T : Trip time (seconds)

K,C,L : Coefficients

G : Input current (Fault current)

Gs : Capture current set

TM : Time multiplier

The overcurrent protection relays operate according to equation (1). When the equation in (1) is taken into consideration, the coefficients according to the relay characteristics and standards are as follows.

Table 1. Table of relay coefficients

Types of Characteristics	Standard	K	L	C
Standard Inverse	IEC	0,14	0,02	0
Very Inverse	IEC	13,5	1	0
Extremely Inverse	IEC	80	2	0
Moderate Inverse	ANSI/IEEE	0,0515	0,02	0,114
Very	ANSI/IEEE	19,61	2	0,491

Inverse				
Extremely Inverse	ANSI/IEEE	28,2	2	0,1217

The coefficients K, L, C vary according to the type of curve used (Standard Inverse, Very Inverse etc.) while calculating the opening time of the overcurrent protection relay as shown in Table 1 (0,14-0,02-0 / 13,5-1-0 etc.).

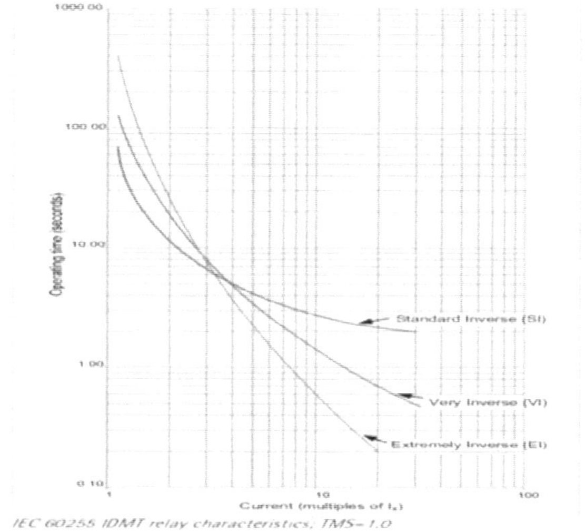

Figure 1. Representation of trip times of the relays according to curve types under IEC standard [5]

If the overcurrent protection relays are adjusted according to the fixed time characteristic, they wait for the set time until the set current is exceeded and send the trip signal. If they are set according to the other characteristics, they operate in such a way that the signal sending time decreases as the fault current increases according to their type [4].

3. Ihsaniye Region Where Sample Application is carried out

The transformer power supplying the settlements in İhsaniye region is about 6,6 MVA. Besides, it also creates a burden since there are spa tourism (Gazlıgöl etc.) and hotels in the region. These loads are controlled by 15 breakers, 15 disconnectors and 15 overcurrent protection relays. In addition to the special lines in the region, there is approximately 145 km of line (pigeon, raven, swallow) with different sections at the OG(34,5 kV) (except for AG).

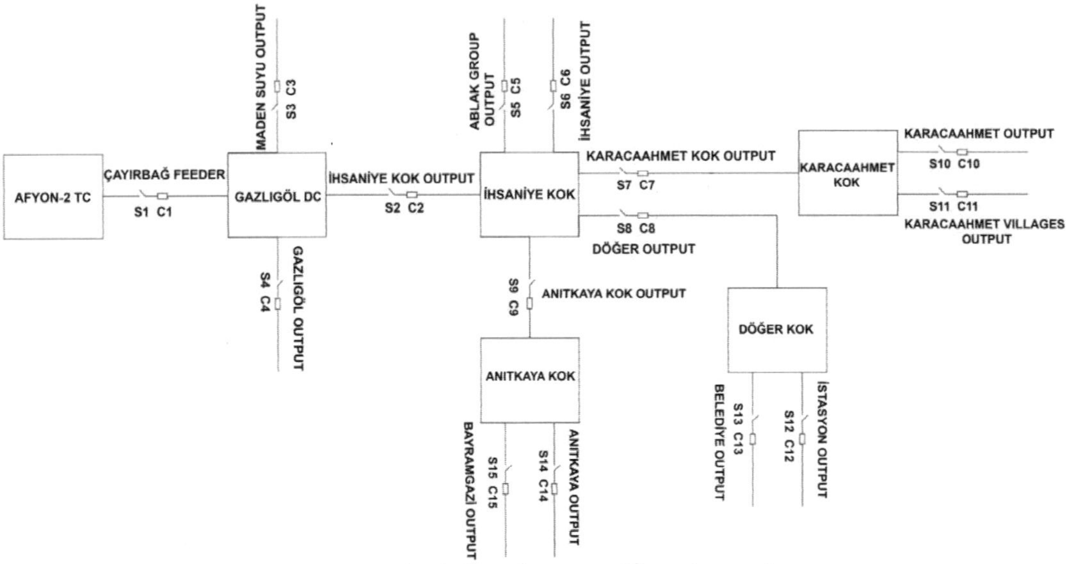

Figure 2. Single line diagram of İhsaniye region

If we refer to the reasons that cause unnecessary energy cuts in the first section of İhsaniye region; the reasons for this energy cut can be very different. For example, technical reasons (Conductor Error, Transformer Error, Trip Signal, Short Circuit Failure and Other Defective Equipment, Isolator Cracking), weather conditions (snowstorm, tropical storms, heavy rainfall, lightning strike, severe wind, solar storm, extreme temperatures), natural disasters (earthquakes, avalanches, tidal waves, landslides and hurricanes), capacity (over-demand and overload of the system), lack of maintenance, animal (snake, cat, raccoon, bird), human (operator error, cut or stealing of conductor and equipment), tree (tree overturning and elongated tree branches), accident (fire, truck or crane crash, traffic accident, drop of the tree branches on the conductor), cut (the raw material used in the generation of electricity cannot be provided) and unknown (other non-recorded reasons for cut) etc. According to the records, the most common cause of energy cut is wind and rain with 31,4%. This factor is followed by equipment failure with 19,9%, ice storm with 11,1%, hose or tropical storm with 10,1%, lightning strike with 8,8%, operator error with 8,5%, fire with 5,6%, voltage drop with 3,9%, earthquake with 1,6% and deliberate reasons (sabotage) with 0,7% [6].

The purpose of relay coordination is to isolate the faulty part from the system with the reaction of all the elements of the system (breaker, relay etc.) as soon as possible when the relay is set with optimum values [7-8-9].

4. Method and Relay Coordination with LabView

Overcurrent protection relays send trip signal to breaker in case of failure. In practice, the nearest relay should be able to detect the fault and the nearest breaker should be able to clean failure as main purposes of the relay coordination in case there is short circuit (or failure). Because it cleans, other breakers must remain in the off position. The tripping time of the relay must be kept as short as possible because the line remains stable and other relay and breakers are not affected by sudden load changes. However, this short retention process must be done precisely that it doesn't affect relay coordination.

- I> : It is the symbol used for delayed overcurrent trip adjustment.
- I>> : It is the symbol used for sudden overcurrent trip adjustment.
- Io> : It is the symbol used for delayed ground overcurrent trip adjustment.

- Io>> : It is the symbol used for sudden ground current trip adjustment

Such problems as opening two breakers and not opening by inappropriate breaker originating from previous miscalculations due to mechanic relay in İhsaniye region (such multipliers as 0,13-0,17 cannot be entered) and lack of suitable current transformers etc. have been tried to be solved. This solution process was done by making simulation program on LabVIEW and considering the field environment. For example, it is seen that the opening time is 150 ms in the catalog information of a breaker, this time increases to 200-220 ms in the field environment (such reasons as 10 years of usage and opening and closing for many times). In this article, a solution (appropriate safety period) is presented by taking most field conditions including them into account.

According to the single line diagram of İhsaniye region, which is shown in Figure 2, because Madensuyu and Gazlıgöl outputs from Gazlıgöl DM, Ablak Group and Central İhsaniye outputs from İhsaniye KÖK (Breaker Measure Box), Bayramgazi and Anıtkaya outputs from Anıtkaya KÖK, Municipality and Station outputs from Döğer KÖK and Karacaahmet and Karacaahmet Villages outputs from Karacaahmet KÖK are the last outputs, they are activated with the value (I>> and Io>>), I>> is 5 times greater than the value I> with a delay of 30 ms and the value Io>> is also 6 times greater than the value Io> with a delay of 30 ms in overcurrent and ground protections.

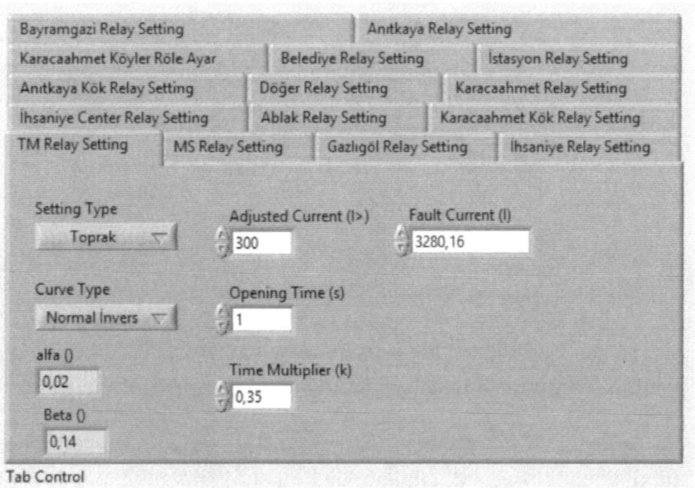

Figure 3. Relay settings of Afyon-2 TM based on LabVIEW

When calculating the I> values (the values of Afyon-2 TM are NI, 300 A, Multiplier 0,35), it is aimed to get the least pulse on conductor and protection elements and to open the breaker by sending the signal to the relay in any fault in the shortest while by considering the loads pulled by every one of exit. For example, according to the equation (1), the relay of Afyon-2 TM waits for 1 s for a fault of 3280,16 A in the overcurrent. The relay of Mineral Water exit from Gazlıgöl DM (NI, 156 A, Multiplier 0,07) waits for 156 ms (approximately 150 ms) for the same fault. Multipliers are calculated on the Labview Program by leaving 275 ms safety time while setting the previous outputs from the last output according to single line diagram.

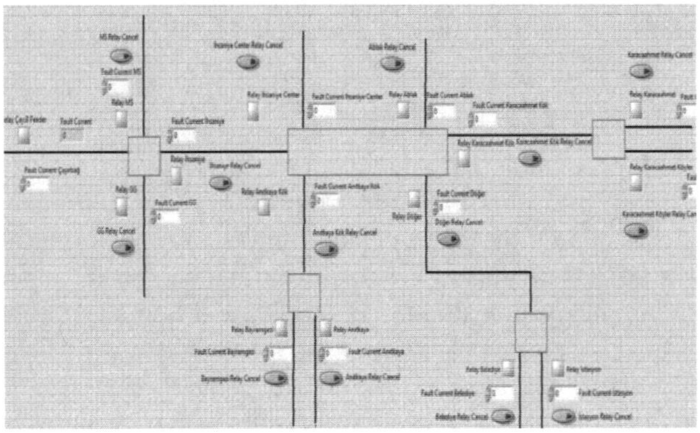

Figure 4. Interface of the program written on LabVIEW for relay coordination.

The malfunctions occurring at any output on the program in Figure 4 are entered in the fault current box. If the entered fault current is higher than the set current, the trip time is calculated automatically according to the curve type and the relay turns on in the calculated time and illuminated the on light for 3 sec (Figure 5).

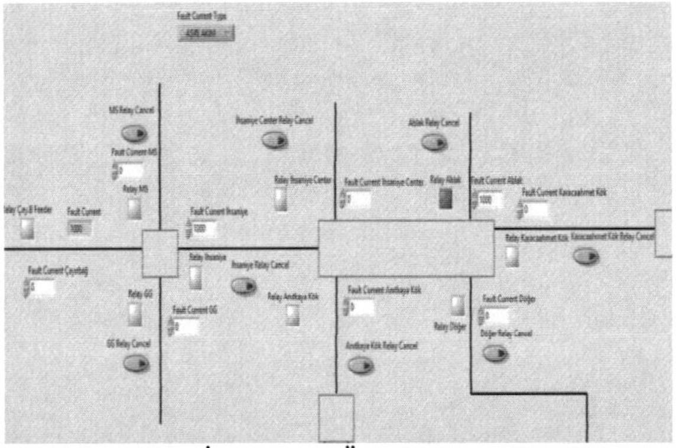

Figure 5. Sample failure from İhsaniye KÖK Ablak exit and activation of the relay

The program has been prepared by considering the actual field events. For example, if the protection elements on a KÖK are disabled (failure of Bar-24, damage to the cable that provides 220 V AC, etc.), the relays will not work and the KÖK will remain unprotected. We can provide this event with the relay cancel button on the program. When the relay cancel button is pressed, the activation and protection of the relay according to the settings of the relay is ensured because the relay right in front of it is activated (Figure 6).

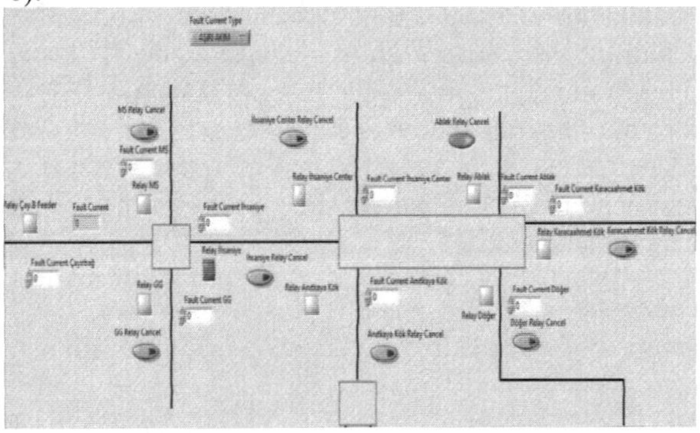

Figure 6. Activation of the previous relay when the relay of İhsaniye KÖK Ablak exit is canceled

5. The Effect of Relay Coordination on the Number of Breaker Trip and Cost

If the appropriate circuit breaker and line section are installed and appropriate relay values are entered after the operation, K9 breaker could have decreased from 300 trips to 156 trips with a difference of 144 and K2 breaker could have decreased from 360 trips to 132 trips with a difference of 228 averagely in a year. In total, it could have decreased from 660 trips to 288 trips with 372 times less power outages (Figure 7).

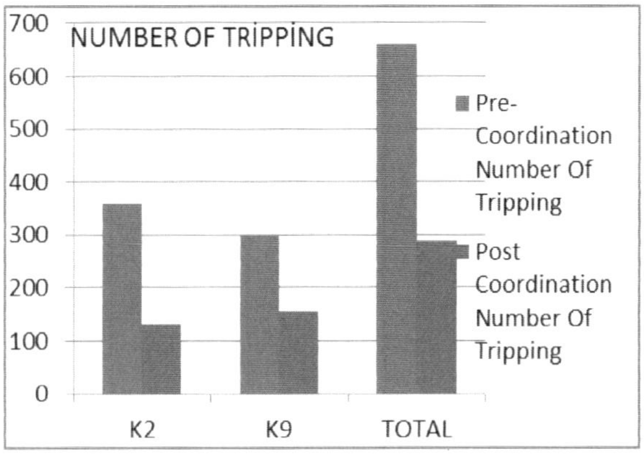

Figure 7. Number of trips before and after coordination (per year)

When the number of opening in the region per year is decreased to that in Figure 7, this has also benefited greatly from the cost. When the unit price of the energy that the energy-selling company received the distribution price is considered for September 2018 (0,137331 TL/kWh = 0,022417 $/kWh), while the previous loss was 162 546,624 TL = 26 532,593 $, the current loss is 63 549,876 TL = 10 373,288 $ with less loss of 98 996,748 TL = 16 159,305 $ (Figure 8) [10].

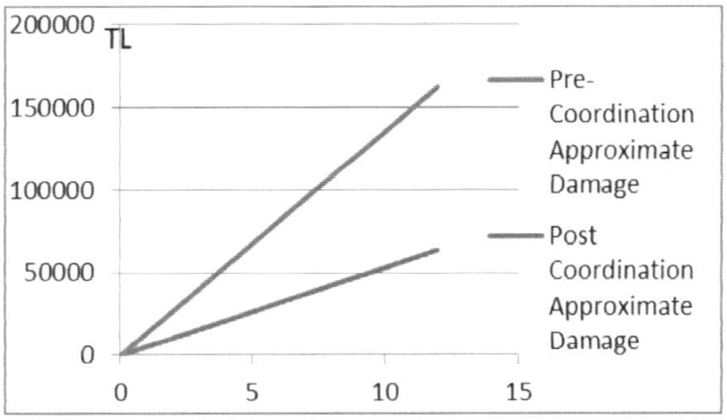

Figure 8. Approximate damage status before and after coordination (per year)

6. Conclusion

In this article, LabVIEW program has been prepared for optimum relay coordination and the decrease of the number of trips after this coordination and the effect of cost have been examined. Coordination was provided by using electronic and analog protection relays in this examination. In this way, the system will be prevented from faults coming from different points of the system.

Disabling the faulty part by intervening to the fault as soon as possible is also important for the efficiency of the system. This is possible with the values set on the relay and the appropriately selected breakers. In this study, it is seen that the breakers have less trips with the values entered on the relay.

In order to make the electrical infrastructure of the plant more reliable and robust, the use of digital protection relays instead of analog protection relays firstly requires the proper parameterization of these relays, the creation of the appropriate selectivity, so of the complete relay coordination. Furthermore, another advantage of using a digital relay instead of an analog relay is that the time multipliers (0,09-0,13-0,17 etc.) that cannot be entered in the formula of relay trip time can be entered more easily in the digital relay.

The study not only makes lines and KÖKs more reliable, but also has many advantages in terms of cost, operation and maintenance for the distribution company.

Acknowledgments

This research is supported by a project from Afyon Kocatepe University Scientfic Research Project Department numbered as 18.KARIYER.249.

References

[1] Öcalan, G. (2005). Realization of protection relay functions by using PLC and SCADA. Master Thesis, Firat University, Institute of Natural and Applied Sciences, Elazığ.
[2] Gençaydın, E., Tanrıöven, M. (2013). A review on safety assessment for protection systems in high voltage transmission systems. Journal of Engineering and Science, Manchester, Sigma 5: 86-102.
[3] Montoya, L.A., Montonegro, D., Ramos, G., (2013). Adaptive protection testbed using real time and hardware-in-the-loop simulation. IEEE Grenoble Conferences (2013), 1-4.
[4] Gürsu, B. (2014). Optimal overcurrent relay coordination in transformer stations with genetic algorithms stopped by punitive function. Journal of the Faculty of Engineering and Architecture, 665-676. Gazi University, Ankara.
[5]http://myelectrical.com/DesktopModules/SunBlog/Handlers/Print.aspx?id=159,30.03.2018
[6] Polat, S., Şekerci, H. (2015). Important electrical failure in the world and our country.
[7] Damchi, Y., Mashhadi, H.R., Sadeh, J., (2016). Optimal Coordination of Distance and Overcurrent Relays Considering a Non-Standard Tripping Characteristic for Distance Relays. The Institution of Engineering and Technology (2016), 1448-1457.
[8] Anupreyaa, K., Karthiga, S.T.C., (2016). Relay Coordination for Distribution System. 2016 Second International Conference on Science Technology Engineering and Management(ICONSTEM), 337-341.
[9] Begbagui, M., Bentarzi, H., Mati, A., (2015). A New Framework of Numerical Distance Relay Using LabVIEW. 4th International Conference on Electrical Engineering (ICEE) (2015), 1-5.
[10] https://altin.in/, 06.10.2018.
[11] Kumar, A., Thakur, M., (2016). Optimal Coordination of Directional Over Current Relays Using a Modified Real Coded Genetic Algorithm: A Comparative Study. Electrical Power and Energy Systems 82 (2016), 484-495.
[12] Begbagui, M., Bentarzi, H., Mati, A., (2015). A New Framework of Numerical Distance Relay Using LabVIEW. 4th International Conference on Electrical Engineering (ICEE) (2015)

IOP Conf. Series: Materials Science and Engineering **533** (2019) 012015 doi:10.1088/1757-899X/533/1/012015

The Analysis and Design of High-power Intrinsic Safe Forward Converter Based on Power-i Technology

Liu shulin[1],Wu hao[1], Zhang yuxi[1], Hu chuanyi[1] and Yan jizhi[1]

[1] College of Electrical and Control Engineering, Xi'an University of Science and Technology, Xi'an, China

e-mail 437373839@qq.com

Abstract. The existing anti explosion measures and the intrinsic safety evaluation methods have become the bottleneck factors that restrict the further improvement of the output power of the native security. Therefore, This article is based on the latest anti explosion standard, the principle of realizing high power output by Power-i technology to suppress the spark discharge energy of switching power supply is analyzed, as well as the composition of the power supply system and its evaluation method. The forward converter is selected as the topology of the high-power local security power supply, according to the technical index of Power-i system, the circuit structure of high power intrinsic safe converter is put forward, and the double protection circuit of overcurrent, overvoltage and short circuit when the output is short circuited and open fault is designed. It has been tested that the output 24V/2A high power intrinsically safe positive converter based on Power-i standard can meet the requirements of explosion protection under type I environment. The experimental results verify the intrinsic safety of the designed high power intrinsic positive shock converter, and reflect the correctness of the theoretical analysis and design method.

1. Introduction

The With the development of coal mine Internet of Things technology, the number of underground safety inspection and monitoring equipment such as sensors and controllers has increased greatly, which has increased the requirements for the output power of intrinsically safe power supplies[1]. Most of the intrinsically safe switching converters currently used in hazardous locations are only a few dozen watts, and there is still much room for improvement.

The earliest intrinsically safe power supply in China is mainly composed of batteries. The current limiting resistor and the Zener diode are used at the output to form an overcurrent and overvoltage protection circuit, and its output power is small[2]. In the middle and late 1970s, scholars began to study series intrinsically safe power supplies (consisting of voltage regulator tubes and transistors) composed of electronic circuits and adding current limiting and voltage limiting links at the output end to improve intrinsic safety measures[3-4]. In the early 1990s, the explosion-proof power supply circuit mainly composed of the power frequency step-down transformer, rectifier circuit, current limiting resistor, three-terminal voltage regulator and output overvoltage and overcurrent protection circuit. In order to solve the self-recovery problem, a thyristor-protected intrinsically safe power supply automatic recovery device is proposed, but a series current detecting resistor is required to limit the output power[5-7]. In the late 1990s, a growing number of research on linear power supplies began to

Content from this work may be used under the terms of the Creative Commons Attribution 3.0 licence. Any further distribution of this work must maintain attribution to the author(s) and the title of the work, journal citation and DOI.
Published under licence by IOP Publishing Ltd

turn to the study of intrinsically safe switching power supplies[6-10], which can improve the efficiency of the whole machine.

In the early days, scholars paid more attention to the influencing factors of minimum ignition energy, drawing the ignition curve, the design of circuit component parameters and its influence on the intrinsic safety performance of the circuit[11]. In recent years, the experts of the domestic and foreign have begun to study multiple protection circuits for intrinsically safe power supplies. By detecting the signal at the moment of circuit failure, the spark energy source is quickly cut off to prevent ignition and explosion, thereby improving the output power of the intrinsically safe power supply. The experimental results show that the perspective of reducing spark discharge time can be considered in order to suppress the spark energy released during circuit failure to improve the effective output power of the intrinsically safe switching converter. Based on the latest power-i technology, a scheme and circuit structure are proposed to improve its output power on the basis of its theoretical research.

2. Intrinsically safe explosion-proof power principle

2.1. Traditional intrinsically safe power supply limits the principle of spark energy

The voltage regulator or overvoltage protection circuit is usually used in the circuit to limit the voltage, and the current limiting resistor or current limiting protection current limiting resistor or a current limiting protection circuit use to limit the current. A circuit with dual protection for auxiliary protection circuits are shown in Figure 1. In the case of a conventional intrinsically safe circuit, the power supply supplies energy to the spark gap for the entire duration of the spark, suppressing spark energy only by limiting the maximum current and maximum voltage.

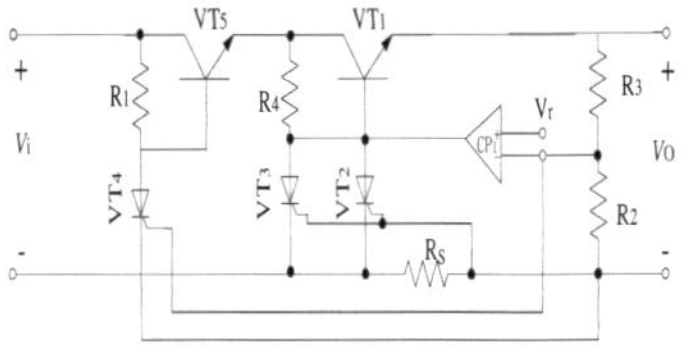

Figure 1. Schematic diagram of the dual protection circuit with auxiliary protection circuit

2.2. Switch intrinsically safe power supply limits the principle of spark energy

When the switching converter detects an output short circuit during normal operation, the protection circuit cut off the switch tube. Taking forward converter as an example as shown in Figure 2, the energy of the input power source is completely cut off, and the energy composition at the output short circuit is mainly derived from the energy of the inductor and capacitor. When designing parameters such as inductance and capacitance, considering that the maximum discharge energy of the inductance and capacitance should not greater than the minimum ignition energy, the output power of the intrinsically safe power supply are still limited.

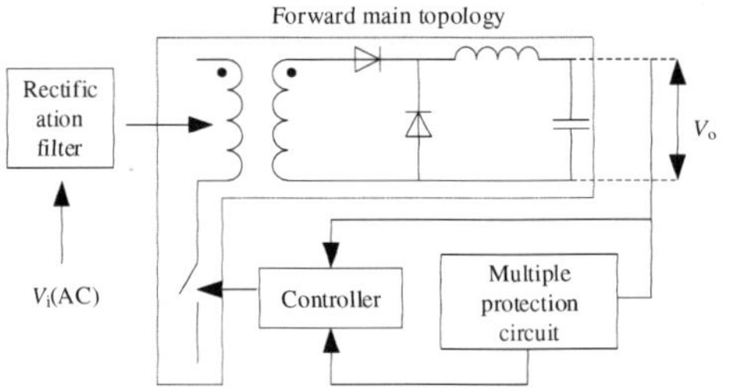

Figure 2. Block diagram of the structure of the intrinsically forward converter

2.3. Switch intrinsically safe power supply limits the principle of spark energy

The protection method of Power-i technology is identifies the fault of the electrical system through the detection circuit. The power supply and the output circuit are quickly disconnected when the fault signal is detected and ensure that the fault spark energy does not detonate dangerous substances. The method can simultaneously cut off the energy stored by the power source, the inductor, and the capacitor, and quickly suppress the spark ignition energy by detecting the current change rate, so that the output power of the intrinsically safe power source with the same parameter can be greatly improved. The block diagram based on the power-i protection circuit is shown in Figure 3.

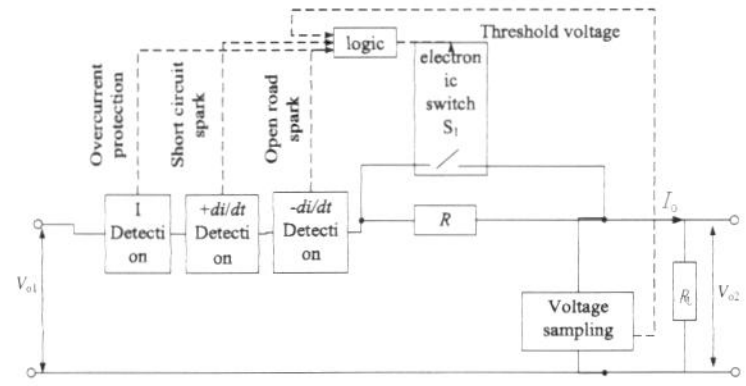

Figure 3. Is based on the power-i protection circuit block diagram

3. Design of high power intrinsically safe converter based on power-i technology

3.1. Design of main parameters

Intrinsically safe switching power supplies have low output power, and at this power level, a simple forward or flyback converter can be considered. Since the peak value of the output voltage and the inductor current of the forward converter is smaller than that of the flyback converter. Therefore, the forward converter topology is used to realize the intrinsically safe power output in this paper. In order to improve the output power of the intrinsically safe converter,it can be realized by connecting the switch tube in series in the output loop of the converter. When the fault occurs, the switch tube is turned off to limit the spark discharge time and thereby reduce the fault spark energy. A high-power intrinsically forward converter circuit structure is designed as shown in Figure 4.

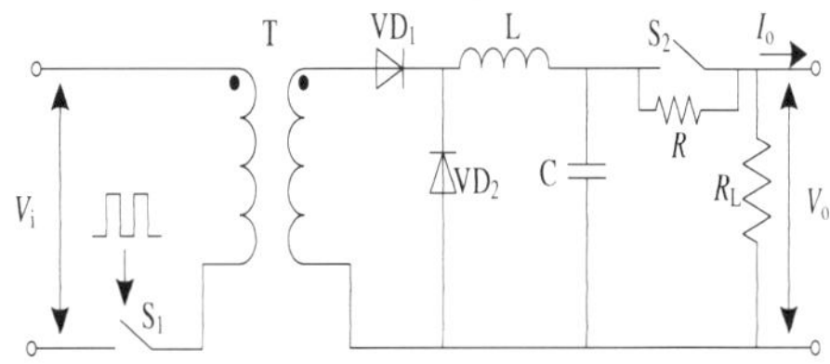

Figure 4. High-power intrinsically safe converter circuit structure

Firstly, the value of the inductor L should be considered so that it operates in CCM mode to reduce the ripple voltage. For the design of the forward converter energy storage inductor L, the CCM and DCM critical inductance $L_{\min}$ corresponding to the maximum input voltage and the minimum load resistance $R_{\mathrm{L,min}}$ can be used as the lower limit of the designed inductance value. That is:

$$L_{\min} = \frac{R_{\mathrm{L,min}}\left(V_{\mathrm{i,max}}/n - V_{\mathrm{o}}\right)}{2fV_{\mathrm{i,max}}/n} \tag{1}$$

From the perspective of optimization design, the actual inductance L should be slightly larger than the minimum inductance value $L_{\min}$.

3.2. Selection of output filter capacitor C

The output filter capacitor can be designed according to specification of the ripple voltage. It is assumed that the value of the inductor satisfies $L>L_{\min}$, and the maximum output ripple of the corresponding dynamic range is the maximum output ripple of the CCM operating mode.

$$V_{\mathrm{PP,max}} = V_{\mathrm{PP,max}}^{(\mathrm{CCM})} = \frac{V_{\mathrm{o}}(1-d)}{8LCf^{2}} = \frac{V_{\mathrm{o}}(1-\frac{V_{\mathrm{o}}}{V_{\mathrm{i}}/n})}{8LCf^{2}} \tag{2}$$

From equation (2), the required minimum output filter capacitance $C'_{\min}$ is

$$C'_{\min} = \frac{V_o(1-d)}{8LV_{PP,\max}f^2} = \frac{V_o(1-\frac{V_o}{V_i/n})}{8LV_{PP,\max}f^2} \tag{3}$$

In practice, considering the parasitic resistance of the electrolytic capacitor and other stray parameters in the circuit, it is necessary to select a filter capacitor larger than the minimum capacitance value to meet the requirements of the output ripple voltage, so the design value of the actual filter capacitor is

$$C_{\min} = \lambda C'_{\min} = \lambda \frac{V_o(1-\frac{V_o}{V_i/n})}{8LV_{PP,\max}f^2} \tag{4}$$

Where λ is the margin factor, generally takes 2~4.
The withstand voltage V_c of the output filter capacitor is

$$V_C = \lambda_1 V_o \tag{5}$$

In the formula, λ 1 is the margin coefficient, which is generally 1.3~3.
According to equations (1) and (3), the maximum output ripple voltage is related to the values of L and C. The minimum capacitance $C_{\min}$ is a function of L and is inversely proportional to the inductance L. Therefore,the larger of the L, the smaller of the $C_{\min}$.

3.3. Design of multiple protection circuits based on power-i technology

A schematic diagram of the multi-protection circuit based on power-i technology is proposed as shown in Figure 5. The monostable flip-flop and the driving resistors R_1, R_2, the transistors VT_1, VT-, and the diode D_1 form a logic module whose main function is respond to the output signal of the detecting module, thereby controlling the turning on and off of the S_1 tube(consisting of PMOS). When the monostable flip-flop input is high, its output is low, the transistor VT2 is turned on, and the parasitic capacitance C_1 of the MOS transistor S_1 charged by D_1 and VT_2. When the C_1 voltage exceeds the threshold voltage of the MOS transistor, he MOS transistor S_1 is turned on, V_{o1} is directly connected to the load R_L through S_1, and the power supply works in Power-i mode. When the monostable flip-flop input is low, its output is high, and the transistor VT_2 is turned off. Then the parasitic capacitance C_2 of the MOS transistor S_1 is discharged through the resistor R_1 and the transistor VT_1, and the C_2 voltage rapidly drops below the threshold voltage, S_1 Turned off, the power enters the shutdown mode.

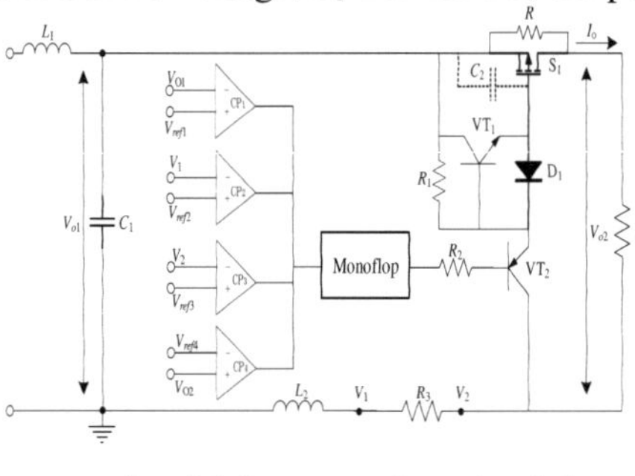

Figure 5. Schematic diagram of multiple protection circuit based on power-i technology

4. Introduction of power-i performance test method for high power intrinsically safe converter

4.1. Response time and evaluation method

The components of general measurement equipment mainly include:

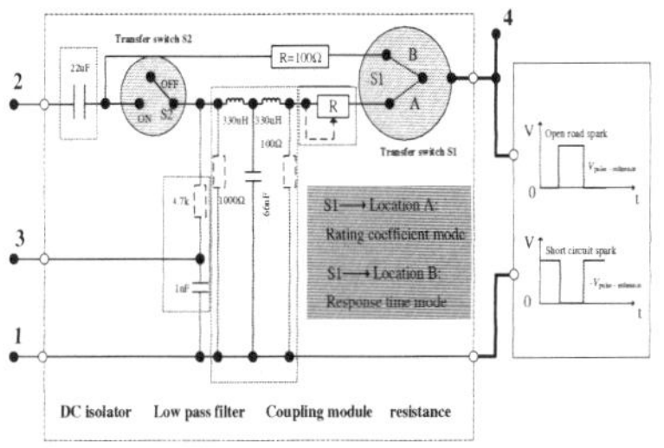

Figure 6. Fundamentals of Power-i Universal Measurement Equipment

Connect the power supply to the universal test equipment shown in Figure 6. The positive pole of the power supply output connect to the 2 pins shown in Figure 6, and the power supply output ground connects to the 1 pin. Turn the switch S_2 of the universal test equipment to the OFF position, and adjust the S_1 to the position B. Start the pulse generator after the power supply works normally, and capture the voltage and current signals at the output through the oscilloscope.

4.2. Transient pulse experimental method

The Power i sourct needs to verify the response capability of the fault signal at the moment of starting under different loads through the transient pulse experiment. The basic layout of the transient pulse test is shown in Figure 7. The Power i sourct transient pulse test Proceed as follows:

When the start button is pressed, the system generates a positive pulse to make the switches ES_1 and ES_2 turn on, and causes the Power i sourct connect with the maximum load. When the current value is 25%, 50%, 75%, and 100% of the maximum output current Io, the reference voltage Vvariable-reference in Figure 7 is adjusted to cause the voltage comparator to be triggered. Press the start button to verify that the designed Power i sourct responds and shuts down quickly, check the output current change with an oscilloscope for this verification.

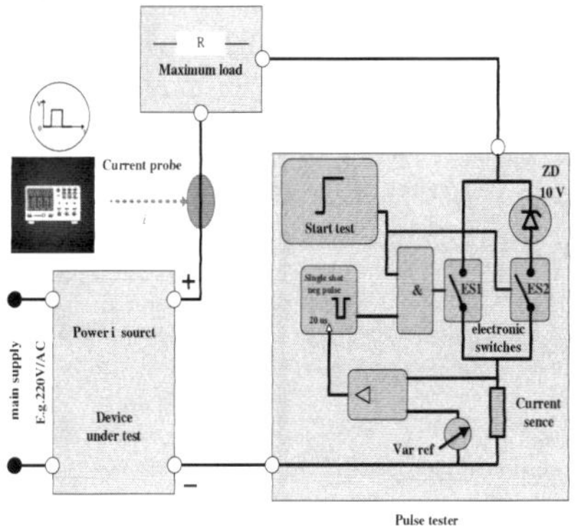

Figure 7. Power-i power transient pulse test equipment

5. Experimental verification

In order to verify the above theoretical analysis, a 24V/2A experimental prototype was produced as shown in Figure 8. The experimental parameters are shown in Table 1.

Table1. Experimental parameters.

parameter	Input voltage (V_i)	Output (V_o)	Transformer ratio (n)	Switch Frequency (f)	Inductance (L)	Capacitance (C)
Setting value	AC220V	24V/2A	4	150kHz	50uH	5uF

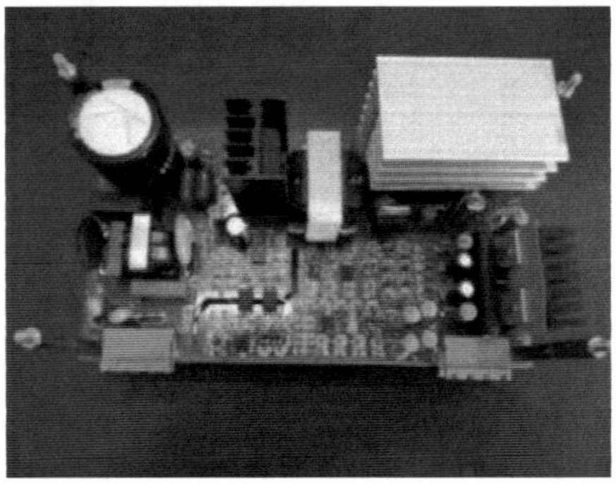

Figure 8. High-power intrinsically forward converter prototype based on power-i technology

5.1. Response time and evaluation coefficient measurement results

After the short-circuit fault occurs, the voltage and current at the output of the power supply respond to the action, and the response time test result is shown in Figure 9. It can be seen that the response time of the designed power supply t power is 1 us.

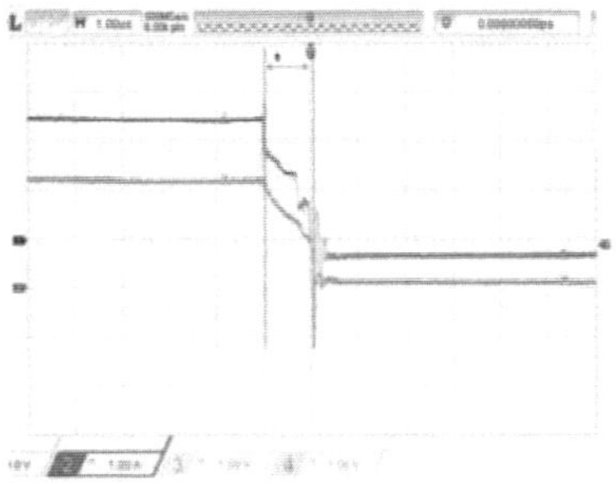

Figure 9. Voltage and current waveform of the output terminal when the power supply responds

In order to determine the minimum value of the evaluation coefficient AF power supply, the maximum value of the $V_{LPF\text{-}limit}$ value during the test is obtained. The $V_{LPF\text{-}reference}$ and $V_{LPF\text{-}limit}$ values are measured according to the above steps, and the experimental waveforms are shown in Figure 10 .

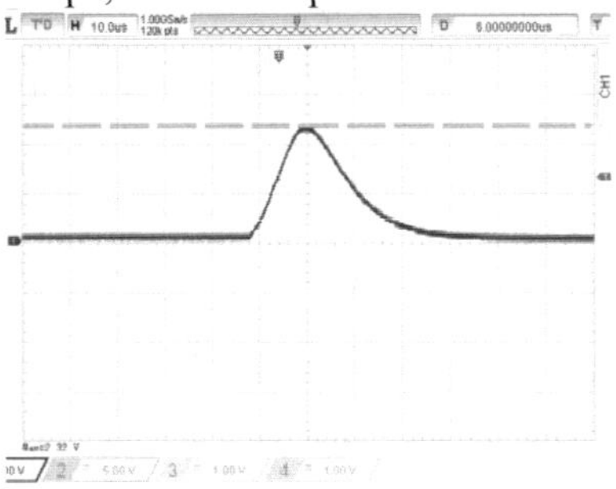

Figure 10. $V_{LPF\text{-}reference}$ voltage waveform

It can be seen that the $V_{LPF\text{-}reference}$ is 2.32V from Figures. 10, and the maximum $V_{LPF\text{-}limit}$ value is 1.24V from Figures. 11. Therefore, it can be known from Equation 6.

$$AF_{\text{power}} = 20\lg\frac{2.32}{1.24} = 5.44 \tag{6}$$

5.2. Transient pulse test results

When the load is 25%, 50%, 75%, and 100% of the rated value, the designed Power i sourct is subjected to the transient pulse test according to the above steps.The output current waveforms measured under different load conditions are shown in Figures 11,12, 13, and 14. According to the

IOP Conf. Series: Materials Science and Engineering **533** (2019) 012015 doi:10.1088/1757-899X/533/1/012015

waveforms ,it can be seen that the output current of the power supply drops to zero within 1us, indicating that the designed high-power intrinsically forward converter can make a quick response for transient pulses under different load conditions.

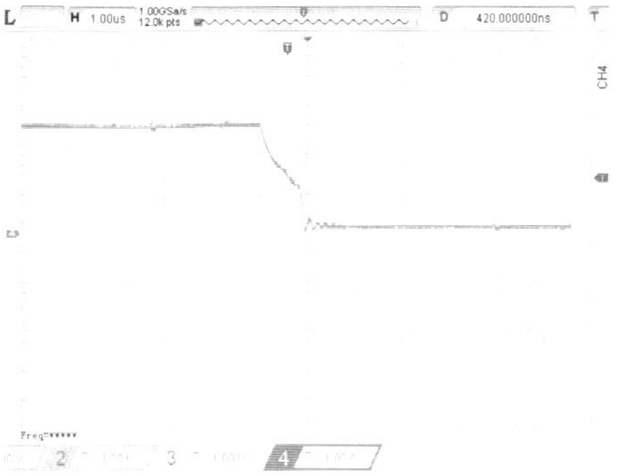

Figure 11. Current response waveform when the load is rated load

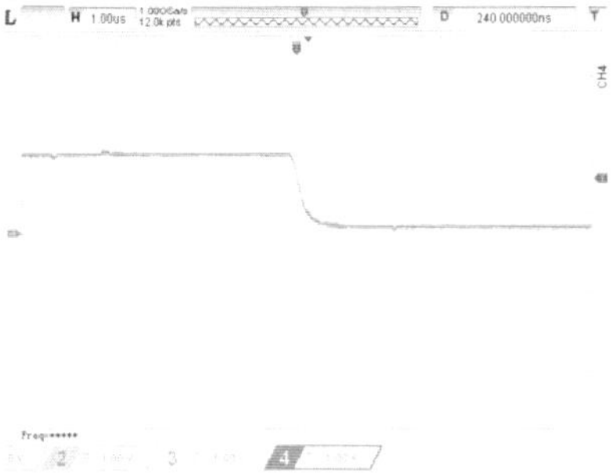

Figure 12. Current response waveform when the load is 75% of the rated load

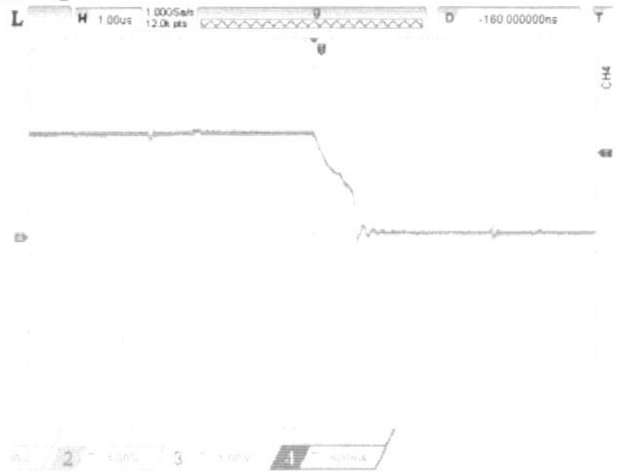

Figure 13. Current response waveform when the load is 50% of the rated load

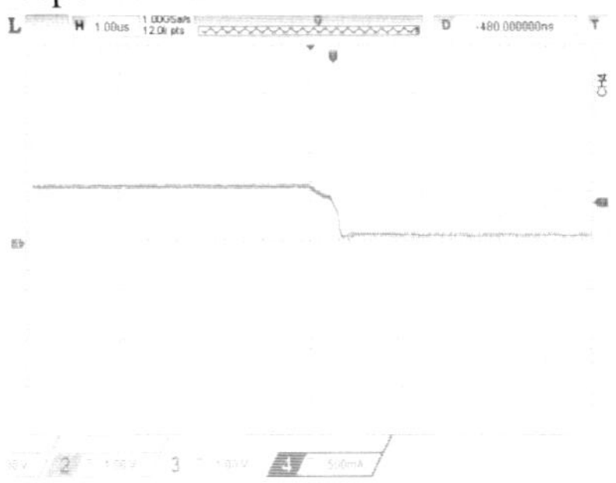

Figure 14. Current response waveform when the load is 25% of the rated load

5.3. Transient pulse test results

The simulation was carried out under the environment of a methane concentration of 8.3%. The converter with this parameter tested under the specified total of 3200 experimental conditions, and no explosion occurred. The experimental results show that the designed high-power intrinsically forward converter meets the intrinsic safety performance index.

6. Conclusion

The principle of the Power-i technology suppresses spark energy to realize high-power output and and evaluation methods and system composition of intrinsically safe power is analyzed in the paper, which lays the theoretical foundation for improving the output power of intrinsically safe power.
Intrinsically safe power used in coal mine environments are divided into primary and secondary intrinsically safe power supplies.primary intrinsically safe power supply is usually explosion-proof and intrinsically safe, which is characterized by large output power and multiple protection circuits;The secondary intrinsically safe power supply is entirely intrinsically safe, with the characteristics of small size and high efficiency and internal intrinsic safety. A high-power output intrinsically forward converter combines Power-i technology is designed, not only with the characteristics of primary intrinsically safe high power and multi-protection, but also secondary intrinsically safe power with small size and high efficiency. The high power output of 48W enable Meet the requirements of the explosion-proof performance of the intrinsically safe power supply under the coal mine.

Acknowledgments
This work is supported by the National Natural Science Foundation of China (51604217,51777167); Shaanxi province technical innovation guidance special plan project(2017CGZH-XNGJ-09)

References

[1] Ming He Important characteristics and realization ways of wisdom mine[J].Industry and Mine Automation, 2018,44(3):31-34.
[2] Mingkai Wang, Zhi Qu, Xiaoyang He, et al. Real time fault monitoring and diagnosis method for power grid monitoring and its application[C].IEEE 2017 Energy Internet and Energy System Integration (EI2), Beijing, China, Sept 28-30, 2017, 1-6.
[3] Yao Hua, Kang Zhao A Mine Explosion-proof and Intrinsically Safe Online Uninterrupted Power Based on Lithium-ionBatteries[J].Instrumentation Techno-logy,2015(7),50-53.
[4] Hyosung Kim. Arcing Characteristics on Low-Voltage DC Circuit Breakers[C].2013 15th European Conference on Power Electronics and Applicati-on(EPE), 2013,Lille:1-7.
[5] S. Erik Reynolds,Mihail Bantic.FMECA for Intrinsically Safe Devices[C].2016 Annual Reliability and Maintainability Symposium,2016,1-5.
[6] Dubaniewicz,Thomas H. DuCarme,Joseph P.Internal short circuit and accelerated rate calorimetry tests of lithium-ion cells: Considerations for methane-air intrinsic safety and explosion proof/flameproof protection methods[J].Journal of Loss Prevention in the Process Industries, 2016(9):575-584.
[7] Yahui Liu, Design of mine-used power supply with super wide input voltage range[J].Industry and Mine Automation,2016,42(6):57-60.
[8] Liebers L.Dynamic arc recognition and termination intrinsic safety without the power limits[A].SICE Annual Conference(SICE)[C].2011:13-18.
[9] Qinghai Meng, Jingjii Wang, Dual Normal Distribution of Arc Discharge Time for Inductive Intrinsically Safe Circuits[J].Transactions of China Electrotechnical Society,2017,32(2):119-124.
[10] Yuting Wang, Shulin Liu, Yibo Ma, Research on Digitization of the Minimum Ignition Voltage Curve of Simple Capacitive Circuit[J]Transactions of China Electrotechnical Society,2014,345-350.
[11] Shulin Liu, Qiang Cui, Yong Li ,Buck converter output short circuit spark discharge energy and output intrinsic safety criterion[J].,Acta Physica Sinica2013, 62(16):430-439.

Influence of electric traction system on power grid harmonics

Meng Tan [1], Dongshan He [2,a], Jun Wen [2], Xiaohong Geng [2] and Sichao Wang [2]

1 State Key Laboratory of Advanced Transmission Technology, Global Energy Internet Research Institute, 102209 Beijing, China
2North China Electric Power University, School of Electrical and Electronic Engineering, 102206 Beijing, China

415254257@qq.com

Abstract. This paper analyzes the traction power supply system and traction substation of electrified railway. Due to the use of a large number of power electronic devices, the electric locomotive has strong nonlinear characteristics. Focusing on harmonics, this paper studies the harmonics generated by AC-DC and AC-DC-AC electric locomotives respectively. Based on this, The mathematical model of the power supply system of electrified traction locomotive is established, and the harmonic model is established. The simulation results are in good agreement with the actual conditions.

1. Introduction

Nowadays, electrified traction locomotive combines lots of new technologies in many fields. It has big transport capacity, high volume, speediness, energy-saving, environment-friendly, and many other advantages [1]. With the rapid development of high-speed railway electrification in China, by the end of 2017, the length of railroad lines in service has reached 127,000km, of which 25,000km is high-speed railway, accounting for 66.3% of the world's total high-speed railway, and ranking first in the world. According to the 13th Five-Year Plan for Railway Development, by the year of 2020, the length of railroad lines in service nationwide will reach about 150,000 kilometres, and the electrification rate will reach about 70%.

Heavy-haul and high-speed railway are the main development directions of railway in China in the future. The conventional external voltage classes of the system are mainly 110kV and 220kV, most of which are 110kV. At present, the major locomotive used in heavy-haul railway in China is AC-DC electric locomotive, whose control mode is phase-controlled rectifier. When getting electric energy from the power grid, the locomotive will generate low-order harmonics of 3, 5, 7, and so on, these will be injected into the power system through the traction network. The single-phase modulation technology--PWM is adopted in AC-DC-AC electric locomotive, thus the third and fifth harmonics are reduced, but the higher harmonics near the switching frequency increase. Harmonics will do great harm to the whole power supply system, mainly: reduce the efficiency of power generation, power transmission and electricity consumption, leading to the increase of electrical loss; make noise, which will affect the normal life of the surrounding residents; and the traction electric network is distributed parameter element, so that resonance may occur when harmonic current passes through the traction network., causing fold increase of the harmonic current, overvoltage, affecting the safe and stable operation of power supply equipment. Therefore, it is of great necessity to conduct in-depth analysis and research on harmonics of electric traction system.

Content from this work may be used under the terms of the Creative Commons Attribution 3.0 licence. Any further distribution of this work must maintain attribution to the author(s) and the title of the work, journal citation and DOI.
Published under licence by IOP Publishing Ltd

At present, the research results on harmonics of electric traction system mainly focus on deriving the harmonic content of a certain locomotive from the mechanism, most of which simplify the locomotive load into harmonic current source; analyzing the influence of electrified traction locomotives on power quality based on the field measured data; establishing the electromagnetic transient simulation models according to the different research objects by simulation software; modelling of the harmonic source and negative sequence source and power flow calculation of harmonic and negative sequence based on the field measured data; the resonance characteristics of traction power supply system, etc. This paper establishes a harmonic model for the power supply system of electrified traction locomotive, and simulates and analyzes the traction conditions. The simulation results show that the model is typical.

2. Harmonic of Electrified traction locomotive

2.1. Traction power supply system

Power systems and transmission lines transfer power from power plants to traction power supply systems. The voltage level of power grids, to which traction power supply systems are connected, is generally 110kV or 220kV. At present, most of general electrified railways in China are connected to 110kV power grids, while high-speed passenger dedicated lines are connected to 220kV power grids [2].

The simple principle of the electrified railway power supply system is shown in Fig.1. In Fig.1, 1 is a three-phase AC high-voltage transmission line; 2 is a traction substation; 3 is a feeder line; 4 is a contact line; 5 is track or ground as a traction current regression channel; 6 is a return line; 7 is an electric locomotive.

In general traction power supply system is composed of traction substation and traction network, in which the traction network includes feeder line, contact line, track circuit, return line and other parts [3].

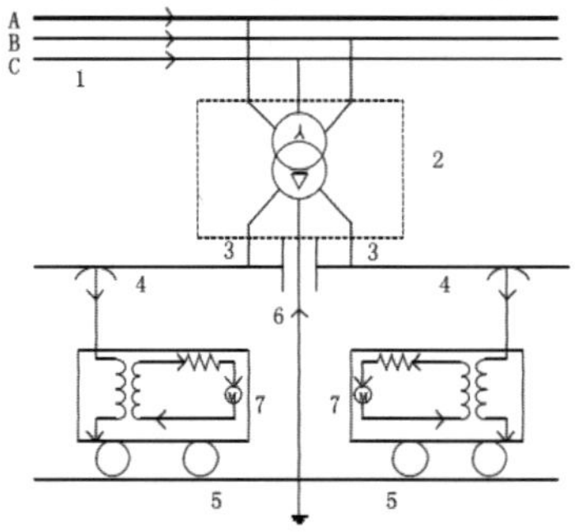

Figure 1. Structure diagram of traction power supply system

In the whole traction power supply system, traction substation plays a key role, and the biggest role of it is to realize varying voltage and phase transformation. It can convert the three-phase high-voltage (110kV or 220kV) power supply of the upper power grid into a low-voltage (25 kV or 27.5 kV)power grid which can supply electric locomotives and meet the needs of locomotive traction. Traction substation is usually installed near the railway, ordinarily using more than one disconnected power to supply electricity, or by a single power to supply power through double circuit transmission line, to raise the efficiency of the power supply system. Combined with the current situation in foreign countries, the key types adopted are: single-phase traction substation, three-phase YN, d11 connected traction substation, three-phase-two-phase traction substation.

The single phase traction substation includes single phase connected traction substation and single phase V, v connected traction substation. Single-phase connected traction substation is simple in structure and low in investment in comparison and can increase capacity utilization greatly, especially when the loads of two power supply arms is consistent, which are mostly used in China at present. As is shown in Fig2, this type of substation is only connected by two phases from the power system, so there is obvious negative sequence current in the power grid. While single phase V, v connected traction substation adopts 2 single phase traction transformers, the connection form is V, v.

IOP Conf. Series: Materials Science and Engineering 533 (2019) 012016 doi:10.1088/1757-899X/533/1/012016

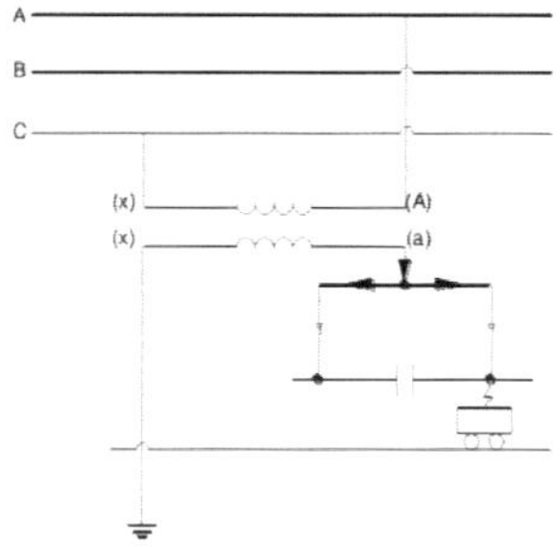

Figure 2. Structure diagram of single phase traction substation

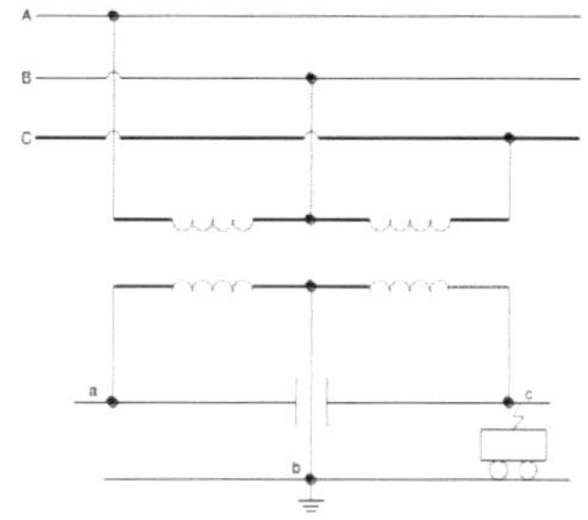

Figure 3. Structure diagram of single phase V, v connected traction substation

2.2. AC-DC electric locomotive

AC-DC electric locomotive first transforms three-phase electric energy into two single-phase electric energy supply arms through three-phase-two-phase traction converter (such as three-phase YNd11 connection and balanced connection transformer), and then transmits electric energy to the train through the traction network of direct power supply or direct power supply with reflux power supply. Converter-driven DC motors, which have nonlinear, mobility and high-power load characteristics, will inject a large number of latent wave and negative sequence current through the traction network and traction transformer network, which will pose a greater threat to the safety and stability of power grid operation.

At present, the main electric locomotive used in Chinese electrified railway is the Shaoshan series electric locomotive, which is developed independently by China.

The harmonic content diagram of SS-4G electric locomotive is shown in Fig.4.

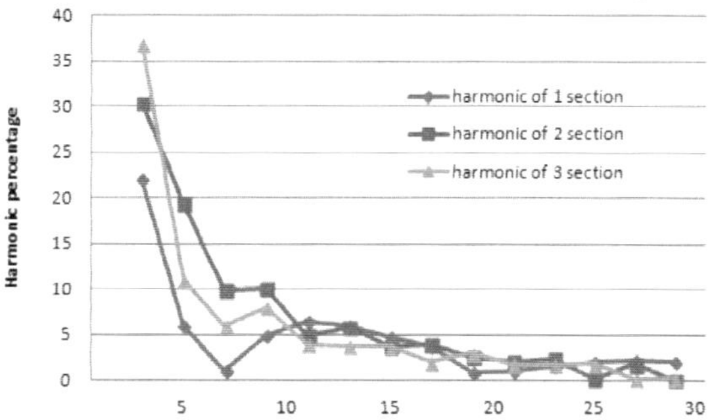

Figure 4. Harmonic content diagram of SS-4G electric locomotive

According to document [4], the measured harmonic currents of 8K AC-DC electric locomotives are shown in Table 1.

Table 1. Measured harmonic currents of 8K AC-DC electric locomotives

order	3	5	7	9	11	13	15
content（%）	21.47	13.98	5.85	5.25	5.28	3.76	3.193
order	17	19	21	23	25	27	29
content（%）	3.45	3.14	2.41	2.02	1.77	1.52	1.23

The spectrum of current on line side is shown in Fig 5.

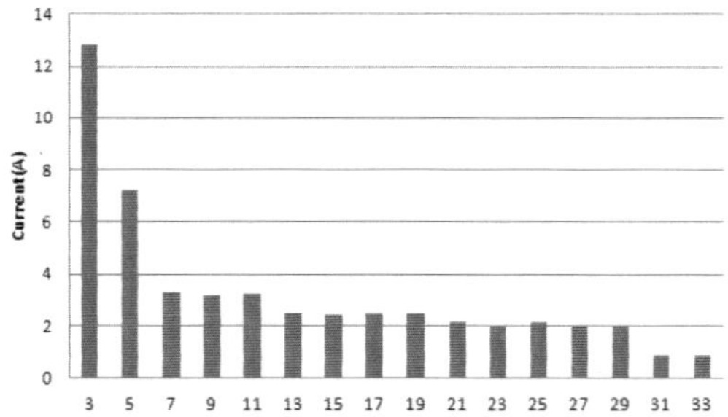

Figure 5. Measured spectrum of current on line side

It can be concluded that the AC-DC electric locomotive mainly generates odd harmonics of 3 and 5 order in the low frequency band. As the order of harmonics increases, the content of harmonics generated by the AC-DC electric locomotives gradually decreases. The 3-order harmonic content is the largest.

2.3. AC-DC-AC electric locomotive

Different from the general speed railway, the traction transformer of the high speed railway adopts V, v connection transformer, and the installation capacity is larger. The traction electric network adopts full parallel AT power supply mode to reduce electromagnetic interference, but the structure is more complex and the power supply interval is longer. The AC-DC-AC electric locomotive, compared with the general speed railway, runs faster and the power is higher, the power factor is close to 1. With PWM rectifier, the harmonic current of AC side is smaller, the low-frequency harmonic component is smaller, low frequency harmonic components are smaller, and the low-order harmonic is improved, but the current frequency band becomes wider, the high-order harmonic distribution is wider.

Table 2. Measured harmonic currents of 8K AC-DC electric locomotives

Type	Voltage level number	Swiching frequency/Hz	Low-order harmonics	High-order harmonics
CRH1	2	450	3、5、7	31、33、···41
CRH2	3	1250	3、5、7	45、47、···55
CRH3	2	1250	3、5、7	95、97、···105
CRH5	2	250	3、5、7	15、17、···25

In China, AC-DC-AC electric locomotives are mainly "Harmony" series of freight electric locomotives, such as HXD1, HXD2, HXD3 etc. And currently the widely used CRH China Railway High-speed is CRH1, CRH2, CRH3 and CRH5 etc.

Table 2 shows the basic data and characteristic harmonic frequencies of various types of Harmony EMUs which are widely used at present.

According to document [4] Fig.6 shows the measured harmonic current content of the net side.

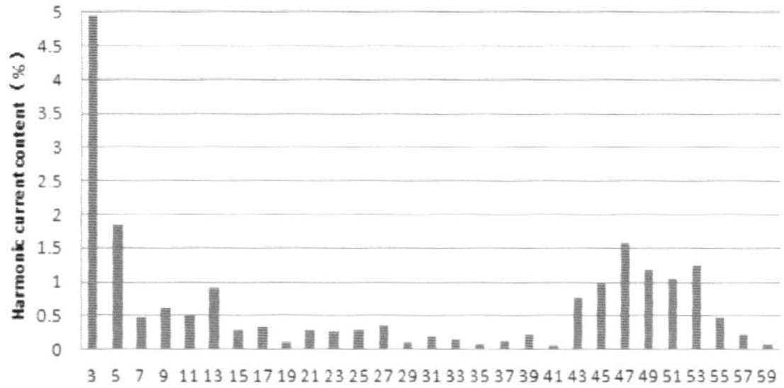

Figure 6. Measured harmonic current content of the net side

Theoretical analysis shows that low-frequency harmonic component of AC-DC-AC electric locomotive should be small, while high-frequency harmonic component large, but in reality the effect is not ideal enough, because the control of low harmonics is subject to certain conditions. However, generally, compared with AC-DC electric locomotive, the high-frequency harmonic components generated by AC-DC electric locomotive are very small.

3. Harmonic model of traction power supply system

3.1. Load harmonic source modelling

Modelling harmonic source of electric traction locomotive load is the basic premise of harmonic analysis and harmonic control of traction power supply system. Previous analysis shows that traction load is a typical high-power harmonic source, which has the characteristics of asymmetry and fluctuation compared with general power system load.

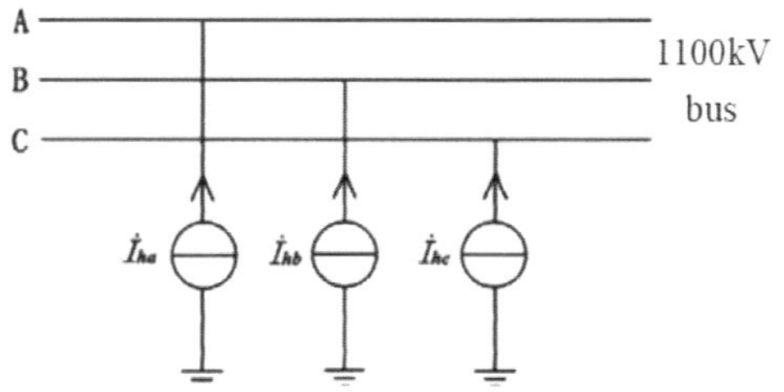

Figure 7. Equivalent structure model of traction power supply system

In the modelling of harmonic source of traction power supply system, the magnitude of DC side current and the waveform of AC side current of traction load are mainly determined by its own characteristics, which are basically independent of AC side grid parameters and can be regarded as harmonic current source [5]. The equivalent structure model of traction power supply system is shown in Fig6. A simplified harmonic current source model is used to describe the harmonic characteristics of traction loads. The real and imaginary parts of each harmonic current in Fig.6 are expressed as functions of fundamental positive sequence voltage, current RMS and phase difference.

According to the working principle of traction load control system of electrified traction locomotive, the traction control system of electric locomotive determines the armature current of traction motor by the speed and level of train running. The voltage equation of traction circuit is shown in equation (1).

$$U_d = C_e \phi n_d + I_a (R_a + R_w) \tag{1}$$

In the formula, R_a is armature resistance, R_w is stray resistance of traction circuit, ϕ is main pole magnetic flux, n_d is traction motor speed. The DC voltage of the rectifier circuit can be determined by equation (1), and according to the working principle of the rectifier circuit, the working state of the rectifier bridge can be determined, and then the trigger angle of the thyristor can be determined. That is, the amplitude and phase angle of harmonic current generated by traction load under this working condition are constant. Therefore, the harmonic generated by traction load of electrified traction locomotive can be determined only by the traction load operation condition.

In order to make the model more practical in engineering and to make the harmonic source model more descriptive under the test of measured data, a three-phase fundamental positive sequence power exchanged between locomotive and grid-side power supply system is proposed as the excitation of the model to evaluate the harmonic characteristics of traction load, and the basic form of the model is given as follows:

$$\dot{I}_h = F_h\left(P_1, Q_1, C\right), h = 1, 2, \cdots N \tag{2}$$

In the formula, $\dot{I}_h$ is the phasor of h-order harmonic current injected into the network side, P_1, Q_1 are the fundamental positive sequence active and reactive power exchanged between locomotive and power supply system; C is a set of model parameters. It can be seen from equation (2) that by this model $\dot{V}_1$ can be obtained without fundamental power flow calculation when solving harmonic current because of the use of three-phase fundamental power as excitation, which brings great convenience to practical engineering application.

3.2. Traction condition

The measured data collected from three 110kV traction substations in reference [6] classify the traction conditions into seven typical categories, while some traction conditions have several sets of data samples, which are listed as events, as shown in Table3.

The data samples of the same type of traction condition are different from each other in the starting time of sampling, sample size and the average power $\overline{P}$ of the locomotive during this period, it gives:

$$\overline{P} = \frac{P_1 + P_1 + \cdots + P_n}{n} \tag{3}$$

In the formula, n is sampling point number; P_n is active power of sampling point n.

Table 3. Classification of traction conditions

Event	Average power (MW)	Traction substation	Traction side load distribution	Operation state
1	10.39	V/V	Two armed	Level
2	9.47	V/V	Single armed	Level
3	3.23	balance	Single armed	Level
4	3.36	balance	Single armed	Level
5	1.72	balance	Single armed	Level
6	1.92	balance	Single armed	Over compensation
7	1.27	Y - Δ	Single armed	Over compensation
8	1.29	Y - Δ	Single armed	Over compensation
9	-6.0	Y - Δ	Single armed	regenerative braking
11	-6.04	Y - Δ	Single armed	regenerative braking
12	-6.12	Y - Δ	Single armed	regenerative braking

3.3. Harmonic source modelling of traction conditions

On the premise of guaranteeing the accuracy of the polynomial model, minimize the order of the polynomial model as far as possible to improve its engineering practicability. Based on the principle that the correlation coefficient is not less than 0.9, the order of polynomial is determined to be 5. For any harmonic on the net side, the positive and negative sequence currents are:

$$\begin{cases} \dot{I}_h^{(1)} = I_{hx}^{(1)} + j\dot{I}_{hy}^{(1)} \\ \dot{I}_h^{(2)} = I_{hx}^{(2)} + j\dot{I}_{hy}^{(2)} \end{cases} \tag{4}$$

In the formula, I_{hx} is real part of sequence current, I_{hy} is imaginary part of sequence current. The harmonic source model expression is given for the real part or the imaginary part of the sequence current.

$$I_h = F_h\left(P_1, Q_1, C\right) = c_1 + c_2P_1 + c_3P_1^2 + c_4P_1^3 + c_5P_1^4 + c_6P_1^5 \\ c_7Q_1 + c_8Q_1^2 + c_9Q_1^3 + c_{10}Q_1^4 + c_{11}Q_1^5 \quad (5)$$

In the formula, P_1, Q_1 are active and reactive power of three phase fundamental wave positive sequence on net side, which is determined by the load level of a given locomotive or measured. $c_1 - c_{11}$ is parameters of the model.

In particular, when the locomotive is stopped, that is, when the active power and reactive power are 0, take c_1 to be 0. In order to identify the parameters of the model, the objective function is given:

$$J = \min\sqrt{\left(I_{hs} - F_h\left(P_1, Q_1, C\right)\right)^T\left(I_{hs} - F_h\left(P_1, Q_1, C\right)\right)} \quad (6)$$

In the formula, I_{hs} is the measured effective value of h-order harmonic current. P_1, Q_1 are measured power. The parameter set C of the polynomial model can be determined only when the minimum value of the objective function J is obtained. The quasi-Newton method is used for parameter identification, and the iteration number is set to 1000. Convergence criterion is $\nabla J\left(C^{(k)}\right) \le 10^{-10}$. Take the reference current $I_B = 10\text{A}$ and the reference power $S_B = 19053.3\text{kVA}$.

Detailed identification results of the real and imaginary parts of the positive and negative sequence components of the third harmonic currents under various operating conditions can be found in reference [6].

4. Harmonic simulation of electrified traction locomotive

After obtaining the identification results of the model parameters of each traction condition, the positive sequence active power and reactive power of the three-phase fundamental wave on the original network side are substituted into the model as excitation, and the model response is obtained. According to reference [6], taking the positive and negative sequence current amplitudes of 3-order harmonic as an example, the model responses of two typical events are compared with the measured data, as shown in Fig.8 and Fig.9.

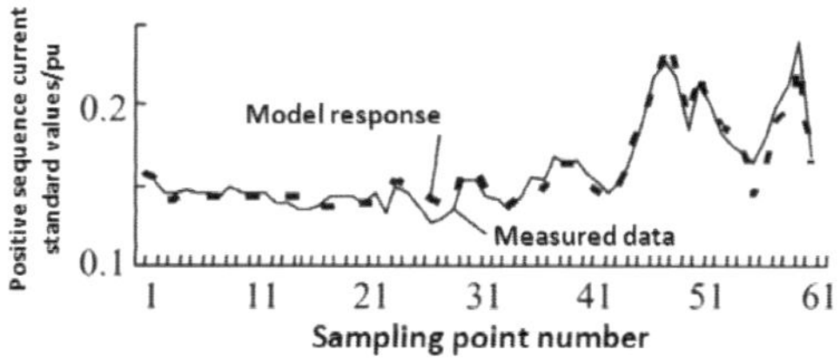

Figure 8. Measurement and model response of event 3

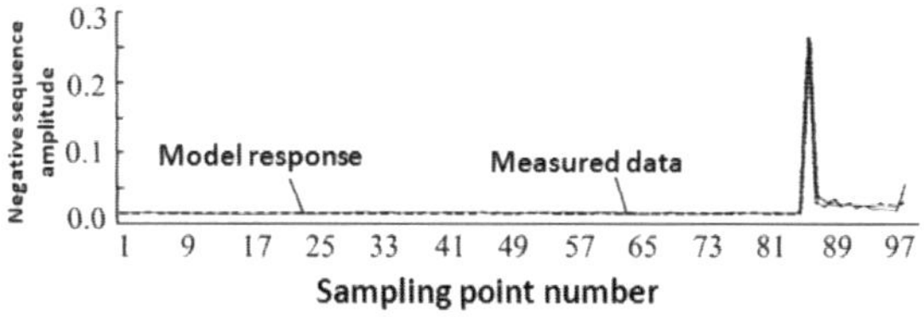

Figure 9. Measurement and model response of event 7

The remaining 30 groups of measured harmonic current and power data under various traction conditions are taken as samples to test the harmonic source model. The maximum average relative error of the 3-order harmonic currents under different traction conditions is

$$\max\sigma = \frac{\sum_{1}^{30}\left|\left(I_{sn} - F\left(P_{1n}, Q_{1n}, C\right)\right)/I_{sn}\right|}{30} = 6.91\% \quad (7)$$

In the formula, I_{sn} .is the real or imaginary part of the 3-order harmonic of positive sequence or negative sequence in group n under a certain working condition. The maximum average relative error of the other harmonics is within 8%. The accuracy of the model is verified.

3-order harmonic model parameters of AC-DC electric locomotive are given here. It is known from the previous section that the amplitude of harmonics generated by AC-DC-AC electric locomotive is very small compared with that of AC-DC electric locomotive because of its PWM control mode. For AC-DC electric locomotive, the 3-order harmonic current value is the largest, so this model is representative.

5. CONCLUSIONS

Firstly, the traction power supply system and traction substation of electrified railway are analyzed in this paper. It is known that the electric locomotive has strong nonlinear characteristics due to the use of a large number of power electronic devices, which brings a lot of power quality problems to the traction network and power system. This paper focuses on harmonics. On this basis, the harmonics generated by AC-DC and AC-DC-AC electric locomotives are studied respectively. It is concluded that compared with AC-DC electric locomotives, the high-frequency harmonics generated by AC-DC electric locomotives are very small. The mathematical model of the power supply system of electrified traction locomotive is established, and the harmonic model is established. The simulation results are in good agreement with the actual conditions. The model is typical and suitable for the follow-up harmonic analysis of various electrified traction locomotives.

References

[1] Li Li. High speed rail traction power supply system harmonic analysis and suppression research [D]. East China Jiaotong University, 2017.
[2] Li Hongqiang. Harmonic analysis of traction power supply system [D]. Southwest Jiao Tong University 2014.
[3] Ju Hao. Study on power quality assessment of traction substations [D]. Xi'an University of technology, 2018.
[4] Sheng Cai Fei. Simulation study on harmonic currents of electric locomotives and EMUs [D]. Beijing Jiaotong University 2009.
[5] Xie Shaofeng, Li Qunzhan, Zhao Liping. Study on Harmonic Distribution Characteristics and Probabilistic Model of Traction Load of Electrified Railway [J].Chinese Journal of Electrical Engineering, 2005, 25 (16): 79-84.
[6] Li Xinran, Sun Qian, et al. Harmonic Source Model for Traction Load of Electric Railway Based on Measured Data [J]. Grid Technology, 2012, 36 (1): 158-162
[7] Yu Xijun, Wang Pengju, Wu Binshun. The influence of traction load of electrified railway on power quality of power grid [J]. China New Communications, 2018, 20 (21): 212.
[8] Zeng Yi, Sun Xiaojia, Pan Ling, Pan Aiqiang, Zhang Peng. Study on resonance of Tibetan Power Grid Considering the influence of electrified railway [J]. Electricity and energy, 2018,3(05): 593-598.

Loss Analysis and Optimization Design of Half-Bridge LLC Resonant Converter

TONG Jun[1], LI Facheng[1], LI Xiang[1]and YANG Xingchen[1]

1 College of Electrical and Control Engineering, Xi'an University of Science And Technology, Xi'an 710054, China

e-mail: 582327086@qq.com

Abstract. The efficiency improvement of switching converters has always been a research hotspot. This paper analyzes the loss of LLC resonant converter including magnetic components loss and Switching loss. For easier analysis, a mathematical model of the switching tube is established, and then derive the formula of the loss. The synchronous rectification technology is used to improve the efficiency of the LLC, and finally established a simulation model to verify the correctness of the technology to improve the efficiency of the converter.

Keywords: LLC Resonant Converter, synchronous rectification technology, MATLAB simulation

1. Introduction

With the rapid development of power electronics technology, the requirements for power supplies in various fields are getting higher and higher. Therefore, high efficiency, small size, high power density, high withstand voltage and high reliability have become hot research circles in the industry. From the perspective of reducing the size of the switching converter, increasing the switching frequency is one of the most effective methods. However, increasing the switching frequency will increase the switching loss and reduce the switching efficiency. This paper analyzes the loss of each part of the LLC converter, which has important significance for improving the efficiency of the converter.

2. The Basic Working Principle of LLC Resonant Converter

The topology of the LLC resonant converter is shown in Figure 1, where Q1 and Q2 are two MOSFET switching transistors. $L_r$is a resonant inductor, which can be replaced by the leakage inductance of the transformer? L_m is the magnetizing inductance of the transformer, Cr is the resonant capacitor, L_r, L_m and C_r form the resonant cavity [1], and their resonant frequencies are:

$$f_1 = \frac{1}{2\pi\sqrt{L_r C_r}} \tag{1}$$

Content from this work may be used under the terms of the Creative Commons Attribution 3.0 licence. Any further distribution of this work must maintain attribution to the author(s) and the title of the work, journal citation and DOI.
Published under licence by IOP Publishing Ltd

$$f_2 = \frac{1}{2\pi\sqrt{(L_r + L_m)C_r}} \tag{2}$$

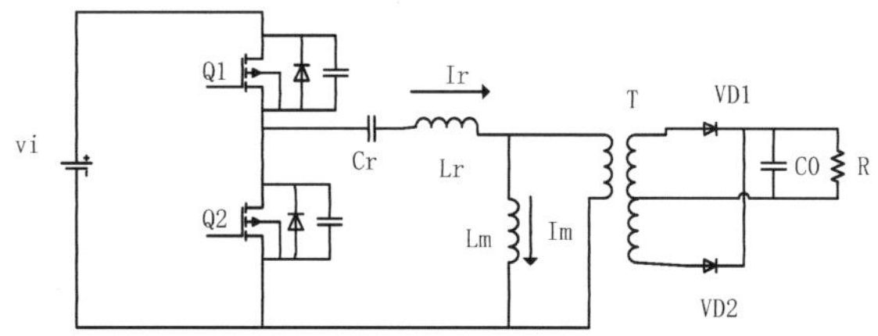

Figure 1. Topology of LLC resonant converter

A switching cycle of the switch can be divided into 8 stages [2], as shown in Figure 2.
Where VQ1 and VQ2 represent the driving pulses of the switching tubes Q1 and Q2, i_r represents the current of the resonant cavity, i_m represents the excitation current of the excitation transformer, and VD1, VD2 and i_{D1}, i_{D2} respectively represent the voltage and current of the secondary rectifier diode.C_1 and C_2 represent the parasitic capacitance of Q1 and Q2.

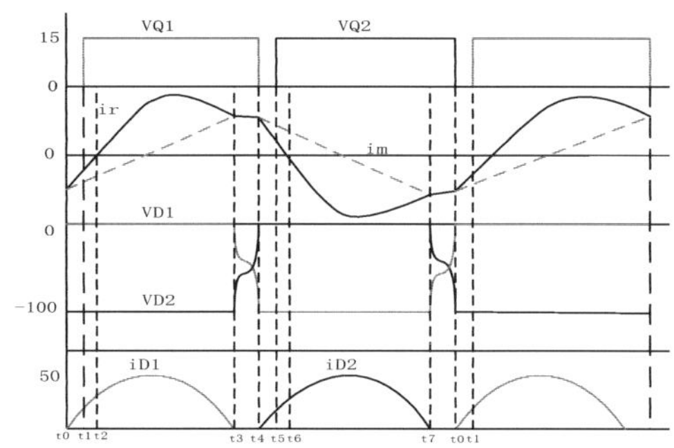

Figure 2. main waveform when $f_2 < f_s < f_1$

Stage 1(t_0<t<t_1):Q2 is turned off at t_0,and then enter dead time, C_1 discharge while C_2 charging, Reverse current $i_r < i_m$, The primary winding voltage of the transformer is forward voltage, and the secondary rectifier diode D1 is turned on.
Stage 2(t_1<t<t_2): The body diode of Q1 is turned on, creating conditions for ZVS of Q1. The voltage of the primary winding of the transformer is forward voltage, and the secondary rectifier diode D1 is turned on. The voltage of l_m is clamped at nV_o and does not participate in resonance. When $i_r$becomes 0, the process ends.
Stage 3(t_2<t<t_3): At time t_2, Q1 is turned on, and $i_r > i_m$, $l_m$is linearly charged at this stage. This process ends until $i_r = i_m$.
Stage 4(t_3<t<t_4): At time t_3, $i_r = i_m$, the transformer has no energy transfer, the secondary rectifier is turned off, and the excitation winding l_m participates in resonance.
Stage 5(t_4<t<t_5): At time t_4, Q1 is turned off and enters the dead time. At this stage, C_1 is charged, C_2 is discharged, $i_r < i_m$, the primary winding voltage of the transformer is reverse voltage, and the secondary rectifier diode D2 is turned on.
Stage 6(t_5<t<t_6): At time t_5, C_2discharge is completed, and the body diode of Q2 is turned on to create conditions for ZVS of Q2. The state of the transformer and rectifier diode is the same as that of the previous stage.
Stage 7(t_6<t<t_7): At time t_6, Q2 is turned on, and the secondary side rectifier D2 is turned on, and the voltage of l_m is clamped at -nV_o and l_m is linearly charged at this voltage. This process ends until $i_r = i_m$.

Stage 8(t_7<t<t_0): At time t_7, $i_r = i_m$, the working state is the same as Stage 4. and then Enter the next cycle

3. Loss Analysis of LLC Resonant Converter

The loss of the LLC resonant converter mainly includes two parts, one is the loss caused by the magnetic component such as the transformer, and the other is the loss caused by the switching tube.

3.1 Loss analysis of magnetic components

Transformer losses are clearly a major part of the loss of magnetic components. It mainly includes iron loss and copper loss. Iron loss refers to the energy lost during transformer magnetization, including hysteresis loss and eddy current loss. Iron loss has a great relationship with core selection. Therefore, the selection of high frequency transformer core material It is very important for the design of the converter.

The core loss P includes hysteresis loss P_h, eddy current loss P_c, and residual loss P_e, that is,

$$P = P_h + P_c + P_e \tag{3}$$

3.1.1 Hysteresis loss. The loss in the static magnetic field is mainly hysteresis loss. The hysteresis loss is proportional to the area enclosed by the hysteresis loop [3], which can be expressed by

$$P_h = \oint BdH \tag{4}$$

The hysteresis loss is generally calculated by the following formula.

$$P_h = K_h f B_m^{1.6} \tag{5}$$

K_h is the proportionality factor and is related to the material.

3.1.2 Eddy current loss. In the dynamic alternating magnetic field, there is eddy current loss, which can be calculated by the following formula:

$$P_c = \frac{\pi^2 f^2 B^2 d^2}{6\rho} \tag{6}$$

Where d is the material density, unit g/cm^3 ρ is the resistivity, the unit is Ω·m.

3.1.3 Residual loss. The residual loss is due to the magnetization hysteresis caused by the post-magnetic effect. That is, when B→0 and f→0, the part of the loss whose total loss is not zero is the residual loss, which is related to the type of material and the excitation condition.

3.2 Power switch tube loss

3.2.1 Loss of the primary side switch of the transformer. Generally speaking, the loss of the power tube is divided into on-state loss and switching loss. The on-state loss is the loss caused by the switch tube when it is turned on, and the switch loss means that the switch tube is turned from on to off or from off to on. The resulting transition loss. In the first section, the primary side switch of the LLC resonant converter can be analyzed to achieve ZVS, so the turn-on loss is not considered. Then, only the on-state loss and the turn-off loss need to be considered.

During time t_0<t<t_3, the voltage across the magnetizing inductance L_m is nV_0, which is clamped by the secondary side without participating in resonance. In this stage, the exciting inductor current i_m can be expressed as:

$$i_m(t) = -I_{m\cdot max} + \frac{nV_0}{L_m} \cdot t \tag{7}$$

At time t_3, the excitation current reaches the maximum value $I_{m\cdot max}$,that is: $i_m(t_3) = I_{m\cdot max}$, at this time $t_3 = \pi/w_r$. Then

$$i_m(t_3) = -I_{m\cdot max} + \frac{nV_0}{L_m} \cdot \frac{\pi}{w_r} = I_{m\cdot max} \tag{8}$$

So $I_{m\cdot max} = \frac{nV_0}{2L_m} \cdot \frac{\pi}{w_r}$

At time t_3, $i_m = i_r$, the voltage across the magnetizing inductance L_m is no longer clamped, and participates in resonance with $L_r$and C_r. Since the value of the magnetizing inductance L_m is much larger than the resonant inductor L_r, the resonant current $i_r = I_{m\cdot max}$ can be considered in the stage of $t_3 < \mathrm{t} < t_4$.

Therefore, the expression of the resonant current $i_r(t)$ in the first half of the cycle is as follows:

$$i_r(\mathrm{t}) = \begin{cases} \frac{V_{in}-nV_0}{Z_r}\sin(w_r t - \theta) & t_0 < \mathrm{t} < t_3 \\ I_{m\cdot max} & t_3 < \mathrm{t} < t_4 \end{cases} \tag{9}$$

where $Z_r = \sqrt{\frac{L_r}{C_r}}$ is resonant network impedance, $\theta = \sin^{-1}(I_{m\cdot max})$.

The on-state loss during Q1 conduction can be divided into two parts, one part is the loss caused by the body diode of Q1, and its size is the product of the on-voltage drop V_d of the body diode and the current i_d. This process occurs at t_1<t<t_2 period, $i_d = i_r$, its loss can be expressed as:

$$P_{ssd} = \int_{t_1}^{t_2} i_d \cdot V_d \, dt \tag{10}$$

The other part is the on-state loss. This process occurs during the period t_2<t<t_3, and it can be expressed as:

$$P_{dt} = \frac{1}{T}\int_{t_2}^{t_3} (i_r(t))^2 \cdot R_{on} \, dt \tag{11}$$

R_{on} is the on-resistance, which can be found in the device manual.

The turn-off loss occurs during the period of t_4<t<t_5. After the Q1 is turned off, the resonant current cannot be rapidly dropped to 0 due to the leakage inductance of the transformer, so loss occurs, and $t_4$enters the dead time, and C_1 is charged at this stage. C_2 discharge, establish its mathematical model as shown in Figure 3:

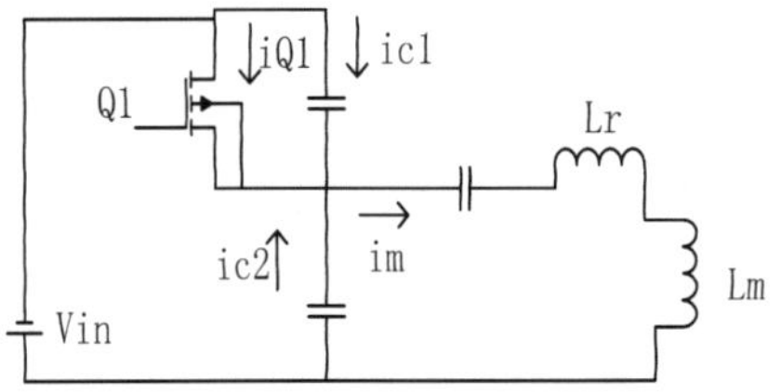

Figure 3. Mathematical model when Q1 is turned off

When Q1 is turned off, the current i_{Q1} decreases linearly, and its size is:

$$i_{Q1}(\mathrm{t}) = i_b - i_b/t_f \cdot t \tag{12}$$

Where i_b is the initial current, the magnitude of which is equal to the current at time t_4, $i_b = I_{m\cdot max}$ and t_f is the fall time. According to the model, you can get:

$$\begin{cases} i_m = i_{c1} + i_{c2} + i_{Q1} \\ i_{c1} = \frac{c1 dV_{c1}}{dt} \\ i_{c2} = -\frac{c2 dV_{c2}}{dt} \\ V_{c1} + V_{c2} = V_{in} \\ C1 = C2 \\ i_{c1} = -i_{c2} \end{cases} \tag{13}$$

we can get $i_{c1} = \frac{i_m}{2t_f} \cdot t$

$$V_{s1}(\mathrm{t}) = V_{c1}(t) = (i_m \cdot t^2)/4C_1 t_f \tag{14}$$

Where V_{c1} V_{c2} is the capacitor voltage, and V_{s1} is the voltage of Q1. The loss of the MOS tube is thus:

$$P = \frac{1}{T}\int_0^{t_f} V_{s1}(t) i_{Q1}(\mathrm{t})\, dt = \frac{{i_m}^2 t^2 {t_f}^2 f}{48C_1} \tag{15}$$

3.2.2 Secondary side rectifier diode loss. Generally, the voltage drop of the rectifier diode is 0.7V, and the conduction time of each diode of the secondary side is π/w_r so the conduction loss of the two rectifiers is:

$$\mathrm{P}_{D1、D2} = 2 * 0.7I\frac{\frac{\pi}{w_r}}{T} = 0.7I = 0.7\frac{V_0}{R_0} \tag{16}$$

The diode rectification conduction voltage drop is about 0.7V. If the secondary side output voltage is as low as a few volts, the loss due to this diode voltage drop becomes very large. The synchronous rectification technology is very suitable for this low voltage and high current occasion. Greatly reduce the loss of the diode [4].

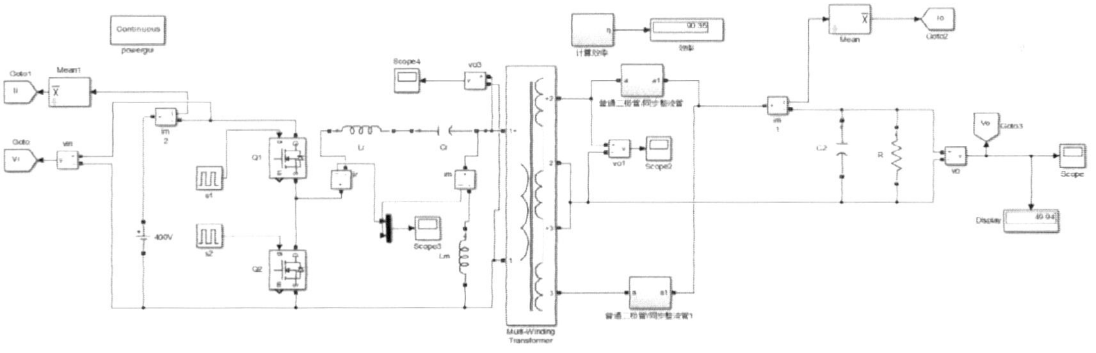

Figure 4. Simulation model of LLC resonant converter

4. Simulation Verification of Secondary Side Synchronous Rectification Technology

The relevant parameters of this simulation are as Table 1:

Table 1. Simulation parameters

Input voltage V_{in}	400	Load R_o	5Ω
Magnetizing inductance L_m	879	Dead time	125nS
Resonant inductor L_r	122	Transformer ratio n	4
Resonant capacitor C_r	$22n$		

The simulation model is established as shown in Figure 4. When the secondary diode is rectified on the secondary side, the output voltage waveform is shown in Figure5. The average output current is $I_o = 9.5A$, and the average input current is $I_{in} = 1.3A$, so the efficiency of normal diode rectification can be calculated.

$$\eta = \frac{V_o I_o}{V_{in} I_{in}} 100\% = \frac{47.6 \cdot 9.5}{400 \cdot 1.3} 100\% = 86.9\% \tag{17}$$

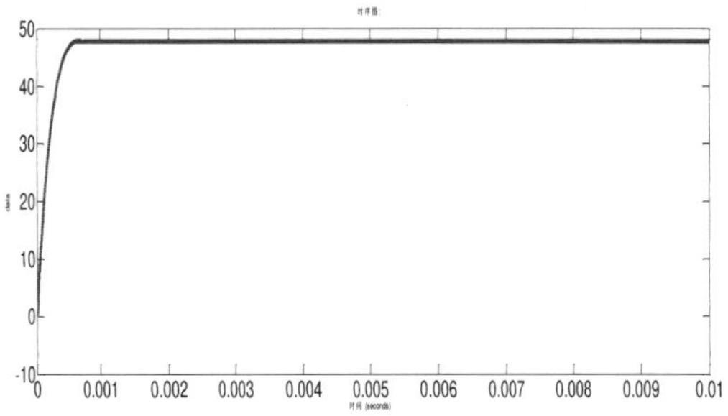

Figure 5. Output voltage waveform when rectifying with a common diode

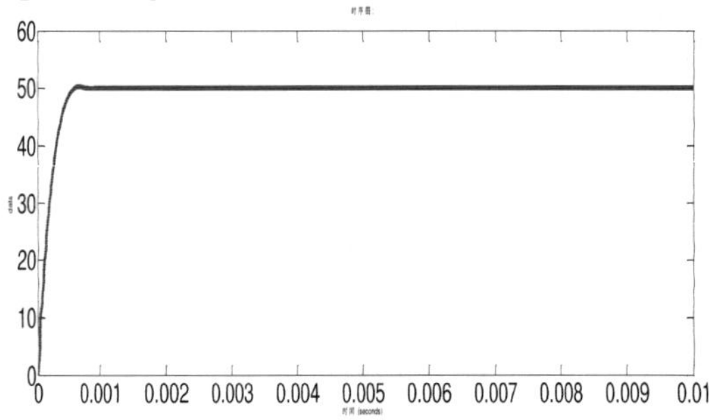

Figure 6. Output voltage waveform when using synchronous rectifier

The output waveform when using the synchronous rectification technique on the secondary side is shown in Fig. 6. At this time, the output voltage $V_o = 49.97V$, the output average current is $I_o = 10.0A$, and the average input current is $I_{in} = 1.38A$ and the efficiency is:

$$\eta = \frac{V_o I_o}{V_{in} I_{in}} 100\% = \frac{49.97 * 10.0}{400 * 1.38} 100\% = 90.5\% \tag{18}$$

It can be seen that the efficiency of the secondary side is greatly improved after using the synchronous rectification technology. The simulation results in i_m and i_r waveforms as shown in Figure 7.

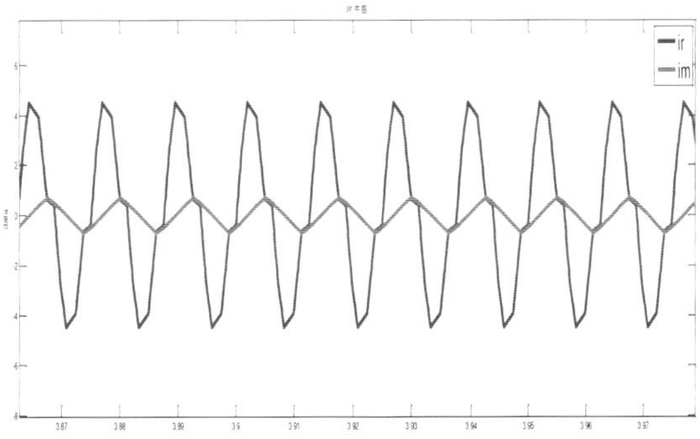

Figure 7. Resonant current i_r and excitation current i_m waveform

5. Conclusion

Nowadays, with the increasing demand for power supply in various fields, it is urgent to improve the efficiency of switching converters. In this paper, the loss of LLC resonant converter is analyzed, including the loss of magnetic component of the transformer and the loss of the switching tube, and a mathematical model is established to analyze the on-state loss and turn-off loss of the switch. At the same time, the LLC resonant converter simulation model is established. and used MATLAB software simulation to verify that the synchronous rectification technology can improve the efficiency of the LLC resonant converter.

References

[1] Designing an LLC resonant half-bridge power converter[R]. Texas instruments power supply design seminar.2010.

[2] Yu Changbin. Research on LLC Resonant Half-Bridge DC-DC Converter [D]. Chongqing University.2007.

[3] Feng Xue. LLC Resonant Converter Loss Analysis and Efficiency Optimization [D]. Beijing Jiaotong University.2017.

[4] Feng Jijun. Mathematical Modeling and Loss Analysis of LLC Resonant Converter [J]. Electronic technology 2016, 29[4]:140-143

Support Vector Machine based on clustering algorithm for interruptible load forecasting

Xiang Yu[1,a], Guangfeng Bu[1],Bingyue Peng[1] ,Chen Zhang[1],Xiaolan Yang[1],Jun Wu[1], Wenqing Ruan[1],Yu Yu[1],Liangcai Tang[1]and Ziqing Zou[2]

1. State Grid Yangzhou Power Supply Company, China
2. School of Electrical Engineering, Southeast University, China

a. Corresponding author: Xiang Yu@18795969086@163.com

Abstract. Accurately forecast interruptible load can help to alleviate the power supply tension during peak load and make scheduling more flexible. Support vector machine (SVM) method which has been widely used in load forecasting usually selects a period of date close to the forecast day without considering the information characteristics of itself. An interruptible load forecasting method based on clustering algorithm is proposed in this paper. This method puts forward a new idea to select the sample of prediction model which takes full account of the weather and date information of the forecast day and solve the problem that the traditional SVM method cannot properly reflect it. In this paper, the principles of clustering algorithm and support vector machine are introduced firstly. Then K-means clustering algorithm is used to classify the historical data, and the support vector machine forecasting model is constructed by using the categories of the forecast day membership. Finally, the prediction is carried out by combining with the actual data. The results show that the prediction accuracy of this method is more than 95%, and it has higher precision.

1 Introduction

As an important component of demand side management (DSM) in power system, interruptible load management (ISM) makes use of the flexibility of users to alleviate the power supply tension during peak load. So as to avoid or reduce the expensive rotating reserve and the generation capacity investment needed to meet the growth of electricity demand, it is conducive to the safe and economic operation of the power system, weaken the impact of market power in the electricity market, and suppress price spikes. Therefore, interruptible load forecasting is an important part of power system energy management. Interruptible load forecasting not only provides a guarantee for the safe and economic operation of power system, but also is the basis of power market scheduling and power supply planning [1-2].

Support vector machine (SVM) algorithm has the characteristics of high fitting accuracy, strong generalization ability and global optimization, and has been applied to load forecasting by some scholars [3-4]. But this kind of traditional load forecasting based on support vector machine usually chooses a period of date close to the forecasting date, and the closer the forecasting date is, the more similar the characteristics of the load forecasting date are. However, even if it is only one or two days apart, the actual load can be very different, for example, the load on the day before the holiday and during the holiday is markedly different, and the adjacent days may also be greatly affected by the weather. Direct prediction can lead to a certain deviation.

Content from this work may be used under the terms of the Creative Commons Attribution 3.0 licence. Any further distribution of this work must maintain attribution to the author(s) and the title of the work, journal citation and DOI.
Published under licence by IOP Publishing Ltd

Therefore, this paper proposes a support vector machine forecasting based on load clustering, which takes both meteorological factors and time factors into account. Firstly, the meteorological data and date information of historical loads are sorted out, then the historical data are clustered by K- means clustering algorithm, and the clustering results are classified according to the similarity of historical loads. The relevant information of the forecast day is obtained by querying the information and multiple historical loads are found in which are the same as the information of the forecast day. Then the support vector machine prediction model is constructed with the historical date of this category. The method is applied to the interruptible load forecasting of an enterprise in Yangzhou, and the higher forecasting accuracy is obtained, which proves the accuracy and superiority of the support vector machine forecasting method based on clustering algorithm.

2 Principle of support vector machine based on clustering

2.1 Clustering algorithm

Clustering is an analysis method which divides a group of data sets into several categories according to the similarity between the set elements [5-6]. Through clustering analysis, the attributes and characteristics of each category can be studied for further processing. Cluster analysis requires that the distance between elements in the same category is small, while the distance between different categories is very large.

2.1.1 Normalization processing

First of all, the clustering should be normalized. Normalization is a simplified way of calculating, that is, a dimensional expression, transformed into a dimensionless expression, become a scalar. Many operations can be handled in this way, not only to ensure the convenience of the operation, but also to highlight the essence of the meaning of the physical quantity [7].

Normalization definition: normalization is to process the data that needs to be processed (by some algorithm) to limit to a certain range you need. First of all, the normalization is for the convenience of later data processing, and the second is to accelerate the convergence of the preservative program when it runs. The specific function of normalization is to induce and unify the statistical distribution of samples. The normalization between 0-1 is a statistical probability distribution, and the normalization is a statistical coordinate distribution on a certain interval.

User load data obtained by power system load measurement devices will vary greatly in numerical range, and these differences will have a great impact on the results of classification. Therefore, the sample data should be normalized before classification in order to eliminate the impact of these differences. The load curves referred to in this paper are all representative daily load curves of users after normalization, so the classification of power system users is transformed into the classification of user load curves.

2.1.2 K-means clustering algorithm

According to the different principles of classification, clustering analysis can be divided into partition clustering method, hierarchical clustering method, density clustering method, grid and model-based method and so on. In this paper, we choose the k-means clustering algorithm in partition method. K-means is an iterative clustering algorithm, in which the objects in the cluster are constantly moved until an ideal cluster is obtained, and each cluster is represented by the average value of the objects in the cluster. Using k-means algorithm to get the cluster, the similarity of objects in the cluster is very high, and the difference between objects in different clusters is also very high.

The main steps of the algorithm are [8]:

(1) Randomly selecting k objects from n data objects as the initial cluster center;

(2) The average value of each cluster is calculated, and the average value is used to represent the corresponding cluster;

(3) According to the distance between each object and the center of each cluster, the cluster is assigned to the nearest cluster;

(4) The second step is to recalculate the average value of each cluster.

This process repeats until it is satisfied that a criterion function is no longer significantly changed or that the clustered objects are no longer subject to change. The criterion function of the general K-means algorithm is the squared error criterion, which is defined as:

$$E=\sum_{i=1}^{k}\sum_{p\in C_i}|P-mi|^{2} \tag{1}$$

Where E is the sum of the mean variance of all the objects in the dataset and the corresponding cluster center, p is the given data object, and m_i is the mean value of the c_i clusters (p and m are both multidimensional). The k-means algorithm is relatively scalable and efficient for large databases, and generally ends in a local optimal solution.

2.2 Principle of support vector machine (SVM)

The support vector machine (SVM) algorithm is based on statistical theory. SVM method is a successful implementation of statistical learning theory [9-10]. It is based on the VC theory of statistical learning theory and the principle of structural risk minimization. According to the limited sample information, the best tradeoff is found between the complexity of the model (the learning accuracy of a specific training sample) and the learning ability (the ability to identify arbitrary samples without errors), in order to obtain better generalization ability. It has the following characteristics:

(1) The SVM principle is implemented, which can minimize the upper bound of generalization error rather than the training error, so it has better generalization performance.

(2) Compared with neural network method, SVM has less free parameters. There are only three free parameters in the SVM algorithm, but there are a lot of free parameters in the neural network, which need to be selected subjectively by experience.

(3) The neural network may not converge to the global optimal solution, so it is easy to fall into the local optimal solution. In SVM algorithm, training SVM is equivalent to solving a quadratic convex programming problem with linear constraints, so its solution is unique, global and optimal.

The load forecasting algorithm based on support vector machine is as follows:

For the sample set $\{x_i, y_i\}_{i=1}^{l}$,, $x_i \in R^n$ as the input variable and $y_i \in R$ as the corresponding output value, a nonlinear mapping $\varphi(x)$ is introduced to map the n -dimensional input and 1-dimensional output sample sets D from the original sample space to the high-dimensional feature space, and the regression function is in the form of the following[11-12]:

$$f(x)=\omega \cdot \varphi(x)+b \tag{1}$$

In the formula: $\omega \in R^n$ is the weight and $\varphi(x)$ is the x mapped eigenvector, b is the threshold. According to the principle of structural minimization, SVM can determine the regression function by minimizing the target, that is:

$$\min \frac{1}{2}\|\omega\|^2 + C\sum_{i=1}^{m}\xi_i^2 \tag{2}$$

$$\begin{aligned} s.t.\ \ & y_i[\omega \cdot \varphi(x_i)+b]=1-\xi_i \\ & \xi_i \ge 0, i=1,2,...,m \end{aligned} \tag{3}$$

Where $C>0$ is the penalty function, ξ_i is the relaxation factor. For the constraint condition, the corresponding multiplier α_i is introduced, and the kernel method is introduced, the form of feature mapping is considered, then the final decision function is given as:

$$\varphi(x) = \sum_{i=1}^{l} (\alpha_i^* - \alpha_i) K(x_i, x) + b \tag{4}$$

The commonly used kernel functions and their parameters are as follows [13-14]:
(1) Polynomial kernel function

$$K(x_i, x_j) = (x^T{}_i x_j)^d \tag{5}$$

(2) Linear kernel function

$$K(x_i, x_j) = (x^T{}_i x_j) \tag{6}$$

(3) Gauss kernel function

$$K(x_i, x_j) = \exp(-\frac{\|x_i - x_j\|^2}{2\sigma^2}) \tag{7}$$

Where $\sigma > 0$ is the width of the kernel function.
(4) Laplace kernel function

$$K(x_i, x_j) = \exp(-\frac{\|x_i - x_j\|}{\sigma}) \tag{8}$$

Where $\sigma > 0$ is the width of the kernel function.
(5) Sigmoid kernel function

$$K(x_i, x_j) = \tanh[\beta(x_i^T x_j) + \theta] \tag{9}$$

As for complex prediction problems, because of the different sources of data, their composition may also be heterogeneous data sets. This leads to the complexity of data distribution. If we only choose a single kernel function SVM model, we may not be able to achieve satisfactory prediction performance. For the common kernel functions mentioned above, they can be divided into global kernel function and local kernel function according to their influence on the regression prediction problem. The typical representative of global kernel function is polynomial kernel function, whose outstanding feature is its strong generalization performance. It can extract the whole situation of the sample data very well, but the disadvantage is that the learning ability is relatively weak. Gauss kernel function, as one of the typical representatives of local kernel function, has good learning ability for samples in a certain distance. The disadvantage is that its generalization ability is relatively poor. Therefore. Considering the merits and demerits of the two kinds of kernel functions, the two kinds of kernel functions are combined according to certain weights to form the final kernel function, which can improve the comprehensive performance of the model.
According to the above analysis, if the weight coefficient of Gauss kernel function K_G is α and the weight coefficient of polynomial kernel K_P is $1\text{-}\alpha$, then the final mixed kernel function after combination is as follows:

$$K = \alpha K_G + (1-\alpha) K_P \tag{10}$$

3 Prediction model

3.1. Load clustering based on k-means clustering algorithm

In this paper, the K-means algorithm is used to cluster the characteristic variables corresponding to 96 points daily load data of interruptible load, and then the load pattern clustering is realized.

Firstly, the influencing factors of interruptible load are analysed [15]. The interruptible load fluctuates greatly because of the weather and date. The selected weather factors are daily maximum temperature, daily average temperature, and the definition of the specific weather conditions, defining rainy days and non-rainy days, rainy days for the category 1, non-rainy days for the category 0. The date factor is reflected as whether it is a holiday or not, the load curve of the ordinary working day and the legal holiday will have great difference, so the definition of the statutory holiday is class 2, the ordinary working day is category.

3.2 Establishment of support vector machine model

From the above analysis, we can see that there are five model parameters to be chosen in this paper, namely regularization parameter C, Gauss kernel width σ 、 The weight coefficient α of the Mixed kernel function K 、 Polynomial coefficient d and ε. Among them, C and σ play a very important role in support vector machine (SVM) algorithm.

C determines the training error and generalization ability. Too small a value will cause underlearning to the training data, and too much will easily cause an overlearning phenomenon to the training data, resulting in deterioration of generalization performance. The suitable value of C should be between 10 and 100. σ reflects the distribution characteristics of the training sample data. If the value of σ is too small, the training set will be overlearned, and the training set will be underlearned if the value of σ is too small. The suitable value of σ should be between 1 and 10.The performance of support vector machine algorithm is insensitive to ε, and the change of ε has little effect on the standard mean square error of training set and test set, so it is not affected by the variation of ε. In general, the number of support vectors decreases with the increase of ε. However, the large value of ε can also reduce the approximation accuracy of the data points, so the value of ε can not be too large. In order to give full play to the advantages of the two kinds of kernels, the mixed kernels take half of each function, that is, α is 0.5. The range of the polynomial degree d is $[0,4]$ .Therefore, this paper selects the parameters of support vector machine C=50, σ =2， ε=0.001.

Therefore, the forecast flow is as follows:

1） Input historical load and historical weather date data;
2） K-means clustering method for load clustering;
3） Get 4 kinds of loads;
4） Query the information of the days to be forecasted and find out the similar load of the days to be forecasted.
5） Load forecasting of SVM with similar date load input;
6） Output prediction results.

The specific flow chart is shown in the following figure:

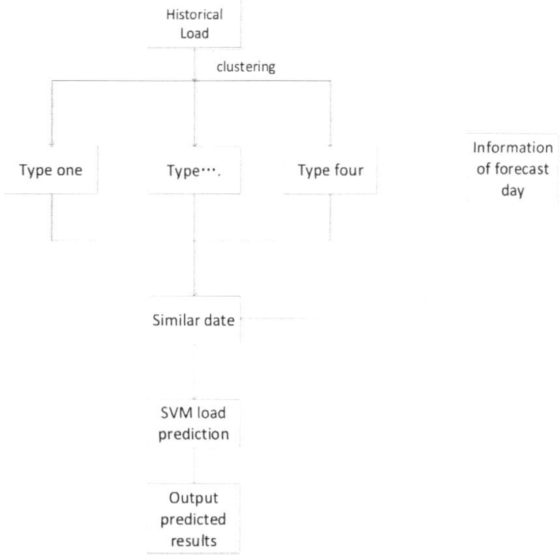

Figure 1. Prediction flow chart

4 Analysis of example

In this paper, the interruptible load of an enterprise in Yangzhou is forecasted by the above methods, and the forecasting results are as follows:

4.1 The analysis of clustering results

Firstly, the daily load data of interruptible load in December 2017 are clustered. Because of the similar temperature in December, the load is divided into four categories according to whether it rains or not, that is, sunny day holiday, rainy day holiday. Sunny working day and rainy working day, get the result below.

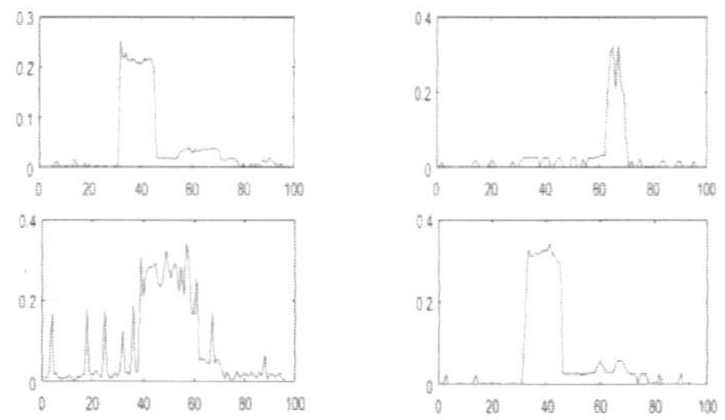

Figure 2. Cluster center curve

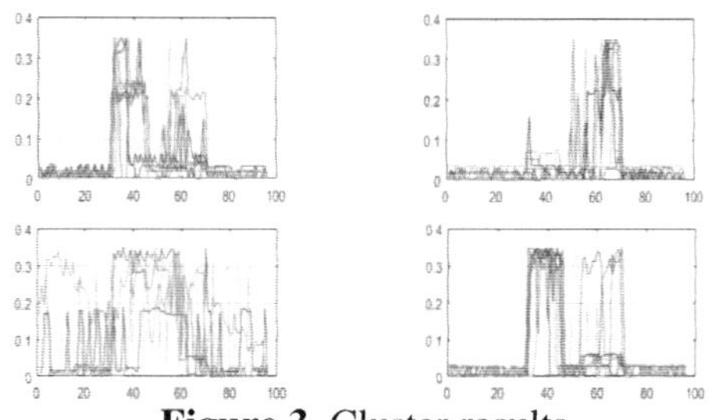

Figure 3. Cluster results

After clustering the load curves with k-means method, all the 31 samples to be clustered are divided into their own categories, and none of them is left out. And basically, the curve in each class has a high similarity with the original cluster center.

4.2 The analysis of prediction results

To forecast the daily load curve of interruptible load on December 31, the basic information of 31 December is searched, the corresponding category is found, and the forecast is carried out with the input data. The forecast results are as follows:

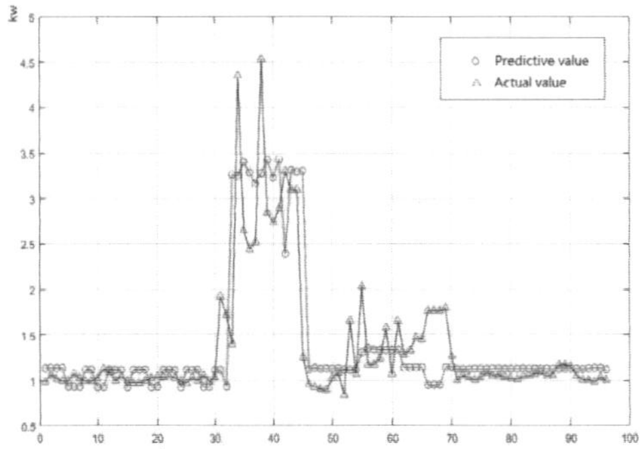

Figure 4. Comparison of forecasted and actual loads as at 31 December

Table 1. Prediction results of 31 December

Times	Actual load (MW)	Forecast load (MW)	Relative error
0:00	1.1422	1.0818	0.053
2:00	1.1189	1.0754	0.039
4:00	0.9267	0.9432	0.018
6:00	1.1131	1.0945	0.016
8;00	3.4834	4.2538	0.249
10:00	3.2275	2.89630	0.103
12:00	1.1237	1.0559	0.060
14:00	1.3532	1.3967	0.032
16:00	1.1429	1.1841	0.036
18:00	1.1237	1.1423	0.016
20:00	1.1263	1.1202	0.005
22:00	1.1346	1.1235	0.009
24:00	1.1189	1.1156	0.003

As can be seen from the results in figure 4 and table 1, the predicted value and the actual value almost coincide in a curve, except that there are differences at a few points. The big difference between the forecast load and the actual load at 8 o'clock may be due to the adjustment of the enterprise's production plan at the early peak. In general, the prediction accuracy of this method is high, up to 95%.

5. Conclusion

There are many factors affecting interruptible load forecasting, and the relationship between them is complex, and the correlation among them is uncertain. Support vector machine (SVM) algorithm can be widely used to establish a nonlinear mathematical relationship between various complex factors

and loads. However, the traditional support vector machine (SVM) method usually chooses a period of date close to the forecast date as a training sample, which can not fully reflect the information characteristics of the forecast day, and the prediction error is large.
Clustering analysis can solve the problem of sample selection well by clustering the similar samples into one group according to the similarity of each sample. Therefore, this paper proposes an interruptible load forecasting method based on clustering algorithm of support vector machine, that is, after clustering, the load similar to the information characteristics of the forecast day to be forecasted is found as a forecasting sample, and then the support vector machine method is used to forecast the load. This method provides a new idea for the selection of forecasting model samples and overcomes the problem that the selection of training date based on traditional support vector machine (SVM) method can not reflect the characteristics of the information of the forecast day. From the forecasting results, it can be seen that the method is accurate and reliable, and it is feasible to be used in interruptible load forecasting.

References

[1] T. Zhang, J.H. Song，X.L. Cheng, "*A Summary of Interruptible Load Study*". NEP **6**，46(2007)
[2] C.Q. Kang, Q. Xia, B.M. Zhang, "*Summary of Power system load forecasting Research and discussion on its Development Direction*". AUTOMAT ELECTRON POWER SYS **28**, 1(2008).
[3] D.F. Zhao, M. WANG, J.S Zhang, "*Short-term load forecasting based on support Vector Machine*"， PRON CHIIN SOC ELECTRICAL ENG **4**, 26(2002).
[4] Y.C. Li，T.J. Fang，E.K. Yu, "Research on support Vector Machine for Short-term load forecasting",（PRON CHIIN SOC ELECTRICAL ENG, 2003）.
[5] J. Wang, S.T. Wang, Z.H. Deng, "*Survey on challenges in clustering analysis research*", CONTROL DECIS,**27**,321(2012).
[6] Q. Wang, C. Wang, Z.Y. Feng, "*Review of K-means clustering algorithm*", EDE **20**，21（2012）.
[7] L.J. Cheng, C. Jing, Y.X Wu," Research *on normalization method of mathematical expression*". J. Zhejiang Univ-TECHNO. 40, 229(2012).
[8] T.B. Zhu, J. Fu, Y.F. Yang, "*Cluster analysis of load characteristic based on electricity consumption information acquisition system*", EM & I **53**, 70 (2016).
[9] V.Vapnik ，S. Golowich，A. Smola." *Support vector method for function approximation*", (MIT，1997).
[10] A.J. Smola, "*Regression estimation with support vector learning ma-chines",* (TS,1996).
[11] S.K. Shevade，S.S. Keerthi，C. Bhattacharyy," *Improvements to SMO algorithm for SVM regression*", IEEE Trans **5**, 1188(2000).
[12] L. Zhang, X. S Liu, H.J. Yin, "*Improvements to SMO algorithm for SVM regression*". IEEE Trans **5**, 566(2006).
[13] G.H. Liu, X. Yang, "*Application of support vector machine based on time sequence in power system load forecasting*" SA **11**, 19(2017).
[14] L. Li , Y.L. Lu, M.M Zhou , "*A New Fuzzy Support Vector Machine Based on Kernel Method*", POWER SYS TECHNO 28, 38(2004).
[15] X.D. Niu, Z.H. Gu, M. Xing, "*Study on Forecasting Approach to Short-term Load of SVM Based on Data Mining* ", PRO CSEE **26**, 6(2006).

IOP Conf. Series: Materials Science and Engineering **533** (2019) 012019 doi:10.1088/1757-899X/533/1/012019

Research on Position Sensorless Control of PMSM Based on Improved SMO

Zhang Guoqing[1,2,3], Xiong Xinhong[3], Zheng Yousheng[3], Zhang Yanhui[1,2], Chen Lianbo[1,2,3], Wu Xinyu[1,2,*] and Feng Wei[1, 2,*]

[1]Guangdong Provincial Key Lab of Robotics and Intelligent System, Shenzhen Institutes of Advanced Technology, Chinese Academy of Sciences, Shenzhen 518055, China
[2]CAS Key Laboratory of Human-Machine-Intelligence Synergic Systems, Shenzhen Institutes of Advanced Technology, Shenzhen 518055, China
[3]School of Logistics Engineering, Wuhan University of Technology, Wuhan 430070, China

Feng Wei: wei.feng@siat.ac.cn; Wu Xinyu: xy.wu@siat.ac.cn

Abstract. The sliding mode observer (SMO) can effectively achieve the permanent magnet synchronous motor (PMSM) to perform position sensorless control, but there are still drawbacks of system chattering and high complexity. In this paper, an improved SMO is established based on hyperbolic tangent function and Kalman filter. The system chattering is effectively suppressed, and the accuracy of position and speed observation error is significantly improved without phase compensation. It is found that the convergence rate of the hyperbolic tangent function has a strong correlation with the system chattering degree and the speed observation error. When a is between 3 and 4, the speed error accuracy of the improved SMO is the optimum situation.

1. Introduction

Permanent magnet synchronous motor (PMSM) are widely used in industrial robots, numerical control machine tools and new energy vehicles as their high inertia ratio, high efficiency and high power density [1]. Conventional PMSM provide feedback signal to the control system with position sensor such as a photoelectric encoder, which not only increases the amount and weight of the servo motor, but also increases the cost. The most severe situation is that the position sensor will be affected by the working conditions and thus reduce its reliability. By contrast, sliding mode observer (SMO) can effectively solve the above mentioned issues [2].

The SMO has good robustness and fast dynamic response [3-4]. However, the conventional SMO can easily cause system chattering due to the discontinuity of the Signum function. In recent years, it is a research hotspot to suppress the chattering of PMSM system based on SMO. Kim et al. [5] used the Sigmoid function instead of the Signum function. Their method could effectively suppress chattering, while still adopted the low pass filter to process the estimated back EMF, which requires phase compensation and increases system complexity. Hou et al. [6] constructed a cascaded SMO to effectively solve the phase lag problem of the low pass filter, but still increased the complexity of the system. Based on the Sigmoid function, Tang et al. [7] used Kalman filter to replace the low pass filter,

Content from this work may be used under the terms of the Creative Commons Attribution 3.0 licence. Any further distribution of this work must maintain attribution to the author(s) and the title of the work, journal citation and DOI.
Published under licence by IOP Publishing Ltd

which can avoid the phase lag and filter out the noise signal, and obtaining the observation curve with better tracking performance. The continuous function that approximates the Signum function is an effective method to suppress system chattering. It is important to explore the influence of the convergence rate of the continuous function on the system chattering and observation error. However, researchers focused on the selection of sliding mode control function and signal processing. Few people have done research in this area. In order to enhance the observation performance of the SMO, we use the hyperbolic tangent function as the control function of the SMO, replacing the traditional low pass filter with the Kalman filter. Based on improved SMO, the convergence coefficient a of the hyperbolic tangent function is changed, and the influence of a value on the PMSM speed estimation value and system chattering is analyzed.

2. Conventional SMO

The mathematical model of the PMSM is

$$\begin{cases} \frac{di_\alpha}{dt} = -\frac{R_s}{L_s} i_\alpha + \frac{1}{L_s} u_a - \frac{1}{L_s} e_\alpha \\ \frac{di_\beta}{dt} = -\frac{R_s}{L_s} i_\beta + \frac{1}{L_s} u_\beta - \frac{1}{L_s} e_\beta \\ e_\alpha = -\psi_f \omega_e \sin\theta_e \\ e_\beta = \psi_f \omega_e \cos\theta_e \end{cases} \tag{1}$$

Where i_α, i_β are the stator current components on the α and β axes; u_α and u_β are the stator voltage components on the α and β axes; e_α and e_β are the back electromotive force components on the α and β axes; R is the stator resistance. L is the stator inductance, ψ_f is the permanent magnet flux linkage, ω_e is the electrical angular velocity, and θ_e is the rotor electrical angle. According to Equation 1, the mathematical model of the conventional SMO can be constructed as:

$$\begin{cases} \frac{d\hat{i}_\alpha}{dt} = -\frac{R_s}{L_s} \hat{i}_\alpha + \frac{1}{L_s} u_a - \frac{k}{L_s} sign(\hat{i}_\alpha - i_\alpha) \\ \frac{d\hat{i}_\beta}{dt} = -\frac{R_s}{L_s} \hat{i}_\beta + \frac{1}{L_s} u_\beta - \frac{k}{L_s} sign(\hat{i}_\beta - i_\beta) \end{cases} \tag{2}$$

Where $\hat{i}_\alpha$ is the current component estimate of the α axis in the stationary coordinate system; $\hat{i}_\beta$ is the estimated value of the current component of the β axis in the stationary coordinate system; k is the sliding mode gain coefficient.

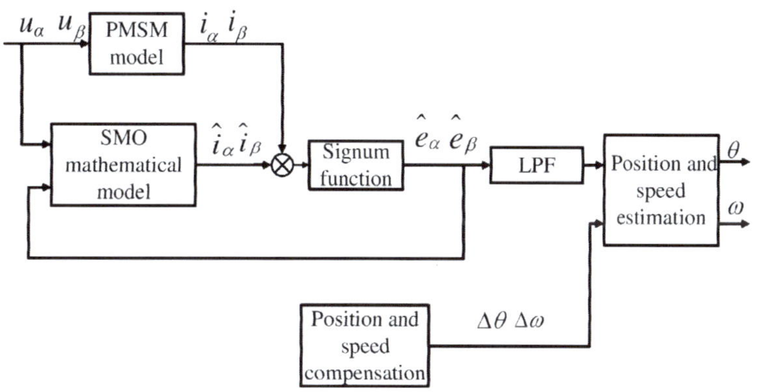

Figure 1. Conventional SMO model

The structure of the conventional SMO is shown in Figure 1. The difference between the estimated current of the SMO and the actual current of the PMSM is used as the sliding surface. Signum function is used as a sliding model control function, and then low pass filter is used to process the estimated back-EMF. The arctangent function is used to calculate the rotor position and motor speed of the PMSM based on the back-EMF. The formula is as follows:

$$\begin{cases} \hat{\theta}_e = \arctan(-\dfrac{\hat{e}_\alpha}{\hat{e}_\beta}) \\ \omega_e = \dfrac{\sqrt{(\hat{e}_\alpha)+(\hat{e}_\beta)}}{\psi_f} \end{cases} \tag{3}$$

The low pass filter can suppress chattering. However, the effect is limited, and the low pass filter will result in phase lag, so phase compensation is required, which increases the complexity of the system. Therefore, an improved SMO is built, and the influence of the convergence rate of the hyperbolic tangent function on chattering is further studied based on this model.

3. SMO Based on Hyperbolic Tangent Function and Kalman Filter (Improved SMO)

3.1. The Design of Improved SMO

The formula for the hyperbolic tangent function is

$$\tanh(ax) = \frac{e^{ax} - e^{-ax}}{e^{ax} + e^{-ax}} \tag{4}$$

Constructing a mathematical model of SMO based on hyperbolic tangent function：

$$\begin{cases} \dfrac{d\hat{i}_\alpha}{dt} = -\dfrac{R_s}{L_s}\hat{i}_\alpha + \dfrac{1}{L_s}u_\alpha - \dfrac{k}{L_s}\tanh(\hat{i}_\alpha - i_\alpha) \\ \dfrac{d\hat{i}_\beta}{dt} = -\dfrac{R_s}{L_s}\hat{i}_\beta + \dfrac{1}{L_s}u_\beta - \dfrac{k}{L_s}\tanh(\hat{i}_\beta - i_\beta) \end{cases} \tag{5}$$

The difference between the estimated current and the actual current is taken as the sliding surface.

$$s(x) = \begin{bmatrix} \hat{i}_\alpha - i_\alpha \\ \hat{i}_\beta - i_\beta \end{bmatrix} \tag{6}$$

According to the stability principle of Lyapunov, the stability condition of the improved SMO as follows:

$$\begin{bmatrix} (\hat{i}_\alpha - i_\alpha) & (\hat{i}_\beta - i_\beta) \end{bmatrix} \begin{bmatrix} \dfrac{d(\hat{i}_\alpha - i_\alpha)}{dt} \\ \dfrac{d(\hat{i}_\beta - i_\beta)}{dt} \end{bmatrix} \le 0 \tag{7}$$

If $k\tanh(\hat{i}_\alpha - i_\alpha) > e_\alpha$, the requirements can be met to achieve stability.

The model of improved SMO is shown in Figure 2. Since the low pass filter is removed, the system no longer needs phase compensation. Compared to Figure 1, the SMO structure is simplified.

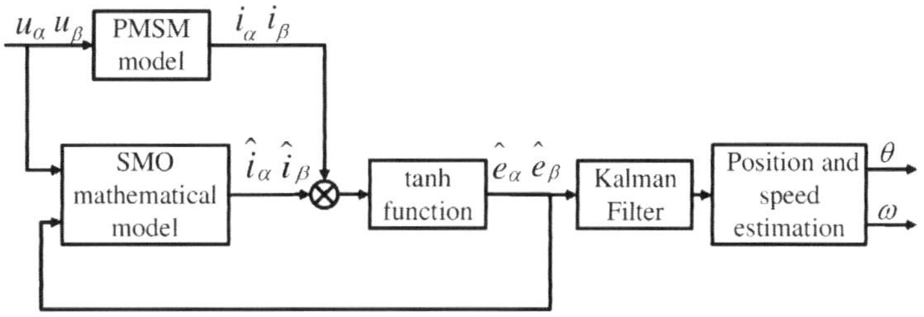

Figure 2. Improved SMO model

The i_d=0 control is used to build the simulation model. The PI regulator is used as the speed controller and current controller. The improved SMO speed observations are fed back to the system inputs. The position sensorless control system structure diagram of PMSM based on the improved SMO as shown in Figure 3.

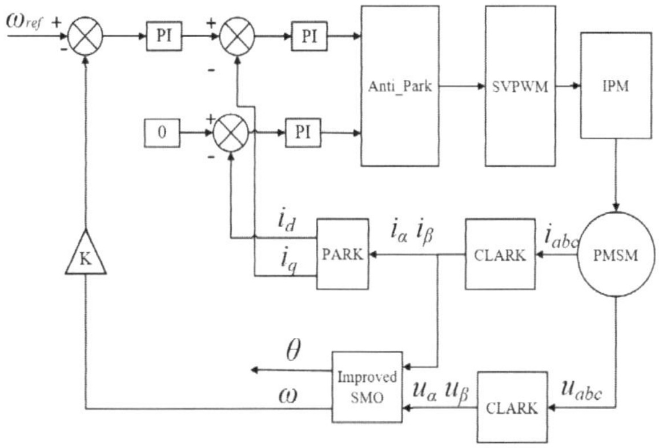

Figure 3. The position sensorless control system of PMSM based Improved SMO model

3.2. The Analysis of Simulation Results

Figure 4 and Figure 5 show the estimated, actual and estimated errors of the rotor speed of the PMSM under the conventional SMO and the improved SMO, respectively. The hyperbolic tangent function convergence coefficient is 1 (a=1), given a rotational speed of 1000r/min with no-load start. It can be observed that the improved SMO effectively suppresses the system chattering. The system provides a load torque of 5 N•m at 0.03s. Both the conventional SMO and the improved SMO show good anti-interference capability. And yet, the observation accuracy of the improved one is greatly enhanced. and its observation error is stabilized at 0.17 r / min.

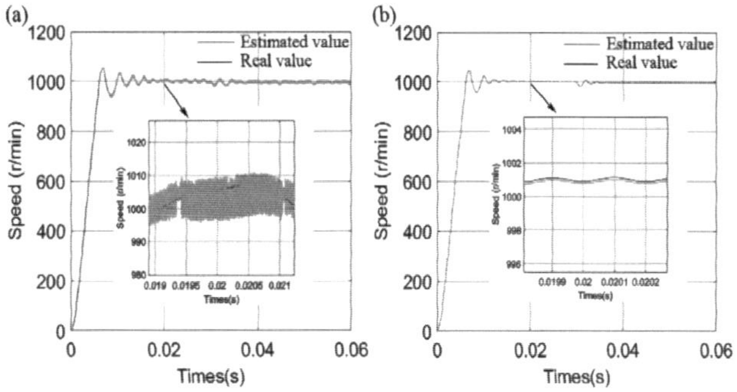

Figure 4. Estimation and actual speed (a) Conventional SMO; (b) Improved SMO

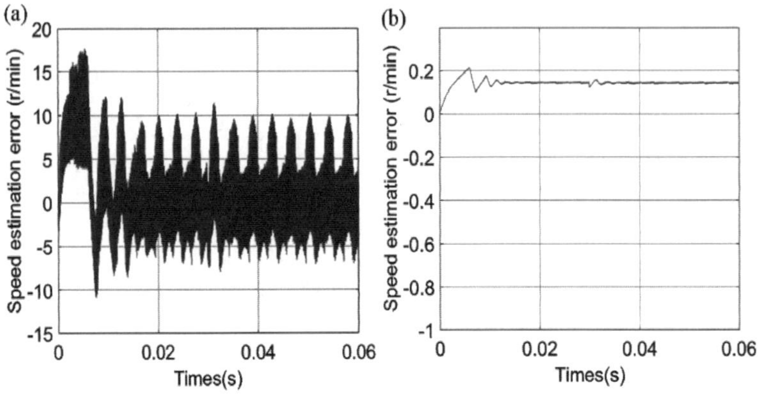

Figure 5. Speed estimation error (a) Conventional SMO; (b) Improved SMO

The speed response can be seen as a time response curve, so it can be evaluated by the time response dynamic performance indicator. It includes rise time (t_r), peak time (t_p), maximum overshoot (σ_p), adjustment time (t_s) and oscillation number (N). These can be calculated from Figure 4. The above indicators are listed in Table 1.

The t_r and t_p of the conventional SMO are nearly equal to the improved one, which means that the two systems have the same rapidity in the startup phase. The σ_p of improved SMO is significantly smaller than the conventional one, and the oscillation number is less, which means that the system runs more smoothly. t_s can reflect the overall rapidity of the system, so there is a obvious advantage in improved SMO.

Table 1. Speed response dynamic performance indicator

	t_r(s)	t_p(s)	t_s(s)	σ_p(%)	N
Conventional SMO	0.00612	0.00678	0.012	5.4	2
Improved SMO	0.00611	0.00667	0.009	4.5	1

Figure 6 illustrates the estimated value and actual value of the rotor position of the PMSM under the conventional and improved SMO. There is a significant position estimation error when conventional SMO is adopted, but the improved SMO has a good position tracking effect. The estimated position curve and the actual position curve are completely coincident, the chattering is weakened, and the curve is smoother. At the moment of motor starting, the conventional SMO has large observation errors and transient severe jitter (indicated by the arrows 1), but the improved one is a good way to avoid this.

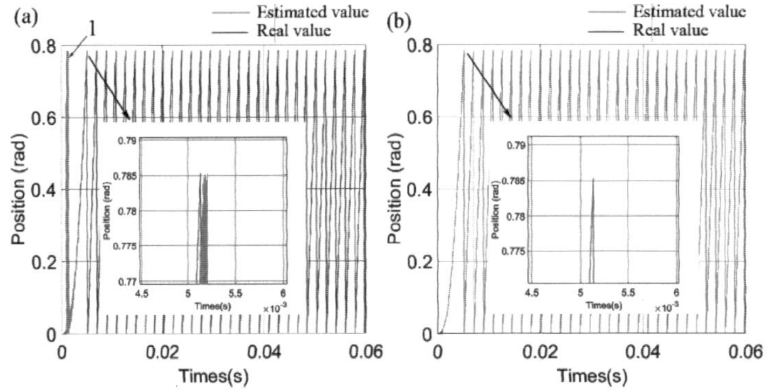

Figure 6. Estimated and actual rotor position (a) Conventional SMO; (b) Improved SMO

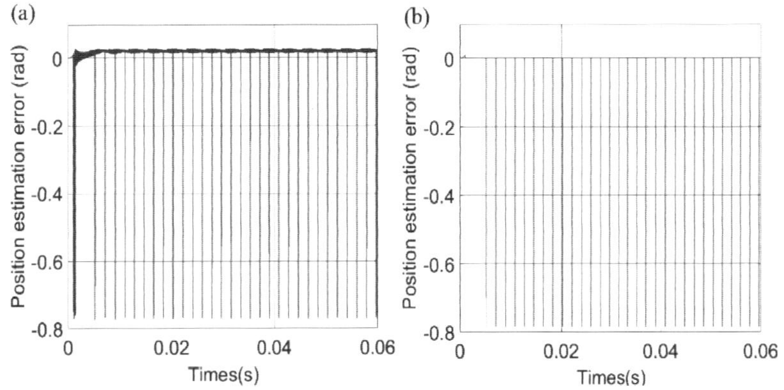

Figure 7. Position estimation error (a) Conventional SMO; (b) Improved SMO

It can be seen from Figure 7 that the position observation error of the improved SMO is 0.000075 rad, and the observation error of the conventional one is 0.025 rad. The improved SMO significantly improves the position observation accuracy. Similar to Figure 6, it can be obtained from the position estimation error curve that the improved SMO eliminates the position estimation error at the starting moment.

PMSM is controlled by three-phase current. Figure 8 shows the three-phase current under different control systems. Compared with conventional SMO, the improved SMO three-phase current curve converges faster, and the curve is smoother, which is closer to the ideal sinusoidal current.

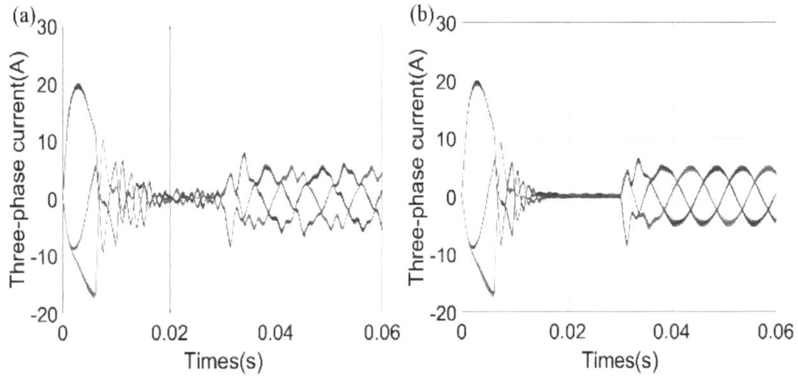

Figure 8. Three-phase current (a) Conventional SMO; (b) Improved SMO

4. Influence of the Convergence Rate of Hyperbolic Tangent Function on Observation Performance

The convergence rate of the hyperbolic tangent function is affected by the factor a. The larger a is, the closer the hyperbolic tangent function is to the Signum function. Figure 9 is the function image of the hyperbolic tangent function.

The SMO established in Section 3 of this paper has a good observation effect. The speed error curve of a=1, 2, 3, 4 is obtained based on this model, as shown in Figure 10. As the a increases, the jitter of the speed error curve is more severe, which means that the chattering of the system is more serious. During the process of a from 1 to 3, the degree of chattering is positively correlated with the value of a, and the observation error is negatively correlated with the value of a. When a=4, the chattering is still increasing, but the speed estimation error increases rather than continues to decrease. In the PMSM control system under the improved SMO, when a=3~4, the speed observation error has the minimum value.

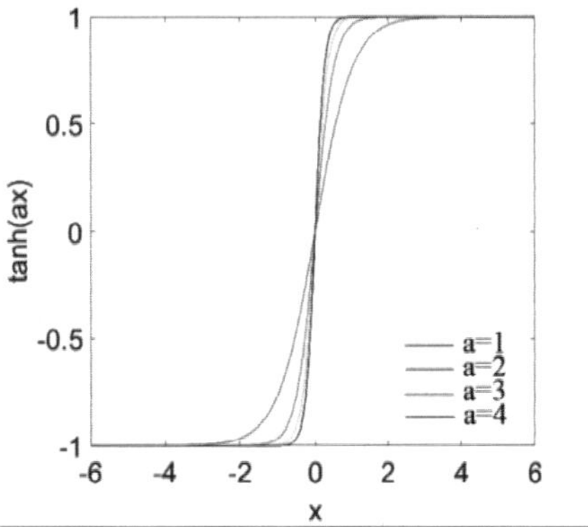

Figure 9. Hyperbolic tangent function image

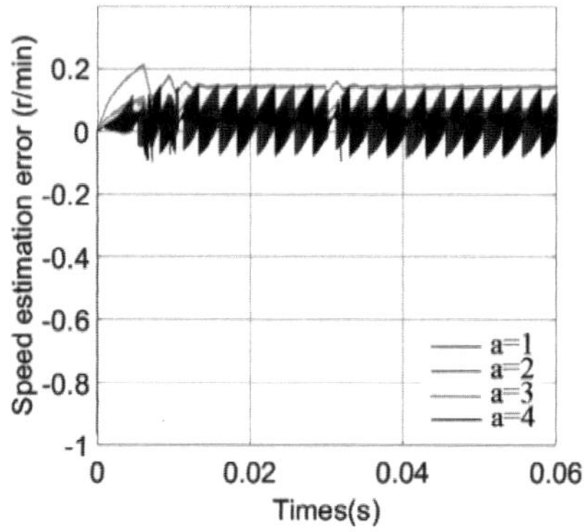

Figure 10. Speed Estimation error

5. Conclusion

In this paper, the hyperbolic tangent function is used as the control function, an improved SMO is established with Kalman filter, which effectively suppresses chattering without phase compensation. The speed observation error of the PMSM based on the improved SMO is 0.17r/min, which has a very high observation accuracy. In terms of position observation, the instantaneous and severe chattering at the start moment is effectively eliminated, and the position observation accuracy at the startup instant is improved. In general, the improved SMO has greatly improved the parameter identification ability, effectively improving the control performance of PMSM. Based on the improved SMO, the influence of the convergence rate of the hyperbolic tangent function on the speed observation error and chattering is studied. The simulation results indicate a strong regularity. This paper only studies under simulation conditions, and the experimental research will be in the future work.

Acknowledgements

This work is financially supported by Guangdong frontier and key technological innovation (NO.2017B090910013), Shenzhen knowledge innovation technology basic research project (JCYJ20170818153048647), national natural science foundation of Guangdong (No.2016A030313177), Shenzhen technology project (JSGG20160301160759264, JSGG20171823140127645).

References

[1] Kumar J, Kumar P S and Rambabu M 2012 *International Conference on Energy. IEEE.* 1-6.
[2] Itoh J I, Nomura N, Ohsawa H 2002 *Power Conversion Conference. IEEE.* **3** 1310-1315
[3] Bernardes T, Montagner V F, Grundling H A and Pinheiro H 2013 *IEEE Transactions on Industrial Electronics.* **61** 1679-1691.
[4] Fakham H, Djemai M, Busawon K 2008 *IET Electric Power Applications.* **2** 353-361.
[5] Kim H, Son J and Lee J 2011 *IEEE Transactions on Industrial Electronics.* **58** 4069-4077.
[6] Hou L M, Wang H Z, Li Y and Zhang H G 2016 *Control & Decision.* **31** 2071-2076.
[7] Tang H, Li H, Lin J 2018 *IEEE Computer Society.*290-293.

Research on torque ripple suppression of BLDCM method based on Cuk converter

TONG Jun[1],LI Xiang[1,a],LI Wenquan[1],LI Facheng[1] and YANG Xingchen[1]

[1]College of Electrical and Control Engineering, Xi'an University of Science And Technology, Xi'an 710054, China

2292441262@qq.com

Abstract. Due to the different manufacturing process and control mode, the commutation torque ripple will occur in the operation of BLDCM(Brushless DC motor). In this paper, the problem of BLDCM torque ripple is analyzed theoretically. In order to suppress the torque ripple during commutation, a method of combining Cuk converter with power inverter circuit is proposed. By changing the bus voltage, the DC bus voltage can be reduced at low speed and increased at high speed, so that the torque ripple of BLDCM can be suppressed at full speed. The simulation results are verified by MATLAB/Simulink, and the simulation results show the effectiveness of the method.

Key words: BLDCM;Cuk converter;Commutation torque ripple suppression;MATLAB simulation

1. Introduction

BLDCM is widely used in many applications due to its high energy density, simple control method, large output torque and stable operation[1].While the motor inevitably generates torque ripple during manufacturing and control, which limits the application of BLDCM in high standard servo field[2,3]. Therefore, suppressing or eliminating torque ripple becomes a problem that must be solved.
In this paper, the combination of Cuk converter and power inverter circuit is used to change the DC bus voltage. By changing the bus voltage,the torque ripple suppression of BLDCM at full speed section is realized.

2. Analysis of working principle of BLDCM

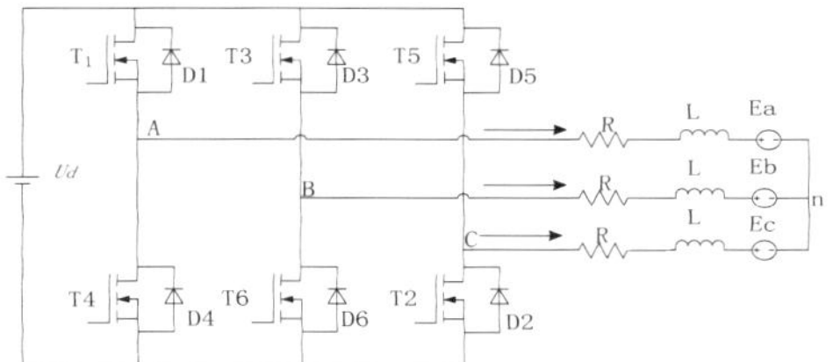

Figure 1. BLDCM circuit structure and driver circuit

Content from this work may be used under the terms of the Creative Commons Attribution 3.0 licence. Any further distribution of this work must maintain attribution to the author(s) and the title of the work, journal citation and DOI.
Published under licence by IOP Publishing Ltd

Figure 1 shows a three-phase Y-connected brushless DC motor and its drive circuit[4].The stator voltage equation, state equation and torque equation are as follows:

$$\begin{bmatrix} u_a \\ u_b \\ u_c \end{bmatrix} = \begin{bmatrix} R & 0 & 0 \\ 0 & R & 0 \\ 0 & 0 & R \end{bmatrix} \begin{bmatrix} i_a \\ i_b \\ i_c \end{bmatrix} + p \begin{bmatrix} L_{aa} & L_{ab} & L_{ac} \\ L_{ba} & L_{bb} & L_{bc} \\ L_{ca} & L_{cb} & L_{cc} \end{bmatrix} \begin{bmatrix} i_a \\ i_b \\ i_c \end{bmatrix} + \begin{bmatrix} e_a \\ e_b \\ e_c \end{bmatrix} \tag{1}$$

$L_{xy}(x, y = a,b,c)$ is the mutual inductance between three-phase windings, p is the differential operator. Ignoring the salient pole effect of the rotor, $L_{aa}=L_{bb}=L_{cc}=L$, $L_{xy}=M$, there is no neutral line and the excitation self-inductance second harmonic can be ignored.therefore: $ia+ib+ic=0$, $Mi_b + Mi_c = -Mi_a$.

$$\begin{bmatrix} u_a \\ u_b \\ u_c \end{bmatrix} = \begin{bmatrix} R & 0 & 0 \\ 0 & R & 0 \\ 0 & 0 & R \end{bmatrix} \begin{bmatrix} i_a \\ i_b \\ i_c \end{bmatrix} + p \begin{bmatrix} L_s & 0 & 0 \\ 0 & L_s & 0 \\ 0 & 0 & L_s \end{bmatrix} \begin{bmatrix} i_a \\ i_b \\ i_c \end{bmatrix} + \begin{bmatrix} e_a \\ e_b \\ e_c \end{bmatrix} \tag{2}$$

Where $L_s = L - M$.

The state equation :

$$p \begin{bmatrix} i_a \\ i_b \\ i_c \end{bmatrix} = \begin{bmatrix} 1/L_s & 0 & 0 \\ 0 & 1/L_s & 0 \\ 0 & 0 & 1/L_s \end{bmatrix} \cdot \left\{ \begin{bmatrix} u_a \\ u_b \\ u_c \end{bmatrix} - \begin{bmatrix} R & 0 & 0 \\ 0 & R & 0 \\ 0 & 0 & R \end{bmatrix} \begin{bmatrix} i_a \\ i_b \\ i_c \end{bmatrix} - \begin{bmatrix} e_a \\ e_b \\ e_c \end{bmatrix} \right\} \tag{3}$$

The back EMF(Electromotive force) waveform and current waveform of brushless DC motor under 120 degree conduction mode of three phase six-state are shown in Figure 2.

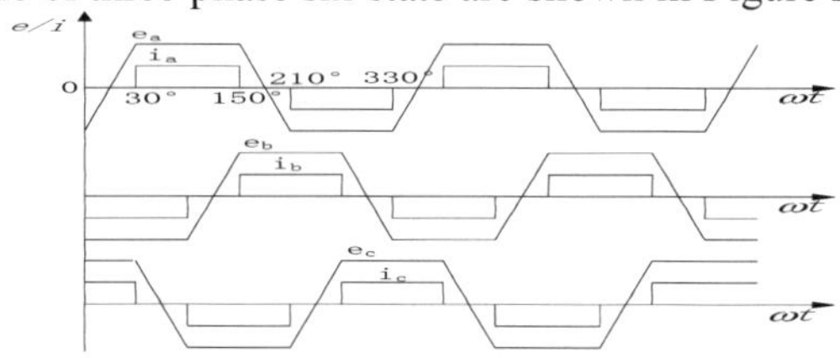

Figure 2. BLDCM back EMF and current waveform

The stator winding current interacts with the magnetic field generated by the rotor permanent magnet to generate electromagnetic torque.

$$T_e = \frac{e_a i_a + e_b i_b + e_c i_c}{\omega} \tag{4}$$

According to Figure 2, if phases A and B are turned on,

$e_a = -e_b = E$, $e_c = 0$, $i_a = -i_b = I$, $i_c = 0$,

$$T_e = \frac{2EI}{\omega} \tag{5}$$

3. Commutation torque ripple analysis of BLDCM

When the motor operated in the three-phase six-state 120° conduction mode, only two-phase windings are turned on at any time. Assume that phases A and C are turned on. $i_a + i_b + i_c = 0$, $E_a = E_b = -E_c = E_m$,the ideal electromagnetic torque of the motor is:

$$T_e = \frac{2E_m I}{\Omega} \tag{6}$$

When commutating,the motor converted from the two-phase windings of A and C to B and C.The current of A-phase winding is freewheeled by diode D4;the current of C-phase winding will gradually increase to a stable value.

Ignoring the influence of stator winding,

$$\begin{cases} L\frac{di_a}{dt} + Ri_a + E_a - \left(L\frac{di_c}{dt} + Ri_c + E_c\right) = 0 \\ L\frac{di_b}{dt} + Ri_b + E_b - \left(L\frac{di_c}{dt} + Ri_c + E_c\right) = U_d \end{cases} \tag{7}$$

Assuming that the values of the winding current of each phase are steady-state values, the current and electromagnetic torque of each phase winding are respectively:

$$\begin{cases} i_a = I - \frac{U_d + 2E_m}{3L} t \\ i_b = \frac{2(U_d - E_m)}{3L} t \\ i_c = -I - \frac{U_d - 4E_m}{3L} t \end{cases} \tag{8}$$

$$T_e = \frac{2E_m}{\Omega}\left(I + \frac{U_d - 4E_m}{3L} t\right) \tag{9}$$

From (9), t_{on} of the conduction phase winding and t_{off} of the off-phase winding during the commutation are respectively:

$$\begin{cases} t_{on} = \frac{3LI}{U_d + 2E_m} \\ t_{off} = \frac{3LI}{2(U_d - E_m)} \end{cases} \tag{10}$$

The key to suppressing the commutation torque ripple is to make the rising rate of the conduction phase current equals to the falling rate of the shutdown phase current[5-6],

$$U_d = 4E_m \tag{11}$$

It can be known that the commutation time in the commutation process is related to the back EMF, and the back EMF is proportional to the motor speed,the current variation law during BLDCM commutation at different speeds as shown in Figure 3:

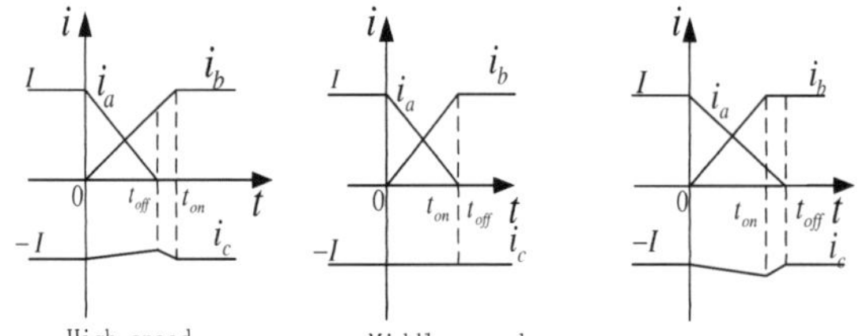

Figure 3. Variation of commutation current at different speeds

When $U_d < 4E_m$, when i_a has dropped to 0, i_b has not reached the steady-state value, as shown in high speed. The torque ripple can be obtained as:

$$T_e = \frac{2E_m I_0}{\Omega} \cdot \frac{U_d - 4E_m}{U_d + 2E_m} \tag{12}$$

When $U_d > 4E_m$,before i_a has dropped to 0, i_b has reached the steady-state value, as shown in middle speed. The torque ripple can be obtained as:

$$T_e = \frac{2E_m I_0}{\Omega} \cdot \frac{U_d - 4E_m}{U_d - E_m} \tag{13}$$

When $U_d = 4E_m$,when i_a falls to $0, i_b$ reaches a stable value,as shown in middle speed.The torque remains constant during the commutation process:

$$T_e = \frac{2E_m I_0}{\Omega} \tag{14}$$

It is the inconsistency of the change rate of the current between the turn-off phase and the conduction phase that causes the current fluctuation in the constant conduction phase and the fluctuation of the commutation electromagnetic torque. When commutation, the DC bus voltage should satisfy the relation $U_d = 4E_m$, and the back EMF is directly proportional to the motor speed.

4. Torque ripple suppression method based on Cuk converter

The topology of the Cuk circuit is shown in Figure 4.Control the power switching tube Q by PWM chopping.By adjusting the difference of the duty ratio D,the DC buck-boost can be realized[7].

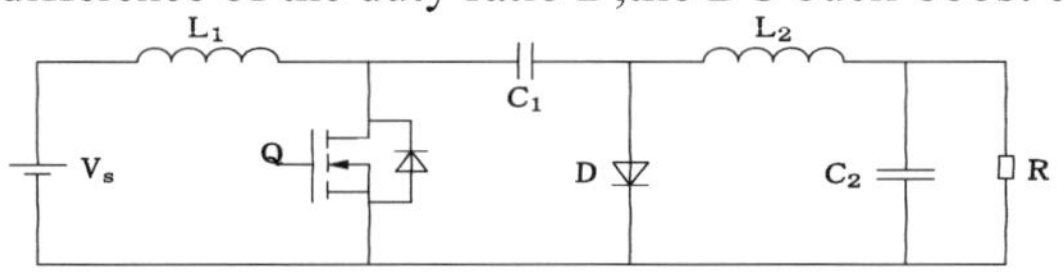

Figure 4. Cuk converter

When $T_{on} = DT_s$,the diode D is turned off. the DC power supply V_s charges the inductor L1 via the power switch tube Q.

$$V_{L1} = L_1 \frac{di_{L1}}{dt} = V_s \tag{15}$$

$$\Delta i_{L1+} = \frac{V_s}{L_1} DT_s \tag{16}$$

C_1 charges C_2 through the switch tube and discharge to the load and the inductor L2.

IOP Conf. Series: Materials Science and Engineering 533 (2019) 012020 doi:10.1088/1757-899X/533/1/012020

$$V_{L2} = L_2 \frac{di_{L2}}{dt} = V_{C1} - V_O \tag{17}$$

$$\Delta i_{L2+} = \frac{V_{C1} - V_O}{L_2} DT_s \tag{18}$$

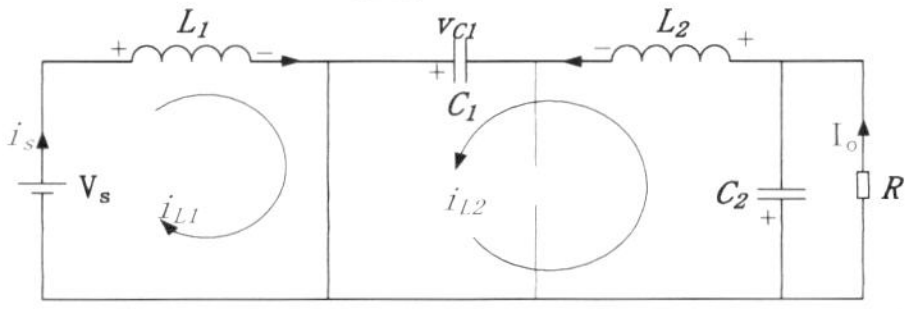

Figure 5. Switching tube Q conduction equivalent circuit

When $T_{off} = (1-D)T_s$, D is forward conducting.Since the inductor current cannot be abruptly changed,when the switch Q is turned off,L1 charges the C1 through D,

$$V_{L1} = L_1 \frac{di_{L1}}{dt} = V_s - V_{C1} \tag{19}$$

$$\Delta i_{L1-} = \frac{V_{C1} - V_s}{L_1} (1-D)T_s \tag{20}$$

At the same time,the current of L2 is supplied to the load through D,

$$V_{L2} = L_2 \frac{di_{L2}}{dt} = -V_O \tag{21}$$

$$\Delta i_{L2-} = \frac{V_O}{L_1} (1-D)T_s \tag{22}$$

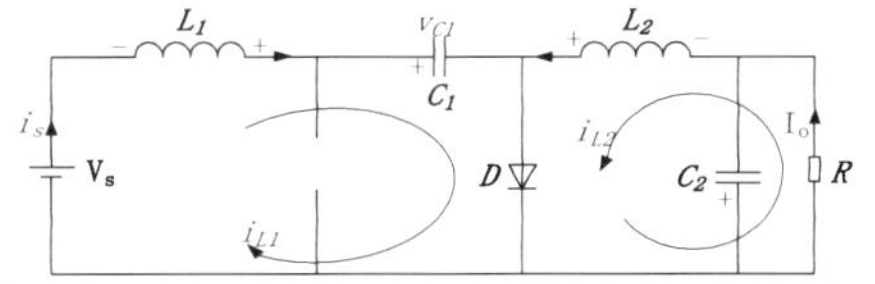

Figure 6. Switching tube Q cutoff equivalent circuit

When the circuit works in a stable operation,the increase of the inductor current in one switching cycle equals to the amount of reduction.

$\Delta i_{L1+} = \Delta i_{L1-}$, $\Delta i_{L2+} = \Delta i_{L2-}$.

$$V_{C1} = \frac{1}{1-D} V_s \tag{23}$$

$$V_O = DV_{C1} \tag{24}$$

$$V_O = \frac{D}{1-D} V_s \tag{25}$$

From (25), by adjusting the duty ratio , the output voltage V_o of the Cuk circuit can be controlled.And the motor back EMF is proportional to the speed:

$$E_m = k_e n$$

(26)

When the power switch Q is turned on,the relationship between duty cycle and speed is:

$$D=\frac{4k_e n}{V_s+4k_e n} \tag{27}$$

It can be seen that the torque ripple generated by the commutation can be suppressed by adjusting the duty ratio D of the Cuk pre-stage conversion circuit.

5. Simulation results and analysis

In order to verify the feasibility of the proposed method,a system simulation model was established on the MATLAB/Simulink simulation platform.The simulation model system is shown in Figure 7.

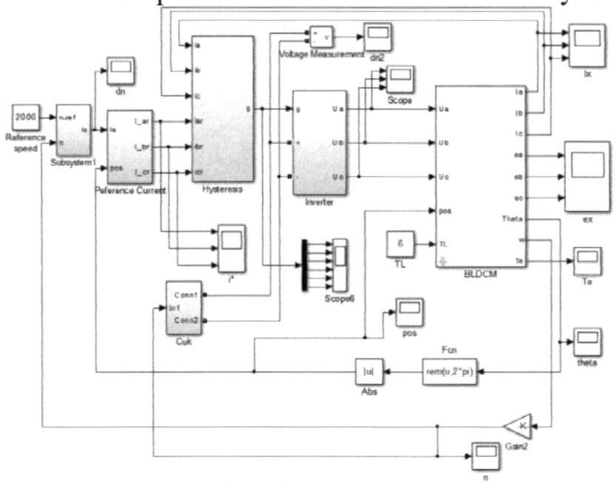

Figure 7. Simulation Model system

The motor load torque is $6N\cdot m$.When the motor speed is 2000r/min,the bus voltage is 48V. When the front stage conversion circuit is not compensated,the bus voltage waveform,the A phase current waveform and the output torque are shown in the following Figures.

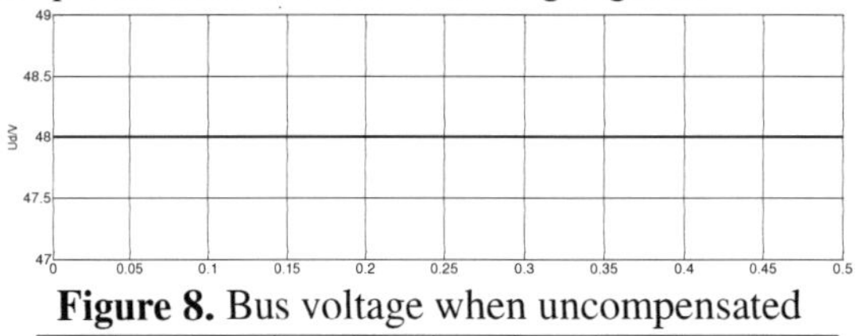

Figure 8. Bus voltage when uncompensated

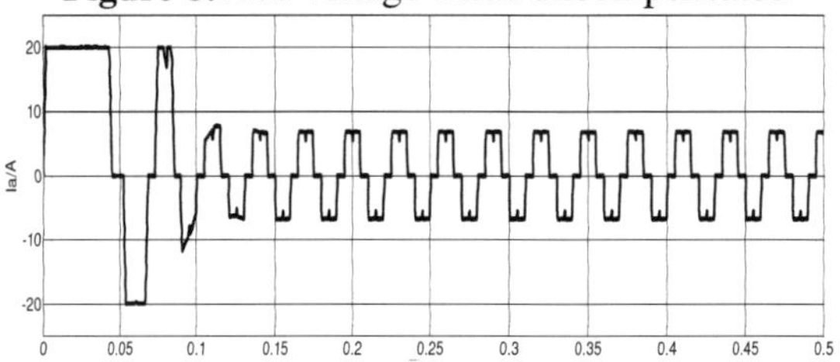

Figure 9. Phase A current waveform without compensation

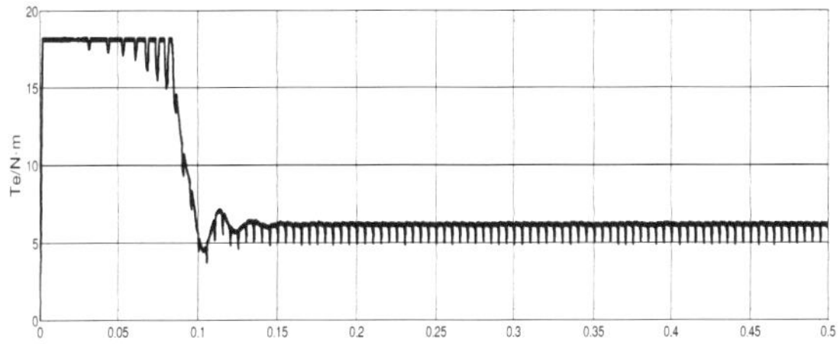

Figure 10. Torque waveform without compensation

After increasing the compensation , the bus voltage increased to 60V,

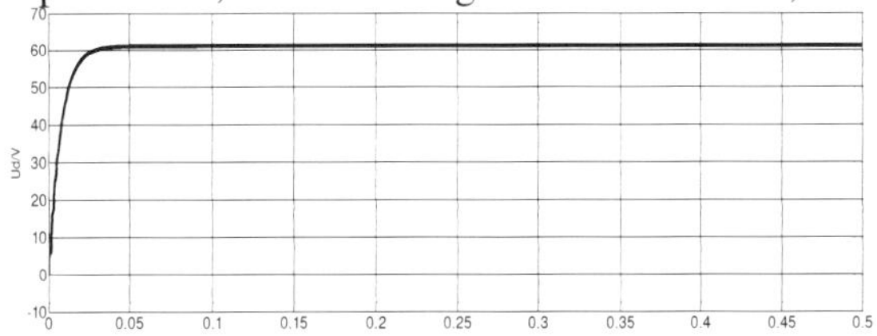

Figure 11. Bus voltage after Cuk conversion compensation

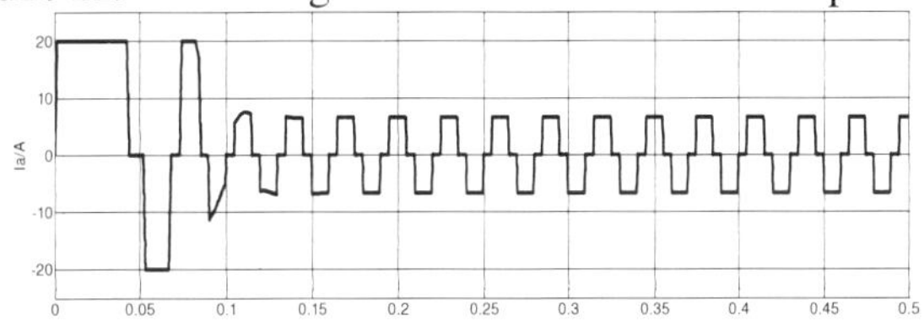

Figure 12. Phase A current waveform after compensation

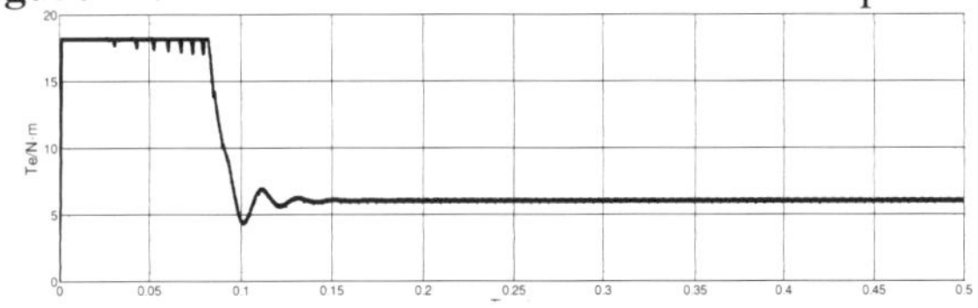

Figure 13. Compensated torque waveform

Comparing the above figures, it can be seen that the pulsation of the flat top portion of A-phase current is suppressed,and the corresponding output torque is smooth, the torque ripple is effectively suppressed.

6. Conclusion

In this paper,the mathematical model of BLDCM and the reason of commutation torque ripple are analyzed.A method combining Cuk converter and power inverter circuit is proposed.By changing the bus voltage,realize torque ripple suppression at full speed.The simulation is verified on the MATLAB platform and the results show the effectiveness of the proposed method.

References

[1] SHI Jian,LI Tiecai. A PWM Strategy to eliminate commutation torque ripple of brushless DC motors[J]. Proceeding of the CSEE,2012,32(24):110.

[2] WANG Dafang, BU Deming,ZHU Cheng, et al. A modulation method to decrease commutation torque ripple of brushless DC motors[J]. Transaction of China Electrotechnical Society,2014,29(25):160.

[3] Carlson R,Lajoie-Mazenc M,Fagunde J C D S. Analysis of torque ripple due to phase commutation in brushless dc machines [J].IEEE Transaction on Industrial Application,1992,28(3):632-638.

[4] SHI Tingna,GUO Yuntao,Peng Song,et al. A new approach of minimizing commutation torque ripple for brushless DC motor based on DC-DC converter[J].IEEE Transaction on Industrial Electronics,2010,57(10):3483.
[5] Ogasawara S and A H. An approach to position sensorless drive for brushless DC motors[J]. IEEE Transactions on Industry Applications, 1991,27(5):928-933.
[6] Matsui N, Takeshita T. A novel starting method of sensorless salient-pole brushless motor: Industry Applications Society Annual Meeting[C]. Denver, CO: IEEE, (1994).
[7] WANG Zhaoan,HUANG Jun.Power Electronic Technology[M].Beijing:Mechanical Industry Press,(2005).

IOP Conf. Series: Materials Science and Engineering **533** (2019) 012021 doi:10.1088/1757-899X/533/1/012021

Innovative Design of the Cooling Topologies for Electric Vehicle Motors

ZHANG ZI-CHAO[1, 2], SONG QIANG[1, a] and BILAL AHMED[1]

1 School of Mechanical Engineering, Beijing Institute of Technology, Beijing, China
2 College of Environmental and Energy Engineering, Beijing University of Technology, Beijing, China

songqiang@bit.edu.cn

Abstract. According to the different cooling medium, the electric motor cooling styles can be divided into liquid cooling and air cooling; the former involves water cooling and oil cooling while the latter comprises natural air cooling and forced air cooling. Whereas, the air cooling system is difficult to meet various performance requirements for the modern Electric Vehicle (EV). Therefore, approximately all EV motors are liquid cooled at present. Moreover with the trend of EV motors to Permanent Magnet (PM) and to high speed, for instance, the maximum working rotated speed from 10,000rpm to 20,000 rpm, the packaging of the motor is minimized at same power while efficiency drops, lifetime shorten and even cause demagnetization due to thermal effects without optimum design. Consequently, the improvement of the liquid cooling system is essential to enhance the efficiency of the EV motor in the future for higher power density. In this perspective, the innovative cooling topologies of electric machines based on the development results of the parts such as lubricating oil and motive seal of electric vehicles are presented in this paper. It solves some problems brought about by predecessors in view of the high efficiency and reliability of the motor with higher energy density.

1. Introduction

For sustainable and green future, it is crucial to reduce the use of an internal combustion engine from the transportation; electric motors ought to be used. Many researchers have been working on the cooling system of the EV motors in recent years. For the cooling of EV motors, investigators are using two basic methods: air cooling and liquid cooling. Air cooling further can be split into natural air cooling and forced air cooling. Since it's relatively low heat transfer coefficient, the electric motors of EV are mostly cooled by a liquid cooling system nowadays that consists of water cooling and oil cooling. Usually, water used in cooling jackets for indirect cooling and oil enter in the internal part of the motor to cool the hot spot directly. The development of the electric motor, particularly PM motor leads to several improvements, for example, the maximum rotated speed rises from 10,000rpm to 20,000 rpm, power density becomes high, mass is reduced, and diameter and axial length become smaller. However, heat generated from the electric motor is increased. Therefore the performance and the lifetimes of these machines are easily affected by the thermal condition [1], and the flow structure has a significant impact on the overall thermal performance of the cooling design [2]. Optimum design of the motor cooling system will surely play the vital role in improving efficiency, prolonging service

Content from this work may be used under the terms of the Creative Commons Attribution 3.0 licence. Any further distribution of this work must maintain attribution to the author(s) and the title of the work, journal citation and DOI.
Published under licence by IOP Publishing Ltd

life, and preventing demagnetization of the permanent magnet as well as the insulation of coils. Furthermore, thanks to the development of the motive seal, high speed bearing and new fluid for e-axle that the key attributes are durability at low viscosity, copper corrosion protection, excellent oxidative stability and thermal management, foam suppression and high speed operation, etc. in this way, we have immense possibilities and suitable conditions for the cooling topologies of the electric motor.

In 2014, Zhang Fengge proposed three cooling schemes combined with air cooling and water cooling for a 1.12MW PM motor[3]. Comparing and analyzing the temperature field, it is concluded that the temperature rise of the rotor is the key point in the design of the cooling system. In 2015, BAE Systems publishes a patent in which the stator is cooled by the oil from the rotated rotor with the cooling flange [4]. Its purpose is to effectively cool the motor and develop the motor performance. An IEEE paper mentions the evolution and cooling methods of electric motor [5]. Christian Paar concluded that specific thermal optimization of the electric motor considering its thermal environment and operating profile permits a better meeting of customer requirements such as packaging, weight, performance, and cost [6]. In the same year, Tesla Motor, Inc. reveals a liquid cooled motor based on pressure and gravity [7], which is characterized by the shape of the oil tray and the design of a drainage plate under the action of gravity to make the oil dispersed in contact with the expected components to avoid the blocking of the openings. It does not depend on the size and quantity of the openings commonly using in only pressure-based distribution. The difficulty is that the dispersal of gravity cannot guarantee the cooling effect of the lower part of the motor. So the lower part may overheat, which puts forward higher requirements for the thermal stability of the stator and rotor materials of the motor.

In 2016, a paper said that the high temperature of rotor produced by losses is one of the important issues for the high-speed electric motor, and it has great significance for the cooling design topologies to enhance the motor's efficiency [8]. Shanghai Dajun Technologies Inc. introduces a new type of oil cooling motor characterized by spraying the wire bag and iron core to overcome the defects of the water cooling motor and improve the power density [9]. Daniel publishes a patent of the liquid cooling motor with the motive seal to set an injector that moves relatively towards the stator winding [10]. The target is to eliminate the coil erosion and the adverse effects of the oil-air mixture on the performance of the oil pump after the spray of cooling fluid. LASSILA shows a liquid cooling motor through the eccentric setting of the injector nozzle and some former ways to make the relative motion between the oil injector and coil [11]. The goal is to avoid corrosion as well as local overheating. In 2017, Zhang Shixiang introduced a utility model of an oil cooling motor [12], which has the slot in the rotor's circumferential direction. Some of the oil enter in the jacket slot to cool the rotor, and the others are injected from the jacket openings onto the stator winding. But the oil is difficult to enter the slot for the pressure of the air gap. The results from Pia Lindh confirm that the proposed direct cooling method is practical also in small machines. Moreover, it offers significant improvements in the thermal management of the machine [13]. BYD develop a utility model of the oil cooling motor [14]. Its characteristic is that the stator shell has a stator oil way and has a shell injection port towards the stator winding. The rotor has a rotor oil path connected with the stator oil way via the endbell oil way. The magnetic isolating plate has an oil channel in a radial direction towards the winding. Its purpose is to optimize the heat dissipation and improve the efficiency of the design. The shortcoming is that the cooling capacity still needs to be optimized and the shaft is likely to have a dynamic imbalance, leading to NVH problem. In 2018, a review on state-of-art modern cooling systems employed for thermal cooling of electric motors provided[15].In summary, the cooling of the electric motor needs to be improved to increase its efficiency by enhancing its cooling mechanism.

2. Analysis of typical patents

2.1. BYD oil cooling system

The utility model is an oil cooling motor. The housing has shell oil cooling channels which connect to external oil supply pipeline through the inlet and communicate with the external oil pipeline through the outlet. The rotor that is provided with the rotor injection ports facing to the stator has a shaft oil way connected with the shell oil channel in which the injection ports are facing to the stator.

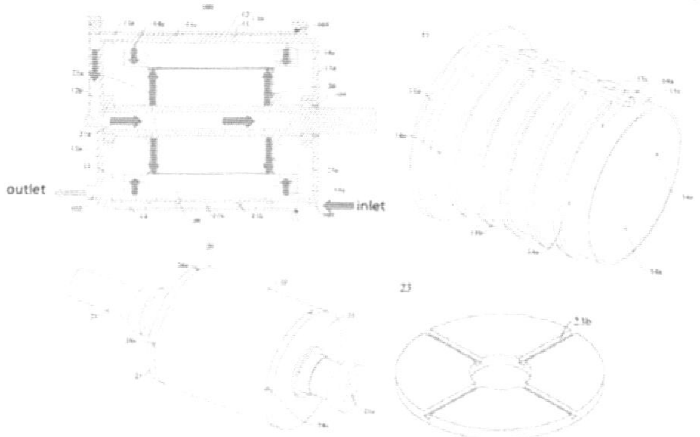

Figure 1. BYD oil cooling system

The cooling topologies can cool both rotor and stator, especially the end parts of the stator which is easy to be overheated. The stator core is cooled through cooling jackets. And direct inner oil cooling is also applied in terms of the stator winding via fixed-point injection of the cooling jacket and the rotating injection of the magnetic isolating plates, which means the stator winding is cooled by forced convection heat transfer with oil. As the rotor with the permanent magnets is sensitive to temperature, two ways are supported. One is the heat conduction to the cooling shaft; another is the heat transfer by cooling channels in the magnetic isolating plates.

However, the limited position of the injection openings and channels cause unsatisfactory heat transfer area. It leads to an inadequate cooling on the stator winding as well as on the end parts of the rotor. For further high-speed motors, the heat generated in the air gap tend to be even larger. This cooling contracture will not work very well. Besides, two isolating magnet plates are placed at both ends of the rotor under certain conditions of oil pressure and the motor speed. It is unavoidable that the flow rate of oil in the channel is not uniform, thereby generating vibration of the dynamic imbalance. So the NVH problem brought by the rotated injection rotor should also be taken into consideration.

2.2. Tesla liquid cooling system

The liquid cooling system includes a manifold is extending above the stator (S) which is receiving the liquid under pressure and having the openings that direct the liquid jets onto the stator and the end parts of the rotor (R). Meanwhile, the trays acquire fluid from the manifold injectors and distribute gravity-fed liquid onto the end parts.

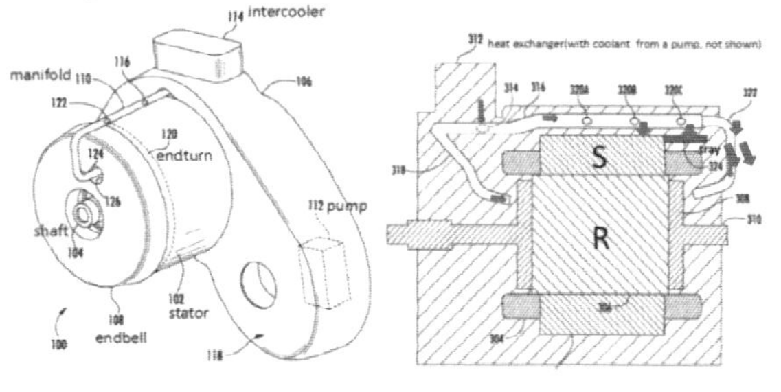

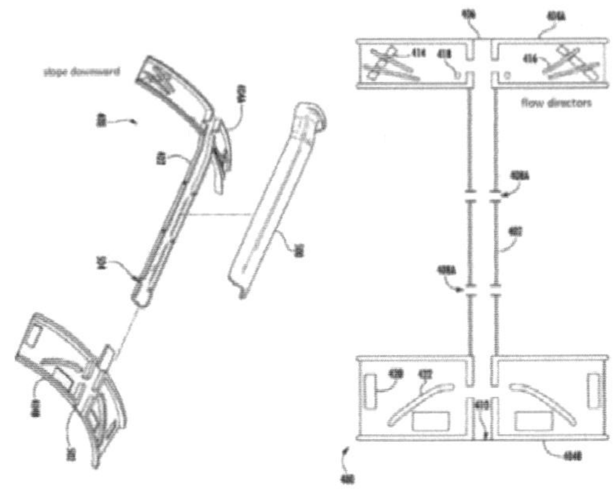

Figure 2. Tesla liquid cooling system

In this cooling topology, both rotor and stator are directly cooled by liquid. In detail, liquid manifold injection dissipates the heat of the stator core by means of the forced convection heat transfer with the coolant, and heat transfer via gravity-fed liquid distribution of the trays and the rotated rotor makes the stator winding cool. The rotor is cooled by heat transfer with the fluid via liquid manifold injection and gravity-fed liquid distribution of the trays. The cooling effect at the end parts of the rotor is relatively good, but the coolant cannot contact with the outer and inner surface of the rotor and thereby reduce the heat dissipation of the rotor. It causes insufficient cooling capacity under heavy load conditions. What is more critical is that despite the good cooling effect at the upper part of the stator winding due to the gravity-fed liquid distribution of the trays that slope downward with flow directors, a similar cooling effect cannot be obtained in the lower part of the stator. It will lead to disequilibrium temperature field then require higher thermal stability of silicon steel sheets, end up with shortening the motor service life.

Tesla believes that a system that uses only pressure-based distribution, with ten or more openings, the openings would need to be relatively small and therefore prone to clogging. However, if such openings were made larger, at low oil flow rates, the coolant may not fill the distribution channel (e.g., a manifold) and the coolant would not reach some parts of the motor. The gravity-fed liquid distribution has its advantages as compared to pressure-based distribution, while along with the development of the special fluid for e-axle, it will partly remove the concerns about the lower limit of the opening size. Consequently, to obtain the most adequate and balanced cooling scheme, various methods of the pressure-based distribution are still not negligible.

3. New cooling topologies

3.1 Brief description

As shown in fig.3, the fluid enter into the shaft cooling chamber from the shaft oil supply chamber formed by the inlet dynamic sealing device placed between the housing inlet and the shaft inlet. Then, the oil enters into the air gap cooling cavity or the stator winding rotary cooler low-pressure cavity to do the next cooling process. Optimum design of cooler and cooling chamber can effectively cool the motor and further enhances the service life of the driving motor.

IOP Conf. Series: Materials Science and Engineering 533 (2019) 012021 doi:10.1088/1757-899X/533/1/012021

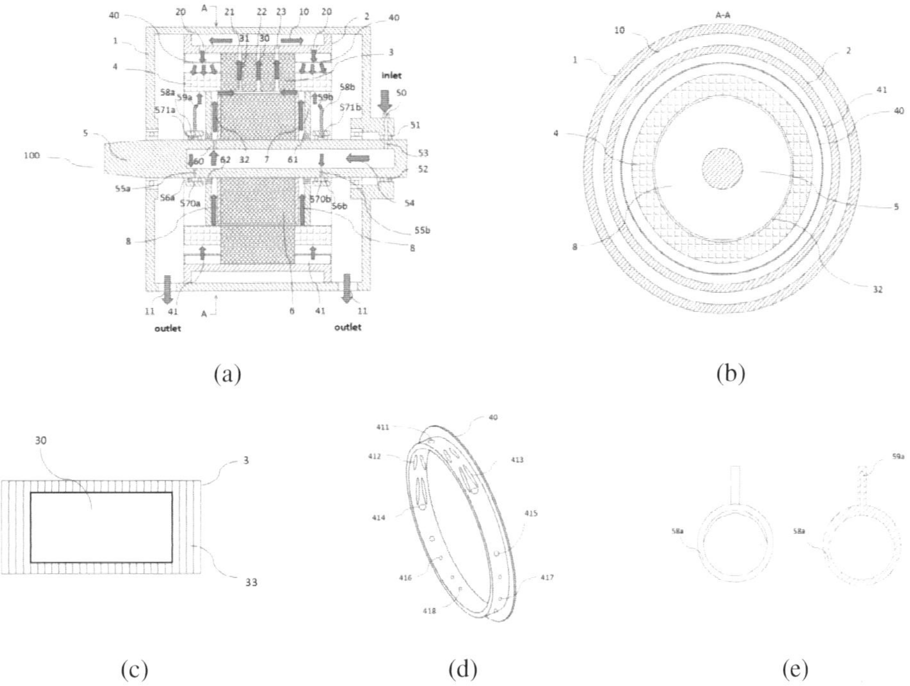

Figure 3. innovative liquid cooling system

Fig.3, (a) provides a schematic diagram of the cooling system. (b) is a schematic diagram of the sectional structure. (c) is the local enlargement structure of the stator core. (d) is a stator winding fixed cooler, (e) is the stator winding rotary cooler 58a and the profile structure of the stator winding rotary cooler 58a.

In fig.3 (a), 10 is the stator core cooling cavity. 2 is the stator core cooling jacket. 20 is the jet of the cooling jacket. 21, 22, 23 are the inlets of the cooling jacket. 30 is the stator core oil channel. 31is the inlet of the stator core. 32 is the air gap cooling cavity. 33 is the silicon steel sheet. 4 is the stator winding. 40 is the stator winding fixed cooler. 41 is the cooler cavity of the stator winding fixed cooler. 411-418 are the stator winding cooler guide openings. 51 is the inlet dynamic sealing device. 52 is the shaft oil supply chamber. 53 is the shaft inlet. 54 is the shaft cooling chamber. 55a and 55b are the stators winding rotary cooler inlet. 56a and 56b are the stators winding rotary cooler dynamic sealing device. 570a, 570b is the stator winding rotary cooler low-pressure cavity. 571a, 571b is the stator winding rotary cooler high-pressure cavity. 58a, 58b are the stator winding rotary cooler. 59a, 59b are the fluid injection port. 60 is the air gap cooling cavity inlet. 61 is the partition bearing dynamic sealing device. 62 is the partition board bearing. 7 is the partition board. 8 is the magnetic isolating plate.

3.2 Cooling principle

The rotor provided in this creative topology takes the heat transfer method via air gap cooling cavity and the means of heat conduction with the shaft cooling chamber. Specifically, two partition boards are mounted on the shaft through the partition bearings integrated with moving seals, and the circumferential surface of which are sealed with the stator forming the air gap cooling cavity, which can enhance the cooling effect of the air gap as well as the end parts of the rotor. Finally, we can get little drag loss but the improvement of overall efficiency and better electromagnetic stability that is the priority for PM motor.In addition, the stator core is made of a plurality of silicon steel sheets, and the sectional shape of the stator core oil channel can be rectangular. The length of the stator core oil channel rectangular section is an integral multiple of the thickness of the silicon steel sheet, for example, it can be 10~20 times the silicon steel sheet thickness. By this approach, not only the cooling loop is optimized but also benefit to reduce torque ripple excitation thereby improving acoustic quality of the electric motor.

Under the action of the centrifugal force of the rotor, the fluid in the air gap cooling cavity can get into the stator core cooling jacket by the stator core oil channels. There are openings of the jacket onto the stator winding fixed coolers. Furthermore, the stator winding fixed cooler is invented by considering that the cooling effect is as equal as possible for the up and down circumference of the stator winding in order to prolong service life. In detailed, the stator winding fixed coolers, sealed with the jacket and stator, have guide openings of different design in the upper and lower circumference to play gravity-fed distribution and pressure based injection respectively, solving the problem of the inhomogeneous cooling intensity of the end parts of the rotor occurred in Tesla motor.

Moreover, in order to strengthen the heat transfer of the stator winding further, the stator winding rotary coolers are provided. The stator winding rotary coolers are dynamic seal with the shaft, forming the stator winding rotary cooler cavity. The design of the coolers make the oil inlet cavity large in volume and small pressure, and the oil outlet cavity has a small volume and high pressure. Some fluid in the shaft cooling chamber will enter into the stator winding rotary cooler low-pressure cavity under the action of centrifugal force and be injected from the stator winding rotary cooler high-pressure cavity. In this way, the cooling chamber or cavity will always be filled with fluid, thus avoiding the NVH problems in BYD motor.

Besides, all the cooler such as the stator winding rotary cooler and the air gap cooling cavity are easy for modularized development of the electric motor cooling system. It is the innovative design of the cooling topologies for EV electric machine that can maintain the efficiency in high speed and heavy load.

4. Conclusion

It is realized that for a sustainable green future, the inevitable trend of automobile electrification has been shown worldwide. To satisfy the need for a lightweight and highly integrated EV, the required motor is to be high speed. Meanwhile, the pursuit of high efficiency makes the PM motor become more and more mainstream. High-speed PM motors need new and more stringent requirements on the cooling and lubrication system. The design of the cooling topologies of the motor is complex. It is perceived that in the past few years, due to heat management of the high speed and high energy density motor, the cooling system is from the indirect water cooling to the direct oil cooling. The new design proposed in this paper is based on the new development results of the parts such as fluid and bearing of electric vehicles. It solves some problems brought about by predecessors in view of the high efficiency and reliability of the motor with higher energy density in the future. However, the calculation of the cooling mass flow rate, the size of the openings and the matching of different working conditions are still under study. For future research, it would be more interesting to perform the analysis of the highly integrated motor and its experimental test for the electric vehicle.

Acknowledgment

This research is support by the national natural science foundation of China (51575041).

References

[1] Han-Kyeol Yeo,Hyeon-Jeong Park,Jung-Moo Seo,Sang-Yong Jung,Jong-Suk Ro,Hyun-Kyo Jung. Electromagnetic and Thermal Analysis of a Surface-Mounted Permanent-Magnet Motor with Overhang Structure [J]. IEEE Transactions on Magnetics, 2017 (6): Article Sequence Number 8203304.

[2] Aurélie Fasquelle, Daniel Laloy. Water Cold Plates Cooling in a Permanent Magnet Synchronous Motor [J]. IEEE Transactions on Industry Applications, 2017 (5): 4406 - 4413.

[3] Zhang fengge, Du guanghui. Temperature Field Analysis of 1.12MW High-Speed Permanent Magnet Machine with Different Cooling Schemes [J]. Transactions of China Electrotechnical Society, 2014, 29 (1): 66-71.

[4] Lassila, Viktor. Method and Device for Liquid Cooling of an Electric Motor: CN 201380056661.X [P]. 2015-07-22.

[5] Chuck Yung. Cool Facts about Cooling Electric Motors: Improvements in Applications That Fall Outside the Normal Operating Conditions [J]. IEEE Industry Applications Magazine, 2015 (6): 47-56.

[6] Christian Paar, Annette Muetze, Hendrik Kolbe. Influence of Machine Integration on the Thermal Behavior of a PM Drive for Hybrid Electric Traction [J]. IEEE Industry Applications Society, 2015 (5): 3914 - 3922.

[7] Gary A. Pinkley, David F. Nelson, William R. Fong, Scott Heines, Edwin M Pearce. Pressurized and Gravity-fed Liquid Cooling of an electric motor: US 20150222162A1 [P]. 2015-08-06.

[8] Ziyuan Huang, Jiancheng Fang, Xiquan Liu, Bangcheng Han. Loss Calculation and Thermal Analysis of Rotors Supported by Active Magnetic Bearings for High-Speed Permanent-Magnet Electrical Machines [J]. IEEE Transactions on Industrial Electronics, 2016 (4): 2027 - 2035.

[9] Zhou Ming, Zhu Lingyu, Zhang Gongming. Oil Cooling Structure for Driving Motor of New Energy Vehicle: CN 201521109578.7 [P]. 2016-06-01.

[10] Engblom, Daniel. Method and Device for Liquid Cooling of Electric Motor: WO 2016159860A1 [P]. 2016-10-06.

[11] Lassila, Viktor. Method and Device for Liquid Cooling of Electric Motor: WO 2016159862A1 [P]. 2016-10-06.

[12] Zhang Shixiang, Zhang Shengchuan, Xia Ji. Motor Rotor Oil Cooling Structure and Motor with the Oil Cooling Structure: CN 206313565 U [P]. 2017-07-07.

[13] Pia Lindh, Ilya Petrov,Ahti Jaatinen-Värri,Aki Grönman; Miguel Martinez-Iturralde, Marco Satrústegui,Juha Pyrhönen. Direct Liquid Cooling Method Verified With an Axial-Flux Permanent-Magnet Traction Machine Prototype [J]. IEEE Transactions on Industrial Electronics, 2017 (8): 6086 - 6095.

[14] Li Huanwei. Oil Cooling Motor and Vehicle: CN 206149098 U [P]. 2017-05-03.

[15] Alberto Carriero, Matteo Locatelli, Kesavan Ramakrishnan, Gianpiero Mastinu, Massimiliano Gobbi. A Review of the State of the Art of Electric Traction Motors Cooling Techniques [J]. SAE Technical Paper, 2018-01-0057.

Parameter Tuning of PID Controller of Servo System Based on Particle Swarm Optimization Algorithm

Chen Lianbo[1,2,3], Xiong Xinhong[3], Zheng Yousheng[3], Zhang Yanhui[1,2], Zhang Guoqing[1,2,3], Wu Xinyu[1,2,a], Feng Wei[1,2,a]

[1]Guangdong Provincial Key Lab of Robotics and Intelligent System, Shenzhen Institutes of Advanced Technology, Chinese Academy of Sciences, Shenzhen 518055, China
[2]CAS Key Laboratory of Human-Machine-Intelligence Synergic Systems, Shenzhen Institutes of Advanced Technology, Shenzhen 518055, China
[3]School of Logistics Engineering, Wuhan University of Technology, Wuhan 430070, China

[a] Corresponding author: wei.feng@siat.ac.cn; xy.wu@siat.ac.cn

Abstract. In order to improve the control precision of permanent magnet synchronous motor servo drive, a Proportional-Integral-Differential (PID) control method based on particle swarm optimization (PSO) algorithm is proposed. Firstly, the mathematical model of the permanent magnet synchronous motor was established. Then the PSO algorithm was used to adjust the parameters of the PID controller. Finally, the simulation experiment was carried out under the MATLAB/Simulink environment to test its performance. Compared with the traditional manual parameters tuning, the PID controller based on PSO algorithm had better performance and better stability.

1. Introduction

Permanent magnet synchronous motor (PMSM) has been widely used in various industrial transmission fields due to its good control characteristics, simple structure, reliable operation, small size and high power factor [1]. The control performance of the PMSM has been improved in the application of vector control. Based on vector control, the PMSM servo system often adopts the double closed loop structure of speed loop and current loop. The speed regulator and current regulator generally use proportional integral (PI) controller [2]. The regulation performance of the PI regulator is closely related to the PI parameters, while the traditional manual tuning depends on the experience and the setting effect is not guaranteed. Therefore, the tuning of PI parameters has become a research hotspot in servo systems.

In order to improve the performance of the PID controller, various scholars have proposed the parameter tuning of PID methods. In order to optimize the effect of the controller, Isidro [3] added BP neural network, and finally achieved satisfactory results. Jimenez [4] proposed a genetic algorithm to optimize the PID controller and applied the strategy to the passive optical network management. The simulation results show that compared to the Z-N tuning algorithm, the adjustment time was reduced by 64% with the above control strategy, showing a higher efficiency. At the same time, the swarm intelligence algorithm has been used to the parameter tuning of the PID controller since the rapid

Content from this work may be used under the terms of the Creative Commons Attribution 3.0 licence. Any further distribution of this work must maintain attribution to the author(s) and the title of the work, journal citation and DOI.
Published under licence by IOP Publishing Ltd

development of the swarm intelligence [5]. Yan et al. [6] applied the wolf group algorithm to circuit fault diagnosis and the calculation results show that the wolf group algorithm has higher precision than the genetic algorithm. Zhao [7] proposed a particle swarm algorithm (PSO) algorithm based on multi-agent strategy for the problems of power system, and achieved good results in reactive power optimization of power system. According to the needs of medical image field, Shi [8]. proposed a new algorithm for medical registration images based on PSO algorithm and hybrid mutual information. Simulation experiments show that the results of the novel algorithm are up to the sub-pixel level, and have strong robustness and adaptability.

PSO algorithm has the simple structure, strong searching ability, high computational efficiency and easy engineering implementation. It has been applied in many related fields. [9] Huang [10] used the PSO algorithm for the path planning of the manipulator, which improved the accuracy of the motion trajectory. Hu [11] applied the PSO algorithm to the hinge position of the auger mechanism of the auger, which significantly improved the performance of the luffing mechanism. Liu [12] used the PSO algorithm to optimize the air-floating frictionless cylinder structure and optimized the cylinder parameters. In order to fully solve the shortcoming of PID controller parameters of PMSM servo system, a mathematical model based on PSO algorithm for PID tuning based on PMSM is proposed. The tuning effect of the PSO algorithm on the parameters of the PI controller was been analysed with the MATLAB/Simulink.

2. Mathematical Model of PMSM based on Vector Control

The dynamic model of typical sinusoidal PMSM in the synchronously rotating reference frame is given by [13]:

$$\begin{cases} i_d = \sqrt{\frac{2}{3}}\left(i_a\cos\theta + i_b\cos\left(\theta - \frac{2\pi}{3}\right) + i_c\cos\left(\theta + \frac{2\pi}{3}\right)\right) \\ i_q = \sqrt{\frac{2}{3}}\left(-i_a\sin\theta - i_b\sin\left(\theta - \frac{2\pi}{3}\right) - i_c\sin\left(\theta + \frac{2\pi}{3}\right)\right) \end{cases} \quad (1)$$

$$\begin{cases} u_d = Ri_d + \frac{d}{dt}\psi_d - \omega_e\psi_q \\ u_q = Ri_q + \frac{d}{dt}\psi_q - \omega_e\psi_d \end{cases} \quad (2)$$

$$\begin{cases} \psi_d = L_d i_d + \psi_f \\ \psi_q = L_q i_q \end{cases} \quad (3)$$

Substituting the eqn. (3) into the eqn. (2), the stator voltage equation can be obtained:

$$\begin{cases} u_d = Ri_d + L_d\frac{d}{dt}i_d - \omega_e L_q i_q \\ u_q = Ri_q + L_q\frac{d}{dt}i_q + \omega_e\left(L_d i_d + \psi_f\right) \end{cases} \quad (4)$$

Where, u_d and u_q are the voltage along d and q axes; i_d and i_q are the stator current along d and q axis; R is the resistance of the stator; ψ_d and ψ_q are the stator flux linkage along d and q axis; ψ_e is the electrical angular velocity; L_d and L_q are the stator self-inductance in the d and q axis, respectively; ψ_f represents the permanent magnet flux linkage.

In the surface-mount PMSM, the stator inductance satisfies L_d=L_q=L_s, so the electromagnetic torque equation is:

$$T_e = \frac{3}{2}p_n i_q \psi_f \quad (5)$$

IOP Conf. Series: Materials Science and Engineering **533** (2019) 012022 doi:10.1088/1757-899X/533/1/012022

In the double closed loop control system with speed loop and the current loop, the actual current of the motor is equal to the given current through the fast current control, so the two current components can be controlled separately. Figure 1 is the schematic diagram of PMSM vector control when using id=0. The main performance of the system is determined by the speed loop, and the quality of the speed loop will affect the performance of the position loop. Therefore, the speed loop PI controller parameters are mainly adjusted in the controller parameter tuning.

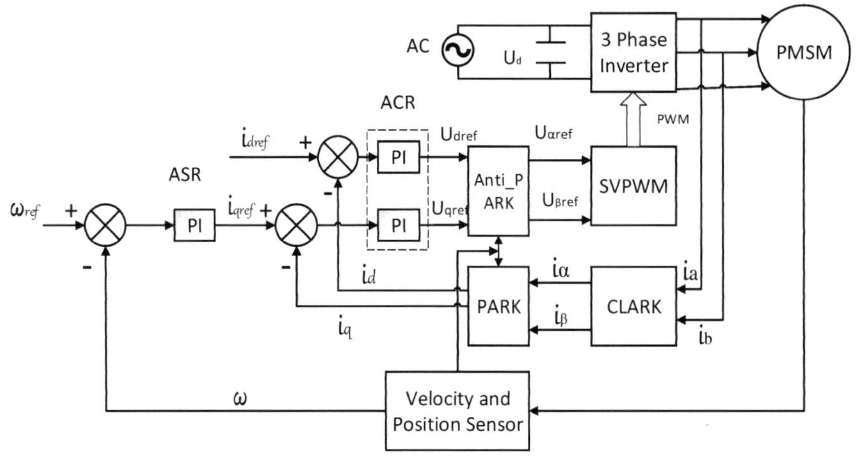

Figure 1. PMSM Vector Control Schematic

3. Particle Swarm Optimization Algorithm

3.1. The Basic Principle of Particle Swarm Optimization Algorithm

Suppose a group of M particles flies at a certain speed in the D-dimensional space. The state attribute of particle i at time t can be expressed as:
Position:

$$x_i^t = (x_{i1}^t, x_{i2}^t, \dots x_{id}^t, \dots x_{iD}^t)^T \tag{6}$$

Speed:

$$v_i^t = \left(v_{i1}^t, v_{i2}^t, \dots v_{id}^t, \dots v_{iD}^t\right)^T \tag{7}$$

Individual optimal position:

$$p_i^t = \left(p_{i1}^t, p_{i2}^t, \dots p_{iD}^t\right)^T \tag{8}$$

Global optimal position:

$$p_i^t = \left(p_{i1}^t, p_{i2}^t, \dots p_{iD}^t\right)^T \tag{9}$$

According to the above information, the speed of particle i at time t+1 can be expressed as:

$$v_{id}^{t+1} = wv_{id}^t + c_1 r_1 \left(p_{id}^t - x_{id}^t\right) + c_2 r_2 \left(p_{gd}^t - x_{id}^t\right) \tag{10}$$

$$x_{id}^{t+1} = x_{id}^t + v_{id}^{t+1} \tag{11}$$

Where: w is the inertia weight; r_1, r_2 are random numbers uniformly distributed in the interval (0, 1); c_1, c_2 are learning factors
The eqn. (10) is mainly composed of three parts: the first part is the last particle velocity with inertia weight, which is the inheritance of the particle to the previous state, the particle's trust to the spontaneous state of motion, and the inertial motion according to its own speed; the second part is the "cognitive" part, it is expressed as the thinking of the particle itself, that is, the particle movement is influenced by the experience of its own movement, so that the particle has a strong local search ability; the third part is the "social" part, indicating that the movement of the particle is affected by the whole The influence of particle swarms, the information sharing between particles and mutual cooperation.

3.2. Algorithm Implementation Flow

In the optimization process of PSO algorithm, an evaluation function is proposed to calculate the fitness value of particle. The fitness value is used to evaluate the performance of particle. The product of the absolute value of error and time is used to the test function in this paper. The expression is as Eqn. 6:

$$F=\int_0^{\infty} t\left|e(t)\right|dt \qquad (12)$$

The process of PSO algorithm is as follows:

1. Produce particle swarms and initialize particle parameters.
2. Calculate the fitness value of the particle according to eqn. (12), and determine the individual optimal value F_{gbest} of the particle and the global optimal value F_{zbest} of the particle swarm.
3. For each particle, compare the fitness value of the historical optimal position with the current particle, and update the optimal position of the individual according to the result.
4. For each particle, compare the fitness value of the global optimal position of the particle group with the fitness value of the current particle, and update the global optimal position according to the result.
5. Update the position and speed of particles.
6. Determine whether the termination condition is currently reached. If reached, the iteration is terminated; otherwise return to 2. The condition for terminating the iteration is that the number of iterations reaches the maximum number of times or the fitness value is less than the convergence precision.

4. Design of Speed Loop PI Controller Based on Particle Swarm Optimization algorithm

The schematic of the PI control system based on PSO algorithm is shown in Figure 2. In the design of the PI regulator, the key is to determine the reasonable values of K_{pw} and K_{iw}. The system has better performance after adjustment by the PI controller. The range of values of K_{pw} and K_{iw} is determined by empirical methods and PSO algorithm is adopted to find the optimal value in this interval.

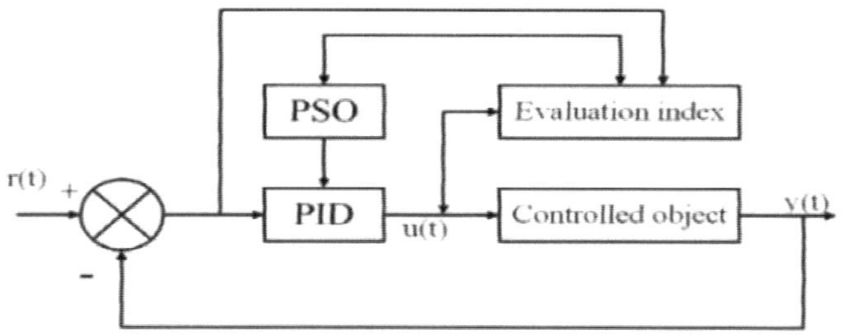

Figure 2. PID Control Schematic Based on PSO

A schematic diagram of the process of tuning the velocity loop PID parameters by using the PSO algorithm is shown in Figure 3.

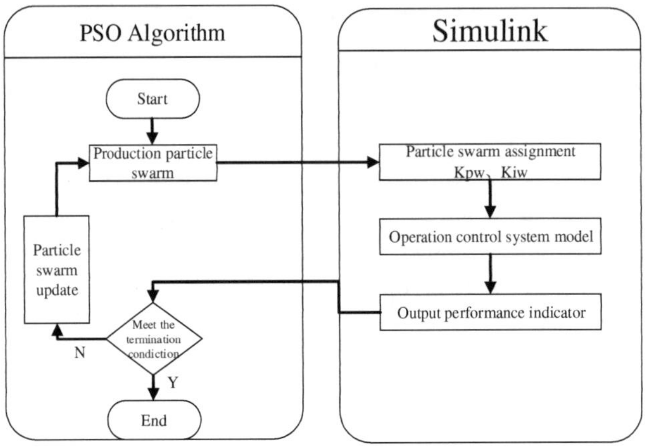

Figure 3. Schematic diagram of the process of setting PID parameters by PSO algorithm

The specific process is presented as follows:

1. A particle swarm with a population size of N=30 is generated within the range of the controller PI parameter, and each parameter of the particle swarm algorithm is initialized. Such as inertia weight, learning factor and number of iterations.
2. In the particle swarm, the position of the particle represents the parameter that needs to be optimized (K_{pw} and K_{iw} in the PI parameter of the velocity loop). According to the range of K_{pw} and K_{iw} value adjustment, the maximum position U_b and the minimum position L_b of the particle position are set, and the maximum velocity v_{max} of the particle is determined thereby.
3. The position information x_i and the velocity information v_i of the particles in the particle group are initialized. And calculate the fitness value F of the particle according to formula (12), set the current position of the particle as the best g_{best} of the individual, and record the best position of all the particles as the global optimal z_{best}.
4. For each particle, the fitness value of the particle is calculated according to equation (12). If the current particle's fitness value F_i is less than F_{gbest}, it indicates that the current position is better than g_{best}. Assign x_i a g_{best} to the latest individual optimal value to complete the individual's best update.
5. Comparing the fitness value of the current particle with the fitness value of the global optimal z_{best}, if the current particle's fitness value F_i is less than F_{zbest}, it indicates that the current position is better than z_{best}. Assign x_i a z_{best} to make it the latest global optimal and complete the global best update.
6. The position and velocity of the particles are updated according to equations (10) and (11). The PI value in the updated particle is substituted into the controller, the system is run, and the corresponding fitness function is calculated.
7. After repeated iterations, it is determined whether the termination condition is currently reached. If the termination condition is reached, the iteration is terminated; otherwise, the loop iteration is continued. The condition for terminating the iteration is that the number of iterations reaches the maximum number of times or the fitness value is less than the convergence precision.

When the PSO algorithm ends the optimization, the optimal result is output. Comparing the optimization results with the performance of the current PI parameters, if it is better than the PI parameters before optimization, the original PI parameters are replaced to achieve the purpose of parameter self-tuning.

5. Comparison of Manual Parameters Tuning and Simulation Analysis Based on PSO

According to the vector control principle of PMSM controller, the parameter tuning simulation model of PMSM controller based on PSO algorithm is explore under MATLAB/Simulink environment. The simulation model is shown in Figure 4. The vector control and adaptive value calculation of PMSM is done in Simulink, while the PSO algorithm is implemented in MATLAB.

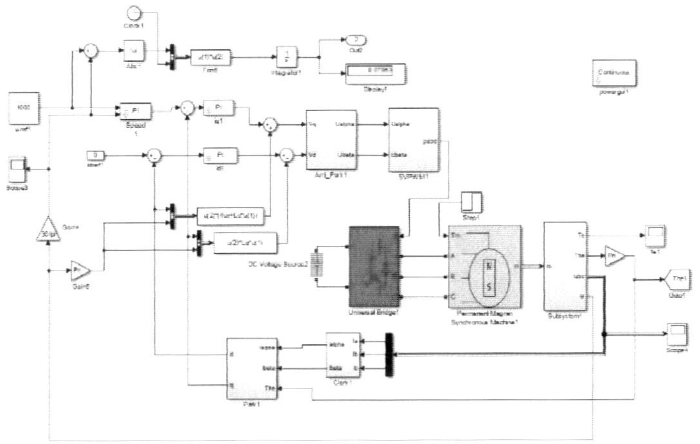

Figure 4. PMSM Simulation Model Based on PSO

After 50 iterations of the PSO algorithm, the parameters of the speed loop PI controller self-tuning parameters K_{pw} and K_{iw} are 0.18 and 3.95 respectively. The servo system PI controller is repeatedly debugged by traditional manual, and the best parameters K_{pw} and K_{iw} are 0.14 and 7. The speed response curves corresponding to two different parameters of the PI controller in two cases is shown in Figure 5.

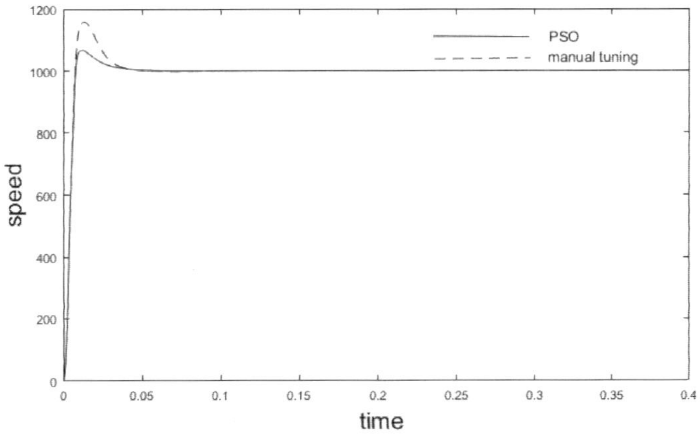

Figure 5. Servo System Speed Response Curve

As can be seen from Fig. 5, the PI controller based on the particle swarm optimization algorithm has a significant reduction in the overshoot of the system, although the response time is not greatly improved compared to the traditional manual parameters tuning method.

In order to describe the difference between the two methods, the performance of the PSO algorithm and the manual parameters tuning are compared from the three aspects of overshoot, rise time and product integral of error and time (ITEA). The overshoot reflects the degree to which the output value deviates from the stable value; the rise time reflects the time required to stabilize the control process area; the ITAE reflects the overall stability of the control process, and the comparison results are shown in Table 1.

Table 1. Permanent Magnet Synchronous Motor Parameters

Setting method	Overshoot/%	Rise/s	ITEA
Trial method	15.8	0.045	0.071
PSO	6.6	0.04	0.040

As shown in Table1, the PSO algorithm has improved over the traditional manual parameters tuning in terms of overshoot, rise time and ITEA.

6. Conclusion

Based on PSO algorithm, an intelligent PID tuning optimization algorithm is proposed. After theoretical analysis and simulation experiments, the PSO algorithm could improve the tuning efficiency of the parameters compared with the traditional manual parameters tuning method. The effectiveness, overshoot, response speed and feasibility of this algorithm have been verified. The dynamic performance and steady state performance of the servo system of the motor has greatly improved.

Acknowledgements

This work is financially supported by Guangdong frontier and key technological innovation (No.2017B090910013), Shenzhen knowledge innovation technology basic research project (JCYJ20170818153048647), National Natural Science Foundation of Guangdong (No.2016A030313177), Shenzhen Technology Project (JSGG20160301160759264, JSGG20171823140127645).

References

[1] Zhang H Y, Xu H P, Fang C and Xiong C 2018 *Transactions of China Electrotechnical Society* **33** 2923-2934 (2018)
[2] Jan R M, Tseng C S, Liu R J *Electric Power Systems Research* **78** 1161-1168 (2008)
[3] Isidro S, Julio R B and Antonio A A *Journal of Food Engineering* **46** 21-29 (2000)
[4] Jimenez T, Merayo N 2015 *Expert Systems With Application* **42** 9211-9229 (2000)
[5] GirirajKumar S M, Jayaraj D and Kishan A R *International Journal of Computer Applications* **19** 12-18 (2010)
[6] Yan X L, Ding P and Ma J *Computer Engineering and Applications* **53** 152-156 (2017)
[7] Zhao B, Cao Y J *Proceeding of the CSEE* **25** 1-7 (2005)
[8] Shi Y X, Qiu T S, Han J and Jin Sheng Chinese Journal of Biomedical Engineering **34** 1-7 (2015)
[9] Zhang G H, Shao Xin Y, Li P G and Gao L *Computers & Industrial Engineering* **56** 1309-1318(2009)
[10] Huang Q J, Zhang K, Song J C and Yu Z L *Journal of Northeastern University* **11** 1636-1641 (2018)
[11] Hu J P, Peng Y M *Chinese Journal of Engineering Design* **5** 561-566 (2018)
[12] Lu Y, Zhao G X *Journal of System Simulation* **9** 3564-3570 (2018)
[13] Kiran B, Kottalil A M and Ananthamoorthy N P *International Journal of Advanced Research in Electrical, Electronics and Instrumentation Engineering* **1** 689-695 (2013)

Fuzzy force control based on grasped objects stiffness of tendon-driven underactuated prosthetic hand system

Luo Huanxin[1,2], **Deng Hua**[1,2,a]

[1]School of Mechanical and Electrical Engineering, Central South University, Changsha 410083, China[2]

[2]State Key Laboratory of High Performance Complex Manufacturing, Changsha 410083, China

[a]Corresponding author: hdeng@ csu.edu.cn

Abstract. The design of an fuzzy force based on grasped objects stiffness method for tendon-driven underactuated prosthetic hand system are addressed in this study. First, a mathematical model of tendon-driven underactuated prosthetic hand is developed for force control. Next, fuzzy force based on grasped objects stiffness was discussed to improve the control performance. Finally, the proposed control strategy for underactuated prosthetic hand is simulated to six different stiffness objects. The results shown that the controller is suit for prosthetic hand grasp objects with different stiffness. The simulation results verify the effectiveness of the force control method based on grasped objects stiffness.

1. Introduction

Underactuated tendon-driven prosthetic hand uses tendon-driven transmission mode, and the driving torque on each knuckle is generated by the pulley on the tendon[1,2]. The compression spring is installed in the finger, which makes the finger have certain coupling and adaptive ability, makes the finger move flexibly and can realize various grasping modes. Upper limb amputees need to maintain a proper and stable grip when using prosthetic hand to grasp objects. Excessive grip force will lead to deformation or even damage of the grasped objects, and too little grip force will lead to sliding[3]. Moreover, the shape, size, stiffness and weight of grasped objects are different, it is difficult to grasp the objects stably[4].

Underactuated tendon-driven prosthetic hand is a highly coupled nonlinear system, and it has uncertainties such as external disturbances and modeling errors. Force control method with good adaptability and strong robustness need to be applied to prosthetic hands. At present, there are mainly PID control, fuzzy control, sliding mode control, hybrid position/control and impedance control applied to the grip force control of prosthetic hand. The conventional force control method assumes that the characteristics of the controlled object are known. However, Upper limb amputees grasp different objects with different properties in daily life, the controller does not know ahead of time what those objects are [5]. The controller in robot manipulators adaptively adjust the feedback gain based on measurement of object stiffness[6,7]. PID Sliding Mode control based on the grasped object stiffness shown to be an effective improvement to the positon control of a prosthetic hand when grasping different objects with different stiffness properties[8].

Content from this work may be used under the terms of the Creative Commons Attribution 3.0 licence. Any further distribution of this work must maintain attribution to the author(s) and the title of the work, journal citation and DOI.
Published under licence by IOP Publishing Ltd

IOP Conf. Series: Materials Science and Engineering **533** (2019) 012023 doi:10.1088/1757-899X/533/1/012023

In order to make the prosthetic hand grasp the flexible object with different stiffness stably and improve the success rate of grasping the flexible object with different stiffness, an fuzzy force controller based on stiffness is proposed in this paper. The model control parameters of the prosthetic hand are adjusted by stiffness identification to improve the control strategy of the grasped object. A fuzzy force controller is designed to control the stability of the prosthetic hand.

2. Underactuated prosthetic hand system

The prosthetic hand is designed with underactuated principle. The underactuated mechanism is compact and finger size is close to that of human hand. As shown in Fig.1, A single finger consists of three joints, and two winding wheels are installed at each joint. One is used for the transmission of rope power. The rope connects the distal knuckle to the winding wheel of the motor. The rope passes through the winding wheel of each knuckle. The other is used to restore the compression and reset of the spring[9].

The finger has three DOF actuated by a DC motor, which flexes the fingers by pulling cables extending to the fingertips. A bidimensional mathematical modle has been developed in order to evaluate the finger dynamic behavior during the reaching phase to the objects[10].

The finger dynamic model can be written starting from the Lagrangian formulation. The equation of dynamics of an underactuated prosthetic finger in Cartesian coordinates can be obtained in the form:

$$M\ddot{\theta} + C\dot{\theta} + D\theta = Bu - lf \tag{1}$$

where M represents the inertia of the system, C is the resistance caused by damping and clearance friction, and D is System stiffness. u is driving voltage of the motor, f indicates the reaction force of the grasped object to the prosthetic hand. The reaction force is mainly related to the position of the contact point.

The contact force between the finger and the environment can be simplified as a constant stiffness spring model[11,12], that is

$$f(t) = K_{ev}x(t) \tag{2}$$

where K_{ev} is the environment stiffness, $x(t)$ is the vertical displacement of the prosthetic finger at the contact point.

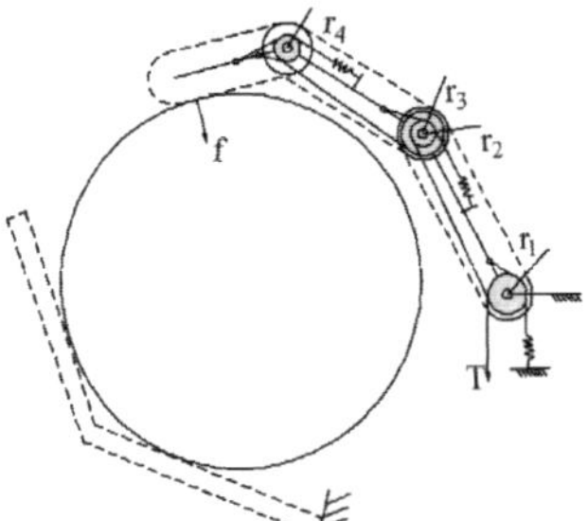

Figure 1. Finger structure.

When the underactuated prosthetic hand contact with the environment, we assume the contact point is the equilibrium position of the prosthetic hand system, so the displacement x can be approximated by :

$$x(t) = l\theta(t) \tag{3}$$

where l represents the distance between the rotating origin of the prosthetic hand joint and the distance between the prosthetic hand and the object.

Integrating Eq. 1, Eq. 2, and Eq.3. , yields

$$\ddot{\tilde{f}}(t) + a_1\dot{\tilde{f}} + a_0\tilde{f} = cu \tag{4}$$

where $a_1 = C/M$, $a_0 = (D + l^2 K_{ev})/M$, $c = BlK_{ev}/M$.

3. Fuzzy controller based on the grasped object stiffness

In the process of prosthetic hand grasping, the position, damping and viscosity coefficient of the objects are different, and there are a little errors in the modeling of friction force in the process of prosthetic hand grasping objects, at the same time, the friction model is constantly changing in the process of prosthetic hand grasping objects. The traditional PID algorithm is used to control the system, which has poor robustness and easy to affect the control accuracy. It is difficult to achieve the requirements of stable grasp. When people grasp objects, it is a fuzzy process to estimate grasping force by experience according to the size, shape and characteristics of the object being grasped. Fuzzy logic has superior ability to deal with uncertain and nonlinear problems. Therefore, many scholars use fuzzy control method to control the grasping force of prosthetic hand. When a prosthetic hand grasps an object, the stiffness of the object has a great influence on the control system. Therefore, an fuzzy control method based on the stiffness estimation of the grasped object is proposed to adjust the parameters of the fuzzy system.

3.1. An fuzzy controller

An fuzzy control[13,14] is designed as shown in Fig.2. The single input is desired force f_d , f is the contact force with environment , x is the displacement.

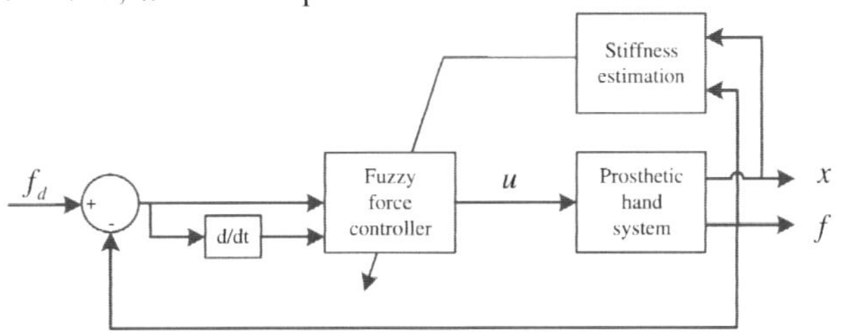

Figure 2. Schematic structure of fuzzy force control system.

The structure of the fuzzy controller is shown in Figure 3, where error $e = f_d - f$, K_e , K_d , K_0 and K_u are ratio coefficient, u is output of controller.

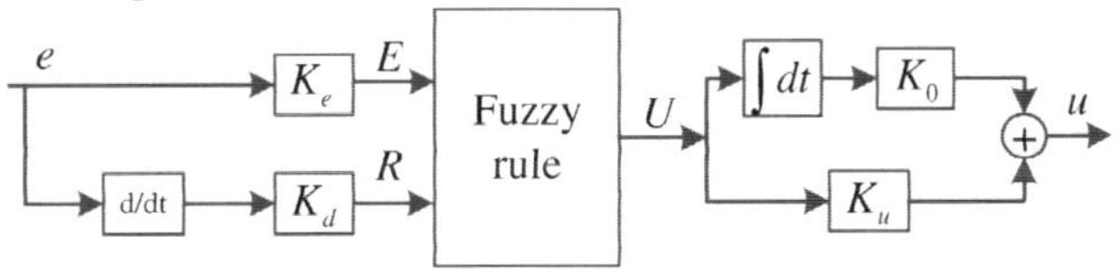

Figure 3. Fuzzy controller.

Thus, finite rules clearly show the saturation effect, and the output model of the rule base becomes

$$u = sat(\sigma) = \begin{cases} \text{sgn}(\sigma) & |\sigma| > 1 \\ g(\sigma) & |\sigma| \le 1 \end{cases} \tag{5}$$

where

$$g(\sigma) = \sigma + (1-\gamma)(kh - \sigma) \tag{6}$$

where $\sigma = K_e E + RK_d$, γ is a nonlinear parameter, h is half of the spread of each input and output membership function. Therefore, the control output can be given as

$$U = U_{pi} + U_{pd} = K_0 \int \tilde{u} dt + K_u \tilde{u} = K_0 \int sat(\sigma) dt + K_u sat(\sigma) \qquad (7)$$

The membership functions of input variable E , R and output variable U are triangular membership function shown in Fig.4. Fuzzy language variables are set as $\{NB, NM, NS, ZO, PS, PM, PB\}$.

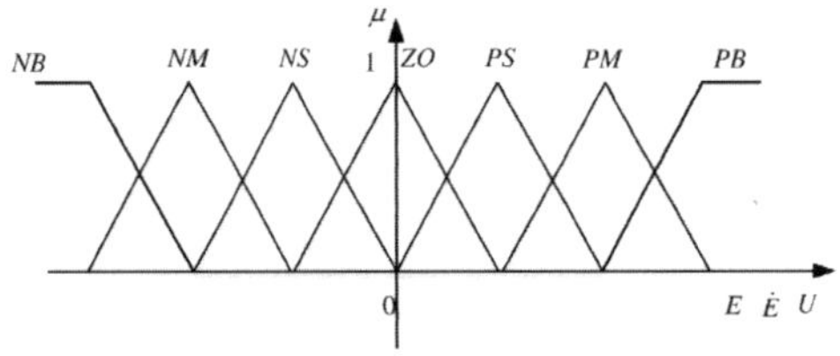

Figure 4. Membership function of fuzzy controller..

The fuzzy rule as shown in Table I, IF E is A_i and R is B_i ,then u is G_{i+j} . A_i is fuzzy rule of input variable E , B_i is fuzzy rule of input variable R , G_i is fuzzy rule of output variable u .

Table1. Rule base of fuzzy controller.

R/E	A_{-3} (NL)	A_{-2} (NM)	A_{-1} (NS)	A_0 (ZO)	A_1 (PS)	A_2 (PM)	A_3 (PL)
B_3 (PL)	G_0 (ZO)	G_1 (PS)	G_2 (PM)	G_3 (PL)	G_3 (PL)	G_3 (PL)	G_3 (PL)
B_2 (PM)	G_{-1} (NS)	G_0 (ZO)	G_1 (PS)	G_2 (PM)	G_3 (PL)	G_3 (PL)	G_3 (PL)
B_1 (PS)	G_{-2} (NM)	G_{-1} (NS)	G_0 (ZO)	G_1 (PS)	G_2 (PM)	G_3 (PL)	G_3 (PL)
B_0 (ZO)	G_{-3} (NL)	G_{-2} (NM)	G_{-1} (NS)	G_0 (ZO)	G_1 (PS)	G_2 (PM)	G_3 (PL)
B_{-1} (NS)	G_{-3} (NL)	G_{-3} (NL)	G_{-2} (NM)	G_{-1} (NS)	G_0 (ZO)	G_1 (PS)	G_2 (PM)
B_{-2} (NM)	G_{-3} (NL)	G_{-3} (NL)	G_{-3} (NL)	G_{-2} (NM)	G_{-1} (NS)	G_0 (ZO)	G_1 (PS)
B_{-2} (NL)	G_{-3} (NL)	G_{-3} (NL)	G_{-3} (NL)	G_{-3} (NL)	G_{-2} (NM)	G_{-1} (NS)	G_0 (ZO)

3.2. *Stiffness estimation method based on force information*

In the contact state, a little increase of grip force will cause the object to tiny deformation. The stiffness equivalent derivative of object is related to the rate of grip force and deformation according to Hooke's law.

When the prosthetic hand contact with the object, there is a relative speed between the finger and grasped objects, so the grasping process is a dynamic contact process. In this paper, a continuous collision force model is used to model the grasp process. The nonlinear continuous collision force model[15] is:

$$f(\delta, \dot{\delta}) = \begin{cases} K\delta^n + C\delta^n\dot{\delta} & \delta \geq 0 \\ 0 & \delta < 0 \end{cases} \qquad (8)$$

where δ is embedded depth, n is exponential coefficient, $\dot{\delta}$ relative velocity at contact point K is local contact stiffness at impact location and C is damping coefficient at impact location.

When the prosthetic hand grasps the object, the grasping speed is very low, and the collision force between the prosthetic hand and the object is very small. In a short time, the deformation of the object

IOP Conf. Series: Materials Science and Engineering **533** (2019) 012023 doi:10.1088/1757-899X/533/1/012023

is very small, that is, the value of $\dot{\delta}$ is close to zero. Therefore, the environmental stiffness can be approximately proportional to the rate of change of the grasping force.

In order to quickly identify the stiffness equivalent of the grasped objects during the contact stage of the prosthetic hand and finger, a force sensor is installed on the prosthetic hand and finger to measure the contact force between the fingertip and the grasped object, and the gradient of the contact force is calculated in a fixed sampling time[16].

$$df = (f_{n+1} - f_n)/T \tag{9}$$

Where $f_n = \frac{1}{m}\sum_{i=1}^{m} f_i$, f_n is contact force with sample time, m is the number of sample point.

The change rate of contact force produced by objects with different stiffness is different. The change rate of contact force produced by objects with large stiffness equivalent is larger than that produced by objects with small stiffness equivalent.

4. Simulation

In order to verify the effectiveness of the proposed fuzzy force controller based on stiffness self-adaptation, we choose objects, soft, medium and hard objects as the grasping objects of the prosthetic hand to simulate, and the grasped objects are simulated by virtual spring. At the same time, the performance of the two controllers, fuzzy controller and stiffness-based fuzzy controller, are tested respectively.

Six kinds of different stiffness objects have been selected as simulation objects, the stiffness of grasp object and desired force are shown in Table 2.

Table2. The parameter of six cases.

Case	$K(N/mm)$	$Desired\ force(N)$
1	0.2	2
2	0.5	3
3	1	4
4	3	5
5	7	7
6	10	8

Case 1: Consider an object with stiffness $k_1 = 0.2N/mm$.

The grasping force f can be derived as $f = k_i dx$, when the prosthetic hand contact with the soft object, the stiffness can be identified and the parameters of the AFLC are automatically selected according to previous design. The designed parameters of AFLC with soft object are as follow: $K_e = 1.8$, $K_r = 1/20$, $K_0 = 60$, $K_u = 1.5$. The transfer function of the system is $G = \frac{15.6}{0.0075s^2 + 0.5s + 34.8}$.

The desired force is 2N. Two kinds of controller performance of force response with $k_1 = 0.2N/mm$ are shown in Fig.5.

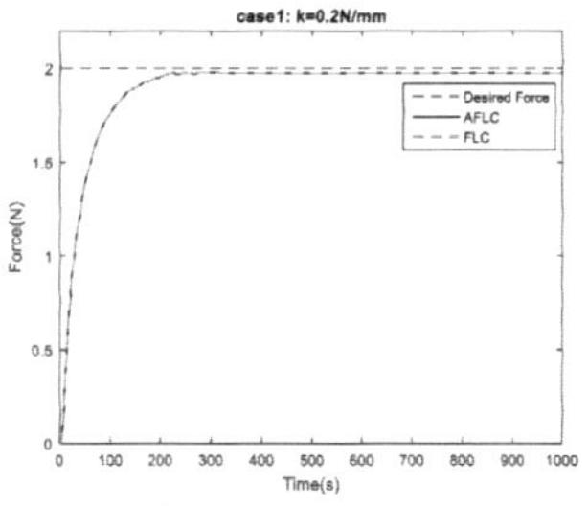

Figure 5. Control performance of controllers for case 1.

Case 2: Consider an object with stiffness $k_2 = 0.5N/mm$.

The designed parameters of AFLC with soft object are as follow: $K_e = 2$, $K_r = 1/20$, $K_0 = 50$, $K_u = 1.2$. The transfer function of the system is $G = \dfrac{39}{0.0075s^2 + 0.5s + 86}$. The desired force is 3N. Two kinds of controller performance of force response with $k_2 = 0.5N/mm$ are shown in Fig.6.

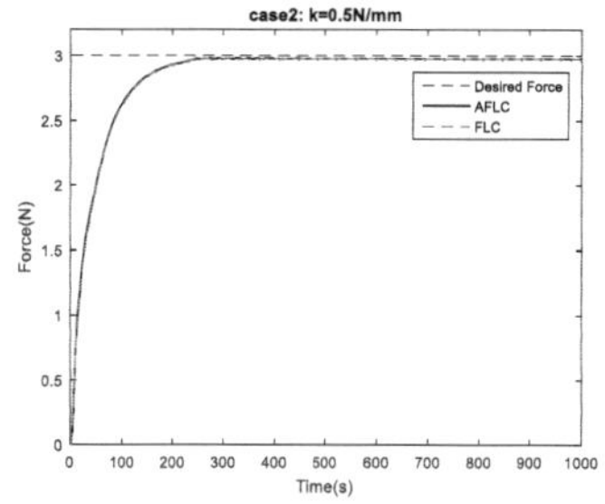

Figure 6. Control performance of controllers for case 2.

Case 3: Consider an object with stiffness $k_3 = 1N/mm$**.**

The designed parameters of AFLC with soft object are as follow: $K_e = 2$, $K_r = 1/30$, $K_0 = 50$, $K_u = 1$. The transfer function of the system is $G = \dfrac{78}{0.0075s^2 + 0.5s + 170}$. The desired force is 4N. Two kinds of controller performance of force response with $k_3 = 1N/mm$ are shown in Fig.7.

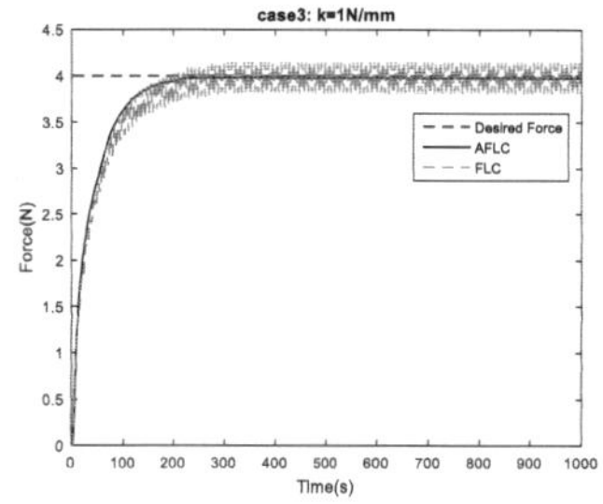

Figure 7. Control performance of controllers for case 3.

Case 4: Consider an object with stiffness $k_4 = 3N/mm$.

The designed parameters of AFLC with soft object are as follow: $K_e = 3$, $K_r = 1/50$, $K_0 = 40$, $K_u = 0.8$. The transfer function of the system is $G = \dfrac{234}{0.0075s^2 + 0.5s + 508}$. The desired force is 5N. Two kinds of controller performance of force response with $k_4 = 3N/mm$ are shown in Fig.8.

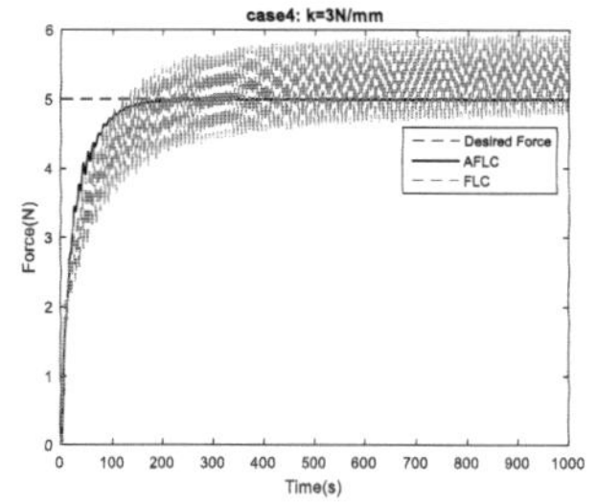

Figure 8. Control performance of controllers for case 4.

The stiffness-based fuzzy controller performance of force response with case 5-6 are shown in Fig .9.

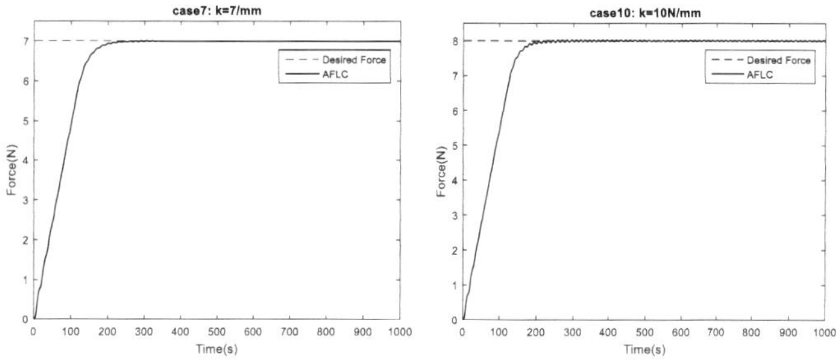

Figure 9. Control performance of controllers for case 5 and 6.

The simulation result in different situtior(Fig. 5-Fig .8) show that the fuzzy force control based on the grasped object stiffness has a good control performance on the prosthetic hand system. By comparison, it is found that the conventional fuzzy controller cannot adapt to the situation of stiffness change. The parameters of fuzzy controller cannot be adjusted according to the change of environment. As shown in Fig. 5, when the design parameters of fuzzy controller satisfy the requirements of grasping soft object, it cannot adapt to medium stiffness object and hard object as shown in Fig. 6 and Fig. 7.

According to the simulation results, the equivalent stiffness of different objects can be divided into three grades and the corresponding control parameters are as shown in Table 3.

Table3. Look-up table of control parameter with different objects.

k	K_e	K_r	K_o	K_u
$0.1 \leq k < 0.5$	1.8	1/20	60	1.5
$0.5 \leq k < 0.8$	2	1/20	50	1.2
$0.8 \leq k < 1.5$	2	1/30	50	1
$1.5 \leq k < 4$	3	1/50	40	0.8
$4 \leq k < 7.5$	4.5	1/60	20	0.1
$7.5 \leq k < 10$	5	1/60	20	0.05

From the simulation results, it can be seen that after touching the object, the controller can judge the stiffness of the object to be grasped and control the object adaptively, so that the object can be grasped stably. The fuzzy force control method is effective for grasping different stiffness objects. The fuzzy control method based on the grasped object stiffness has the advantages of small overshoot, short adjustment time and small steady-state error.

5. Conclusion

The fuzzy force control method based on the grasped object stiffness is presented to control underactuated prosthetic hand. The model of the prosthetic hand system is established for the control system. Six environment stiffness are simulated by the presented control method, the simulation results demonstrate the effectiveness of fuzzy force control based on the grasped object stiffness. The simulation show that the fuzzy force control method based on the grasped object stiffness is effective for grasping different stiffness objects.

References

[1] Huang H, Jiang L, Zhao D W, et al. The Development on a New Biomechatronic Prosthetic Hand Based on Under-actuated Mechanism. *Ieee/rsj International Conference on Intelligent Robots and Systems. IEEE*, 3791 (2004)

[2] Tenim S, Vicatos G. UNDERACTUATED PROSTHETIC HAND:, WO 2016005871 A1[P]. (2016)

[3] Deng H, Zhang Y, Duan X G. Wavelet Transformation-Based Fuzzy Reflex Control for Prosthetic Hands to Prevent Slip. *IEEE Transactions on Industrial Electronics*, 99 (2017).

[4] Engeberg E D, Meek S G. sliding mode control of grasped object slip for prosthetic hands *Ieee/rsj International Conference on Intelligent Robots and Systems. IEEE*, 4174 (2011)
[5] Winges S A, Eonta S E, Soechting J F, et al. Effects of object compliance on three-digit grasping. *Journal of Neurophysiology*, **101**,2447(2009)
[6] Burn K, Short M, Bicker R. Nonlinear Fuzzy Force Control Techniques Applied to Robots Operating in Uncertain Environments. *Journal of Robotic Systems*, **20**,391 (2003)
[7] Guldner J, Dawson D M, Qu Z. Hybrid control for the tracking of rigid-link electrically-driven robots. *Advanced Robotics*, **9**,331 (1995)
[8] Andrecioli R, Engeberg E D. Grasped object stiffness detection for PID sliding mode position control of a prosthetic hand. *IEEE Ras & Embs International Conference on Biomedical Robotics and Biomechatronics. IEEE*, 526-531 (2012)
[9] Luo H, Hu R, Deng H. Force control of an underactuated prosthetic hand based on sliding mode with exponential reaching law. *International Conference on Advanced Electronic Science and Technology*. (2016)
[10] Luo H, Duan X, Deng H. Sliding mode impedance control of an underactuated prosthetic hand. *IEEE International Conference on Information and Automation. IEEE*,726 (2014)
[11] Engeberg E D, Meek S. Improved grasp force sensitivity for prosthetic hands through force-derivative feedback. *IEEE transactions on bio-medical engineering*, **55**,817 (2008)
[12] B. Massa,S. Roccella, M Carrozza, and P. Dario,"Design and development of an underactuated prosthetic hand, *IEEE Int Conf Robot. Autom*, **4**, 3374 (2002)
[13] Duan X G, Deng H, Li H X. A Saturation-Based Tuning Method for Fuzzy PID Controller. *IEEE Transactions on Industrial Electronics*, **60**,5177 (2013)
[14] Duan X G, Li H X, Deng H D H. A simple tuning method for fuzzy PID control. *IEEE International Conference on Fuzzy Systems, 2008. Fuzz-Ieee. IEEE*, 271 (2008)
[15] Zhang Y, Chu L, Zhuo W, et al. Grasping force control preventing excessive deformation for underactuated prosthetic hands. *Advanced Information Management, Communicates, Electronic and Automation Control Conference. IEEE*,:1747 (2017)
[16] Zhang Y, Duan X G, Deng H. A Method to Modify Initial Desired Force of Prosthetic Hand Based on Stiffness Estimation. *Intelligent Robotics and Applications. Springer International Publishing*,348 (2015)

IOP Conf. Series: Materials Science and Engineering 533 (2019) 012024 doi:10.1088/1757-899X/533/1/012024

Speed Estimation of Asynchronous Motor Based on Model Reference Adaptive Method and On-line Identification of Rotor Time Constant

TONG Jun[1], SHI Lei[1] and LIU Li-juan[2] and Fan Chen-yue[1]

[1]College of Electrical and Control Engineering, Xi'an University of Science And Technology, Xi'an 710054, China

[2]Electric Power Science Research Institute, State Grid Gansu Electric Power Company, Lanzhou 730050, China

Corresponding author: 976701775@qq.com

Abstract. In this paper, the principle of model reference adaptive method (MRAS) is analysed and studied. Aiming at the shortcomings of installing speed sensor to collect the speed signal and feedback to the control system, a method for speed estimation and parameter identification of asynchronous motor without speed sensor is presented, And taking the model reference adaptive method (MRAS) as the main line, the Popov super stability theory is used to derive and analyse the adaptive law. Finally, the proposed method is verified by MATLAB software and hardware circuit experiments.

1. Introduction

Nowadays, electric motors play an important role in national production. Whether it is mining or transportation, they are inseparable from motors. Such as coal mining machines, high-speed rail and airplane and so on. This makes it particularly important to improve the performance of the motor. Since the advent of the electric motor, many scientists have been working on the structure of the motor and the control method in order to make the motor perform better in practical applications. The birth of the first DC motor in the 19th century opened the door to motor applications. Due to its good working performance, simple mathematical model, and simple operation, it was active in the industrial field until the middle of the 20th century. In the 1970s, the industrial field put forward higher requirements for the use environment of the motor, single-unit capacity, and voltage level. The DC motor could not be satisfied [1], which promoted the rapid development of the asynchronous motor, making the AC electric drive have a wide and stable speed range. High accuracy, safety and stability, high reliability, easy maintenance and many other advantages. In some industrial and mining enterprises, mining locomotives often use slip frequency control with speed closed loop in order to get better working efficiency and control performance. The detection of speed becomes an indispensable part.

The core of the speed estimation is the value of some parameters of the asynchronous motor. For example, the stator rotor resistance of the motor, self-inductance, mutual inductance. The acquisition

Content from this work may be used under the terms of the Creative Commons Attribution 3.0 licence. Any further distribution of this work must maintain attribution to the author(s) and the title of the work, journal citation and DOI.
Published under licence by IOP Publishing Ltd

IOP Conf. Series: Materials Science and Engineering **533** (2019) 012024 doi:10.1088/1757-899X/533/1/012024

of these parameter values is difficult to obtain directly, so it is often obtained by parameter identification.
The asynchronous motor speed estimation and parameter identification proposed in this paper are based on the Model Reference Adaptive Method (MRAS). Because the model reference adaptive method has the characteristics of simple algorithm and high accuracy compared with the traditional speed estimation and parameter identification method, so it is favored by researchers at home and abroad.

2. The basic principle of model reference adaptive method

The model reference adaptive method can be divided into three structures according to the output connection method. Among them, the output parallel model reference adaptive method (also commonly referred to as output error method) is most common, and the basic structure is shown in Figure 1.

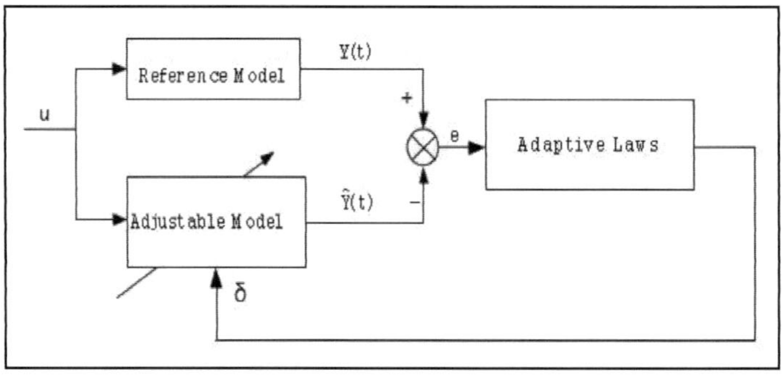

Figure 1. basic structure.

The reference model is an already determined value, equivalent to a reference value. The adjustable model is related to the parameter to be estimated, the output value is uncertain, and it needs to be adjusted continuously. The outputs of the two models are y(t) and respectively, which have the same physical meaning. The difference between the two outputs is the error signal e, which is fed back to the adjustable model by adaptive law. When the output of the two models is the same, the error signal e tends to 0.At this time, it is considered that the parameter to be estimated converges to the true value, and the identification result is obtained [2][3].
It can be seen from Fig. 1 that the design of the adaptive law is related to the whole system, so the design of the adaptive law becomes the focus. The design of the adaptive law in this paper is based on the Popov hyper stability theory. The parameter optimization method was the first method used to design the MRAS system. The MIT method has a simple structure and does not need to set state variables. However, the MIT solution cannot guarantee the stability of the system, therefore, the stability check must be performed at the end. Currently, there are few applications. So the Popov method is used in this paper.

3. MRAS method based on rotor flux linkage

Analytical derivation of reference models, tunable models, and adaptive laws [4].
Equation (1) is the voltage equation of the asynchronous motor in the two-phase stationary α-β coordinate system:

$$\begin{bmatrix} u_{s\alpha} \\ u_{s\beta} \\ u_{r\alpha} \\ u_{r\beta} \end{bmatrix} = \begin{bmatrix} R_s + L_s p & 0 & L_m p & 0 \\ 0 & R_s + L_s p & 0 & L_m p \\ L_m p & \omega_r L_m & R_r + L_r p & \omega_r L_r \\ -\omega_r L_m & L_m p & -\omega_r L_r & R_r + L_r p \end{bmatrix} \begin{bmatrix} i_{s\alpha} \\ i_{s\beta} \\ i_{r\alpha} \\ i_{r\beta} \end{bmatrix} \tag{1}$$

The flux linkage equation is:

$$\begin{bmatrix}\psi_{s\alpha}\\ \psi_{s\beta}\\ \psi_{r\alpha}\\ \psi_{r\beta}\end{bmatrix}=\begin{bmatrix}L_s & 0 & L_m & 0\\ 0 & L_s & 0 & L_m\\ L_m & 0 & L_r & 0\\ 0 & L_m & 0 & L_r\end{bmatrix}\begin{bmatrix}i_{s\alpha}\\ i_{s\beta}\\ i_{r\alpha}\\ i_{r\beta}\end{bmatrix} \tag{2}$$

From $u_{s\alpha}, u_{s\beta}$ of formula (1) and $\psi_{r\alpha}, \psi_{r\beta}$ of formula (2) :

$$\begin{cases}\psi_{r\alpha} = \dfrac{L_r}{L_m}\left[\int (u_{s\alpha} - R_s i_{s\alpha})dt - \sigma L_s i_{s\alpha}\right]\\ \psi_{r\beta} = \dfrac{L_r}{L_m}\left[\int (u_{s\beta} - R_s i_{s\beta})dt - \sigma L_s i_{s\beta}\right]\end{cases} \tag{3}$$

In the formula, σ is the magnetic flux leakage coefficient, and $\sigma = 1 - L_m^2/(L_s L_r)$ .Equation (3) describes the relationship between the rotor flux change rate and the induced electromotive force, called the voltage model.

The rotor side of the asynchronous motor forms a closed loop, which can be regarded as a short circuit. Let $u_{r\alpha} = u_{r\beta} = 0$, from $u_{r\alpha}$, $u_{r\beta}$ of equation (1) and $\psi_{r\alpha}$, $\psi_{r\beta}$ of equation (2) of flux linkage we can obtain:

$$\begin{cases}p\psi_{r\alpha} = -\dfrac{1}{T_r}\psi_{r\alpha} - \omega_r\psi_{r\beta} + \dfrac{L_m}{T_r}i_{s\alpha}\\ p\psi_{r\beta} = -\dfrac{1}{T_r}\psi_{r\beta} + \omega_r\psi_{r\alpha} + \dfrac{L_m}{T_r}i_{s\beta}\end{cases} \tag{4}$$

Equation (4) represents the functional relationship between $\psi_{r\alpha}$, $\psi_{r\beta}$ and $i_{s\alpha}$, $i_{s\beta}$, which is the current model.

Substituting the updated value of the correlation amount in the formula (4) into the following:

$$\begin{cases}p\hat{\psi}_{r\alpha} = -\dfrac{1}{T_r}\hat{\psi}_{r\alpha} - \hat{\omega}_r\hat{\psi}_{r\beta} + \dfrac{L_m}{T_r}i_{s\alpha}\\ p\hat{\psi}_{r\beta} = -\dfrac{1}{T_r}\hat{\psi}_{r\beta} + \hat{\omega}_r\hat{\psi}_{r\alpha} + \dfrac{L_m}{T_r}i_{s\beta}\end{cases} \tag{5}$$

Where $\hat{\omega}_r$ is the speed estimate, $\hat{\psi}_{r\alpha}$ and $\hat{\psi}_{r\beta}$ are the rotor flux estimates.

The error formula for defining the rotor flux linkage is:

$$\begin{bmatrix}e_{r\alpha}\\ e_{r\beta}\end{bmatrix}=\begin{bmatrix}\psi_{r\alpha}\\ \psi_{r\beta}\end{bmatrix}-\begin{bmatrix}\hat{\psi}_{r\alpha}\\ \hat{\psi}_{r\beta}\end{bmatrix} \tag{6}$$

Substituting $p\psi_{r\alpha}$, $p\psi_{r\beta}$, $p\hat{\psi}_{r\alpha}$, $p\hat{\psi}_{r\beta}$ and respectively into the rotor flux error formula can be obtained (7):

$$\dot{e} = p\begin{bmatrix}e_{r\alpha}\\ e_{r\beta}\end{bmatrix} = Ae - W_\omega \tag{7}$$

It is obtained by Popov's hyper stability theory, V=Ge, where the value of G is related to system stability. In order to simplify the calculation, take G=I, then $V = e$

$$\hat{\omega}_r = \int_0^t f_1(e, t, \tau)d\tau + f_2(e, t) \tag{8}$$

Substituting V , W_ω and $\hat{\omega}_r$ into Popov integral inequality and decomposing:

$$\begin{cases} \eta_1(0, t_0) \geq -\gamma_1^2 \\ \eta_2(0, t_0) \geq -\gamma_2^2 \end{cases} \tag{9}$$

Where γ_1^2 and γ_2^2 are finite positive numbers with the same properties as γ^2.

When $k_i > 0$, inequality (10) is always established.

$$\int_0^{t_0} k_i \dot{x}(t)x(t)dt = \frac{k_i}{2}[x^2(t_0) - x^2(0)] \geq -\frac{k_i}{2}x^2(0) \tag{10}$$

For $\eta_1(0,t_0)$, the value is as follows:

$$\gamma_1^2 = \frac{k_i}{2}x^2(0) \tag{11}$$

$$\dot{x}(t) = e_{r\beta}\hat{\psi}_{r\alpha} - e_{r\alpha}\hat{\psi}_{r\beta} \tag{12}$$

$$k_i x(t) = \int_0^t f_1(e, t, \tau)d\tau - \omega_r \tag{13}$$

Then $\eta_1(0, t_0) \geq -\gamma_1^2$ can be guaranteed. Calculated by equations (12) and (13) we can get:

$$f_1(e, t, \tau) = k_i(e_{r\beta}\hat{\psi}_{r\alpha} - e_{r\alpha}\hat{\psi}_{r\beta}) \tag{14}$$

For $\eta_2(0, t_0)$, if the integrand is greater than zero, then inequality $\eta_2(0, t_0) \geq -\gamma_2^2$ must be true, so take:

$$f_2(e, t) = k_p(e_{r\beta}\hat{\psi}_{r\alpha} - e_{r\alpha}\hat{\psi}_{r\beta}) \tag{15}$$

Substituting $f_1(e,t,\tau)$ and $f_2(e,t)$ into equation (8), the rotating speed adaptation law is

$$\hat{\omega}_r = k_i\int_0^t(\psi_{r\beta}\hat{\psi}_{r\alpha} - \psi_{r\alpha}\hat{\psi}_{r\beta})d\tau + k_p(\psi_{r\beta}\hat{\psi}_{r\alpha} - \psi_{r\alpha}\hat{\psi}_{r\beta}) \tag{16}$$

Switch to the S domain, which is:

$$\hat{\omega}_r = (k_p + \frac{k_i}{s})(\psi_{r\beta}\hat{\psi}_{r\alpha} - \psi_{r\alpha}\hat{\psi}_{r\beta}) \tag{17}$$

4. On-line identification of rotor time constant

The online identification of the rotor time constant is based on the MRAS method, and the commonality of the rotational speed adaptive law derivation is that the reference model and the adjustable model are the same. While the difference is that the two adaptive laws are different[5].

Error equation of state related to the rotor time constant:

$$\begin{aligned} \dot{e} &= p\begin{bmatrix} e_{r\alpha} \\ e_{r\beta} \end{bmatrix} \\ &= \begin{bmatrix} -\frac{1}{T_r} & -\omega_r \\ \omega_r & -\frac{1}{T_r} \end{bmatrix}\begin{bmatrix} e_{r\alpha} \\ e_{r\beta} \end{bmatrix} - \left(\begin{bmatrix} \frac{1}{\hat{T}_r} - \frac{1}{T_r} & 0 \\ 0 & \frac{1}{\hat{T}_r} - \frac{1}{T_r} \end{bmatrix}\begin{bmatrix} \hat{\psi}_{r\alpha} \\ \hat{\psi}_{r\beta} \end{bmatrix} - \left(\frac{1}{\hat{T}_r} - \frac{1}{T_r}\right)\begin{bmatrix} L_m i_{s\alpha} \\ L_m i_{s\beta} \end{bmatrix}\right) \\ &= Ae - W_T \end{aligned} \tag{18}$$

If the nonlinear time-varying link satisfies the Popov inequality, the system will approach stability. For the sake of convenience, directly ask for the adaptive law of $\frac{1}{\hat{T}_r}$:

IOP Conf. Series: Materials Science and Engineering 533 (2019) 012024 doi:10.1088/1757-899X/533/1/012024

$$\frac{1}{\hat{T}_r} = \int_0^t f_1(e, t, \tau)d\tau + f_2(e, t) \tag{19}$$

Substituting $V = e$, W_T ,and $\frac{1}{\hat{T}_r}$ into the Popov integral inequality and decomposing, calculated:

$$f_1(e, t, \tau)=f_2(e, t) \tag{20}$$

Substituting $f_1(e,t,\tau)$ and $f_2(e,t)$ into $\frac{1}{\hat{T}_r}$ and switching to the S domain, will get the following formula:

$$\frac{1}{\hat{T}_r} = (k_p + \frac{k_i}{s})[(L_m i_{s\alpha} - \hat{\psi}_{r\alpha})(\psi_{r\alpha} - \hat{\psi}_{r\alpha}) + (L_m i_{s\beta} - \hat{\psi}_{r\beta})(\psi_{r\beta} - \hat{\psi}_{r\beta})] \tag{21}$$

5. Simulation and experimental verification

In order to verify the feasibility and accuracy of the method proposed for motor speed estimation and parameter identification, a simulation model is built on the basis of MATLAB software. System simulation diagram shown in Figure 2.

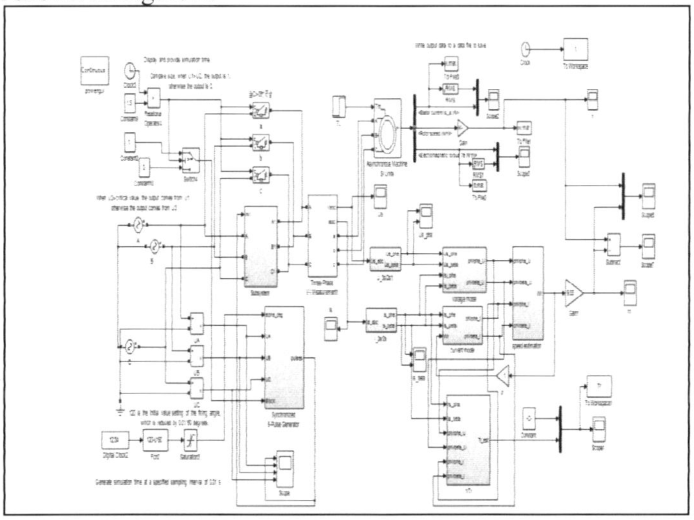

Figure 2. system simulation diagram.

Asynchronous motor was used for the simulation, and the relevant parameters are as follows:

Table1. Asynchronous motor related parameters.

Rated power P_N	1.1KW	Stator resistance R_s	5.27 Ω
Rated speed nN	2850 r/min	Rotor resistance Rr	5.07 Ω
Rated current I_N	2.5A	Stator leakage inductance Lls	0.0304H
Rated voltage U_N	380V	Rotor leakage inductance Llr	0.0298H
Rated frequency f_N	50Hz	Mutual sense Lm	0.394H
Number of pole pairs P	1	Moment of inertia J	0.02kg·m^2

5.1. *MRAS method based on rotor flux linkage*

In the simulation, according to the change trend of the whole speed under different values of kp and ki, as shown in Fig. 3, the PI parameter value is determined after multiple adjustments: kp=100,ki=22000. and the simulation results under no load are shown in Fig. 4 It can be seen that the estimated rotational speed waveform and the actual waveform are basically coincide, and the rotational speed of the motor is relatively large when the motor is started, when the motor is fully started, the error is basically zero.

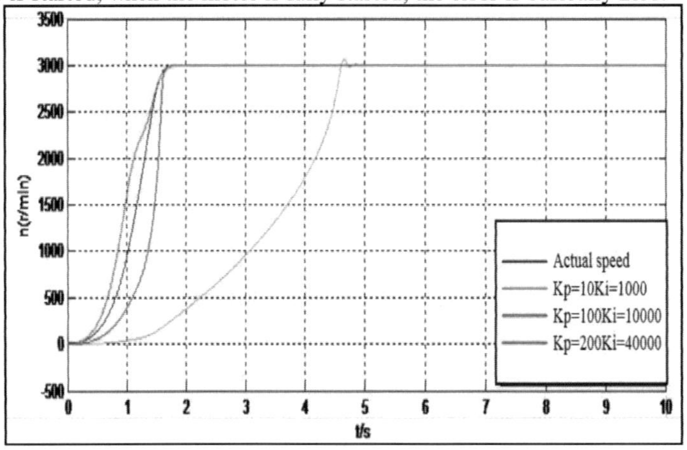

Figure 3. Comparison of estimated waveforms with actual waveforms in three different cases.

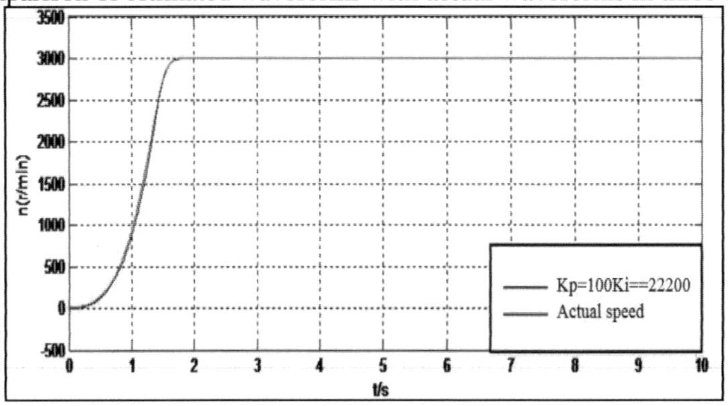

Figure a. Speed waveform.

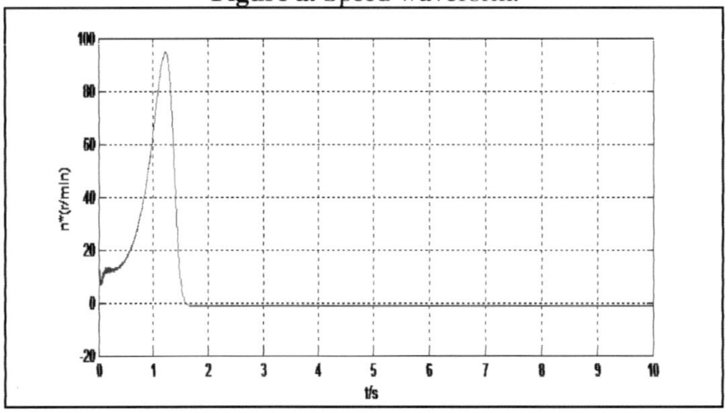

Figure b. Speed error waveform.

Figure 4. No-load speed waveform.

5.2. *Rotor time constant online identification simulation*

The simulation module constructed according to the derived rotor time constant adaptive law[6] is shown in Fig. 5.

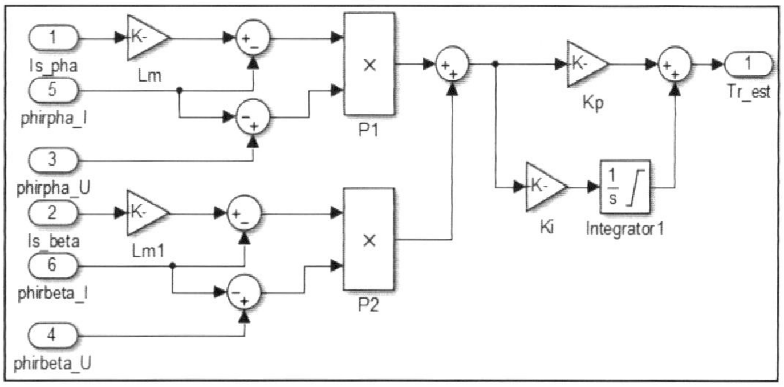

Figure 5. Rotor time constant adaptive law module simulation diagram.

Figure 6 shows the simulation results of online identification. Similarly, take different values for kp and ki, observe the trend, and adjust repeatedly to get kp=0.7, ki=39. The reciprocal of the rotor time constant calculated according to Equation 1/ Tr= Rr /Lr is 11.958 s. It can be clearly seen from Fig. 6(a) that 1/Tr can be stabilized to the actual set value in 1s~2s, and the convergence speed is fast. It can be seen from Fig. 6(b) that the difference between the actual value and the estimated value is small, which proves that the online identification result is in good agreement with the actual one.

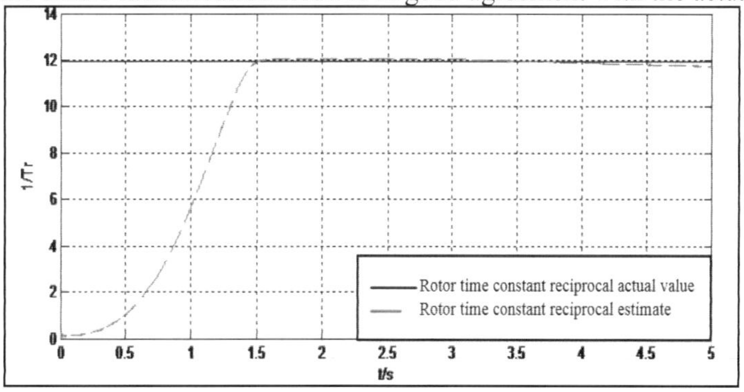

Figure a. 1/Tr online identification waveform.

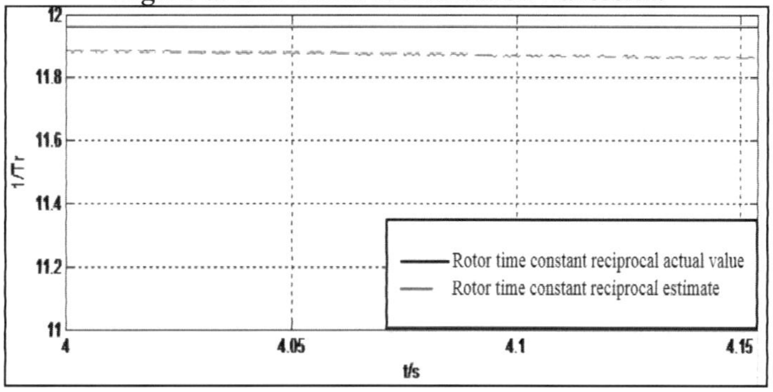

Figure b. 1/Tr online identification of steady-state amplified waveform..

Figure 6. A online identification results.

Based on the above theoretical basis and simulation analysis, a related hardware experiment platform was built. The motor parameters used in the test platform are shown in Table 1. The motor is started under no-load conditions. The voltage and current waveforms are shown in Figure 7. The three curves represent the voltage rms, the current rms, and the current instantaneous value.

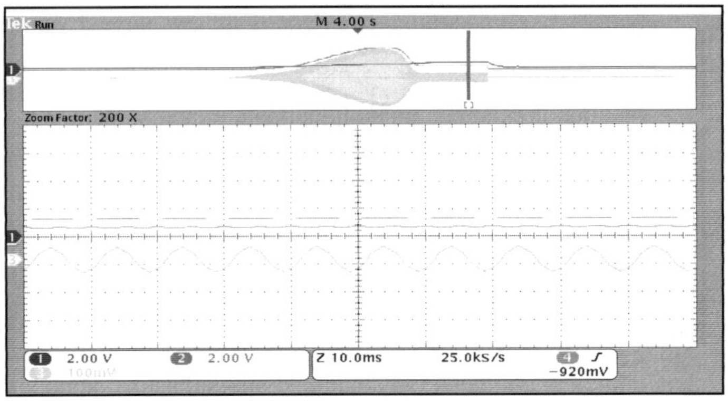

Figure 7. No-load starting voltage and current waveform.

Figure 8 shows the comparison between the actual speed and the estimated speed waveform. The actual speed of the asynchronous motor is indicated by the blue line, the actual speed is collected by the installation speed sensor and obtained after conversion; the red line indicates the estimated speed.

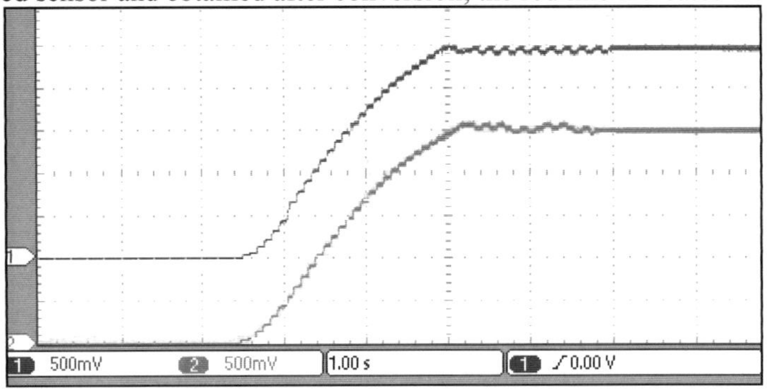

Figure 8. Comparison of actual speed and estimated speed waveform.

6. Conclusion

Based on the model reference adaptive method, this paper deeply studies the method of asynchronous motor speed estimation and parameter identification. The adaptive laws of rotational speed and rotor time constant are deduced respectively, and the relevant simulation models are built. Compared with the actual values, the estimated values of rotational speed and rotor time constant are more accurate, and the feasibility of the method is verified. Good results have also been achieved on the hardware experimental platform, providing a basis for practical application in the future.

Acknowledgments

At the completion of this thesis, I would like to express my sincere gratitude and most sincere respect to the instructor, Teacher Tong Jun. The unswerving scientific spirit of the teacher, the profound knowledge, the work attitude of the facts, the generous and modest spirit, and the spirit of selfless dedication have always inspired me to constantly learn knowledge and pursue the realm of life. Inexhaustible, there are too many people who have helped me, sincerely thank all those who care and help me, thank them for giving me the power to overcome myself, transcend myself, and constantly pursue, thank you.

References

[1] Chen Boshi. Electric Drag Automatic Control System——Motion Control System [M]. Beijing:Mechanical Industry Press, 2006.

[2] Han Zhengzhi, Chen Pengnian, Chen Shuzhong. Adaptive Control [M]. Beijing: Tsinghua University Press, 2014.

[3] GC Goodwin, KS Sin. Adaptive Filtering Prediction and Control[J].Prentice-Hall, 2009, 5(4):21-21.
[4] Wang Chuan, Song Shenmin, Zhang Fuen, et al. Indirect Adaptive Fuzzy Controller Based on Superstability Theory[J]. Journal of Harbin Engineering University, 2006, 27(2): 238-242.
[5] Yang Yan, Song Guojie. Teaching Design on the Concept of Lyapunov Stability[J]. Curriculum Education Research, 2015(14): 145-146.
[6] Chen Wei, Yan Wenlong, Lu Haifeng. A Speed Identification Method of Asynchronous Motor Based on MRAS[J]. Advanced Technology of Electrical Engineering and Energy, 2006, 25(2): 52-55.Luigi T.De Luca, Propulsion physic

IOP Conf. Series: Materials Science and Engineering **533** (2019) 012025 doi:10.1088/1757-899X/533/1/012025

Design of five axis welding robot based on leadshine control system

Zhao Yongguo[1], Li Minglei[2], Wang Peng[1], Han Yuanda[3]

[1]Institute of Automation,Qilu University of Technology(Shandong Academy of Science),Shandong Provincial Key Laboratory of Robot and Manufacturing Automation Technology,Jinan,China
[2]School of Electrical Engineering of Qilu university of technology(shandong academy of science), Jinan,China
[3]Shandong Naikan Intelligent Equipment Co.,Ltd,Dezhou,China

qd_lml@163.com

Abstract. The five-axis welding robot based on leadshine control system can accomplish various welding tasks in industry, and can accomplish arc welding and linear welding and other welding types. When welding, the machine has small chattering and high welding accuracy. In this paper, the D-H linkage coordinate system and kinematics forward and inverse solutions of the robot are analyzed, and the reachable space of the robot is calculated by Monte Carlo method. The hardware system of the robot is also summarized in more detail in this paper. Using SimMechanics tools to model the robot. Then using MATLAB function to validate the model, recording the difference between theoretical data and simulation data, and judging the correctness of the theory according to the difference.

1. Introduction

As we all know, the industrial robot manipulator plays an important role in today's industrial production. The terminal manipulator is programmed to move along the prescribed Cartesian space[1-2]. In order to obtain the trajectory of the end manipulator, its inverse kinematics must be solved. For welding robot, the most important part is to analyze the kinematics of the robot and then to model and simulate the robot[3]. However, for many welding robot mechanisms, the inverse kinematics can not get the solution of the joint angle, or there are many solutions, which brings many problems to the path planning of the welding robot[4-5].

In this topic, the hardware connection of the robot is completed and the teaching device is compiled.After analyzing the D-H link coordinate system[6], the forward and inverse kinematics solutions are obtained. It is convenient to plan the path of robots later[7]. In solving the inverse kinematics of the robot, the algebraic method and the geometric method are combined, and the robot model is modeled and simulated by the tools of MATLAB and SolidWorks[8]. Monte Carlo method is applied to display the motion space of the robot[9-10]. After several cycles of the end position, the points are displayed in three-dimensional space, so that the reachable space of the end of the robot can be perceived intuittively. At the same time,using matlab function to validate the robot model established by SimMechanics toolbox, and determine the correctness of the forward and inverse solutions of the robot by error.

Content from this work may be used under the terms of the Creative Commons Attribution 3.0 licence. Any further distribution of this work must maintain attribution to the author(s) and the title of the work, journal citation and DOI.
Published under licence by IOP Publishing Ltd

IOP Conf. Series: Materials Science and Engineering 533 (2019) 012025 doi:10.1088/1757-899X/533/1/012025

2. Design of Welding Robot

Firstly, this paper will introduce the whole control system of the robot in detail, as shown in Fig. 1 is the control connection diagram of each module of the five-axis welding robot in this project. The control system can complete the motion control and manual teaching function of the robot. Among them, the motion controller is the core, coding various instructions and controlling the movement.

Other paragraphs are indented (BodytextIndented style).

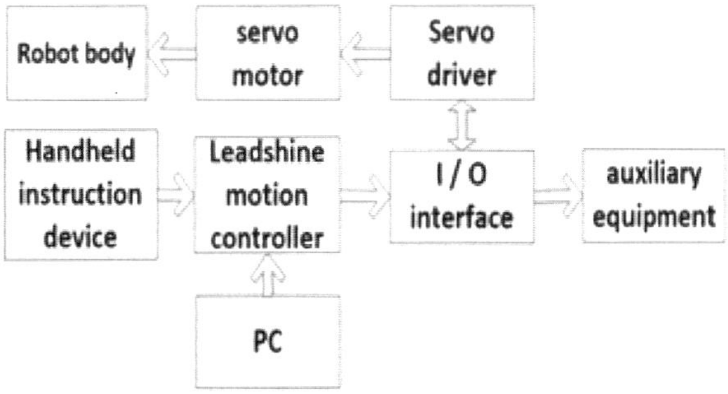

Figure 1. Control chart of robot system

Because before modeling and simulation, we must have a more detailed understanding of the various mechanical structures and parameters of the robot, in order to ensure the accuracy of the later data, so this paper introduce the robot ontology used in this project by means of charts.

Table1. A slightly more complex table with a narrow caption.

Performance	quota
Outline size and weight	1940mm×1240mm×610mm; 150kg;
Maximum uniaxial speed	1#: ≥112°/s; 2#: ≥112°/s; 3#: ≥ 750mm/s; 4#: ≥850°/s; 5#: ≥120°/s
Rated load	6kg
Welding speed per minute	Vmin≤450mm/min; Vmax≥5000mm/min
Basic motion control mode	Computer entry mode
work space	Rmax: 1420±10mm; Hoax: 650±10mm
Trajectory repeatability	Under the no-load speed of 9000 mm/min, the robot runs 500 times along the set trajectory, and the deviation between the trajectories does not exceed 0.15 mm.

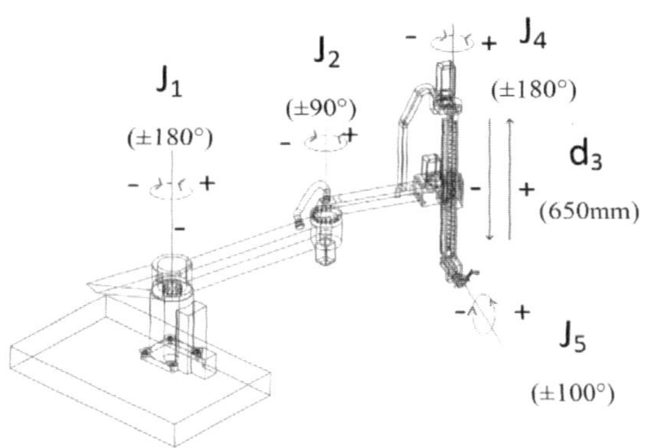

Figure 2. Robot parameter diagram

IOP Conf. Series: Materials Science and Engineering **533** (2019) 012025 doi:10.1088/1757-899X/533/1/012025

According to the above figure, it can clearly seen that the robot belongs to a series structure robot with five joints, four of which are rotating joints and one is moving joints. In addition, when analyzing the welding joints of the robot in Chapter 3,it is necessary to establish a coordinate system to analyze the fixed welding joints trajectory.
The hardware control system and robot ontology are as follows:

Figure 3. Physical diagram of the hardware connection and robot ontology map

3. Kinematics analysis of robot

In this chapter, analyzing the forward and inverse solutions of the robot in detail. The forward solutions can help us get the position and attitude of the end of the robot conveniently, and the inverse solutions can help us get the joint variables of each joint according to the position and attitude. Firstly, the D-H linkage coordinate system is established according to the robot. Because it involves the end-effector welding joint, this paper build it in two parts. But because the coordinate system established in the welding joint is fixed, we will not do more research once.
The D-H link coordinate system is shown below. The origin of 0 and 1 coordinate systems coincide, and the origin of 3, 4 and 5 coordinate systems coincide.

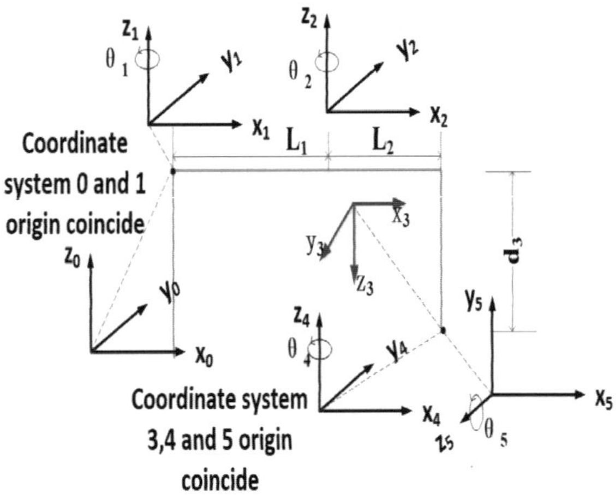

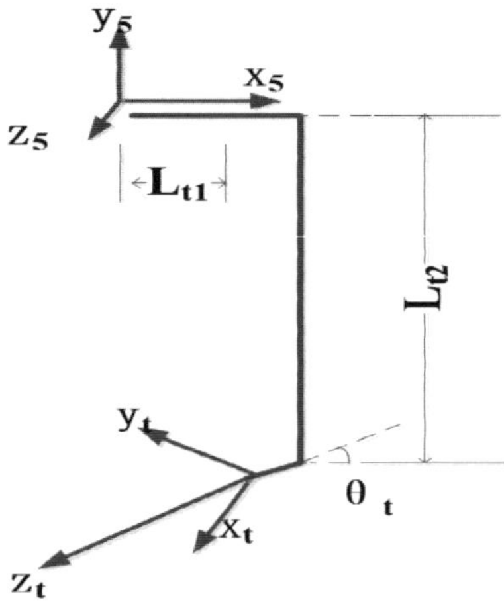

Figure 4. Connecting Rod Coordinate System of Robot Body and Welding Joint

Z_t is aligned with the normal end of the welding torch, pointing to the workpiece. X_t and Z_5 are in the same direction, the origin is at the end of the welding torch.
The following table is an introduction to the parameters of D-H connecting rod. And the positive and negative solutions will be displayed.

Table 2. D-H connecting rod parameters

i	a_{i-1}	α_{i-1}	d_i	θ_i	offset
1	0	0	0	θ_1	0
2	L_1	0	0	θ_2	0
3	L_2	180°	d_3	0	0
4	0	180°	0	θ_3	0
5	0	90°	0	θ_4	0

The following formula is a demonstration of the positive solution of kinematics.

$$ {}_5^0T={}_1^0T{}_2^1T{}_3^2T{}_4^3T{}_5^4T = \begin{bmatrix} T_{5\_11} & T_{5\_12} & T_{5\_13} & T_{5\_14} \\ T_{5\_21} & T_{5\_22} & T_{5\_23} & T_{5\_24} \\ s\theta_5 & c\theta_5 & 0 & -d_3 \\ 0 & 0 & 0 & 1 \end{bmatrix} \tag{1} $$

$$ \begin{aligned} T_{5\_11} &= c\theta_1 c\theta_2 c\theta_4 c\theta_5 - c\theta_1 s\theta_2 s\theta_4 c\theta_5 - s\theta_1 c\theta_2 s\theta_4 c\theta_5 - s\theta_1 s\theta_2 c\theta_4 c\theta_5 \\ T_{5\_12} &= c\theta_1 s\theta_2 s\theta_4 s\theta_5 - c\theta_1 c\theta_2 c\theta_4 s\theta_5 + s\theta_1 c\theta_2 s\theta_4 s\theta_5 + s\theta_1 s\theta_2 c\theta_4 s\theta_5 \\ T_{5\_13} &= c\theta_1 c\theta_2 s\theta_4 - s\theta_1 s\theta_2 s\theta_4 + c\theta_1 s\theta_2 c\theta_4 + s\theta_1 c\theta_2 c\theta_4 \\ T_{5\_14} &= L_1 c\theta_1 + L_2 c\theta_1 c\theta_2 - L_2 s\theta_1 s\theta_2 \\ T_{5\_21} &= c\theta_1 c\theta_2 s\theta_4 c\theta_5 + c\theta_1 s\theta_2 c\theta_4 c\theta_5 + s\theta_1 c\theta_2 c\theta_4 c\theta_5 - s\theta_1 s\theta_2 s\theta_4 c\theta_5 \\ T_{5\_22} &= -c\theta_1 c\theta_2 s\theta_4 s\theta_5 - c\theta_1 s\theta_2 c\theta_4 s\theta_5 - s\theta_1 c\theta_2 c\theta_4 s\theta_5 + s\theta_1 s\theta_2 s\theta_4 s\theta_5 \\ T_{5\_23} &= c\theta_1 s\theta_2 s\theta_4 + s\theta_1 c\theta_2 s\theta_4 + s\theta_1 s\theta_2 c\theta_4 - c\theta_1 c\theta_2 c\theta_4 \\ T_{5\_24} &= L_1 s\theta_1 + L_2 c\theta_1 s\theta_2 + L_2 s\theta_1 c\theta_2 \end{aligned} \tag{2} $$

Then the formula for welding head is analyzed.

$$ {}_t^5T = \begin{bmatrix} R_Z(\theta_t)R_Y(-90^\circ) & \begin{matrix} L_{t1} \\ -L_{t2} \\ 0 \end{matrix} \\ 0\ \ 0\ \ 0 & 1 \end{bmatrix} = \begin{bmatrix} 0 & -s\theta_t & -c\theta_t & L_{t1} \\ 0 & c\theta_t & -s\theta_t & -L_{t2} \\ 1 & 0 & 0 & 0 \\ 0 & 0 & 0 & 1 \end{bmatrix} \tag{3} $$

$$ {}_t^0T={}_5^0T{}_t^5T \tag{4} $$

Next we will analyze the inverse kinematics solution. When $x_{5_o}{}^2 + y_{5_o}{}^2 = (L_1 + L_2)^2$, θ_1 and θ_2 have one sets of solutions. Here applying the geometric solution.

$$ \begin{cases} \theta_1 = A\tan 2(y_{5_o}, x_{5_o}) \\ \theta_2 = 0 \end{cases} \tag{5} $$

When $(L_1 - L_2)^2 \le x_{5_o}{}^2 + y_{5_o}{}^2 < (L_1 + L_2)^2$, θ_1 and θ_2 have two sets of solutions.

$$\begin{cases}\theta_1 = A\tan 2(y_{5o}, x_{5o}) - A\cos(\dfrac{x_{5o}{}^2 + y_{5o}{}^2 + L_1{}^2 - L_2{}^2}{2L_1\sqrt{x_{5o}{}^2 + y_{5o}{}^2}}) \\ \theta_2 = A\cos(\dfrac{x_{5o}{}^2 + y_{5o}{}^2 - L_1{}^2 - L_2{}^2}{2L_1L_2})\end{cases}$$
$$\begin{cases}\theta_1 = A\tan 2(y_{5o}, x_{5o}) + A\cos(\dfrac{x_{5o}{}^2 + y_{5o}{}^2 + L_1{}^2 - L_2{}^2}{2L_1\sqrt{x_{5o}{}^2 + y_{5o}{}^2}}) \\ \theta_2 = -A\cos(\dfrac{x_{5o}{}^2 + y_{5o}{}^2 - L_1{}^2 - L_2{}^2}{2L_1L_2})\end{cases} \quad (6)$$

Before solving θ4 and θ5, we first set the analytical formula of the tool coordinate system as follows. Among them, ${}^0_tT = {}^0_GT$.

$${}^0_GT = \begin{bmatrix} g_{11} & g_{12} & g_{13} & g_{14} \\ g_{21} & g_{22} & g_{23} & g_{24} \\ g_{31} & g_{32} & g_{33} & g_{34} \\ 0 & 0 & 0 & 1 \end{bmatrix} \quad (7)$$

According to θ_1、 θ_2 and θ_t, I can get θ_4 and θ_5 analytic formula.

$$\begin{cases}\theta_4 = A\tan 2(g_{11}c(\theta_1+\theta_2) + g_{21}s(\theta_1+\theta_2), \\ \quad\quad g_{11}s(\theta_1+\theta_2) - g_{21}c(\theta_1+\theta_2)) \\ \theta_5 = A\tan 2(-g_{32}s\theta_t - g_{33}c\theta_t, g_{32}c\theta_t - g_{33}s\theta_t)\end{cases} \quad (8)$$

So two sets of inverse solutions can be obtained. Secondly, because ±180° is actually a position in programming, we all consider these four joint variables to be within the range of (-180° ~ +180°].

4. Model simulation

This chapter will introduce the simulation of a five-axis welding robot. Firstly, using Monte Carlo method to visually display the workspace of the robot, and then using SolidWorks to model the robot and cooperate with MATLAB to verify whether the forward and inverse solutions in Chapter 3 are correct. Finally, I will make an error in the theory and simulation.

4.1. Robot workspace

In this figure, showing the three-dimensional view of the workspace as well as its face view, overhead view and side view.

Since MATLAB is mainly used for simulation, this paper have shown the following steps:

1) Generating random variables of joints and moving pairs.

2) The joint variable value is substituted in the wrist coordinate system relative to the 0 coordinate system.

3) Search for random points in each tier and column.

4) Ten new random points are generated near each searched boundary point, and the generated data are stored in the matrix of the boundary point in sequence.

IOP Conf. Series: Materials Science and Engineering 533 (2019) 012025 doi:10.1088/1757-899X/533/1/012025

5) Find the next boundary point and repeat the first four steps with the boundary point until the end of the cycle number or the random points of the boundary change little.

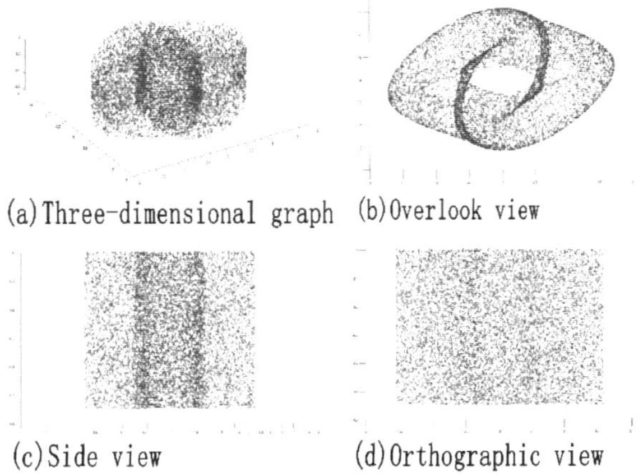

Figure 5. Motion space illustration

4.2. Modeling and combining with MATLAB to verify the correctness of positive and negative solutions.

The quality of SolidWorks modeling determines the degree of integration when the model matches MATLAB in the later stage. Through the actual measurement of the robot, the three-dimensional views of each workpiece and the overall framework of the robot are drawn, and then the three-dimensional views are presented in SolidWorks.

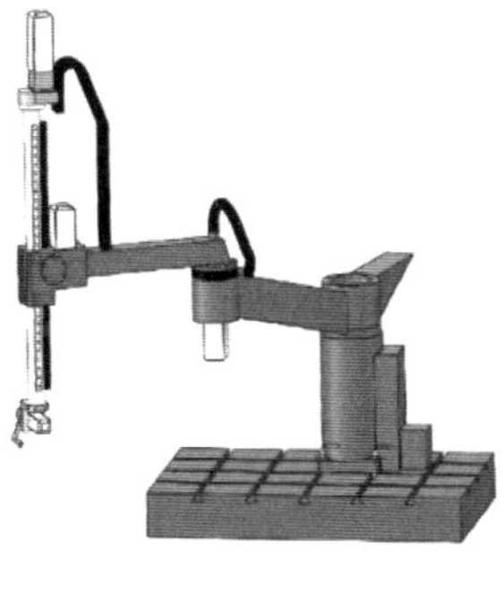

Figure 6. Three dimensional simulation of robot

The robot model is imported into MATLAB and connected with the model established by SimMechanics. Then, the joint angle obtained by theoretical calculation is input into the model, and then the model is moved to see if the final movement is beyond the workspace. If all arrival positions are within the workspace after further verification, it shows that the previous theoretical calculation is correct and conforms to the structure of the robot.

When the model is established, the types of joints (moving or rotating joints) are considered firstly, and then the parameters of each manipulator are defined in the program, the corresponding input parameters are set and the model is run. The obtained data are stored in the corresponding modules for later processing and comparison.

Specific model building is shown in Figure 7.

IOP Conf. Series: Materials Science and Engineering **533** (2019) 012025 doi:10.1088/1757-899X/533/1/012025

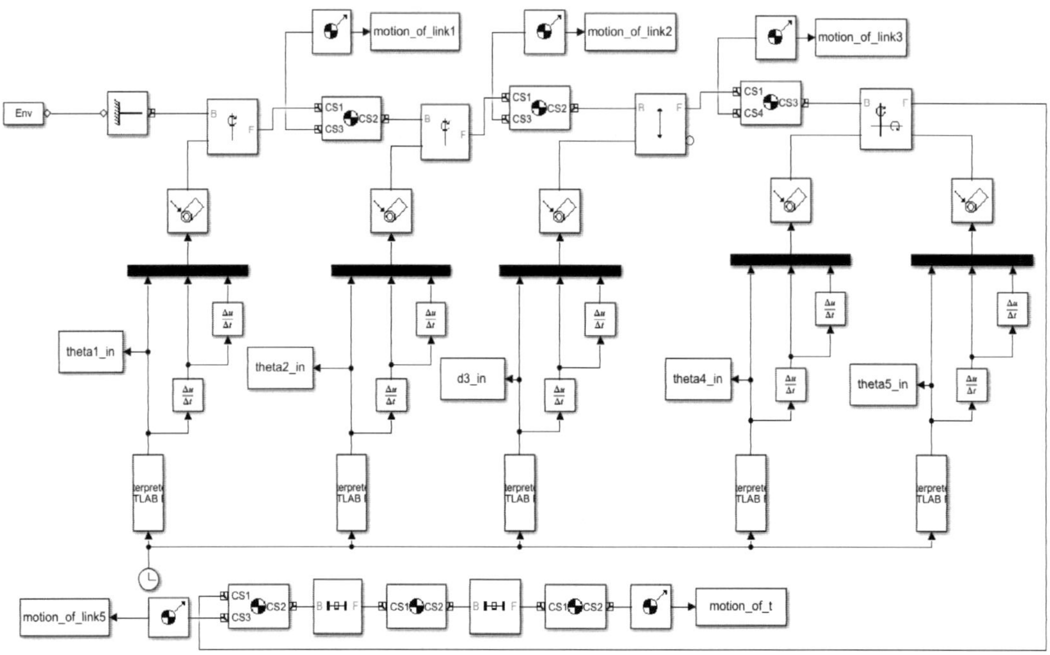

Figure 7. Establishment of simulation model

4.3. Errors between simulation and theory

Firstly, we find a circular arc curve in the workspace, fix 10 points on the arc curve, get the coordinates of these 10 points through SolidWorks, and calculate the theoretical inverse solution through the position of these 10 points. Then the theoretical inverse solutions of these 10 points are brought into the model established by SimMechanics to get the position of the end effector, and then five joint variables are solved again through the end position. Finally, the five joint variables are compared.

point	theta_1	theta_2	d_3	theta_4	theta_5
1	-1.20477548284713	2.08525753107086	-0.86804	0.690314278571168	-1.110223e-16
2	-1.13168742779630	2.19452067827626	-0.86804	0.507963076314937	-1.110223e-16
3	-0.992150789991377	2.29473098772504	-0.86804	0.268216129061239	-1.110223e-16
4	-0.761207939378694	2.09968063987005	-0.86804	0.232323626303544	-1.110223e-16
5	-0.809282653405590	1.90445099283349	-0.86804	0.475627987366995	-1.110223e-16
6	-0.823382330926091	1.69723018994388	-0.86804	0.696948467777112	-1.110223e-16
7	-0.813043682659518	1.53331965839445	-0.86804	0.850520351059966	-1.110223e-16
8	-0.756865314679825	1.92457220109808	-0.86804	0.403089440376637	-1.110223e-16
9	-0.806485718806543	1.79309617486831	-0.86804	0.584185870733132	-1.110223e-16
10	-1.17425961283125	2.12864745151033	-0.86804	0.616408488115820	-1.110223e-16
1	-1.20374023682013	2.08572303567916	-0.86804	0.688813527935867	-1.110223e-16
2	-1.13419915326121	2.19391062395399	-0.86804	0.511084856102110	-1.110223e-16
3	-0.997232225061644	2.30177328901575	-0.86804	0.266255262840789	-1.110223e-16
4	-0.761207939378694	2.09968063987005	-0.86804	0.232323626303544	-1.110223e-16
5	-0.812664215986286	1.91396309907767	-0.86804	0.469497443703510	-1.110223e-16
6	-0.823382330926091	1.69723018994388	-0.86804	0.696948467777112	-1.110223e-16
7	-0.817426200945147	1.54250214450581	-0.86804	0.845720383234234	-1.110223e-16
8	-0.756376435993254	1.92299897470173	-0.86804	0.404173788086420	-1.110223e-16
9	-0.810434827746754	1.80299640725351	-0.86804	0.578234747288141	-1.110223e-16
10	-1.17664723526852	2.13294320724135	-0.86804	0.614500354822067	-1.110223e-16

Figure 8. Theoretical and simulation errors

5. Conclusion

In this topic, we have completed the basic kinematics analysis of the welding robot, obtained the correct forward and inverse solutions, and used SolidWorks and MATLAB to model the welding robot. The SimMechanics tool is used to simulate the motion and get the corresponding simulation data. According to the data analysis and theoretical analysis, the gap between simulation and theory is very small.

In future planning, hoping that the welding robot can complete more complex welding operations by changing the mechanical structure of the welding joint.

6. References

[1] Dahari M , Tan J D . Forward and inverse kinematics model for robotic welding process using KR-16KS KUKA robot[C]// International Conference on Modeling. IEEE, 2011.
[2] Jadran Lenarčič, Andreja Košutnik. Approximate calculation of robot inverse kinematics applied to arc welding[C]// International Symposium on Experimental Robotics I. Springer-Verlag, 1989.
[3] Jeon Y B , Kim S B , Park S S . Modeling and Motion Control of Mobile Robot for Lattice Type Welding[J]. Ksme International Journal, 2002, 16(1):83-93.
[4] Shaoke C , Xuan F , Hanyang L , et al. Trajectory planning for five-axis welding robot based on CAD/CAM technology[J]. Modern Manufacturing Engineering, 2017.
[5] Lu-Gang C , Xue-Liang P , Xi-Wen X U , et al. Trajectory Planning of Welding Robot Based on ADAMS[J]. Journal of Jiangnan University, 2011.
[6] Fayong G , Tao M , Jianghai Z . Problems and Improvement of D-H Method for Establishing Connecting Rod Coordinate System[J]. China Mechanical Engineering, 2014, 25(20):2711-2714.
[7] Liang-Yu L I , Yang L , Tian-Qi W , et al. Operation space trajectory planning of welding robot based on rational B-splines curve[J]. Journal of Tianjin Polytechnic University, 2015.
[8] Cheraghpour F , Vaezi M , Jazeh H E S , et al. Dynamic modeling and kinematic simulation of Stäubli© TX40 robot using MATLAB/ADAMS co-simulation[C]// IEEE International Conference on Mechatronics. IEEE, 2011.
[9] Zhizhong L , Liu. Improvement on Monte Carlo Method for Robot Workspace Determination[J]. Transactions of the Chinese Society for Agricultural Machinery, 2013, 44(1):230-235.
[10] Bin F U , Fei-Fei H U , Lin C , et al. The Solution of Robot Manipulator's Collision Free Workspace Based on Monte Carlo Method[J]. Modular Machine Tool & Automatic

Novel Experiment Design for Unmanned Aerial Vehicle Controller Performance Testing

Cherub Dim, Francis Nabor, Giancarlo Santos, Martin Schoeler, Alvin Chua*

Mechatronics Research Laboratory, Mechanical Engineering Department, De La Salle University, Manila, Philippines

*corresponding author: alvin.chua@dlsu.edu.ph

Abstract. Unmanned Aerial Vehicles (UAV) are becoming popular to the innovating world. This paper presents a novel experiment design to determine the performance of UAV control system. The hardware platform developed is composed of a quadcopter with a multipurpose autopilot system. While the experimental design considers both quantitative and qualitative aspects of the controller. The qualitative components dealt with the hardware compatibility, acceptability and license. On the other hand, the quantitative component measured the accuracy of waypoints, altitude hold and loiter performance of the UAV. The results shows that the new experiment design platform was able to compare two popular autopilot software in terms of performance and acceptability.

1. Introduction

Unmanned Aerial Vehicles (UAV), also known as drones, are becoming popular more and more. They provide humans with different applications similar but not limited to: mapping, delivery, and entertainment. The UAV could be driven remotely, or be programmed on autopilot. These UAV/drones vary through the hardware configuration in the platform: Multi-Rotor, Fixed-Wing, Single-Rotor, and Fixed-Wing Hybrid. Multi-Rotor drone contains multiple motor which drives the platform, an advantage would be high-precision flight, quick throttle changes, and heavy-weight lifting capability as compared with other types of drones. The disadvantage of multi-rotor drone is flight endurance-due to multiple motors, and speed limitation due to the rotors configuration[1][2][3]. Drone control software is in the heart of any type of UAV. The implemented controller to drive the drone will define its performance and capability to perform the specified tasks. Currently, most drones are already equipped with control software. However, some drones are driven by closed source or proprietary control software. These software are exclusively developed and distributed by the manufacturer. Without the ability to modify the drone's software, the drone's application will be limited only to what was pre-programmed by the manufacturer.[4][5][6] Furthermore, whenever there are occurrences of bugs and errors on the software, the patches and updates will be dependent on the software's manufacturer. Hence, open source UAV software are progressing wherein the user can modify and develop the program to suit the vehicle's application. One of the advantage of open source software is the development of the software is dependent on the community wherein innovative ideas, progresses and bug fixes are accomplished for the application of the autopilot software and is available for further improvement. Moreover, open source UAV software are freely distributed, and are available at the user's dispersal. There are many variations of open sourced UAV software; each equipped with its own capability and methods of implementations[7]

Content from this work may be used under the terms of the Creative Commons Attribution 3.0 licence. Any further distribution of this work must maintain attribution to the author(s) and the title of the work, journal citation and DOI.
Published under licence by IOP Publishing Ltd

This paper presents a new experimental design that is able to analyze the performance of an autopilot software. A quadrotor platform together with a multipurpose controller was used as the testbed of the experimental design. The experiment design focuses on the qualitative and quantitative aspects of the flight control system. Two well-known open source autopilot software in the market were chosen as test case to validate the experimental design.

2. Experimental Platform

The control system of a quadcopter plays a key role in its performance. The precise control of the attitude and altitude states of the aerial vehicle is very critical in different applications. This section describe the frame and the electronic components chosen for the experimental platform (see Figure 1). The frame of the quadcopter chosen is the DJI F450 Frame due to its proven design and functionality. The following are the other components: ReadyToSky 2212 920KV, ReadyToSky 30A Brushless ESC, Propellers, Dinogy 11.1V 5000mAh, RadioLink AT 10 II, 3DRobotics Radio Telemetry, and Ublox Neo-M8N GPS.

Figure 1. Experimental Platform

Two autopilot softwares are chosen in the testing namely: PX4[8][9] and Ardupilot[10]. These are the most widely used open source flight controller available in the market. They use different ground stations as shown below. For Arudpilot, they use Mission Planner (see Figure 2) while PX4 uses QGroundControl (see Figure 3).

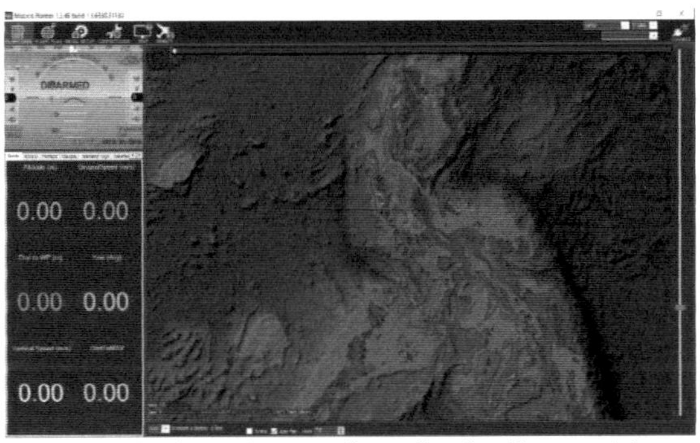

Figure 2. Mission Planner (http://ardupilot.org/planner/)

Figure 3. QGroundControl (http://qgroundcontrol.com/)

3. Methodology

The methodology flowchart was shown in Figure 4. This starts with the design and assembly of the quadcopter hardware as shown in the previous section. This was then followed by the software integration of Ardupilot and PX4 into the quadcopter flight controller. The experiment design developed considers both the qualitative and quantitative aspects of the flight control system. Finally, actual testing was done.

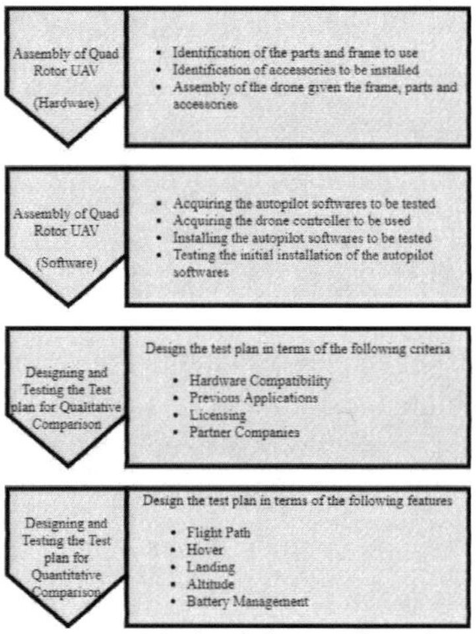

Figure 4. Methodology Flowchart

The design and testing of the research focuses on the test plan in qualitative and quantitative analysis. Qualitative Analysis accounts for the immeasurable data that couldn't be gathered from the testing. While the Quantitative analysis will account for the actual data that could be acquired using the experimental platform for actual data analysis.

3.1. Qualitative Analysis

Qualitative analysis includes the following: hardware compatibility, user experience, licensing, and partner companies. First, the hardware compatibility accounts for various particular operating system or device management software for the unmanned aerial vehicle in both ArduPilot and PX4 applications. Second, the user experience accounts for the different aspects of the interface which includes the calibration process, parameters guideline, PID tuning, and user interface (UI). These open source software have their own unique UI that help distinguish one from the other by having features and modes that each software is capable of doing within the interface. Also, the licensing plays a crucial role to the researcher so that it can be modified and redistributed. Finally, partner companies shows the support of different companies to the software adding value to its existence and improvements in the future.

3.2. Quantitative Analysis

On the other hand, the Quantitative Analysis relies on mathematical and statistical method to evaluate data points. Quantitative Analysis focuses on the Pre-Test such as GPS and Barometer Calibration, Flight Path in terms of time and accuracy, Loiter, Return to Launch, Altitude Control, and Battery Management.

Calibrating the barometer and GPS are needed to ensure that the data gathered with respect to altitude was accurate. The accuracy of the flight path for each software was tested and how fast the software can complete the run of the desired flight path. This is to determine which software has the better capabilities of following configured paths that will have low percentage errors while being the fastest to complete the run. Considering that all the devices and instruments attached to the drone are correctly calibrated. Figure 5 shows the flight path accuracy test points.

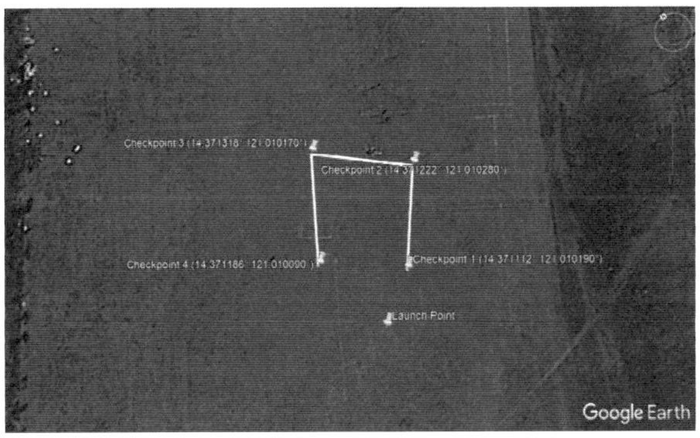

Figure 5. Flight Path (Accuracy and Time)

The stability of the drone during Loiter mode at a certain altitude was tested. This is to determine which software has the better capabilities of holding the latitude and longitude at a certain altitude with minimal errors. Figure 6 show the Loiter flight path.

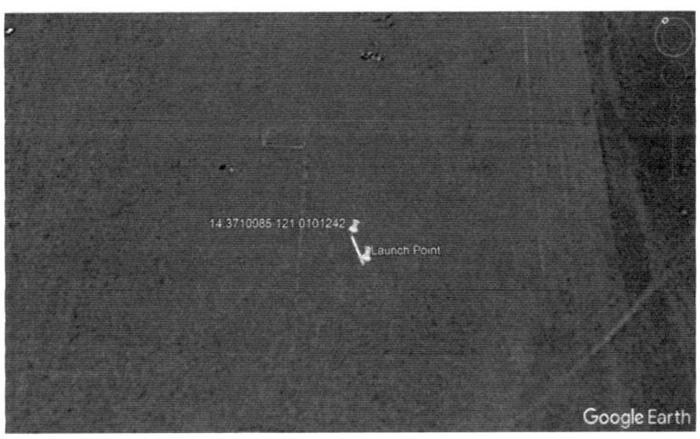

Figure 6. Loiter Flight Path

The return to launch (RTL) experimental procedure below was used to test the efficiency of the quadcopter to go back to its home position. The test will focus on the duration and deviation of the GPS coordinates the drone landed on to the original location. Figure 7 shows the RTL flight path.

Figure 7. Return to Home Flight Path

The altitude elevation accuracy of each software was tested. This is to determine which software has the better capabilities of following desired altitude elevations. Figure 8 shows the altitude control flight path.

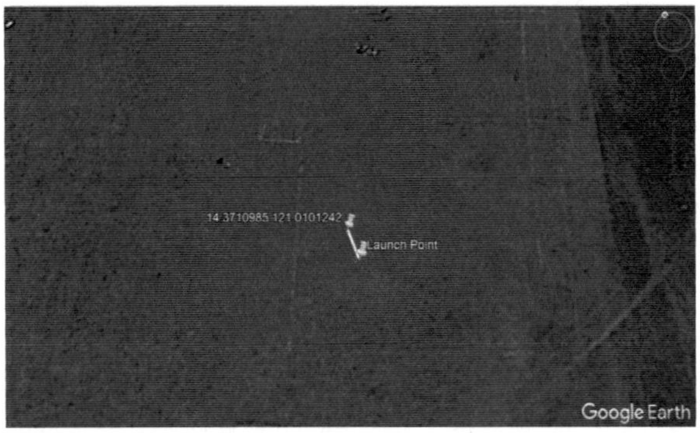

Figure 8. Altitude Hold Flight Path

The battery management of the software will be tested to determine how the two software consume battery in order to accomplish a certain task. Figure 9 shows the flight path.

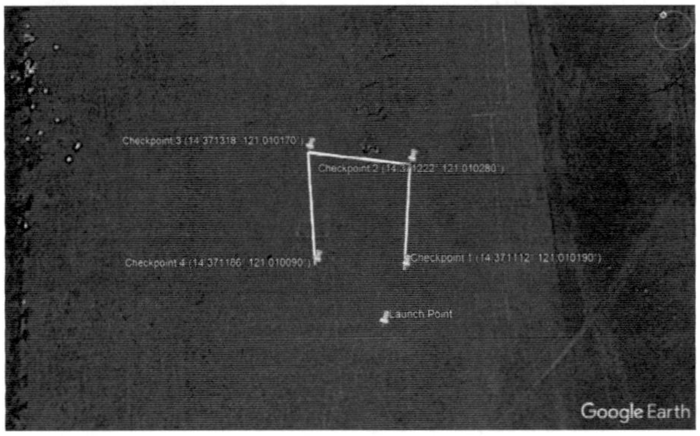

Figure 9. Battery management flight path

4. Results

After performing the experiment design and testing of the flight paths, the data gathered from the test was constructed and analyzed. Both ArduPilot and PX4 were compared qualitatively and quantitatively.

4.1. Qualitative Analysis

As seen from the items listed in their respective websites, it can be observed that ArduPilot is compatible to more hardware as compared to PX4 especially with closed source hardware with 11 different types of off-the-shelf drones to PX4's 3 compatible closed source hardware. Based on Table 1, it can be concluded that ArduPilot focuses more on their software being compatible to different kinds of open and closed source

Table 1. Hardware Compatibility Tally

ArduPilot		**PX4**	
Open Source Hardware	10	Open Source Hardware	13
Closed Source Hardware	11	Closed Source Hardware	3
Total	**21**	**Total**	**16**

ArduPilot has more partner companies for each software developer compared to PX4 comparing 39 to 24. Numerically, ArduPilot is greater than PX4 in number of partner companies, but the partners of PX4 are bigger companies weighing over the 39 partner companies of ArduPilot.

The licensing of the two software is very different from one another wherein these licenses focuses on two different scenarios. ArduPilot licensing, under the GNU General Public Version 3, has a focus on user experience and distribution of businesses and closed source hardware in which they must notify the developers or beta testers of any modifications or bugs in the software as well as notifying the end users that the software is open source. PX4, on the other hand, focuses more on the proper distribution of the software in which the copyright notice, list of conditions, and disclaimer must first be assented along with prior copyright holder and contributors.

Overall, the calibration process of PX4 is better than the calibration process of ArduPilot as it is more interactive and easier to understand as it uses pictures and graphics in order for the user to know what should be done next. PX4 also has a better calibration process because all components that need to be calibrated could be done within the UI.

For the Parameters changed, the use of ArduPilot is easier because unlike PX4 wherein the PID and other parameters had to be manually tuned, ArduPilot could be used immediately after uploading the software on to the drone as it already works smoothly right out of the box. The only parameter that had to be modified in ArduPilot was the function wherein the motors would not activate when the drone was armed before the flight.

Moreover, Ardupilot, again has a better user interface even if it has a cluttered interface and a not so streamlined interface. Ardupilot has more features available at its disposal than PX4 which is less refined with only limited features and flight modes. Another important feature that ArduPilot has is that the user can download and analyze data within the software and export it into a Matlab file where as PX4 requires the use of external software.

4.2. Quantitative Analysis

The barometer and the GPS module were tested. It was found that the average error of the barometer reading is 6.7% for Ardupilot and 2.6% for PX4. For the GPS module the average error for Ardupilot is 0.87% while PX4 is 0.24%.

The flight path accuracy tests shows that PX4 is better in checkpoint accuracy as compared to Ardupilot to which this difference could be attributed to the response of the software as the mission is being performed. Arudupilot registered an average error of 0.7% while PX4 had a 0.55% error. Figure 10 shows a sample run of the flight path accuracy data while Figure 11 show the flight path accuracy t-test. The t-test done showed that the resulting displacements of both software controllers are significantly different from one another. This would attributed to the difference in response and its respective performance in coordinates calculations.

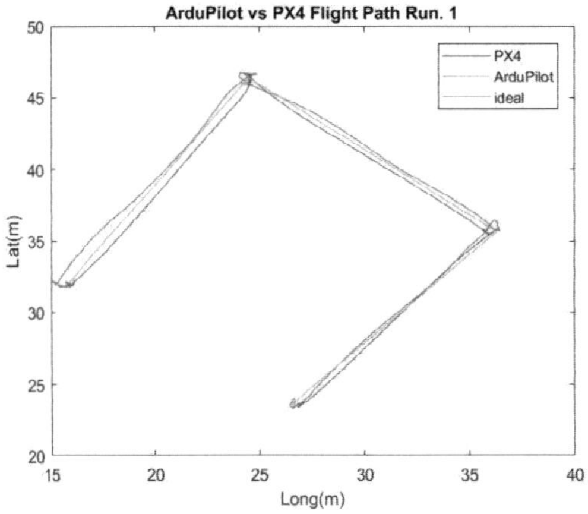

Figure 10. Flight Path Accuracy (Run 1) Plot

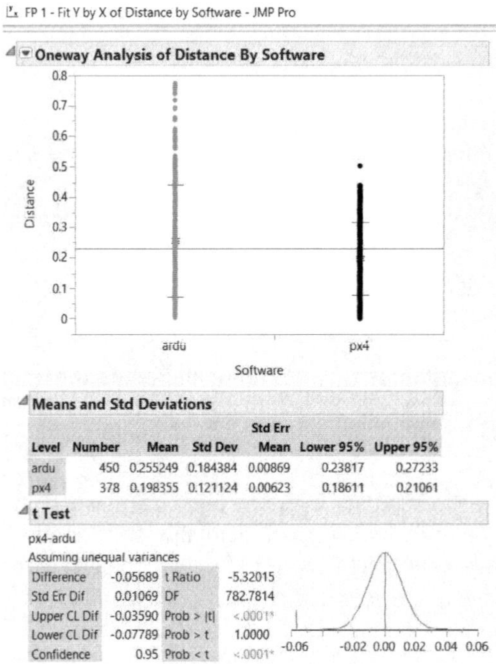

Figure 11. Flight Path Accuracy (Run 1) T-test Results

The flight path time test shows that PX4 performs the mission faster than Ardupilot by an average of 0.476 seconds. Voltage measurement were also checked and were observed to have insignificant effect on time completion of the mission. Table 2 shows that flight path time data summary.

IOP Conf. Series: Materials Science and Engineering 533 (2019) 012026 doi:10.1088/1757-899X/533/1/012026

Table 2. Flight Path Time Data Summary

	ArduPilot		PX4	
Run	**Time**	**Voltage**	**Time**	**Voltage**
1	17.09 s	12.2 V	17.30 s	12.2 V
2	17.34 s	12.0 V	16.08 s	12.1 V
3	17.04 s	11.8 V	16.85 s	11.9 V
4	17.06 s	11.7 V	16.56 s	11.8 V
5	17.02 s	11.6 V	16.38 s	11.7 V
Average	**17.11 s**		**16.634 s**	

From Figure 10, ArduPilot have points wherein there are notably large displacements. In terms of speed, PX4 is also faster than ArduPilot in accomplishing the RTL mode as by an average of 9.3 seconds, at a set height of 15 meters. ArduPilot slows down the decent at the 10 meter mark unlike PX4 wherein the speed of the decent is consistent throughout the RTL mode.

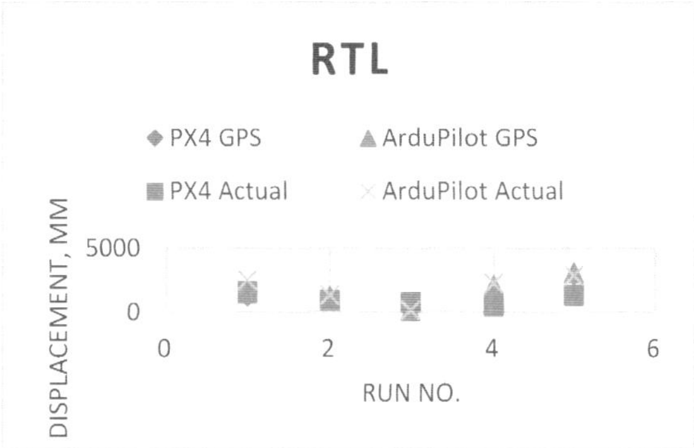

Figure 12. Return to Launch Graph Summary

Based from the data generated for the battery test, ArduPilot consumes more battery in-flight than PX4. This means that PX4 can have a longer flight time than ArduPilot due to its battery consumption in-flight. This also means that there is a distinctive difference between the two software in terms of how the sensors communicate with each other and how it accomplishes missions.

The PX4 has a better performance in altitude control as compared to the capabilities of ArduPilot. The data gathered in Figure 11 shows that PX4 can elevate to different heights as accurately as possible unlike ArduPilot which elevates at heights with a large discrepancy to the desired altitude. It can also be said that ArduPilot has the better capability of holding the height it has elevated to as it is shown in the graph that it is more stable unlike PX4 which has spikes in the graphs below.

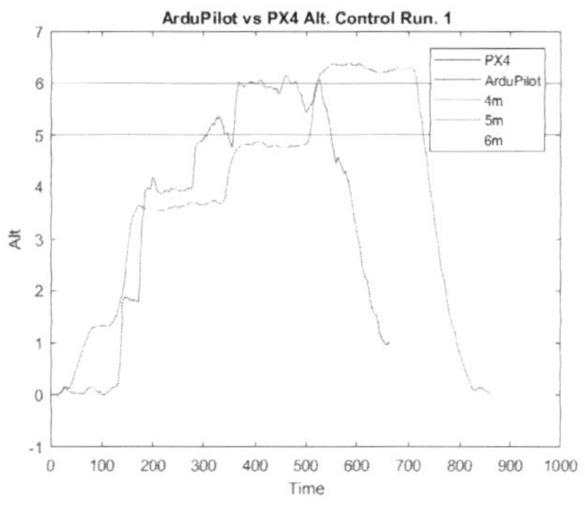

Figure 13. Altitude Control Test Run 1

From the final tally in Tabel 3 done for the quantitative tests, PX4 is better between the two-autopilot software. The tests conducted shows that PX4 tends to be more accurate in attaining a desired value especially on runs that concern altitude control and flight path accuracy. On battery management test, PX4 was only marginally better than ArduPilot, While on RTL test Ardupilot has more displacement error However, ArduPilot performed better in loiter test. On the other hand, deducing from the qualitative findings of both software, ArduPilot is shown to be better in hardware compatibility and user experience. While PX4 is better in licensing and partner companies.

Table 3. Final Tally Results

Qualitative Test	**ArduPilot**	**PX4**
Hardware Compatibility	X	
Licensing		X
Partner Companies		X
User Experience	X	
Quantitative Test	**ArduPilot**	**PX4**
Flight Path Accuracy		X
Flight Path (Time)		X
Altitude Control		X
Loiter	X	
Return to Launch (RTL)		X
Battery Management		X

Conclusions

The paper was able to present a novel experiment design for UAV performance testing. A test case was done comparing two popular autopilot software namely: ArduPilot and PX4. The selection of the components of the experimental platform used in the study was based on availability, costing, and compatibility to different autopilot software. The integration and implementation of each software were done in accordance to each specifications. Hence, it became possible for the designed flight test to be uniformly conducted to each software.

The qualitative comparison of both software shows that the two software are better in some aspects. Ardupilot was better in terms of hardware compatibility and user interface. While PX4 in terms of licensing and partner companies.

On the other hand, the quantitative test results show that PX4 is better than Ardupilot., PX4 faired better in most of the tests which involved accuracy in attaining a desired value, such as altitude and checkpoints. However, in loiter tests ArduPilot performed better. These results were obtained by extraction of flight data log from each software that were statistically analyzed and used to plot different graphs.

The researchers recommend that further performance testing of other UAV controllers be done using the novel experimental method as a strategy. This would provide a benchmark to compare future controllers for its performance.

Acknowledgments

The authors would like to acknowledge the funding support of DOST-PCIEERD and the Mechanical Engineering Department of De La Salle University for the assistance in the conduct of the study.

References

[1] Aarons, T. D. (2011). Development and Implementation of a Flight Test Program for a Geometrically Scaled Joined Wing SensorCraft Remotely Piloted Vehicle. Masters Theses [17872],

[2] Choi, H., Geeves, M., Alsalam, B., & Gonzales, F. (2016). Open source computer-vision based guidance system for UAVs on-board decision making. 2016 IEEE Aerospace Conference, Pages: 1-5, DOI: 10.1109/AERO.2016.7500600.

[3] DiCesare, A., Gustafson, K., & Lindenfelzer, P. (n.d.). Design Optimization of a Quad-Rotor Capable of Autonomous Flight. Massachusetts: Worcester Polythechnic Institute

[4] Liang, O. (2017). Quadcopter PID Explained. OscarLiang.com.

[5] Garcia Carillo, L., Rondon, E., Sanchez, A., Dzul, A., & Lozano, R. (2011). Stabilization and Trajectory Tracking of a Quad-Rotor Using Vision. Journal of Intelligent & Robotic Systems, Volume 61, pp 103-118.

[6] Guido, G., Gallelli, V., Rogano, D., & Vitale, A. (2017). Evaluating the accuracy of vehicle tracking data obtained from. International Journal of Transportation Science and Technology 5 (2016) 136–151, https://doi.org/10.1016/j.ijtst.2016.12.001.

[7] Hoffmann, G., Rajnarayan, D., Waslander, S., Dostal, D., Jang, J., & Tomlin, C. (November 2004). The Stanford Testbed of Autonomous Rotorcraft For Multi Agent Control (Starmac). In In The Proceedings Of The 23rd Digital Avionics System Conference (pp. Pp. 12.E.4/1-10). Salt Lake City.

[8] PX4 autopilot. (2017). Retrieved from Pixhawk Autopilot: https://pixhawk.org/modules/pixhawk

[9] Team, P. D. (2017). Architectural Overview. Retrieved from PX4 Development Guide: https://dev.px4.io/en/

[10] Team., A. D. (2016). Learning Ardupilot. Retrieved from Ardupilot: http://ardupilot.org/dev/docs/learning-ardupilot-introduction.html

IOP Conf. Series: Materials Science and Engineering **533** (2019) 012027 doi:10.1088/1757-899X/533/1/012027

Development of AUV path planner based on unstable mode

Mikhail Medvedev[1] and Boris Gurenko[1]

[1] Southern Federal University, Research Institute of Robot Engineering and Control, 347922, Taganrog, Russian

boris.gurenko@gmail.com, medvmihal@sfedu.ru

Abstract. In this article, development of an autonomous underwater vehicle (AUV) path planner based on unstable mode is considered. The path planner is based on the bionic approach and does not require pre-mapping, it reduces requirements for the sensor subsystem of AUV. In the path planner, we use a method for obstacle avoidance in which an unstable mode is realized in a separate dynamic link. The output of the dynamic link corrects the desired angles of the AUV orientation. It allows setting the unstable mode only at the planning level, and at the regulatory level, a control system works in a stable mode. In addition, this approach allows planning the direction of obstacle avoidance. Detailed description of the proposed path planner and results of the study in Matlab are provided.

1. Introduction

Many studies are devoted to the development and research of control systems for autonomous underwater vehicles [1-8]. One research task in the development of such systems is the development of path planners using intelligent technologies [9], because AUVs should be able to operate autonomously in an environment with obstacles, the location of which is not indicated on the map. There are a lot of intelligent methods for obstacles avoidance, among which we can note the method are based on unstable mode [10]. This method is based on the bionic approach and does not require pre-mapping, it reduces requirements for the sensor subsystem of AUV. The main idea of the method is as follows: a bifurcation parameter β is introduced into the regulator structure. The value of the bifurcation parameter β depends on the distance to an obstacle. If the distance from the AUV to an obstacle is greater than the allowable distance R*, the bifurcation parameter $\beta = 0$ and desired AUV trajectory is stable. If this distance is less than the allowable distance R*, then $\beta \neq 0$ and the desired trajectory of the AUV becomes unstable, it leads to a change in the real trajectory of the AUV movement.

It is important to note that, on the one hand, the introduction of the bifurcation parameter directly to the controller provide an instantaneous reaction of an AUV to an obstacle that appears on the path. On the other hand, unstable trajectories of motion can lead to the exit of the AUV coordinates to technologically acceptable boundaries, to loss of controllability, to abrupt switching of control signals on actuators. In addition, as shown in [10], in this case, a direction of obstacle avoidance is not controlled. In this research, we propose a method for obstacle avoidance in which an unstable mode is realized in a separate dynamic link. The output of the separate dynamic link corrects the desired angles of the AUV orientation. It allows setting the unstable mode only at the planning level, and at the regulatory level, the control system functions in a stable mode. In addition, this approach allows planning the direction of obstacle avoidance.

Content from this work may be used under the terms of the Creative Commons Attribution 3.0 licence. Any further distribution of this work must maintain attribution to the author(s) and the title of the work, journal citation and DOI.
Published under licence by IOP Publishing Ltd

2. Control system with the path planner developing

To represent the mathematical model of the AUV, we use two rectangular coordinate systems, movable and fixed, shown in Figure 1.

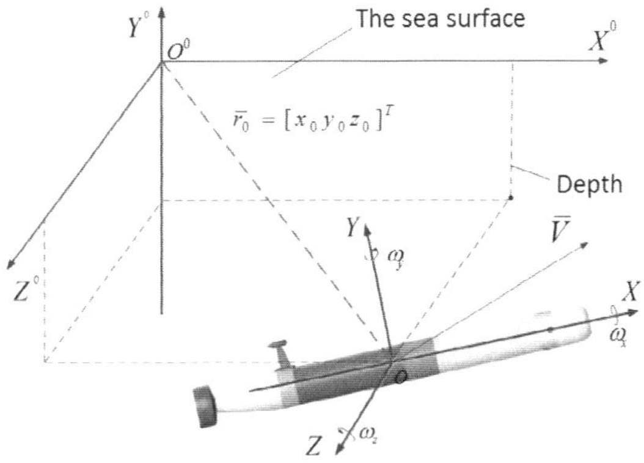

Figure 1. Coordinate system K0 (OX 0 Y 0 Z 0) and K (OX YZ)

Let mathematical model of an AUV is based on well-known solid-state equations. We define a mathematical model of an AUV as a system of nonlinear differential equations and present it in the following vector-matrix form [11]:

$$\dot{Y} = \begin{bmatrix} R_v & 0_{3x3} \\ 0_{3x3} & R_\omega \end{bmatrix} X, \tag{1}$$

$$M\dot{X} = F_u(X, Y, \delta, l, t) + F_d(P, V, W) + F_v(G, A, R), \tag{2}$$

$$T_{actuator}\dot{\delta} + \delta = \Psi_{actuator}(U, \delta), \tag{3}$$

where $T_{actuator}$ is the diagonal (m×m)-matrix of actuators time constants; $\Psi_{\text{actuator}}(\text{U}, \delta)$ is the m-vector of nonlinear functions of right parts of the actuator equations; δ is the m-vector controlled coordinates (engine angles and trusts); U is the control m-vector generated by AUV control system depending on the arrangement of the actuators, $\text{X} = \begin{bmatrix} \text{R}_\text{V} \\ \text{R}_\omega \end{bmatrix}$ is the m-vector of internal coordinates (coordinates state); M is the (m×m) matrix of weight and inertia parameters whose elements are the mass, moments of inertia, added masses; $F_u(X, Y, \delta, l, t)$ is the m-vector of control forces and moments, here l is the vector of design parameters; $F_d(P, V, W)$ is the m-vector of nonlinear elements AUV dynamics; $F_v(G, A, R)$ is the m-vector of measured and unmeasured external disturbances; $\text{Y} = \begin{bmatrix} \text{P} \\ \Theta \end{bmatrix}$ is the n-vector of an AUV position and orientation in the coordinate frame *K(OXYZ)* relative to the base coordinate frame $K^0(OX^0Y^0Z^0)$. Procedure for the conversion of the vector $F_u(X, Y, \delta, l, t)$ into vector U was described in detail and can be found in [11, 12].

To develop control algorithms of AUV, the method [13] is used, which allows solving positional and trajectory problems uniformly. To do this, we define trajectory and velocity vector forms:

$$\Psi_v = \begin{bmatrix} V_x \\ V_y \\ V_z \end{bmatrix} + \begin{bmatrix} -V_k \\ 0 \\ 0 \end{bmatrix} = 0 \quad \Psi_{tr} = \begin{bmatrix} \varphi \\ \delta \\ \gamma \end{bmatrix} + \begin{bmatrix} -\varphi^* \\ -\delta^* \\ 0 \end{bmatrix} = 0, \tag{4}$$

where V_k is the desired speed of the AUV, $\delta^* = \tan^{-1}[(y_k - y_0)/(x_k - x_0)]$, $\varphi^* = \tan^{-1}[(z_k - z_0)/(x_k - x_0)] - \tan^{-1}(V_z/V_x)$.

Then, we define the desired behavior of a closed-loop system as follows:

$$\dot{\Psi}_v + T_1\Psi_v = 0, \tag{5}$$

$$\ddot{\Psi}_{tr} + (T_2 + T_3)\dot{\Psi}_{tr} + T_2 T_3 \Psi_{tr} = 0, \qquad (6)$$

where T_1, T_2, T_3 are positive definite diagonal matrices.

$$F_u = -M \begin{bmatrix} E_{3x3} & 0_{3x3} \\ 0_{3x3} & R_\omega \end{bmatrix}^{-1} \begin{bmatrix} T_1 * \Psi_v \\ \dot{R}_\omega W + (T_2 + T_3) R_\omega W + T_2 T_3 \Psi_t \end{bmatrix} - -F_d - F_v \qquad (7)$$

To avoid obstacles, we introduce into the position-trajectory controller the bifurcation parameter in the following form [10]:

$$\beta = |R - R^*| - (R - R^*), \qquad (8)$$

where R is the distance to obstacles measured by the sensor; R^* is an allowable distance to obstacles, i.e., the distance on which you want to start the process of obstacle avoidance, | | - the module number, it is an absolute value. It is easy to show that if $R \geq R^*$, then $\beta = 0$. If the distance from the AUV to an obstacle is less than allowable, $R < R^*$ then $\beta = -2(R - R^*)$. Given the fact that $(R - R^*) < 0$, $\beta > 0$.
Than we introduce into the structure of the regulator an additional dynamical link of the following form:

$$\dot{z}_i = -(T_{zi} - \beta)\dot{z} + \beta, \qquad (9)$$

where z is an additional variable of the i control channel, T_{zi} is a setting parameter for the i control channel.
We introduce the parameter z into the trajectory vector form in the following way:

$$\Psi_{tr} = \begin{bmatrix} \varphi \\ \delta \\ \gamma \end{bmatrix} + \begin{bmatrix} -\varphi^* \\ -\delta^* \\ 0 \end{bmatrix} + AZ = 0,$$

where $A = \begin{bmatrix} \alpha_1 & 0 & 0 \\ 0 & \alpha_2 & 0 \\ 0 & 0 & \alpha_3 \end{bmatrix}$, α_i – can take the values +1 or -1, thus defining the direction of the obstacle avoidance, $Z = [z_1 \quad z_2 \quad z_3]^T$ is the state vector of additionally dynamical link (9). The direction of the obstacle avoidance can be chosen, for example, from the angle between the direction vectors to the target and to the obstacle or using other intellectual planning methods [9].

3. Closed-loop control system research

To research of the developed controller, we use the following parameters of the AUV [11] in the mathematical model (1), (2):

- AUV mass m_{AUV} = 272.1985 kg;
- Matrix of moments of inertia: $Jm = \begin{bmatrix} 3.3482822 & -0.1472487 & 0 \\ -0.1472487 & 231.9484349 & 0 \\ 0 & 0 & 231.9922593 \end{bmatrix}$,
- Added mass: λ(1,1) = k11 x mAUV; λ (2,2) = k22 x mAUV; λ(3,3) = k33 x mAUV; λ (4.4) = k44 x Jm(1,1); λ(5.5) = k55 x Jm(2.2); λ(6,6) = k66 * Jm(3,3), where the remaining elements of the matrix are zero,
- k11 = 0.02; k22= 1.2; k33 = 1; k4 = 0; k55 = 1.05; k66 = 1.1) ;
- AUV volume V = 0.26183 m3 ;
- AUV length L = 3.7 m;
- AUV area S = 0.07162 m 2;
- Center Masses position xt = (0, -0.1, 0).

The AUV propulsion and steering system consists of a propulsion engine that generates the thrust P1, horizontal thruster P2 and vertical thruster P3, in accordance with Figure 2.

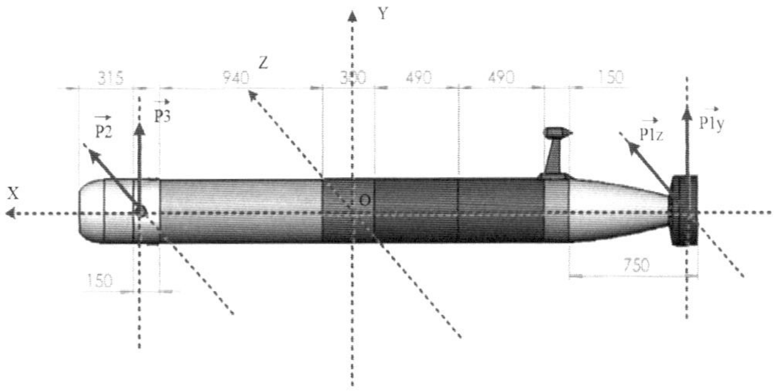

Figure 2. Propulsion system of the AUV

Projection of thrust, created by the main propulsion engine -P_{1x}, P_{1y}, P_{1z}. Forces are created by thrusters: horizontal - P_2, vertically - P_3.

Distance along the axes to the point of application of forces:

OX:	OY:	OZ:
$x_{md} = -1.9;$	$y_{md} = 0;$	$z_{md} = 0;$
$x_{hz} = 1.28;$	$y_{hr} = 0;$	$z_{hr} = 0;$
$x_{vr} = 1.18;$	$y_{vr} = 0;$	$z_{vr} = 0;$

Equations of forces and moments acting on the AUV:

$$\begin{cases} F_{ux} = P_{1x} \\ F_{uy} = P_{1y} + P_3 \\ F_{uz} = P_{1z} + P_2 \\ N_{ux} = 0 \\ N_{uy} = -P_{1z}x_{md} + P_2 x_{hz} \\ N_{uz} = -P_{1y}x_{md} + P_3 x_{vr} \end{cases} \quad (10)$$

The system (10) is a direct transformation of the forces and moments acting on the apparatus. We write the inverse transformation:

$$P_{1x} = F_{ux}$$
$$P_{1y} = \frac{x_{vr}}{x_{vr} + x_{md}} F_{uy} - \frac{1}{x_{md} + x_{vr}} N_{uz}$$
$$P_{1z} = \frac{x_{hz}}{x_{hz} + x_{md}} F_{uz} - \frac{1}{x_{hz} + x_{md}} N_{uy}$$
$$P_2 = \frac{x_{md}}{x_{md} + x_{hz}} F_{uz} + \frac{1}{x_{hz} + x_{md}} N_{uy}$$
$$P_3 = \frac{x_{md}}{x_{md} + x_{vz}} F_{uy} + \frac{1}{x_{md} + x_{vz}} N_{uz}$$

The sensor for detecting obstacles is a forward-looking sonar. Sensor model is given as a ray that comes from the nose of AUV. If there is an obstacle in the path of the AUV, i.e. ray intersects the obstacle; sensor model calculates the distance to the obstacle R. If the distance R is less than the radius of sensor action $R^s = 50$ meters, then the sensor model returns the distance to the control system. We simulated the closed-loop control system of the AUV. During the simulation, the underwater vehicle should move from the point (0; 0; 0) to the point (200; 100; 200) at a speed of 5 m/s. On the AUV path there are two obstacles with a radius of 15 m. Obstacle №1 has coordinates (50, 35, 50), obstacle №2 - (120, 70, 120). Modeling results with the motion trajectory of the AUV in the environment with two obstacles are shown in Figures 3 and 4.

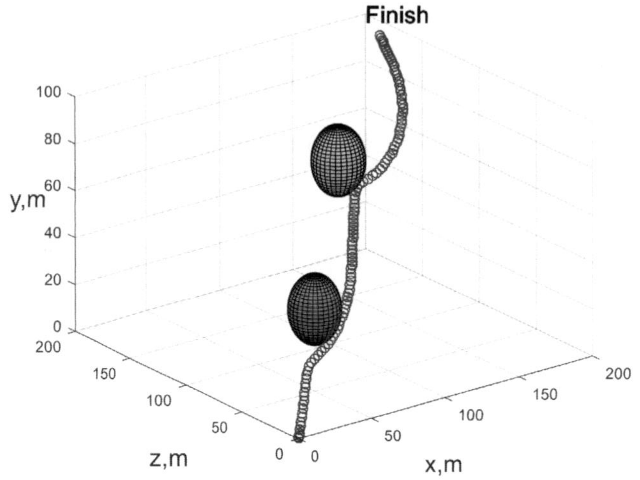

Figure 3. Trajectory of the AUV movement in three-dimensional space in the environment with two obstacles

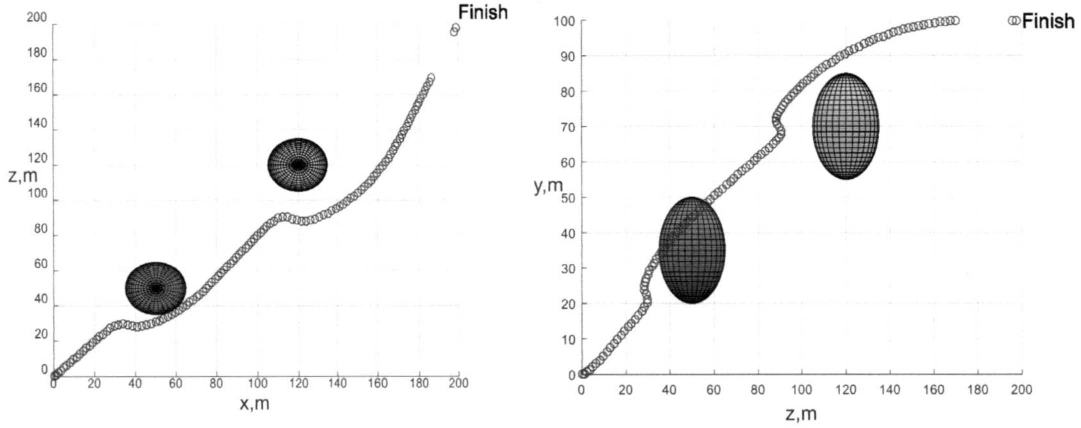

Figure 4. Trajectory of the AUV movement - top view (left) and side view (right)

The additional dynamic system was used only in the course control channel; the value of the parameter α_1 is one. Figure 5 shows how the parameter z_1 changes.

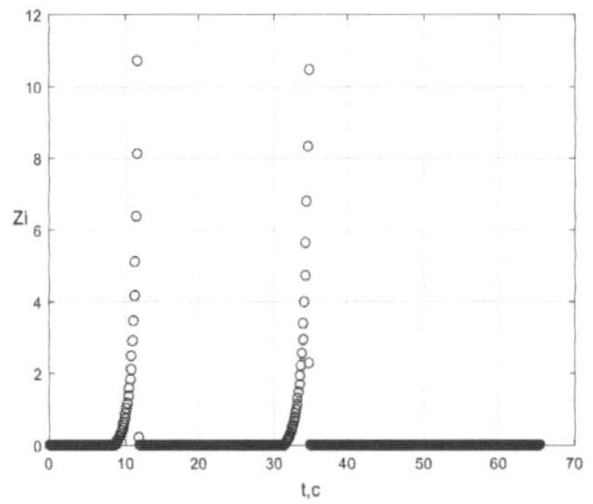

Figure 5. Parameter z_1 in the course control channel

Changing of AUV velocity and control forces are shown in Figure 6.

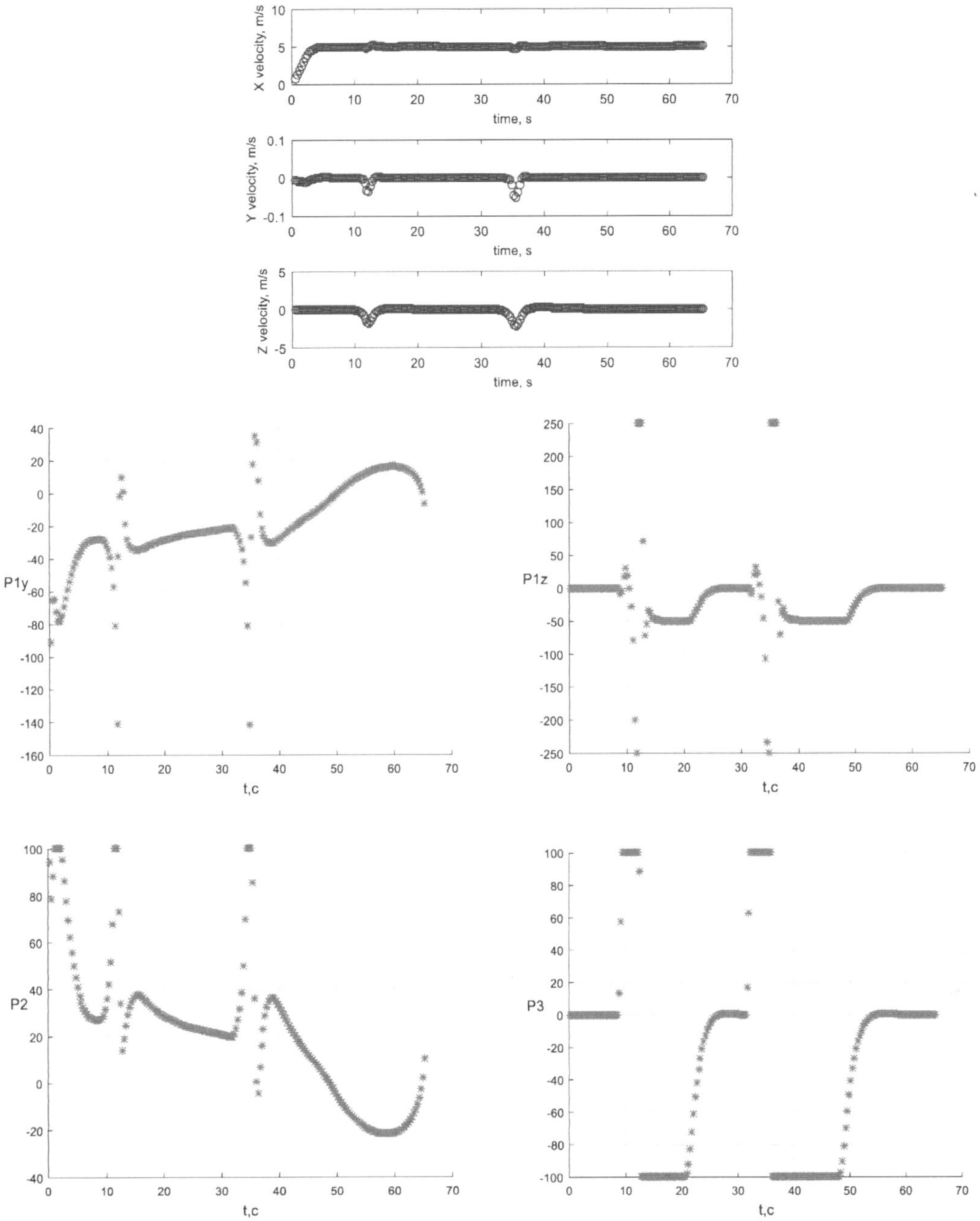

Figure 6. History of AUV speed and control forces

4. Conclusion

In this paper, we propose a hybrid method of path planning of AUV which based on an unstable mode. Description of the method and way of using in AUV control system are given. This method is based on the bionic approach and does not require pre-mapping, it reduces requirements for the sensor subsystem of AUV. We proposed a method for obstacle avoidance in which an unstable mode is realized in a separate dynamic link. The output of the separate dynamic link corrects the desired angles of the AUV orientation. It allows setting the unstable mode only at the planning level, and at the regulatory level, the control system functions in a stable mode. In addition, this approach allows planning the direction of obstacle avoidance.

The method proposed in this article can be compared with the method of artificial potential fields [14,15]. As we known, in the method of potential fields, attractive and repulsive forces are functions of the coordinates

$$F_{attr} = F_{attr}(y), F_{rep} = F_{rep}(y), \tag{11}$$

where $F_{attr}(y)$ is the resultant of attractive forces; $F_{rep}(y)$ is the resultant of repulsive forces. In the method proposed in this article, attractive and repulsive forces are functions of the coordinates, velocities and accelerations of the mobile object and the bifurcation parameter β. They are formed as a solution of a differential equation of the form

$$F_{ar}(y, \dot{y}, \ddot{y}, \beta) = 0, \tag{12}$$

where $F_{ar}(y, \dot{y}, \ddot{y}, \beta)$ is a function, depending on the value of the parameter , which is attractive or repulsive. Obviously, the using of Eq. (12) gives more opportunities to take into account the velocities and accelerations of the mobile object to avoid obstacles. Also, due to the sensitivity of the unstable solutions of equation (12) to the initial conditions, the using of unstable trajectories of motion gives a greater variability of trajectories to avoid obstacles, increases the probability of passing of environment with moving obstacles.

Acknowledgments
The reported study was funded by RFBR according to the research project № 18-38-00441\18.

References

[1] Akyildiz, I.F., Pompili, D., Melodia, T. Underwater acoustic sensor networks: Research challenges // Ad Hoc Networks, Volume 3, Issue 3, May 2005, Pages 257-279, DOI: 10.1016/j.adhoc.2005.01.004

[2] Robert S. McEwen; Brett W. Hobson; Lance McBride; James G. Bellingham Docking Control System for a 54-cm-Diameter (21-in) AUV // IEEE Journal of Oceanic Engineering Year: 2008, Volume: 33, Issue: 4 Pages: 550 - 562, DOI: 10.1109/JOE.2008.2005348)

[3] Leonard, N.E., Paley, D.A., Lekien, F., Sepulchre, R., Fratantoni, D.M., Davis, R.E. Collective motion, sensor networks, and ocean sampling // Proceedings of the IEEE Volume 95, Issue 1, January 2007, № 4118466, Pages 48-74., DOI: 10.1109/JPROC.2006.887295

[4] Rudnick, D.L., Davis, R.E., Eriksen, C.C., Fratantoni, D.M., Perry, M.J. Underwater gliders for ocean research // Marine Technology Society Journal, Volume 38, Issue 2, June 2004, Pages 73-84

[5] Aguiar, A.P., Hespanha, J.P. Trajectory-tracking and path-following of underactuated autonomous vehicles with parametric modeling uncertainty // IEEE Transactions on Automatic Control Volume 52, Issue 8, August 2007, Pages 1362-1379, DOI: 10.1109/TAC.2007.902731

[6] Stilwell, D.J., Bishop, B.E. Platoons of Underwater Vehicles: Communication, Feedback, and Decentralized Control // IEEE Control Systems Volume 20, Issue 6, December 2000, Pages 45-52, DOI: 10.1109/37.887448

[7] Lapierre, L., Soetanto, D. Nonlinear path-following control of an AUV // Ocean Engineering Volume 34, Issue 11-12, August 2007, Pages 1734-1744, DOI: 10.1016/j.oceaneng.2006.10.019

[8] Mac, T.T., Copot, C., Tran, D.T., De Keyser, R. Heuristic approaches in robot path planning: A survey (2016). Robotics and Autonomous Systems. V 86, pp. 13-28.

[9] Pshikhopov V. Path Planning for Vehicles Operating in Uncertain 2D Environments // eBook ISBN: 9780128123065, Paperback ISBN: 9780128123058, Imprint: Butterworth-Heinemann,

1st February 2017, 312 p.
[10] V.Kh. Pshikhopov, M.Yu. Medvedev PATH PLANNING AND CONTROL OF VEHICLES IN 3D ENVIRONMENT USING UNSTABLE MODES // Izvestiya Sfedu. Engineering Science, SECTION VI. MANAGEMENT SYSTEM, 01-02, 2017, pp 146-158.
[11] B. Gurenko, "Mathematical Model of Autonomous Underwater Vehicle," Proc. of the Second Intl. Conf. on Advances In Mechanical and Robotics Engineering - AMRE 2014, pp. 84-87, 2014, doi:10.15224/ 978-1-63248-031-6-156
[12] Fedorenko R., Gurenko B. Autonomous Underwater Vehicle Mathematical Model and Simulator // Proceedings of the 4th International Conference on Control, Mechatronics and Automation (ICCMA' 16). - pp. 56-60. - doi: 10.1145/3029610.3029639
[13] Pshikhopov, V., Medvedev, M Position-Path Control of a Vehicle (Book Chapter))// Path Planning for Vehicles Operating in Uncertain 2D Environments, 1 February 2017, Pages 1-23, ISBN: 978-012812306-5;978-012812305-8
[14] Khatib O. Real-Time Obstacles Avoidance for Manipulators and Mobile Robots, (1986) Int. Journal of Robotics Research. Vol. 5. (1). Pз. 90–98.
[15] Khatib O. Real-time obstacle avoidance for manipulators and mobile robots (1985) IEEE Int. Conf. Robotics and Automation, pp. 500 –505.

IOP Conf. Series: Materials Science and Engineering **533** (2019) 012028 doi:10.1088/1757-899X/533/1/012028

A Satisficing Game Theoretic Approach to Multi-Aircraft Collision Avoidance

Weiye Wang

School of Automation, Beijing Information Science and Technology University, Beijing, China

afatfox@163.com

Abstract. We propose a distributed approach to multi-aircraft real-time conflict resolution based on satisficing game theory which considers reducing flight collisions and improving economic efficiency. In our conflict game model, aircraft as exclusively self-interested agent are assigned different statuses by a priority ranking mechanism proposed. According to the state of one's own priority, they take others' preferences into consideration to choose the most satisficing conflict resolution strategy from policy set including speed and heading modification strategies. The simulation results on the random flight scenario of two concentric circles show that our method which adopts speed and heading adjustment to avoid conflicts is promising in terms of safety and system efficiency.

1. Introduction

Along with the development of air transportation industry, flight conflicts are becoming increasingly prominent as aircraft density in airspace increases dramatically. Collision avoidance plays a vital role in order to ensure the airspace security and improve the capacity of airspace. For conflict resolution, the objective is to provide suitable maneuvers to avoid a predicted conflict. Much research has been devoted over the last decade on conflict resolution tasks.

A review of research in this area of conflict resolution is presented in a study[1]. Short-term conflict resolution using genetic algorithm(GA)[2,3] and optimal control theory[4,5] has been paid attention extensively. These methods are large in computing capacity and can't predict conflict within 5 minutes. Force fields approach[6] is explored to resolve real-time conflict problem. However, the kind of algorithm is not always guaranteed to generate flyable trajectories. Tomlin et al.[7,8] propose non-cooperative two-person zero-sum game theory to design controllers for safety specifications. Krozel et al.[9] introduce the distributed real-time conflict resolution algorithms. A myopic strategy selects the alternative that requires the smallest heading adjustment to resolve collisions in pairs. Archibald et al.[10] adopt a satisficing game theory to resolve real-time aircraft collisions. The maneuvering strategies of conflict resolution are only heading modification, and the flight speed is constant.

In our paper, we present a decentralized approach to multi-aircraft real-time conflict resolution based on the satisficing game theory. The approach adopts flight velocity vector adjustments as maneuvering strategies of collision avoidance focused on the two dimensional case. A resolution maneuver set is obtained for each player by selectability and rejectability. The proposed approach is illustrated in simulation of the random flight scenario of two concentric circles. It is showed that our approach could real-time resolve multi-aircraft conflicts efficiently and robustly.

Content from this work may be used under the terms of the Creative Commons Attribution 3.0 licence. Any further distribution of this work must maintain attribution to the author(s) and the title of the work, journal citation and DOI.
Published under licence by IOP Publishing Ltd

The mathematic formulation of multi-aircraft conflicts and real-time resolution methodology is detailed in the second part. In the third part, we describe the numerical simulation and results. The paper is concluded in the last part.

2. Collision Avoidance Model

2.1. Conflict Prediction

We assumes that each aircraft is aware of flight critical information pertaining to the others within a given detectable radius R_D and A is a set of all aircraft operating in the airspace where there are M aircraft, and any aircraft a_i satisfies $a_i \in A$. In order to illustrate the flight collision, circular protected zone of aircraft, namely near miss radius R_{NM} and collision radius R_C, satisfies $R_{NM} >> R_C$, as shown in figure 1. When the distance of current locations between a_i and a_j represented as $d(i, j)$ is smaller than or equal to R_C, it will lead to a collision. If $R_C < d(i, j) \leq R_{NM}$, a near miss event will occur. Collisions are weighted more heavily than near misses.

Figure 1. Example of near miss zone and collision zone. Yellow represents near miss zone and red represents collision zone, with $R_{NM} >> R_C$.

The conflict prediction method used in our study is based on searching for the closest point of approach (CPA) between two aircraft within look-forward time window $T_w = 8\,\text{min}$, which can be mathematically obtained by linear predictions according to their current action, as shown in figure 2. We define $d_{\min}(i, j)$ to represent the minimum distance at CPA between a_i and a_j, and $d_{CPA}(i, j)$ is the distance from a_i 's current position to a_i 's CPA between a_i and a_j. If $d_{\min}(i, j) < R_{NM}$ and $d(i, j) < R_D$, there are potential collisions between a_i and a_j to be denoted as a conflict vector $w_{ij} = (d_{\min}(i, j), d(i, j)) \in W^2$, where W^2 is the conflict space of pair-wise aircraft. It is supposed that E_i is a set of all the potential conflict with a_i represented as $E_i = \{w_{ij} \mid j = 1, \cdots, i-1, i+1, \cdots, M\}$.

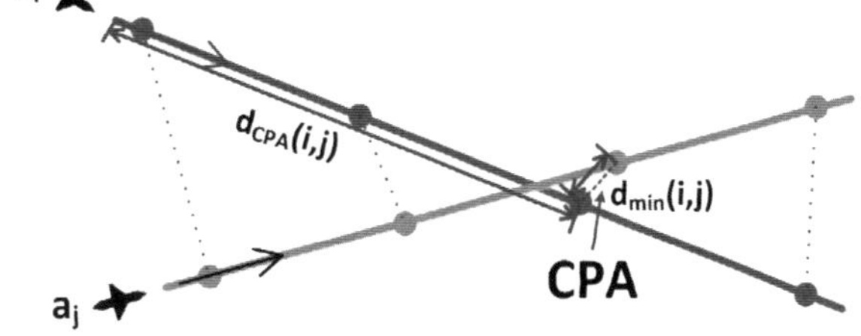

Figure 2. Representation of a conflict. It shows the position of CPA between a_i and a_j, the minimum predicted distance $d_{\min}(i, j)$ at CPA, and the distance $d_{CPA}(i, j)$ from a_i 's current position to a_i 's CPA between a_i and a_j.

In games, the playing pure strategy $s_i^n(n=1,\cdots,N)$ of individual $a_i(i=1,\cdots,M)$ is from the discrete set of optional strategies $S_i(s_i^n \in S_i)$, where N is the total number of a_i's optional strategies, namely the dimensions of S_i . a_i possesses a mapping from different strategy combinations $(s_i^n, s_1^n, \cdots, s_{i-1}^n, s_{i+1}^n, \cdots, s_M^n)$ to rejectable and selectable payoffs. The mapping from strategies space to rejectable and selectable payoffs space can be mathematically constructed as $H: S_i \times S_j \to W^2$, namely the mapping from strategies space to conflicts space.

2.2. Priority Ranking Mechanism

A priority ranking mechanism is introduced to represent individuals' different social status before games. The rules of defining ranking are as follows: firstly, those individuals within 5nmi away from the destination possess higher ranks than all the others; secondly, in each subset, individuals with greater current flight delay cause higher ranks; thirdly, individuals in each subset are allocated ranks respectively by their current flight time, with longer flight time bringing higher ranks; finally, individuals in the same subset with the same delay and flight time are ranked according to the remaining planned flight time, with larger remaining time resulting in higher ranks. It is supposed that the priority ranking mechanism leads to a unique priority for each individual, and that rank orderings in different games are consistent from the perspective of all individuals.

In a round of games, individual a_i determines its neighbors of game according to the potential collisions set E_i and individuals' priority ranking. The individuals with higher ranking will ignore potential conflicts with ones of lower priority and satisfy its own interests absolutely, while ones with lower ranking will preferentially consider potential conflicts and cost its own interests to benefit superior ranking individuals and the whole group. Therefore, the set of a_i's game neighbors is all individuals with higher ranking than a_i in E_i, and represented as E'_i.

2.3. Conflict Resolution

In our study, each aircraft chooses its maneuvering strategies of conflict resolution according to its priority ranking, game neighbors, and selectable and rejectable payoff. Aircraft adopt flight velocity vector changes as the flight strategies to avoid collisions. At each time step, each aircraft selects one resolving strategy from five speed and five direction options as its behavior at the next time step. We assumed that heading and speed changes are instantaneous. The five direction modifications include straight flight, moderate turns left and right (2.5°), and sharp turns left or right (5°). The five speed modifications include flight according to planned speed, acceleration ±5% based on planned speed, and acceleration ±10% based on planned speed[11].

2.3.1. Rejectability. The details of computing the rejectable payoff are as follows. Let s_j^c signify a_j's current flight action. The rejectable payoff function is represented as

$$P_R(s_i^n) = \frac{1}{1 + \sum_{a_j \in E_i'} F(s_i^n, s_j^c)} . \tag{1}$$

The function $F(s_i^n, s_j^c)$ is defined as

$$F(s_i^n, s_j^c) = \begin{cases} 2\alpha, & if\ d_{\min}(i,j) \le R_C \\ \alpha, & if\ R_C < d_{\min}(i,j) \le R_{NM} \\ 0, & otherwise \end{cases} , \tag{2}$$

where α is defined as

$$\alpha = \begin{cases} (2 - \frac{d_{\min}(i,j)}{R_{NM}})(\frac{1}{d_{CPA}(i,j)})^{\beta}, & if\ d_{CPA}(i,j) \le 3R_{NM} \\ (\frac{1}{d_{CPA}(i,j)})^{\beta}, & otherwise \end{cases}, \tag{3}$$

where the parameter β is experimentally tuned variable, set as $\beta = \frac{2}{3}$. If there are conflicts in every optional strategy, the maximum rejectable options will be the ones with the most distant conflicts. Those strategies leading to maximum rejectable payoff function $P_R(s_i^n)$ are seen as safe strategy subset $\bar{S}_i = \{s \mid s = \arg\max_{s_i^n \in S_i} P_R(s_i^n)\}$, $\bar{S}_i \subset S_i$. If there is only strategy in $\bar{S}_i$, aircraft a_i will execute the only one as its behavior at the next time step. If the number of the elements in $\bar{S}_i$ is more than one or there is no conflict for all strategy option, aircraft's action decisions will depend on selectability.

2.3.2. Selectability. Most flight missions require that the aircraft reaches the destination as timely as possible, so selectability that reflects goal achievement according to flight mission requirements is influenced by the required time of arrival. a_i selects the strategy to satisfy the required time of arrival to the greatest extent possible and minimize the flight delay. We assume that a_i's current location coordinate is represented as $L_i^c = (x_i^c, y_i^c)$, and its planned location coordinate at the next time step is $L_i^p = (x_i^p, y_i^p)$. a_i's possible location at the next time step is determined by selected strategy s_i^n from the safety subset $\bar{S}_i$, represented as $L_i^c + \vec{s_i^n}$, where $\vec{s_i^n}$ is the vector form of s_i^n to denote flight speed and direction. Then the selectable payoff function is defined as

$$P_S(s_i^n) = \exp(-\mid L_i^c + \vec{s_i^n} - L_i^p \mid), \tag{4}$$

where $s_i^n \in \bar{S}_i$. The planned location coordinate at the next time step is defined as

$$L_i^p = \begin{cases} L_i^c + \frac{\overline{L_i^c L_i^d}}{T_i^p - T_i^c}, & if\ T_i^p > T_i^c \\ L_i^d, & otherwise \end{cases}, \tag{5}$$

where $L_i^d = (x_i^d, y_i^d)$ is the coordinate of a_i's destination, T_i^p is its planned flight time, T_i^c is its current flight time, and $\overline{L_i^c L_i^d}$ is the vector from its current position L_i^c to its destination L_i^d. The strategy with maximum selectable payoff function $P_S(s_i^n)$ will be executed by a_i in the next time step, specifically,

$$s_i^{next} = \arg\max_{s_i^n \in \bar{S}_i} P_S(s_i^n). \tag{6}$$

3. Simulation Results and Discussion

Our simulation environment is similar to the one used in some previous studies[9.10]. In the airspace, all aircraft cruise at the same altitude and at the same planned speed $v_p = 500mph$. Each aircraft selects a strategy according to the rules described in previous section and updates its heading, speed and position, the new information is then broadcasted to all the others within its detectable zone. For all aircraft in the airspace, it is supposed that the value of the detectable radius is same, set as $R_D = 50nmi$. To maintain calculation accuracy, the frequency of updates of resolution strategy is once each second in our simulation.

3.1. Evaluation Metrics

In order to evaluate our approach to collision avoidance, appropriate metrics will be used to guarantee that both safety and efficiency objectives are met. Safety and system efficiency evaluation concerns are similar to the ones used in other studies[10].

The most important metric is flight safety for any method of conflict resolution. Two types of separation violations are reported: system collisions and system near misses. The collision event described in previous section is the case of $R_C \leq 900ft$ between any two aircraft. System collisions is defined as

$$C = \frac{\sum_{m=1}^{T_S} C_m}{T_S}, \tag{7}$$

where C_m is the number of collisions at mth time step and T_S is total flight time of whole system. When $R_{NM} \leq 5nmi$ between any two aircraft, the near miss event happens. System near misses is defined to be similar to system collisions,

$$NM = \frac{\sum_{m=1}^{T_S} NM_m}{T_S}, \tag{8}$$

where NM_m is the number of near misses at mth time step.

Conflict resolution maneuvers would lead to deviating from aircraft's projected paths, consuming more resources, and bring greater economic losses. System efficiency reflects the deviation to which the aircraft are able to follow their projected paths to destinations. A good method of collision avoidance should maintain the high system efficiency while ensuring safe flight. The fight efficiency of aircraft a_i is defined as $EFF_i = \frac{T_i^p}{T_i}$, where T_i^p is a_i's projected flight time along its projected paths, and T_i is a_i's actual flight time to reach its destination. Then system efficiency is represented as

$$SE = \frac{1}{M}\sum_{i=1}^{M} EFF_i . \tag{9}$$

Additionally, the standard deviation of efficiency is used as an evaluation of fairness of flight delays, represented as

$$SDE = \sqrt{\frac{1}{M}\sum_{i=1}^{M}(EFF_i - SE)^2} . \tag{10}$$

3.2. Simulation Results

In our study, the simulation process is divided into three types according to different maneuvering strategies: only speed changes, only heading changes, and velocity vector changes.

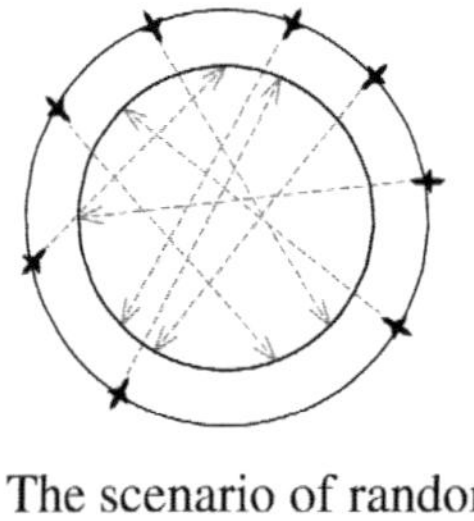

Figure 3. The scenario of random flights.

In order to identify the robust of our model, a random traffic patterns scenario[9] is used. In the random flight scenario, there are two concentric circles. The radii of the outer circle and the inner circle are 60nmi and 50nmi respectively. Each aircraft starts at a random point on the outer circle and arrives at a random point on the inner circle as its destination, and its projected path must pass through the inner circle, as shown in figure 3. In this scenario, conflicts only occur in the inner circle. The inner circle represents the test airspace and the 10nmi ring buffer guarantees that no possible collision exists in the inner circle at initial period of generating aircraft. Because of the randomness of the scenario, it is a great examination of collisions avoidance algorithms with diverse random geometries. When each simulation run starts, a new aircraft emerges with approximately 5s intervals until the upper limit of the number of aircraft is reached in the concentric circles. The upper limit of the number of aircraft denotes an associated traffic density μ in this scenario. When an aircraft arrives at their destinations, a new aircraft is generated to replace it with the same start and end. The simulation results that run for up to 60 minutes are collected and averaged over 20 runs at each group of data.

Table 1. The result for random flights scenario in the case of only speed changes

μ	Evaluation Metrics			
	Near Misses	**Collisions**	**SE**	**SDE**
3	0	0	1	0
6	0.02409771	0	1	0
9	0.170067375	0.000222099	1	0
12	0.3285564	0.000111049	0.9983245	0.00085069
15	0.6291387	0	0.9983005	0.00135001
18	0.98722465	0.000166574	0.9981165	0.0018207
20	1.37991535	0.000333149	0.998004	0.00211189
22	1.3602859	0.000277624	1	0
24	2.0086165	0.000402555	0.997895	0.00246308
26	1.9560525	0.000277623	0.997534	0.00349094
28	2.7481115	0.000499722	0.9979945	0.00228207
30	2.977541	0.000652416	0.99711	0.00434006
32	3.6777205	0.001096613	0.996751	0.00579429

Table 2. The result for random flights scenario in the case of only heading changes

μ	Evaluation Metrics			
	Near Misses	**Collisions**	**SE**	**SDE**
3	0	0	1	0
6	0.023806223	0	0.9999628	0.001636319
9	0.16461992	0	0.99222205	0.018389865
12	0.38903991	0	0.9951527	0.008949775
15	0.70884775	0	0.9830611	0.042024318
18	1.01013355	0.000083287	0.98135935	0.043837987
20	1.22247345	0	0.97308325	0.053015962
22	1.7353	0.00059689	0.96891295	0.059064045
24	2.5912115	0.000624653	0.9544303	0.084839645
26	2.6465705	0.001110492	0.95361955	0.08323377
28	3.9120355	0.005496944	0.9381217	0.0871449
30	4.968198	0.006538036	0.91691315	0.110845485
32	6.4531805	0.012395889	0.89199145	0.122369525

Table 1 shows the simulation results in the case of only speed adjustments as maneuvering strategies. The number of near misses and collision events increase and system efficiency doesn't change much in the vicinity of $SE = 1$ as the traffic density μ grows. In the case of only heading changes, the simulation results for collision avoidance are exhibited in table 2. When μ is increased, the number of near misses and collisions grows and system efficiency lowers. Comparing the two methods, the strategy of alone heading changes can improve the performance of flight safety slightly, whereas its system efficiency is sacrificed as the cost. The simulation results for velocity vector adjustments as resolution strategies are exhibited in table 3. The near miss event increases and system efficiency decreases slightly with growing μ. It is noted that the strategy of speed and heading adjustments greatly improves the performance of flight safety in contrast to two others, and its system efficiency is slightly less than one of only speed changes as resolution strategies.

Table 3. The result for random flights scenario in the case of velocity vector change

μ	**Evaluation Metrics**			
	Near Misses	**Collisions**	**SE**	**SDE**
3	0	0	1	0
6	0	0	1	0
9	0.0843987	0	1	0
12	0.3658218	0	1	0
15	0.5918375	0	0.9997827	0.0054207
18	0.98057	0	0.9935034	0.022086
20	1.3594895	0	0.9862456	0.0384128
22	1.6042109	0	0.980139	0.0481082
24	1.7628547	0	0.9808618	0.0460986
26	2.0192789	0	0.977478778	0.0596417
28	2.728651	0	0.9596697	0.0815321
29	2.964632	0.0003054	0.961747	0.081553
30	3.145503	0	0.9624844	0.075867
32	3.339284	0.0009717	0.9538982	0.1006478

In order to prove the high efficiency of our algorithm, a contrast experiment is shown in figure 4. The two contrastive methods are Archibald's full satisficing algorithm[10] and Krozel's decentralized look-ahead approach[9]. In the scenario of random flights, the contrastive results confirm that our approach which uses speed and heading adjustments as game strategies offers significantly better performance for system efficiency in contrast to two others.

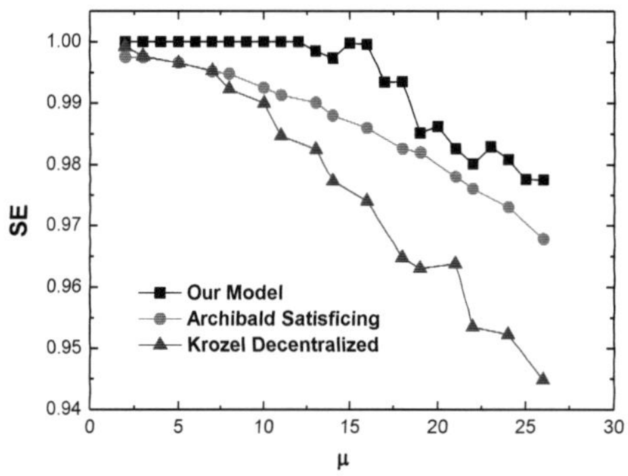

Figure 4. A comparison of system efficiency among three algorithms including our method, Archibald's full satisficing algorithm and Krozel's decentralized look-ahead approach.

4. Conclusion

Further developments of collision avoidance are required for a better management of flight conflicts with the increasing air traffic densities of the next decade. In the study, we present a decentralized approach to multi-aircraft real-time conflict resolution based on the satisficing game theory. The approach adopts speed and heading modification as maneuvering strategies of collision avoidance. The simulation results prove our approach can offer good performance of safety and highly efficient flight. Most importantly, the reported collision events and system efficiency demonstrate that the approach of velocity vector adjustments is promising in the case of high density autonomous flight.

Acknowledgments

This work is supported by Beijing Natural Science Foundation(Grant No. 4174092).

References

[1] J. K. Kuchar and L. C.Yang, IEEE Trans. Intell. Transp. Syst., vol. 1, no. 4, pp. 179-189, Dec. 2000.
[2] S. Mondoloni and S. Conway, AIAA Guidance, Navigation, and Control Conference & Exhibit, Montreal, Canada, AIAA 2001-4054, pp. 11, Aug. 6-9, 2001.
[3] X. Guan, X. Zhang, J. Wei, I. Hwang, Y. Zhu, and K. Cai, International Journal of Systems Science, Volume 47 Issue 9, Pages 1995-2008, July 2016
[4] L. Pallottino, E. M. Feron, and A. Bicchi, IEEE Trans. Intell. Transp. Syst., vol. 3, no. 1, pp. 3-11, Mar. 2002.
[5] Y. Gao, X. Zhang, X. Guan, 2012 International Symposium on Instrumentation & Measurement, Sensor Network and Automation, pp. 310-313, Aug. 2012
[6] K. Zeghal, AIAA Guidance, Navigation and Control Conference, Boston, 1998.
[7] C. J. Tomlin, I. Mitchell, and R. Ghosh, IEEE Trans. on Intell. Transp. Syst., vol.2, no. 2, pp. 110-120, Jun. 2001.
[8] C. J. Tomlin, J. Lygeros, and S. S. Sastry, Proceedings of the IEEE, vol. 88, no. 7, pp. 949-970, Jul. 2000.
[9] J. Krozel, M. Peters, K. D. Bilimoria, C. Lee, and J. S. B. Mitchell, Presented at the 4th USA/Eur. Air Traffic Manag. R&D Seminar, Santa Fe, NM, Dec. 2001.
[10] J. K. Archibald, J. C. Hill, N. A. Jepsen, W. C. Stirling, and R. L. Frost, IEEE Transactions on Systems, Man, and Cybernetics-Part C: Applications and Reviews, vol. 38, no. 4, pp. 510-521, Jul. 2008.
[11] G. Granger, C. Allignol, and N. Durand, 9th USA/Europe Air Traffic Management Research and Development Seminar (ATM-2011), 2011.

IOP Conf. Series: Materials Science and Engineering **533** (2019) 012029 doi:10.1088/1757-899X/533/1/012029

Connected Automated Guided Vehicles in a Smart Container Terminal

Mohamed K. Watfa, Khaldon Al Karmadi

Faculty of Engineering and Info. Sciences, University of Wollongong in Dubai, UAE

MohamedWatfa@uowdubai.ac.ae

Abstract. As container terminals aim for more efficient operation and higher productivity, automation is making major breakthroughs all over the world. Container terminals are increasingly adopting automated solutions to meet the challenge of larger ships, taller cranes and bigger call sizes. Having the right design criteria and taking the correct decisions early in the project is crucial in order to meet the expected cost and performance targets. Therefore, designing an automated container terminal is no longer just a matter of acquiring the right port equipment and integrating the equipment with basic terminal operating systems (TOSs), it is becoming more of designing and implementing a complete smart automated system. Smart automation would enable Automated Guided Vehicles (AGVs) used in container terminals to automatically adapt to changing circumstances, optimizing performance and output. Most of existing research work considered pre-defined AGVs paths using different techniques without considering free path AGVs solutions that can potentially increase transportation efficiency This research work uses the concept of Internet of Things (IoT) to increase transportation efficiency of the Automated Guided Vehicles (AGVs) and reduce vessel discharging time in an automated container terminal. In addition, the research paper proposes an algorithm to increase efficiency in the automated container terminal and achieve the research objective. Finally, simulation results were used to prove the positive effects of using the proposed algorithm in decreasing vessel discharging time compared to existing benchmarks and optimum solutions.

1. Introduction

Containerization, the stowage of regularly or even irregularly shaped freight in sealed, reusable boxes with standardized dimensions is one of the most important cargo moving techniques developed in the 20th century[1]. Countries are investing not only in expanding terminals capacity, but as well in advanced container terminal technologies to increase operations efficiency and reduce vessel discharging time. With the development of material handling and information technology, a number of terminals, such as Europe Combined Terminal (ECT) in Rotterdam are employing automated container-handling equipments to satisfy the growing demands and lower the labour costs [2]. Automated container terminal can be divided to three main operations as follows: 1- Quay Cranes (QCs), 2- Automated Guided Vehicles (AGVs) and 3- Automated Stacking Cranes (ASCs). Quay Cranes (QCs) are responsible for loading and discharging containers from and to vessels on the berth. These QCs are semi-automated cranes which are operated by operators at a remote control room. Crane operator will load and discharge containers according to the stowage plan that the port will receive before vessel berthing to specify the sequence and location of each container in the ship. The

Content from this work may be used under the terms of the Creative Commons Attribution 3.0 licence. Any further distribution of this work must maintain attribution to the author(s) and the title of the work, journal citation and DOI.
Published under licence by IOP Publishing Ltd

Automated Guided Vehicles (AGVs) are responsible to take containers from QCs and transport them to the storage yard during the discharging operation and from storage yard to QCs during the loading operation. These AGVs are driverless vehicles which are moving based on pre-defined paths. Finally, the Automated Stacking Cranes (ASCs) handle the containers received from the AGVs when they arrive to storage yard. Scheduling and controlling these kinds of automated equipments is a very demanding and important task.

Most of the existing AGVs researches assume fixed paths for AGVs movements to reduce traffic complexity and easier vehicles control. On the other hand, using non fixed paths for AGVs movements can increase transportation efficiency, but it will increase traffic complexity and will create new controlling problems. GPS based AGVs will replace the fixed paths AGVs and will be able to reach a destination much faster than the pre-defined paths vehicles which will add more efficiency to the automated container terminal. On the other hand, implementing these free travel AGVs will add more traffic complexity and collision problems and those problems should be considered in the terminal more seriously. However, using sensor networks, Vehicle-to-Vehicle Communications and Internet of Things can help solve those AGVs traffic problems, increase terminal efficiency and therefore reduce vessel discharging time.

2. Related Work

Many researchers studied different operational aspects in the container terminal from different angles. For example, Luo and Wu [2] proposed a mixed integer programming model and genetic algorithm for AGVs' dispatching and storage allocation optimization with the objective of minimizing vessel berthing period in a fully automated container terminal. Huang, Yan and Wang [3] proposed a mixed integer programming model to optimize scheduling of non-automated container terminal resources by minimizing vessel waiting time and reducing terminal energy consumption. According to Xin et al. [4], the control of automated container terminals is complex since Quay Cranes (QCs), Automated Guided Vehicles (AGVs) and Automated Stacking Cranes (ASCs) interact intensively for transporting containers, while collision avoidance of equipment must be ensured. Many research works also focused on AGVs routing, dispatching and collision avoidance algorithms in an automated container terminal. Routing is the process of choosing a static path and specifying the number of trips for each AGV in the terminal. Dispatching is the process of assigning the right AGV to the right container request. The dispatching process is considered as a dynamic problem where an AGV is assigned in real time depending on the location of the vehicle and other real time data from the terminal environment. Most work in this area use integer programming models with some heuristic approaches and concluded that using AGV's for multiple vessels is much more efficient than specifying for single vessel operation [5]. For the dispatching process, research works have studied different types of vehicles using both static and dynamic dispatching approaches with the objective to minimize total operation time or total crane waiting time [6,7]. Other notable works in this area include Kim and Bae [8] who proposed a static and dynamic AGVs dispatching based on vehicle initiated look-ahead algorithm. The suggested dispatching method showed higher efficiency than other dispatching methods such as, shortest travel distance or time, revised shortest imminent operation and earliest due date. AGVs deadlock and collision detection was also studied in the literature including the work of Kim, Jeon, and Ryu [9] who proposed to divide the traffic area in the terminal to grid-blocks and use reservation graphs during the development and validation stages allowing AGVs to be on several grid-blocks. Zeng and Hsu [10] used a grid based solution which partitioned the terminal area to blocks and pre-defined paths. In addition, their method used discrete time segments for AGVs routing to avoid dangerous interactions between vehicles for collision avoidance. On the other hand, the use of Internet of Things was studied extensively in the supply chain and logistics industry but from different perspectives. Ports globally face considerable challenges calling for research on innovative solutions with minimal reliance on manual interventions and controls. As a result, the authors in [11] proposed an RFID based smart solution for Mombasa container terminal which considers the operations of containers which are entering, exiting or still in the terminal. There are many researchers studied the effects of Internet of Things on the transportation sectors and how to increase car traffic efficiency suggesting different methodologies for connected cars localization and distributed fault diagnosis using Vehicle to Vehicle communications [12].

3. Methodology and Framework

The operations of the automated container terminal can be divided in to three stages. The first stage is responsible for the operations of Quay Cranes for discharging/ loading containers from/to the vessel. The second stage focuses on AGVs operations moving containers from QCs to ASCs in the discharging process and from ASCs to QCs during the loading process. In addition, the second stage plans the scheduling of AGVs to generate collision free path for each AGV. Stage 3 handles the operations of ASCs stacking containers in their storage slots. Furthermore, in all of the three stages there are two levels of controlling decisions. In this research paper, we will be focusing on the two levels of stage 2 only which include AGVs operations. The higher level is responsible for the scheduling process of AGVs and assigning a specific job to a specific AGV (Discrete Events Operations). The higher level controller consists of a stage controller for each stage and a supervision controller which acts similar to a central brain of all the three operations stages as shown in Figure 1.

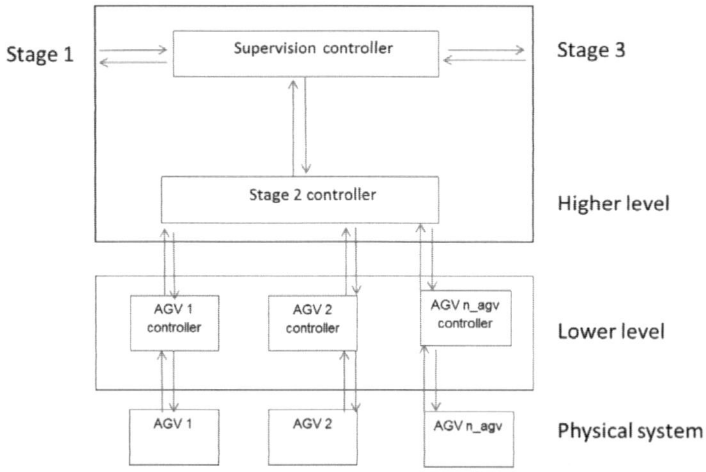

Figure 1. Stage & Level Control Architecture

3.1. High Level Algorithm

The higher level algorithm is responsible for dividing the time intervals and assigning sequences of each job in all the three stages. In addition, the higher level algorithm depends on some assumptions as follows: 1) It considers the discharging operation only. 2) The containers' origins in the vessel are known and are ready for processing at time 0. 3) Each container takes different operation times during the QCs, AGVs and ASCs stages. 4) The capacity of each QC, AGV and ASC is equal to 1 container (1 TEU). The higher level algorithm is divided into five main functions which control the processes of assigning resources and sequences for each container. The main function will control the overall algorithm and acts as the supervision controller responsible to manage the three stages of operation. The main function will declare the number of containers, number of QCs resources, number of AGVs resources and number of ASCs resources in the operation problem. Then, it will start executing the container jobs one by one giving priority to available food containers by calling the food_first function.

- AssignQC function will assign a QC for each container and it will give priority to food containers to get served from the QC which has the least number of jobs in the queue for a less waiting time. Based on the area location of the food container, it will be assigned either to the corresponding QC in that area or the two neighbour QCs to the left or right based on the least number of jobs on their queues. On the other hand, non-food container will be served using the corresponding area QC only.

- AssignASC function will assign ASC for each container. The container will be assigned to the corresponding ASC or the one to the left or the one to the right depending on the least number of jobs in their queues. This process of assigning ASC will be used regardless of the type of cargo inside the container.
- AssignAGV function will assign an AGV for each container in the ship. The function will assign a free AGV for each container based on the nearest AGV which has the shortest distance to the container loading point. The function will calculate the distance between the AGV's current position and the container loading point which is considered as the point of origin of the AGV operation. The function will calculate the distance for all the AGVs and it will choose the one with the shortest distance to serve the container.

3.2. Low Level Algorithm

The lower level considers the movements of AGVs, obstacles detections and collision avoidance between AGVs and other static obstacles. AGVmove function will begin by calculating the number of zones needed to be crossed during transportation from the current location to the point of origin and from the point of origin to the AGV destination. Then, the function will call obstacle_detection function to check if there is an obstacle in front of the AGV and will call the collision_avoidance function otherwise it will move one zone foreword and it will repeat this process to move until it reaches the destination. Also, in the automated container terminal there are two types of obstacles static or dynamic obstacles. Static obstacles are the non-moving obstacles such as cranes and the other infrastructure in the container terminal as shown in Figure 2(a,b). On the other hand, dynamic obstacles are the other AGVs which are moving and interacting with each other using common paths and intersections. To manage these interactions Vehicle -to- Infrastructure and Vehicle -to- Vehicle Communications were used to assure smooth and safe transportation in the automated container terminal. Vehicle -to- Infrastructure communication will help AGVs detect static obstacles by sharing the exact location of all the static obstacles in the automated container terminal. This is done by uploading a detailed map with all exact dimensions and locations of static obstacles to a central cloud system connected to all local AGV systems. In addition, the infrastructure will broadcast a message to all surrounding AGVs to alert them from a possible collision. Vehicle -to- Vehicle Communication will be used to detect dynamic obstacles while an AGV is moving. There will be a local system in every AGV and this system will contain a 1 meter accuracy GPS and a radar sensor to act as a two layer safety sensor. Every AGV local system will share its real time location with other surrounding local systems and the central cloud system to allow other AGVs to read this important information. In addition, the radar sensor, which surrounds the AGV, will act as the second layer of safety to detect any obstacle surrounding it. So, these two layers of safety will assure correct detection of obstacles by analyzing the two readings and giving a better decision. Furthermore, Obstacle_detection function will read and analyze the readings from the GPS and the radar sensors. Based on these readings, the function will decide if there is an obstacle on the way of the AGV or not. The function will return TRUE to the AGVmove function if any of the two sensors detect an obstacle and FALSE if no obstacle is detected. The Collision_avoidance function will act to avoid any possible accidents between the AGV and a static obstacle or the AGV and a dynamic obstacle. The AGV will stop if another AGV is occupying the wanted zone and it will wait until the other AGV releases the zone. Furthermore, the AGV will slightly turn left or right if there is a static obstacle in the way to avoid crashing. Logically, as the number of containers increases, the computational time will increase as well. The time complexity of the algorithm with n containers is O(n2). The higher and lower levels algorithms will work together to allocate resources for each container and assure safe transportation. By having efficient resource allocations and transportation, we will have faster operations which would therefore result in reducing vessel discharging time in an automated container terminal.

IOP Conf. Series: Materials Science and Engineering **533** (2019) 012029 doi:10.1088/1757-899X/533/1/012029

4. Experiments and Simulation

To prove the efficiency of the suggested methodology, we use a typical container terminal map in Figure 2 that shows the environment of the operations and respective coordinates. This container terminal map or reference system also used by [4] assumes 5 QCs for discharging containers from the ship, 5 AGVs to transport containers from the QCs to the ASCs and 5 ASCs for 5 stacking areas so each ASC will be responsible for one stacking area. The AGV will take the container from the origin transfer point which is the QC discharge point to the destination point which is the ASC loading point as shown in Figure 2.

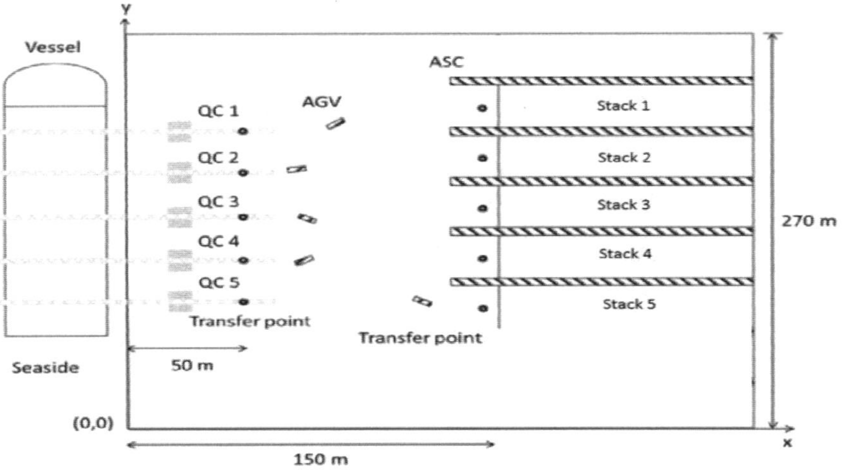

Figure 2. The automated container terminal reference system.

Some assumptions and parameters for the experiments include the following:

- The vessel stowage width is equal to 8 TEUs.
- The maximum distance between the QC interchange point and a container is 100 meters.
- The container terminal yard area is 150 m x 270 m.
- Each stacking location has a volume of (36 TEU length x 10 TEU width x 6 TEU height).
- The maximum speed (velocity) for: QCs is 4 m/s, AGVs is 6 m/s and ASCs is 4 m/s.
- The maximum acceleration (in $(m/s)^2$)for QCs is 0.4, AGVs is 1 and ASCs is 0.4.
- Each QC or AGV or ASC will handle 1 TEU only
- The initial positions of AGVs are loading positions, and of QCs are discharging positions.
- The QC handling time of each container depends on its position in the ship.

Arena simulation software was used in the following experiments to measure the efficiency of the proposed framework and will focus on the following indicators:

- Total Discharging Time: the completion time for handling all containers which leave the ship.
- QC average operation time per container: the average time that the container will spend in the QC stage from waiting in the QC queue until it gets discharged to the AGV.
- AGV average operation time per container: the average time that an AGV will spend starting from the requesting time to delivering the corresponding container to the final destination.
- ASC average operation time per container: the average time that the container will spend in the ASC stage from waiting in the ASC queue until it reaches the final location in the stacking area.
- Average Waiting Time: the average time that a container will wait in the queues during all stages of operation.

Comparison experiments were conducted between the proposed algorithm in this paper and a benchmark experiment in the literature [4] using the same simulation parameters including the same constraints and number of resources. The results shown are the average of five simulation runs.

- ***Discharge Time***: Three experiments were conducted using 1) 5 QCs, 10 AGVs and 8 ASCs, 2) 4 QCs, 8 AGVs and 6 ASCs and 3) 3 QCs, 6 AGVs and 5 ASCs respectively to compare the overall discharge time between the proposed algorithm and the benchmark. Results are shown in Figure 3(a). The results show 13%, 19% and 18% improvement on the discharge time between the proposed algorithm and the benchmark for the three different scenarios with an overall average improvement of 17%.
- ***Average Operation Time***: Using similar scenarios as in the previous experiments, the average operation time was compared between the two algorithms as shown in Figure 3(b). The benchmark algorithm expectedly scored a better result on AGV average operation time compared with the proposed algorithm. This slight increase in the AGV average operation time of the proposed algorithm is because of the extra collision avoidance logic that has been applied to the proposed algorithm. This collision avoidance logic is to assure smooth transportation without any accidents between the AGVs which are using common paths and intersections while moving.
- ***Resource Utilization***: Another comparison experiment was conducted between the proposed algorithm using same number of resources as the benchmark (5 QCs, 10 AGVs and 8 ASCs) and the proposed algorithm using fewer resources (5 QCs, 5 AGVs and 5 ASCs). The results show a very small difference (+3 seconds) in the total discharging time between the two scenarios (365 seconds vs. 368 seconds). Even with the 3 extra seconds, the total discharging time with fewer resources is still lower than the benchmark which has 420 seconds. Therefore, the proposed algorithm with the new configuration will save 5 extra AGVs and 3 extra ASCs while still resulting in a better total discharging time. However, the result of AGV average operation time was slightly more in the proposed algorithm with the fewer resources (+2 seconds). This small increase in the AGV average operation time is because of the fewer available number of AGVs in the automated container terminal. With fewer numbers of AGVs, the response time of the selected AGVs will increase.
- ***Lower Bound Stage Analysis***: Another comparison experiment was conducted between the proposed algorithm and the optimum possible solution as an ideal case scenario. The optimum solution is the ideal situation when there is an assumption of zero waiting time for each container and zero transportation waiting time. There is 161 seconds difference between the proposed algorithm and the optimum solution total discharge time, and this difference is a direct result of the container waiting time. To achieve this optimum result, no shared resources should be considered in the automated container terminal which makes it a very expensive solution and it leads to a very low utilization of resources. In addition, a detailed experiment was conducted to compare between the proposed algorithm and the optimum solution for every stage separately. As shown in Figure 3(c), the proposed algorithm resulted in 175.4 seconds, 29 seconds and 19.3 seconds in the QC stage (stage 1), AGV operations (Stage 2) and ASC (Stage 3) respectively compared to 34.2 seconds, 15.5 seconds and 12.8 seconds for the optimum solution. These savings are a result of the lack of waiting time in Stage 1 and 3, and the assumption of having an AGV ready on the loading point for each container in Stage 2. These results serve as a lower bound and assist us in further improving our proposed algorithms to achieve optimal results.

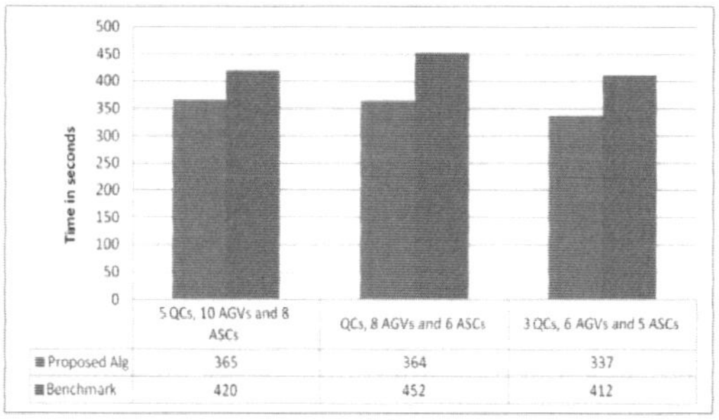

(a)

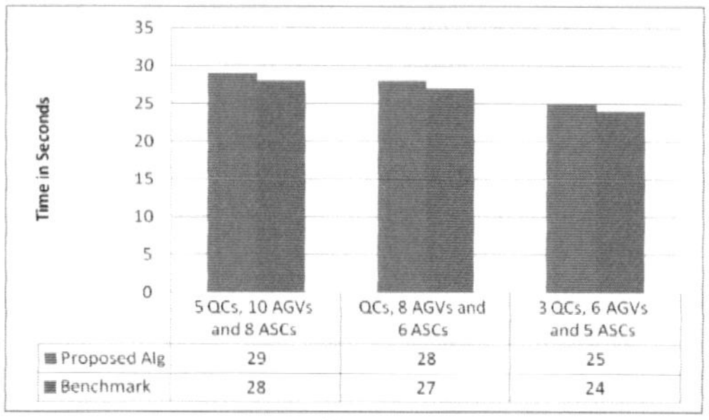

(b)

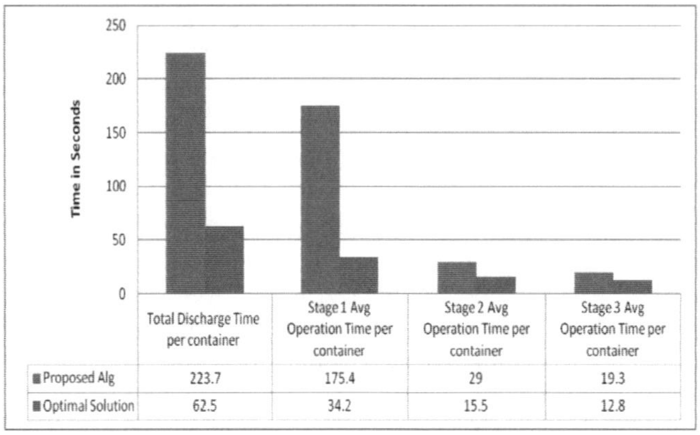

(c)

Figure 3. (a) Total Discharge Time Comparison for the three different configurations. (b) Average Operation Time Comparison (c) Comparison between the proposed algorithm and optimum solution (Stages 1, 2 and 3)

5. Conclusions

This paper studied the three stages of operations in the automated container terminal and proposed a smart algorithm to increase its efficiency using sensor networks. The objective of proposed algorithm is to decrease the total discharging time of a vessel which will increase the efficiency of the automated container terminal assuring smooth traffic in the automated container terminal. The proposed algorithm was proven to work with fewer resources than the benchmark and achieve better results. Future work includes real world deployment in UAE container terminals and will cover different scenarios of different containers sizes to be able to handle 40 feet containers and 2x20 feet containers at the same time.

References

[1] UNCTAD, (2011). (United Nations Conference on Trade and Development) secretariat. Review of Maritime Transport 2011, United Nations publication.

[2] Luo, J., & Wu, Y. (2015). Modelling of dual-cycle strategy for container storage and vehicle scheduling problems at automated container terminals. Transportation Research Part E: Logistics and Transportation Review, 49-64.

[3] He, J., Huang, Y., Yan, W., & Wang, S. (2015). Integrated internal truck, yard crane and quay crane scheduling in a container terminal considering energy consumption. Expert Systems with Applications, 2464-2487.

[4] Xin, J., Negenborn, R., Corman, F., & Lodewijks, G. (2015). Control of interacting machines in automated container terminals using a sequential planning approach for collision avoidance. Transportation Research Part C: Emerging Technologies, 377-396.

[5] Zeng, Q., Yang, Z., & Lai, L. (2009). Models and algorithms for multi-crane oriented scheduling method in container terminals. Transport Policy, 16, 271–278.

[6] Zhang, L. W., Ye, R., Huang, S. Y., & Hsu, W. J. (2005). Mixed integer programming models for dispatching vehicles at a container terminal. Journal of Applied Mathematics & Computing, 17(1–2), 145–17.

[7] Xing, Y., Yin, K., Quadrifoglio, L., & Wang, B. L. (2012). Dispatch problem of automated guided vehicles for serving tandem lift quay crane. Transportation Research Record, 79–86.

[8] Kim, K. H., & Bae, J. W. (2004). A look-ahead dispatching method for automated guided vehicles in automated port container terminals. Transportation Science, 38(2), 224–234.

[9] Kim, K. H., Jeon, S. M., & Ryu, K. R. (2006). Deadlock prevention for automated guided vehicles in automated container terminals. OR Spectrum, 28, 659–679.

[10] Zeng, J., & Hsu, W. J. (2008). Conflict-free container routing in mesh yard layouts. Robotics and Autonomous Systems, 56, 451–460.
[11] Siror, J., Huanye, S., & Dong, W. (2011). RFID based model for an intelligent port. Computers in Industry, 795-810.
[12] Homchaudhuri, B., Pisu, P., & Ozguner, U. (2015). Secure Vehicle Localization and Cruise Control for Connected Vehicles. IFAC-PapersOnLine, 1192-1197.

Feasibility Study of Point Type Equipment for Detecting Train Transponder System

Fei Chen, Jingyang Lv

Beijing University of Posts and Telecommunications, China

chenfei1993t@126.com

Abstract. In the development of railway transportation, safety has always been the most concerned issue. Balise is the core content of high-speed railway train control system, it is one of the most basic unit of information storage and transmission, control and status information necessary for the train. It is affected by many factors in the process of operation, so it is unavoidable to have problems of malfunction. At present, staff are required to operate the transponder to repair or update message information. In order to meet the needs of daily maintenance, the scheme of reading and writing operation of transponder by dynamic train detection is proposed and its feasibility is analyzed. This paper designs the software part of the mobile detection system by analyzing the working status information of different versions of transponders and tests it. The test results show that the dynamic reading and writing operation scheme is feasible.

1. Introduction

Ensuring the safety of train operation is an eternal subject of railway transportation. Especially under the circumstances of increasing axle load, large traffic volume and increasing speed of trunk lines, it is urgent to ensure safety.

In the process of train operation, in order to ensure the safety of driving, it is necessary to obtain correct control information and state information. The CTCs-2 (Chinese Train Control System-2) section is a point-to-link system based on the track circuit of the point transponder to transmit the train operation control information. It indicates that the role of point passive transponder (also known as fixed information transponder[1], this paper is abbreviated as transponder) is crucial for the train to run safely. transponders can send effective data under the action of on-board equipment during the train running, so it should be set up in every block partition of the railway section.

At present, there is no train that can dynamically read and write the transponders so as to update the information and test the related key signals. The current maintenance work of transponder is to update the data read or rewritten by the handheld equipment of railway staff[2]. The staff will not repair the transponder until it is reflected that the ATP on board causes the train to stop. Therefore, the key link of timely discovery of equipment failure hidden danger is missing in daily staff maintenance. It adds uncertainties to safe transportation and disrupts normal transportation order.

For this reason, the work of dynamic detection of transponder is considered to be added to the existing electric test vehicle. Use of electrical test vehicle to meet the safety requirement of daily maintenance of transponder[3]. By means of dynamic detection and data analysis of transponders, it is possible to

Content from this work may be used under the terms of the Creative Commons Attribution 3.0 licence. Any further distribution of this work must maintain attribution to the author(s) and the title of the work, journal citation and DOI.
Published under licence by IOP Publishing Ltd

reduce personnel, improve detection accuracy and efficiency. This method can effectively reduce transport accidents caused by transponder faults and provide reliable guarantee for transport safety.
The structure of this paper is as follows: section 2 analyzes the working methods and features of existing balises, providing a basis for the feasibility study of dynamic detection of trains; section 3 carries out the theoretical study of the feasibility of the system; section
4 carries out actual simulation and realization of the theoretical study of feasibility, verifies the practical feasibility of the theory and verifies the function; section 5, conclusions.

2. Transponder System

2.1. Dot Passive Transponder

Transponder is a point device that has a high amount of stored information and can achieve high rate transmission. It has high reliability and can send key information to the vehicle system at high speed. The main functions include: achieve absolute positioning of trains, provide variable train access information, store important information including line profile data, road and bridge locations, long-term or temporary speed limits and assist train control, etc[4]. Moreover, It is an electromagnetic device that uses the principle of electromagnetic coupling to realize communication between vehicles and vehicles. It does not need an external power supply, and its power supply and system operating clock are obtained by sensing the signal antenna sent by the external device.

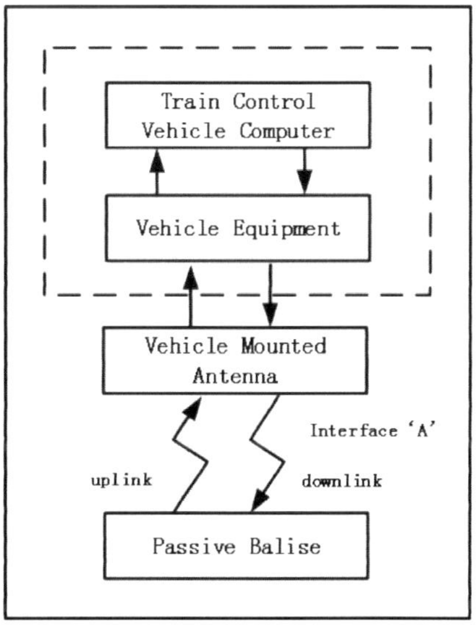

Figure 1. Passive transponder interface diagram.

Figure 1 is an interface diagram of passive transponder. In the transponder system, the communication between passive transponder and the vehicle antenna is performed through A interface, and the A interface can be divided into the following sub- interfaces:
A1 interface: transponder uplink data transmission interface;
A2 interface: vehicle antenna downlink data transmission interface;
A4 interface: vehicle antenna downlink energy supply interface;
In general, we need to perform two operations on the transponder[5]: reading the information mode and writing the information mode. The read operation mode is operated by two interfaces including A1 and A4. During the running of the train, when the vehicle antenna passes over the transponder, the transponder receives the A4 interface signal from the vehicle antenna and converts it into working energy, then continuously transmits the 1023 bits transponder message data through the Al interface. The write operation mode is that the A2 and A4 interfaces work. During the running of the train, when

the vehicle antenna passes over the transponder, the transponder receives the A4 interface signal from the vehicle antenna and transmits 1023 bits of message data to the transponder through the A2 interface cycle.

If the train is to be completed by reading and rewriting the transponder on the section where the ground transponder is distributed, it is required to perform a read- write-read three steps operation. Therefore, it is determined that the transponder is in a normal working state or a programmed state, so its working status is a very important factor. Therefore, the next step is to analyze the actual working condition of the transponder.

2.2. *Transponder Working Analysis*

The passive transponders arranged on the railway now can be roughly divided into two versions: the European standard transponder and the National standard transponder[6]. During the reading operation, The 4M signals sent by the transponder of the two versions are FSK signals with message information, and there is no difference. The difference between the two versions is that the signal sent by the 4M signal is not the same after the write operation is completed. One is to send a new fsk signal with message information that has been written, and the other is to send a single frequency signal. The performance of the 4M signal of the two versions of the transponder are shown in Figure 2 and Figure 3.

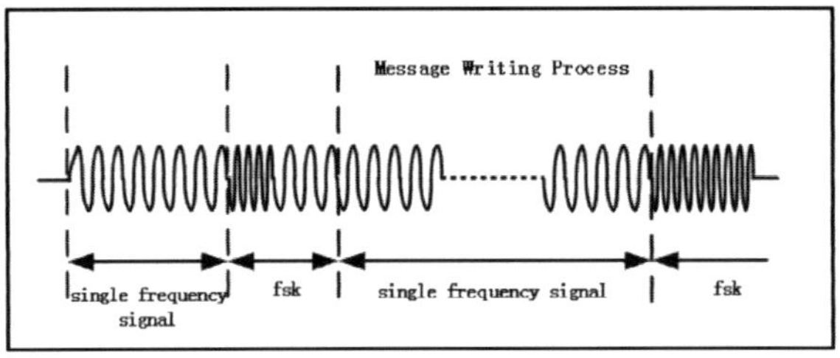

Figure 2. The state of 4M signal of the European standard transponder during write operation.

As shown in figure 2, the European standard transponder will send a signal of 3.951MHz representing the "0" character after the completion of the write operation. The demodulation module of FPGA is also in the working state during the write operation, monitoring the data transmission of the transponder's uplink link.

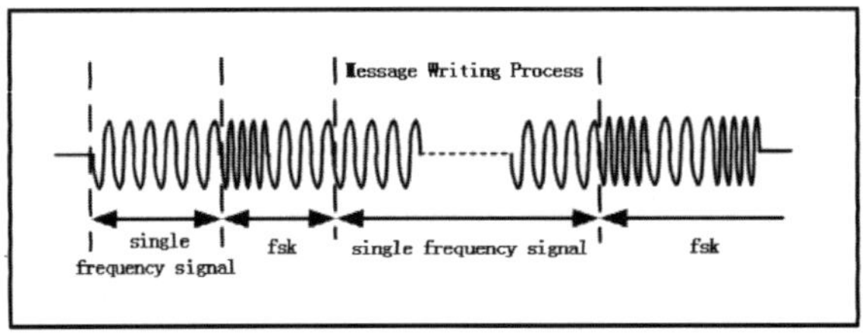

Figure 3. The state of 4M signal of the Nation standard transponder during write operation.

As shown in figure 3, after the write operation, the National standard transponder will automatically send the message of FSK signal modulation which has just write. The demodulation module of FPGA in the write operation is also in working state to monitor the change of the data, when it finds that the single frequency signal is converted into the fsk modulated signal, it means that the transponder's writing process is successful. Because of this feature, the transponder can be directly read for the second time, and the written message can be compared immediately.

Realize the read-write-read operation of the transponder during the running of the train, before using the timer to control the switching between the write operation and the second read operation[7]. After our analysis of the two versions of the transponder, we can directly use the different features of the two versions to reduce the time consuming between operation switching.

3. Feasibility Studies

3.1. Operation Time Analysis
In the previous chapter, the transponder and the analysis of its work were introduced. In our study, while the In other words, the train should be accurate and fast to update the transponder set.

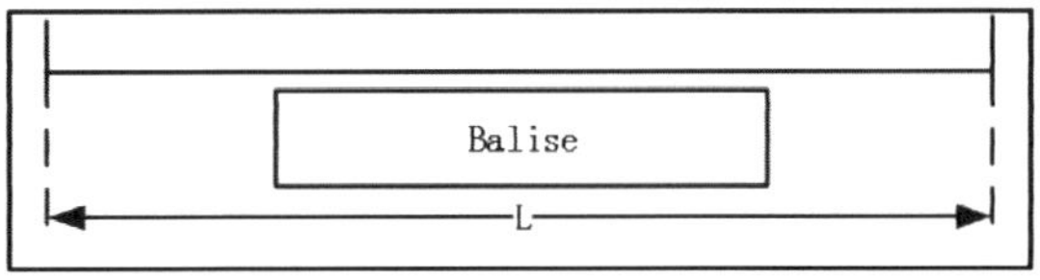

Figure 4. Dynamic read-write transponder system application environment.

The L[8] marked in figure 4 represents the range of action of the train when it passes over the transponder. After measurement,the formula is as follows:

$$L = 0.7\ (m) \tag{1}$$

Here, we assume that T represents the time when the train can transmit data when it passes the scope of action of the transponder, that is, the maximum allowable processing time of read-write-read operation is realized. Then, to guarantee the time of normal operation, the formula should be established.

$$t \ll T \tag{2}$$

We assume that the train is traveling at a speed of Vkm/h, then we can get the value of time at this velocity.

$$\begin{aligned} T &= \frac{L(m)}{V(km/h)} = \frac{0.7(m)}{V(1000(m)/3600(s))} \\ &= \frac{0.7(m)\times 18(s)}{V\times 5(m)} = \frac{12.6}{5V}(s) \end{aligned} \tag{3}$$

Through the above analysis, we have learned that to make the train as fast as possible, we should shorten the processing time of read-write-read operation as much as possible, that is, complete the operation process of the transponder was being read, written and read while running, the train speed should be kept as fast as possible. transponder as soon as possible when passing through the working range of the transponder.

3.2. System Design
The realization of the operation of reading, writing and reading on the transponder during the train running can be described as follows:

1. The upper computer enters the transponder message group information in the driving road before driving, and reads the corresponding message information at the corresponding position when passing each transponder;
2. The upper computer gives the signal of FPGA starting to work (enabling the signal to be pulled down to make FPGA start to work). If the train receives any operation failure or fails to receive successful feedback within a range of 70cm, the upper computer will report error; If the FPGA completes the corresponding operation and gives feedback, the energy signal will be raised;

3. When the upper computer gives the read operation instruction, the FPGA sends 27M signals to supply power;
4. The FSK signal sent by the receiving transponder is demodulated and other operations are performed. After the reading operation, write operation is performed;
5. During the writing operation, the analysis in chapter 2 can monitor the change of 4M signal to determine whether the writing operation is successful or not. After the success, the second reading operation can be performed;
6. The whole operation process ends after the completion of the read back and the second read operation.

The specific working process is shown in figure 5, where above the dotted line is part of the operation mechanism of the upper computer, and below the dotted line is the internal operation mechanism of the lower machine FPGA.

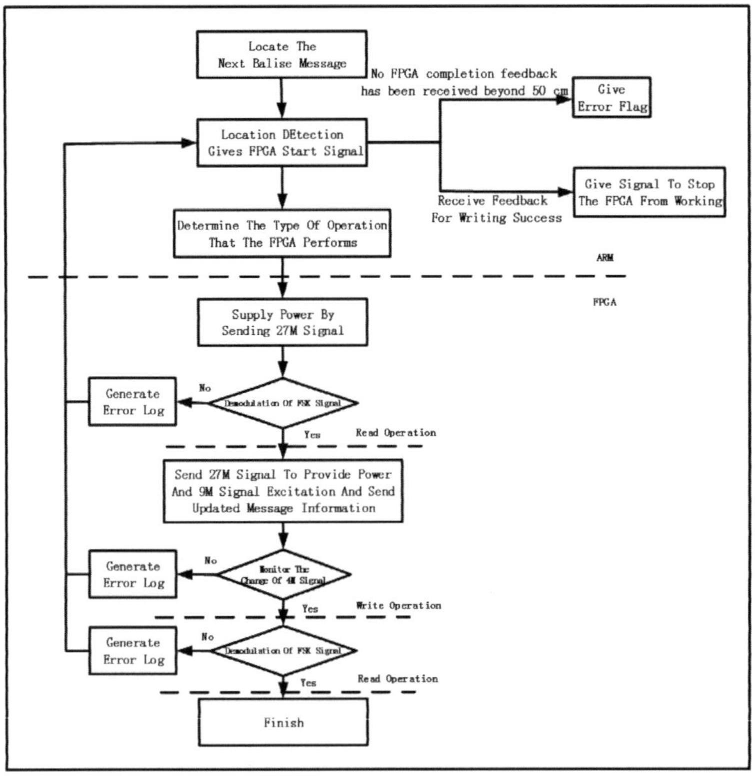

Figure 5. Flow chat of system design.

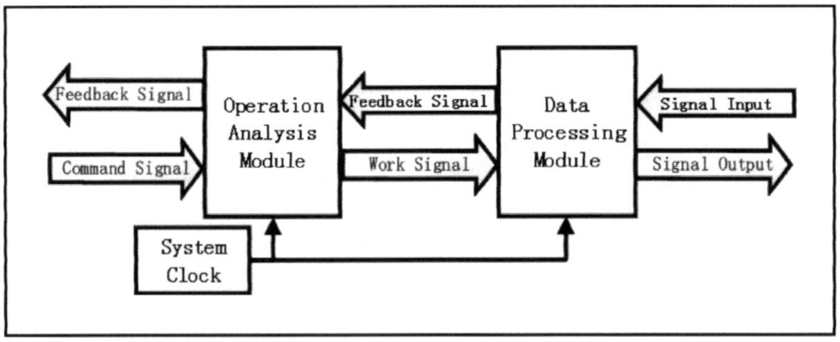

Figure 6. Modular design.

3.3. Modular Design

The software part of the system end is designed to be implemented using the EP3C25F256 chip of Cyclone III series of Altera company, and the wiring layout is conducted on the Quartus II software platform[9].The software part of the system is mainly realized by two parts: operation analysis module and data processing module, as shown in figure 6.

After receiving the data, the data module determines the working state of the system according to the rules of the established operation instructions. According to the corresponding working status, the data processing module sends the corresponding working signal to the data processing module, and the data processing module performs the reading or writing operation. If the module does not get the data feedback information, the working log is automatically generated and the corresponding signal is given to the operation analysis module.

In the process of writing the program code, attention should be paid to distinguish the changes of different control signals in different operation instructions, and avoid the control signal confusion caused by wrong operation. If there is wrong operation, the next instruction can be executed after the operation.

3.4. Software Partial Results

The resource occupancy after the integration of the software part is depicted in figure 7.The logical unit occupied 20%, storage unit utilization 59%. This resource occupancy meets the design requirements.

Flow Status	Successful - Thu Nov 01 15:06:38 2018
Quartus II Version	11.0 Build 157 04/27/2011 SJ Full Version
Revision Name	bept
Top-level Entity Name	bept
Family	Cyclone III
Device	EP3C25F256I7
Timing Models	Final
Total logic elements	8,070 / 24,624 (33 %)
Total combinational functions	6,622 / 24,624 (27 %)
Dedicated logic registers	5,009 / 24,624 (20 %)
Total registers	5009
Total pins	68 / 157 (43 %)
Total virtual pins	0
Total memory bits	359,412 / 608,256 (59 %)
Embedded Multiplier 9-bit elements	28 / 132 (21 %)
Total PLLs	2 / 4 (50 %)

Figure 7. Dynamic read-write transponder system application environment.

4. Performance Test

In order to verify the feasibility of the design and test the performance, a time test is needed. First, the code is downloaded to the existing hardware board platform, and then the operation time is displayed by the interactive display system with the host computer. According to the test time to determine whether the train speed can meet the demand. The structure of display time is tested by the serial assistant on the PC side.

First, the read operation is performed, as shown in figure 8. @ symbol stands for the shortcut key instruction for the read operation, read time 2819 refers to the single reading time of 2.819ms. After several tests, it was found that the values were between 2.8ms-3.8ms.

```
[17:50:04.123] @
[17:50:08.411] read time 2819
[17:50:12.475] Reverse_state  0
[17:50:14.891] CRC Re 427460A6, CRC Cal 427460A6
[17:50:17.092] 90 00 7F 80 00 00 3F A0 0B 96 40 30 02
C0 00 7F FF FF FF FF FF FF FF FF FF FF FF FF FF FF FF
FF FF FF FF FF FF FF FF FF FF FF FF FF FF FF FF FF FF
FF FF FF FF FF FF FF FF FF FF FF FF FF FF FF FF FF FF
FF FF FF FF FF FF FF FF FF FF FF FF FF FF FF FF FF FF
FF FF FF FF FF FF FF FF FF FF FF FF FF FF FF FF FF FF
FF FF FF FF
```

Figure 8. Single read operation test.

Second, perform an integral operation of the system, as shown in figure 9, The system performs two reads and one writes for a total of 73ms.

```
************************
read write time test
A_port Read
A_port Read
read time 10ms
write back read time 63ms
total consuming time 73ms

*****************************
```

Figure 9. System operation time test.

According to the results of system test in figure 8 and figure 9, combined with our analysis, the running speed of the train can be calculated as 34.5km/h.

5. Conclusion

With the heavy railway transportation tasks, The requirement of safety is getting higher and higher. At the same time, it has been a research topic to reduce personnel access, improve detection accuracy and efficiency of detection, and reduce transportation accidents caused by fault of balise, so as to provide reliable guarantee for transportation safety.

This paper first analyses the working methods and characteristics of the existing version of transponder, and the analysis results provide a basis for the feasibility study of dynamic train detection. Then, through the theoretical study of feasibility, the actual simulation and implementation are carried out. After performance testing, although the train is running slowly, the starting point of considering the equipment is to reduce human costs, basically meet the requirements, and there is room for further improvement.

Acknowledgments

I would like to express my gratitude to all those who helped me during the writing of this thesis.

First and foremost, I gratefully acknowledge the help of my supervisor, Mr. Lv Jingyang, who has offered me valuable suggestions in the academic studies. In the preparation of the thesis, she has spent much time reading through each draft and provided me with inspiring advice. Without her patient instruction, insightful criticism and expert guidance, the completion of this thesis would not have been possible.

Last but not least, I owe much to my friends and classmates for their valuable suggestions and critiques which are of help and importance in making the thesis a reality.

References

[1] Ying Yaping, Xu Jianfeng, Chen Yujun. FPGA Design and Implementation of 2FSK Modulation and Demodulation System [J]. Journalof Zhejiang University of Technology, 2010, 38(3): 282-285.

[2] Wang Minzhi. In-depth understanding of Altera FPGA application design [M]. Beijing: Beijing Aerospace Publishing House. 2014.

[3] Xu Ning. Research on the dynamic characteristics of the European standard transponder on-board equipment and the suitability test equipment under high-speed conditions [C]// Institute of Communication Signals, China Academy of Railway Sciences. 2010.

[4] Wang Kun. Research on dynamic detection of transponder in CTCS-2 train control system [D]. 2015.

[5] ZHU Lin-fu, ZHAO Hui-bing, LIANG Di. Modeling and Optimization of Vehicle Transponder Programming Antenna [J]. Journal of the China Railway Society, 2016, 38(4): 44-50.

[6] Cui X, Yu D, Zhang X. A 2-level FSK demodulator for digital-IF receiver[C]//Electron Devices

and Solid-State Circuits, 2007. EDSSC 2007. IEEE Conference on. IEEE, 2007: 1175-1178.
[7] Devgan A, Kashyap C. Block-based static timing analysis with uncertainty[C]//Proceedings of the 2003 IEEE/ACM international conference on Computer-aided design. IEEE Computer Society, 2003: 607.
[8] Sklyarov V, Skliarova I, Barkalov A, et al. Synthesis and Optimization of FPGA-based Systems [M]. Springer Science & Business Media, 2014.
[9] Gustat H, Herzel F. Integrated FSK demodulator with very high sensitivity [J]. Solid-State Circuits, IEEE Journal of, 2003, 38(2): 357-360.

Location and Capacity Planning of Electric Vehicles Charging Piles

Yi Shimin[1], Sun Yunlian[2, a], Zhang Xiaodi[2], Wu Ying[2], Hu Jinlei[3], Zou Qiwu[3], Xie Xinlin[3], Fu Bin[3]

1Guangdong Power Grid Co., Ltd., Guangzhou 510620， China
2Wuhan University, School of Electrical Engineering, Wuhan 430074, China
3Qingyuan Power Supply Bureau of Guangdong Power Grid Co., Ltd., Qingyuan 511500， China

3344147131@qq.com

Abstract. As the planning and construction of electric vehicle charging pile plays a decisive role in the promotion of electric vehicles, this article puts forward a planning method considering the cooperation of different types of charging piles and analyzes the Residential and social charging piles. On the basis of determined number of charging piles in residential area, the planning of social charging piles is analyzed from the demand of charging considering the unbalance of electric vehicles' promotion, the location of charging piles considering the minimum charge distance and the capacity of charging piles considering the situation that EVs don't charge every day. And then the processes of planning are determined. Finally, the planning method is applied to a Xincheng planning in ZhuoZhou which could illustrate the feasibility of this method. Compared with other method, the method of this article pays more attention to the actual situation from the load forecast to the overall planning in the area, thus the result is more realistic and in line with the planning requirements.

1. Introduction

Due to its great advantages in environmental protection, energy saving and emission reduction, electric vehicles have already begun to be promoted in major cities in China. In the next few years, domestic pure electric vehicles will occupy a larger market in the automotive sector [1-3]. With the increase in the number of electric vehicles, the demand for charging infrastructure is also increasing. The construction of charging piles has become an important factor restricting the development of electric vehicles. Charging pile planning has become a hot issue [1, 4].

In the current research, the charging pile planning is mostly based on site selection and capacity. It is mainly based on the existing urban pattern[5], traffic flow information[6] and distribution network structure[7], considering the physical area division and The power network structure and other factors constrain the planning of charging piles and achieved great results. Literature [8] proposed a two-step optimization location method for the location of electric vehicle charging stations in the city, analyzed the charging requirements of the road sections and the urban traffic network, and comprehensively assessed the location of charging stations. The literature [9] analyzed the traffic. And power construction constraints, determine the location of the charging piles, and based on energy equivalence to predict load, determine the capacity of the charging piles and complete the site selection of the

Content from this work may be used under the terms of the Creative Commons Attribution 3.0 licence. Any further distribution of this work must maintain attribution to the author(s) and the title of the work, journal citation and DOI.
Published under licence by IOP Publishing Ltd

charging piles; literature [10] considers traffic flow, urban roads and other factors, and The grid constraints are added to the constant volume of the electric vehicle charging piles, and the constraints of the node voltage amplitude and access to the grid power are taken into account to determine the charging pile power. The literature [11] determines the site of the charging station based on the urban road and uses the queuing theory to determine The number of chargers. However, the current research mainly focuses on social charging piles, and does not consider the complementary relationship between household charging piles and public charging piles, which affects the overall planning.

Under the above background, this paper proposes a charging pile planning method in which the charging pile interacts with the public charging pile in the residential area. Firstly, it analyzes the influence of the unbalanced development of the electric vehicle on the planning and determines the charging demand; then, the social charging pile is given as auxiliary charging. The location of the facility was determined based on the location of the charging pile. Finally, the impact of the charging time of the electric vehicle on the charging load was considered to determine the capacity of the charging pile. Finally, the location of the charging facilities in Zhuozhou Xincheng in 2025 was taken as an example to illustrate the proposed method.

2. The whole charging pile planning steps

The "Guiding Opinions on Accelerating the Construction of Electric Vehicle Charging Infrastructure" issued by the State Council at the end of 2015 pointed out that the construction of charging facilities should be based on the user-owned parking spaces and the special charging facilities installed in the unit parking lots, and public parking facilities should be built. Charging facilities are auxiliary to form a three-dimensional electric vehicle charging infrastructure system. This requires that the demand for electric vehicle charging be met at the city level through the planning of electric vehicle charging piles in the community and charging piles outside the public parking lots outside the community.

In view of this, this paper proposes a three-dimensional charging pile planning method, which divides the charging pile into two types: the charging pile in the community and the public quick charging pile. The community charging piles mainly provide the slow charging service for fixed users. Due to the relatively fixed location of parking spaces, the relatively concentrated charging time, and the long time for slow charging, the charging piles in the residential area are mostly dedicated for buses and need to be based on the penetration rate of electric vehicles. Construction within the community.

The public charging pile is mainly used to meet the charging requirements caused by the imbalance of the charging pile's supply and demand. It needs to pass through factors such as unbalanced supply and demand of the charging pile, characteristics of the vehicle trip, mileage, battery capacity, power consumption per hundred kilometers, whether it is a daily charge, and other factors. Conduct the analysis. For unbalanced charging requirements, model the objective function with the shortest charging distance and determine the location of the common charging pile. Considering the situation of multiple days and one charge, the charging load within the charging pile coverage is predicted so that the capacity of the public charging pile needs to be met. The charging demand of the electric vehicle at the time of maximum charging load determines the number of charging piles. The planning process of public quick charging piles is shown in Figure 1.

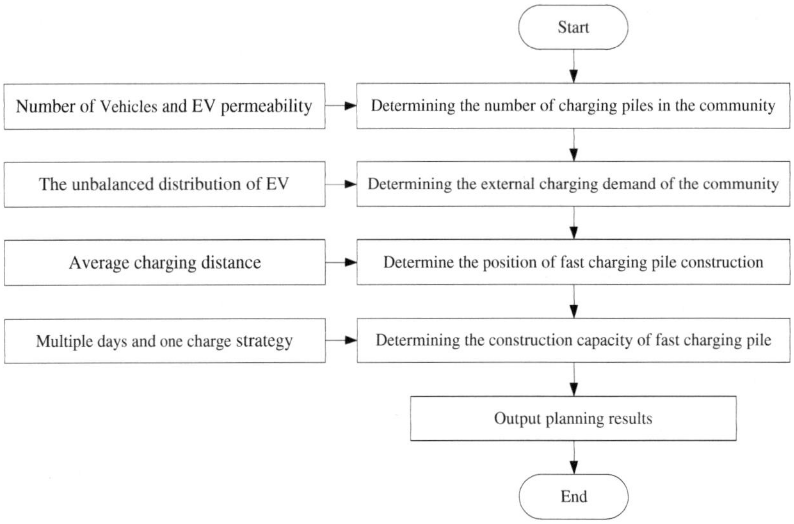

Figure 1. The flow chart of fast charging piles planning.

3. Charging pile planning method

3.1. Determination of charge requirements in consideration of unbalanced distribution of electric vehicles

The number of electric vehicles can be determined by the number of parking spaces planned by the community, the utilization rate of parking spaces, and the penetration rate of electric vehicles. According to the guidance, slow charge piles must be built at a ratio of 1:1. However, the unbalanced distribution of electric vehicles will cause the actual number of electric vehicles in the community to deviate from the predicted value, resulting in a conflict between supply and demand between electric vehicles and charging piles. Given the current domestic housing situation, it will be more difficult to park vehicles in other districts for charging for a long time or to purchase parking spaces in other communities. This requires the construction of a public charging pile to meet this demand for charging. For each cell, the distribution of electric vehicles is not balanced, and Monte Carlo simulation can be used to determine the charging requirements.

Assume that there is a car in the community and the probability of the electric car is P, and the vehicle is determined according to the number of electric vehicles in the community. The above process was repeated 10,000 times and statistics were taken to determine the distribution of the distribution of electric vehicle numbers.

For a normal distribution, there is a 99% probability that the random variable is distributed within the interval $[\mu - 3\sigma, \mu + 3\sigma]$, so the confidence limits can be used to estimate the upper and lower limits of the number of electric vehicles. Among them, the difference between the upper limit of the number of electric vehicles and the average value is the maximum unbalanced number of vehicles that may occur is the charging demand. In the case of a certain average sample size, the maximum number of unequal vehicles in the overall area is:

$$Q = \frac{1}{2} \times 3\sum_{i=1}^{N} \sigma_i \tag{1}$$

N is the number of Residential areas in the area, and σ_i is the variance of the statistical results of the number of EV in the residential area.

3.2. Location planning of public charging pile based on minimum charging distance

Charging distance is the distance from the electric car to the nearest charging pile. According to the charging demand and charging distance of each community, the optimal location and service scope of public charging pile construction are determined.

3.2.1. Charging pile position planning model. If a certain area contains a plurality of charging pile construction points that can be used to charge an electric vehicle, the electric vehicle in the area always selects the charging point that is nearest to its own construction point, which divides the entire area into multiple small sub-domains. Each small area is the range of load requirements that the charging pile should meet within the corresponding construction site, that is, its service scope.

(1) Objective function.

The planning goal of the electric vehicle charging pile in a certain range is that the sum of the weighted total distances of the individual cells reaching the charging pile is the smallest. The objective function is:

$$f(x) = \min\left(C_i \sum_{i=1}^{n} \sum_{j=1}^{m} \min S_{i,j}\right) \tag{2}$$

$\min S_{i,j}$ is the distance from the i th cell to the nearest one of the m charging and construction points, C_i is the number of electric vehicles in the i th cell, and m is the number of electric vehicle construction points.

(2) Constraints

a. Public charging pile location constraints

The public charging piles are mainly built in public parking lots, so the location of the selected public charging piles must be within the scope of public parking lots, that is, the location of public charging piles:

$$(x, y) \in A \tag{3}$$

Where A is the parking location coordinate set.

b. Parking lot constraints

Since the number of parking lots is limited, the number of parking lots needs to be constrained, that is, the number of parking lots is constrained:

$$m \leq M \tag{4}$$

M is the total number of parking lots.

3.2.2. TLBO optimization algorithm. The TLBO algorithm simulates the process of teacher teaching and student learning, and achieves optimization through the two stages of "teaching" and "learning". This method has many advantages such as high computational accuracy, fast convergence speed, small computational complexity, good global convergence for nonlinear optimization, and no constraint on algorithm parameters. It can solve optimization problems by setting the number of populations and termination conditions [12-14]. In view of the fact that the optimization goal in the paper is single-objective and requires a fast convergence speed, the TLBO algorithm is used to solve the optimization problem. In this paper, taking the coordinates of the proposed site as a variable, the minimum total charging distance is the optimization goal; the algorithm flow is shown in Figure 2.

3.3. Taking into account the determination of the charging capacity for multi-day charging

One-day charge limits the user's freedom to reduce the difficulty of calculation, but it will cause the waste of the battery. Literature [15] proposes an orderly charging strategy for electric vehicles based on the needs of individual consumers and the willingness of users. According to the National Travel Survey (England 2014), respondents who traveled less than 2 times per day reached 59.91%, and 98.25% of respondents had fewer than 5 daily trips. . For each trip, less than 8 kilometers of travel accounted for 66% of the total travel. According to relevant survey data, 84.25% of users have daily mileage of less than 60 km [16, 17]. When electric vehicle cruising range is 300 km, the daily power consumption of these users only accounts for 20% of vehicle battery capacity. Charging mode will cause a great waste of battery capacity, and does not meet the habits of most users. Therefore, it should consider the situation of multiple days and one charge.

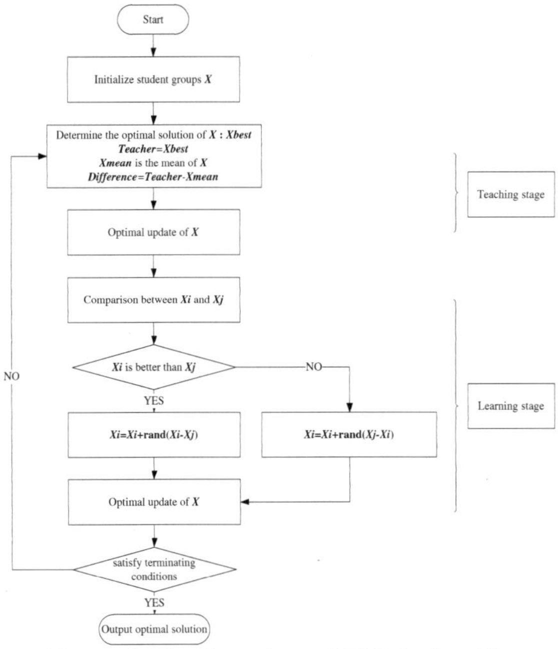

Figure 2. The flow chart of TLBO algorithm.

3.3.1. Electric vehicle charging judgment basis. Considering the case of multi-day charging, it is necessary to determine whether or not the electric vehicle is charged every day. Based on the initial charging power of different electric vehicles, it is determined whether or not the battery is charged on the day, and the charging probability is set to pc.

$$p_c = \begin{cases} 0 & x > 90\% \\ f(x) & 50\% < x \le 90\% \\ 1 & 0 < x \le 50\% \end{cases} \tag{5}$$

Where f (x) is the charge probability function, and its value is determined by the remaining charge x, which is set to satisfy Equation 6:

$$f(x) = \begin{cases} 0.2 & 80\% < x \le 90\% \\ 0.4 & 70\% < x \le 80\% \\ 0.6 & 60\% < x \le 70\% \\ 0.8 & 50\% < x \le 60\% \end{cases} \tag{6}$$

3.3.2. Determination of charging miles for electric vehicles.

The mileage of an electric vehicle is directly related to its charging load. If the electric vehicle is set to charge once a day, the daily mileage is the required charging mileage. When a multi-day charge is involved, the electric mileage should be the total mileage since the last charge. Therefore, the total mileage needs to be calculated. The total mileage is S:

$$S = (((S_n \times w_n + S_{n-1}) \times w_{n-1} + S_{n-2}) \times w_{n-2} + \ldots + S_0) \times w_0 \tag{7}$$

Among them, Sn is the mileage of the electric vehicle in the first n days and wn is the total mileage of the corresponding days. It can be determined by Formula $w_n = f(p_n)$.

4. Analysis of case calculation

Taking the planning of the charging piles of the high-speed railway Xincheng in Hebei Province as an example, the Xincheng is in the planning and construction stage. There are 52,940 parking spaces in the residential area, divided into 41 areas, and the utilization rate of parking spaces in the residential area of Xincheng is expected to reach 90% by 2025. The penetration rate of electric vehicles is 20%. According to the 20% of the number of parking spaces, each community shall reserve the construction of the exchange of slow charging piles and construct DC fast charging piles in public parking lots.

4.1. Charge demand determination

According to the method proposed in Section 2.1 of the article, the maximum number of unbalanced electric cars in each district can be obtained by judging the cars in the community. The results obtained are shown in Table 1.

Table1. Unbalanced number of vehicles (UNOV).

UNOV	No.	UNOV	No.	UNOV
29	15	29	29	15
32	16	25	30	31
24	17	17	31	29
19	18	16	32	26
19	19	24	33	27
19	20	21	34	26
16	21	22	35	27
19	22	0	36	28
18	23	0	37	36
18	24	0	38	35
16	25	0	39	28
15	26	0	40	18
23	27	12	41	23
15	28	17		

4.2. Charging pile site selection

Taking the charging demand of each cell as the weight and minimizing the total charging distance as the optimization goal, the TLBO optimization algorithm is used to determine the optimal construction location of the public parking lot. According to the actual project requirements, the number of available parking lots is 2 to 5, and the construction site location and division are shown in Figure 3. The total charging distance and the number of parking lots are shown in Table 2.

Figure 3. Different construction point location.

Table2. Charge distance of different construction points.

Construction points number	Charge distance/km	Total distance reduction percentage
2	516.2	----
3	401.3	22.25%
4	347.2	13.49%
5	308.3	11.20%

According to the percentage change of the total distance reduction, project construction needs and investment restrictions, select option (b) to build 3 public charging parking lots as the final planning plan. This scheme has a smaller total charging distance, and at the same time, on the basis of the same purchase cost of the charging pile, taking into account the smaller reconstruction cost of public parking lots, and is suitable for application in the charging pile planning of the high-speed railway Xincheng in Zhuozhou City, Hebei Province.

4.3. Determine the charging pile power

Considering the situation of multi-day charging, forecast the charging demand within the service range, use the Monte Carlo method to predict the charging load curve of fast charging piles in public parking lots, and determine the number of charging piles to meet the charging requirements within the range. In this article, 24 hours a day is divided into 96 time points, and the arrival time of vehicles in the community is shown in Equation 12.

$$y = 0.595e^{-\frac{(x-73.77)^2}{2.17^2}} + 0.405e^{-\frac{(x-77.67)^2}{3.77^2}} \tag{8}$$

The statistics of the number of trips per sunrise are shown in the table 3.

Table3. The statistical results of Daily travels.

Travel times	Percentage	Travel times	Percentage
0	6.63%	6	3.49%
1	17.86%	7	1.16%
2	34.70%	8	0.41%
3	22.99%	9	0.16%
4	4.72%	10	0.01%
5	7.86%	11	0.01%

5. Conclusion

This paper proposes a method for considering the location and constant volume of charging piles that are complemented by different types of charging piles. The planning of charging piles and public charging piles in the residential area is considered separately, and the following research work is mainly completed.

(1) The classification of electric vehicle charging piles has been proposed, and a charging demand determination method that takes into account the unbalanced distribution of electric vehicles has been proposed.

(2) The optimization model of the charging pile location targeting the shortest charging distance of the charging user is proposed, and the TLBO algorithm is applied to the optimization calculation of electric vehicle charging pile location.

(3) Proposed to take into account the charging load of electric vehicles for several days

With the prediction method, the results obtained are closer to the actual situation than the original setting.

Through the research on the layout planning of the charging facilities for the high-speed railway Xincheng in Hebei Province, the practical application of this method is introduced. The current charging pile planning system is still not perfect, and more is at the theoretical level. With the increasing number of electric vehicles entering major cities and the increase in charging data for electric vehicles, this kind of three-dimensional sub-type planning method will also be further improved.

References

[1] YANG Xi,DENG Jianshen,LI Hongfeng,FANG et al. Design and research on public service and interactive platform in electric vehicle[J].Power System Protection and Control,2016, 44(10):137-144.

[2] Mohsenian-Rad A, Wong V W S， Jatskevich J, R et al. Autonomous demand-side management based on game theoretic energy consumption scheduling for the future smart grid[J] . IEEE Trans. on Smart Grid, 2010, 1(3) 320-331.

[3] LU Xinyi， LIU Nian， CHEN Zheng， et al. Multi-objective optimal scheduling for PV-assisted charging station of electric vehicles [J]. Transactions of China Electrotechnical Society, 2014, 29(8):46- 56.

[4] XU Qingqiang, KOU Yinggang, MA Jianwei, et al. Research on typical design scheme of charging/battery swap infrastructure for electric vehicle[J].Power System Protection and Control,2015, 43(13):118-124.

[5] LIU Zhipeng, WEN Fushuan, XUE Yusheng, et al. Optimal siting and sizing of electric vehicle charging stations [J]. Automation of Electric Power Systems， 2012， 36(3): 54- 59.

[6] WU Chunyang, LI Canbing, DU Li, et al.Typical schemes of electric vehicle charging infrastructure connected to grid [J]. Automation of Electric Power Systems, 2010， 34(24) : 37-45.

[7] WANG Hui, WANG Guibin, ZHAO Junhua, et al. Optimal planning for electric vehicle charging stations considering traffic network flows [J]. Automation of Electric Power Systems，

2013, 37 (13): 63- 69.

[8] GAO Yajing, GUO Yandong. Optimal location of urban electricvehicle charging stations using a two-step method [J]. Electric Power, 2013, 46(8): 143-147.

[9] GUO Chunlin, XIAO Xiangning. Planning method and model of electric vehicle charging infrastructure [J] . Automation of Electric Power Systems, 2013, 37(13): 70- 75.

[10] HUANG Xiaoqing, YANG Hang, CHEN Jie , et a1. Optimal planning for electric vehicle charging stations based on life cycle cost and quantum genetic algorithm[J]. Automation of Electric Power Systems, 2015, 39(17): 176-182(in Chinese).

[11] ZHAO Shuqiang, LI Zhiwei. Optimal Planning of Charging Station for Electric Vehicle Based on PSODE Algorithm [J]. Journal of North China Electric Power University,2015, Vol. 42, No. 2: 1-7.

[12] Rao R V, Savsani V J, Vakharia D P. Teaching-Learning-Based Optimization: An optimization method for continuous non-linear large scale problems[J]. Information Sciences, 2012, 183(1):1-15.

[13] Satapathy S C, Naik A, Parvathi K. Weighted Teaching-Learning-Based Optimization for Global Function Optimization[J]. Applied Mathematics, 2013, 04(3):429-439.

[14] Rao R V. Applications of TLBO Algorithm and Its Modifications to Different Engineering and Science Disciplines[M].// Teaching Learning Based Optimization Algorithm. 2016.

[15] DANG Jie, TANG Yi, NING Jia, et al. A strategy for distribution of electric vehicles charging load based on user intention and trip rule[J]. Power System Protection and Control,2015,43(16):8-15.

[16] Darabi Z, Ferdowsi M. Aggregated impact of plug-in hybrid electric vehicles on electricity demand profile [J]. IEEE Transactions on Sustainable Energy, 2011, 2(4):501-508.

[17] ZHANG Dexin, CHEN Jinchuan, LIU Ying. Daily travel characteristics and the development strategy of private motor vehicles in Beijing [C]// Proceedings of the6th International Conference of Transportation Professionals. Dalian: Press of Dalian University of Technology,2006.(in Chinese).

Prediction of Bearing Remaining Useful Life based on Mutual Information and Support Vector Regression Model

Wentao Sui[1], Dan Zhang[2,a], Xiaomei Qiu[1], Wei Zhang[1] and Lin Yuan[1]

1.School of Mechanical Engineering, Shandong University of Technology, 255049 Zibo, China

2.School of Electrical & Electronic Engineering, Shandong University of Technology, 255049 Zibo, China

Corresponding author: zhangdan_sdut@163.com

Abstract. In order to evaluate the degradation state of the mechanical equipment and master the information of the remaining useful life (RUL) of the bearing accurately, this paper presents a method for predicting the remaining useful life of bearings based on mutual information (MI) and support vector regression (SVR) model. The proposed method includes two steps of online and offline, the offline step is used to build a degradation model of the bearing by learning , the online step uses the degradation model to predict the remaining useful life. By analyzing the experimental data of bearing full lifetime degradation, the results show that the method can effectively simulate the bearing degradation process and predict the remaining useful life of the bearing.

1. Introduction

Prognostics and health management (PHM) of industrial systems is the core activities of intelligent maintenance such as condition based maintenance (CBM) and predictive maintenance (PM). It includes state monitoring, fault detection, fault diagnosis, fault prediction and decision support. Therefore, fault prediction and health management of components can improve machine availability, reliability and safety. The purpose of fault prediction and health management of rotating machinery is not only to detect faults, but also to predict the time when the machine can run safely and realize its functions.

At present, many literatures are focused on bearing fault diagnosis [1-3] ,the study of remaining useful life prediction has gradually attracted the attention of scholars [4]. Kang [5] use a combination of multiple evaluation criteria to analyze the validity of features, select the effective features accurately and comprehensively. Shen [6] proposed a new method of remaining useful life prediction based on relative characteristics by using the relative root mean square value which is not affected by bearing individual differences to evaluate the bearing performance degradation law. Lei [7] took wavelet packet decomposition technology to extract the frequency band energy ratio of vibration signal as the sensitive index of bearing degradation. The running trend of bearing was predicted by choosing the radial basis function and appropriate parameters. In the above methods the calculation complexity is increased by using multiple evaluation criteria to select features, while the bearing fault information contained in a single indicator is too little to accurately predict the bearing remaining useful life. Therefore, this paper uses mutual information method [8] to find a few effective features which can

Content from this work may be used under the terms of the Creative Commons Attribution 3.0 licence. Any further distribution of this work must maintain attribution to the author(s) and the title of the work, journal citation and DOI.
Published under licence by IOP Publishing Ltd

represent a large number of original features, it provides more remaining useful life information, and then predict the bearing remaining useful life better.

To solve the above mentioned problems, this paper proposes a method of mutual information (MI) for nonlinear feature reduction, and combines nonlinear support vector regression to construct health indicators to evaluate the health status of bearings to predict the remaining useful life.

2. Feature extraction and reduction

2.1. Feature Extraction And Processing

One of the main difficulties in predicting the remaining useful life of bearing is how to establish the characteristic indicators representing the degradation state from the effective feature. In order to accurately predict the bearing remaining useful life, this paper selects several characteristic indicators from time domain and frequency domain as the input sample of the model. These characteristic indicators reflect the running trend and development condition of bearing to some extent, and select the fault sensitive feature to predict the bearing life, as shown in Table 1.

Table 1. The features description.

Feature	Feature name
Time domain	peak value、variance、root mean square、skewness index、kurtosis index、waveform index、peak value index、pulse index、margin index
Frequency domain	mean frequency、frequency center、Standard deviation frequency、Root mean square frequency、kurtosis frequency

Feature processing includes two parts.

(1) The original vibration signals of the bearing contain more noise, the curves drawn by the extracted features will have many "burrs", deal with the sliding average of the features to in order to obtain a relatively smooth curve and reduce noise.7 points slip average is shown in formula (1).

$$x_n^{MA}=\begin{cases}\frac{1}{n+3}\left(x_1+\cdots+x_{n+3}\right) & n\leq 3\\ \frac{1}{7}\left(x_{n-3}+\cdots+x_{n+3}\right) & 4\leq n\leq N-3\\ \frac{1}{N-n+4}\left(x_{k-3}+\cdots+x_N\right) & N-2\leq n\leq N\end{cases} \quad (1)$$

Where xn is the original characteristic sequence, x_n^{MA} is the new sequence after the slip average, N is the total number of vibration data.

(2) In order to reduce the influence of large differences of characteristic variables on the performance of the model, the extracted feature is normalized.

$$X_{norm}=(Y_{max}-Y_{\min})(X-X_{\min})/(X_{\max}-X_{\min})+Y_{\min} \quad (2)$$

Where Xnorm is the result of normalization, Ymax= 1, Ymin= 0, X is the eigenvalue, Xmax is the maximum feature value, Xmin is the minimum feature value.

2.2. Feature Reduction

Mutual information is a basic concept in information theory and a measure of statistical correlation between two random variables. It is usually used to describe the statistical correlation between two systems, or the amount of information contained in another system, as a measure of the amount of information that systems provide to each other. Mutual information is used to measure the degree of correlation between two variables, representing the part with the same information between two variables. Give the random variables X and Y, their marginal probability distribution and joint probability distribution are respectively p(xi), p(yi) and p(xi,yi), the mutual information is defined as:

$$I(X;Y) = E[I(x_i, y_j)] = \sum_{x_i \in X} \sum_{y_j \in Y} p(x_i, y_j) \log \frac{p(x_i, y_j)}{p(x_i) p(y_j)} \tag{3}$$

Where p(xi) is the occurrence probability of xi. When X and Y are irrelevant variable or independent variables, the I(X;Y)=0, it means that there is no information contained between the two variables; conversely, the higher the degree of correlation between the two variables, the bigger the mutual information value, the more the same information is included.

In this paper, the mutual information between each input sample feature and the bearing remaining useful life are calculated, preset a threshold. When the mutual information value and the correlation coefficient absolute value are greater than the threshold value, the feature is retained, otherwise it is eliminated.

3. SVR prediction theory

Support vector machine is a learning algorithm which is first proposed by Vapnik to deal with small samples. Support vector regression has been successfully applied to various machine learning problems, especially regression problems, sunspot frequency prediction and drug discovery. In this paper, support vector regression is used to study the nonlinear degradation model of bearings. The objective of support vector regression is to estimate the relationship between input and output random variables on the assumption that the joint distribution of input and output variables is completely unknown.

Support vector machines are mainly divided into two types: support vector classification (SVC) and support vector regression (SVR), SVR is the most common application of support vector machines. The model created by SVR technology only depends on a subset of the training data because the cost function constructed by the model ignores all training data close to the prediction threshold of the model. Regression estimation can be formally transformed into a inference problem of the function y=f(x), the training set X= {(xi, di), i=1,···, l } is given, $x_i \in R_n$ is the input variable, $d_i \in R$ is the predictive value and l is the number of the training set. The regression function is

$$f(x) = \sum_{i=1}^{l} \left(\alpha_i - \alpha_i^* \right) k\left(x_i, x \right) + b \tag{4}$$

Where $k\left(x_i, x\right)$ is the kernel function, b is the bias threshold, α_i and α_i^* are the Lagrangian multiplier, $\left(\alpha_i - \alpha_i^*\right)$ can get by solving the following optimal problem.

$$\sum_{i,j=1} \left(\alpha_i - \alpha_i^*\right)\left(\alpha_j - \alpha_j^*\right) k\left(x_i, x_j\right) + \varepsilon \sum_{i=1}^{l} \left(\alpha_i + \alpha_i^*\right) - d \sum_{i=1}^{l} \left(\alpha_i - \alpha_i^*\right) \tag{5}$$

$$s.t. \quad \sum_{i=1}^{l} \left(\alpha_i - \alpha_i^*\right) = 0 \qquad \alpha_i, \alpha_i^* \in [0, C]$$

Where ε is insensitive factor, d is the order of kernel function and C is penalty factor.

4. Prediction process

Prognostics and health management is the core activity of condition based maintenance and predictive maintenance. Predicting remaining useful life can be done by three main methods: model-based, data-driven and hybrid prognostics.

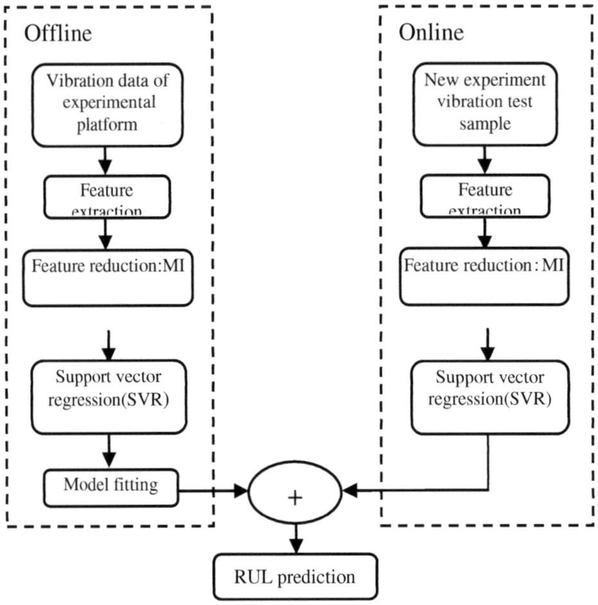

Figure 1. Flowchart of remaining useful life prediction.

The model-based prediction results are more accurate, but they are more difficult to implement because physical modeling is not a simple task in most applications. Data-driven fault prediction relies on the data provided by the sensor to extract features, which are used to build remaining active life models. This method is easy to implement, but the prediction results are not as accurate as the former, which provides a compromise between accuracy and complexity. The hybrid method combines the advantages of the first two methods, but also has these disadvantages.

The method presented in this paper belongs to data-driven, the framework of life prediction is shown in figure 1.In the off-line step, the extracted features of bearing vibration signal are reduced by mutual information method. The combined features are taken as the input of SVR model, and the ratio of current remaining useful life of bearing to full life is model output.

5. Experimental verification

5.1. Experiment platform

PRONOSTIA experimental platform is used to test and verify bearing health assessment, fault diagnosis and fault prediction models, as shown in figure 2. The experimental platform can accelerate the bearing degradation in only a few hours and collect the real experimental data describing the bearing degradation law. The design of the PRONOSTIA platform can provide the bearing degradation data under various operating conditions [9].

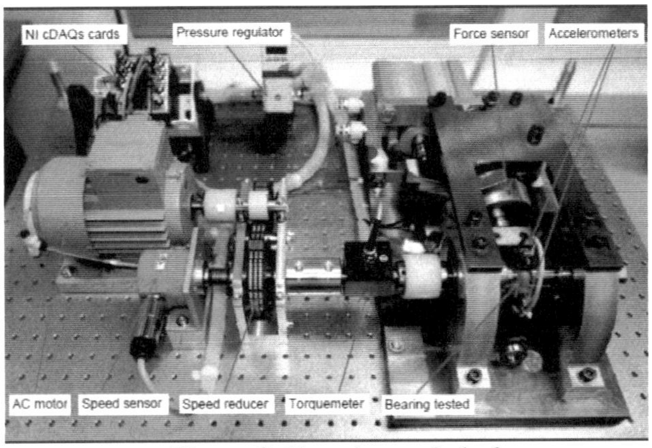

Figure 2. Experimental platform

The degradation of bearing is mainly realized by two kinds of sensors installed on the experimental platform: acceleration sensor and temperature sensor. The inner ring of the bearing rotates with the shaft, the outer ring is fixed, and the acceleration sensor and temperature sensor are mounted on the outer ring of the bearing. The two accelerometers are installed in the horizontal and vertical position of the outer ring of the bearing, and the vibration information is collected from the horizontal direction and the vertical direction respectively. The sampling frequency of acceleration sensor and temperature sensor is 25.6 kHz and 0.1 Hz respectively.

On this platform, the bearing life acceleration experiments are carried out under three different working conditions, and the rotational speed and load information of each condition is shown in Table 2.

Table 2. Working condition information.

Condition	Speed(RPM)	Loading(N)
Condition 1	1800	4000
Condition 2	1650	4200
Condition 3	1500	5000

5.2. The results and analysis

In this paper, mutual information method is used to reduce the features extracted and the correlation coefficient method (RC, Relation Coefficient) [10] is used to compare and analyze with the former. The two methods are used to calculate the mutual information value and correlation coefficient between each input sample feature and the bearing remaining useful life respectively. A threshold value is preset. When the absolute value of mutual information value and correlation coefficient is greater than the threshold value, the feature is retained, otherwise it is eliminated. With a threshold of 0.89, the selected features and predictive indicators of the two methods are shown in Table 3.

Table 3. Feature and prediction re sult indicators value

method	feature	MI or RC	MSE	R
MI	Margin indicator	0.9002	0.0115	0.9267
	Standard deviation frequency	0.9449		
	Kurtosis frequency	0.9522		
RC	Standard deviation frequency	-0.8963	0.0357	0.8977
	Root mean square frequency	-0.8934		
	Kurtosis frequency	-0.8971		

In this paper, the full life data of bearing 1_1 in working condition 1 is used as the training set to establish a prediction model to predict the remaining useful life of bearing 1_3. The sensitive feature combination after bearing reduction is used as the input of the SVR training model and the ratio of remaining useful life to full life is used as the output at present. The prediction results of remaining useful life based on two feature reduction methods are shown in fig.3 and fig.4 respectively. The horizontal coordinate is the acquisition time and the vertical coordinate is the percentage of the remaining useful life.

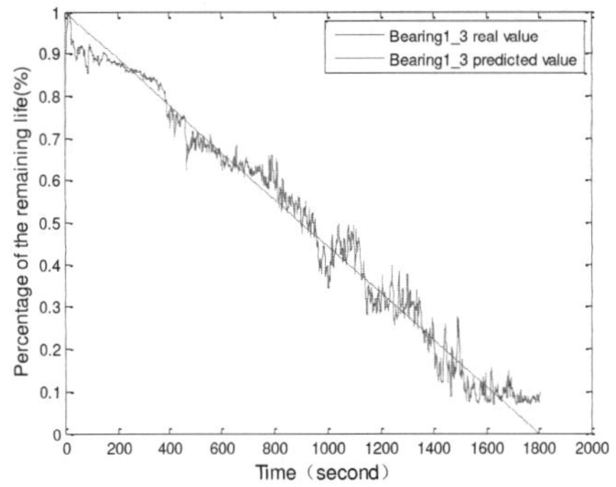

Figure 3. RUL prediction result based on MI

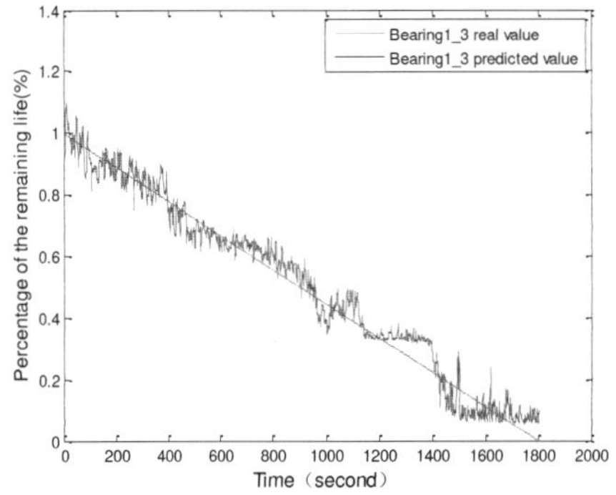

Figure 4. RUL prediction result based on RC

It can be seen from the above two diagrams that there is a certain error between the prediction result and the actual value of the bearing remaining useful life, the predicted value fluctuates up and down near the actual value, and the overall trend of the prediction is consistent with the actual value. From Table 3, the squared correlation coefficient (R) of the prediction result in the model established by the feature reduction of MI method is 0.9267, which is higher than the R value of the model established by the feature reduction of the RC method, the mean square error (MSE) of the model is 0.0115, less than that of the latter. The advantages of the feature reduction method proposed in this paper in predicting the remaining useful life are proved.

6. Conclusion

Aiming at the difficulty of predicting the bearing remaining useful life, the feature reduction method of bearing signal is improved, a prediction model of remaining useful life based on mutual information and support vector regression is proposed. Firstly, reduce the extracted bearing signal features, select several effective features by mutual information method, which form the support vector regression model of sample set with remaining useful life, then the model is used to predict the remaining useful life of other bearings.

Acknowledgments
This research was supported by Shandong Provincial Natural Science Foundation (China, approved grant number ZR2016EEM20, ZR2016FL15).

References

[1] W.T. Sui, D. Zhang, W.Wang. Fault Diagnosis of Rolling Element Bearings based on EMD and MKD [J]. Journal of Vibration and Shock, 2015, 34(9): 55-59+64
[2] W.P. Liu, Y.Q. Liu, S.P. Yang, X.H. Gu. Fault diagnosis of rolling bearing based on typical correlated kurtogram[J].Journal of Vibration and Shock,2018, 37(8):87-92
[3] D. Zhang, W.T. Sui, Q.J. Guo. Condition Monitoring of Rolling Element Bearings Using Optimal Gabor Filters[J]. Journal of Beijing University of Posts and Telecommunications, 2016: 39(2):103-107
[4] T Benkedjouh, K Medjaher, N Zerhouni, S Rechak. Remaining useful life estimation based on nonlinear feature reduction and support vector regression[J]. Engineering Applications of Artificial Intelligence,2013, 26: 1751–1760
[5] Sh.Q. Kang, L.Q. Ye, Y.J Wang, J.B. Xie, Mikulovich V I. Remaining useful life prediction of rolling bearing based on MCEA-KPCA and combined SVR[J]. Journal of Electronic Measurement and Instrumentation,2017,31(9): 1365-1371
[6] Zh.J. Shen, X.F. Chen, Zh.J. He, Ch. Sun, X.L. Zhang, Zh.W. Liu. Remaining useful life Predictions of Rolling Bearing Based on Relative Features and Multivariable Support Vector Machine[J].Journal of Mechanical Engineering, 2013, 49(2): 183-189
[7] Y.G. Lei, X.H. Wu, Ch. Wu, J. Lin. Research On Trend Prediction of Rolling Element Bearings Based on Relevance Vector Machine[C]// National Conference on Equipment Monitoring, Diagnosis and Maintenance, National Equipment Failure Diagnostic Conference and 2014 National Equipment Diagnostic Engineering Conference. 2014
[8] J. Li, J. Li, Sh. Y. Zhang, H.B. Shu. Mechanical Fault Method Using EEMD Algorithm and Mutual Information Method[J].Journal of Huaqiao University(Natural Science), 2018, 39(1): 7-13
[9] Nectoux P, GOURIVEAU, Medjaher K ,et al. Pronostia: An experiment platform for bearings accelerated degradation tests[C].Proceedings of IEEE International Conference on Prognostics and Health Management, 2012: 1-8
[10] Y.F. Zhen, H.J. Shi. Attribute reduction algorithm based on relation coefficient and conditional information entropy[J]. Computer Engineering and Applications, 2011, 47(16):26-28

IOP Conf. Series: Materials Science and Engineering **533** (2019) 012033 doi:10.1088/1757-899X/533/1/012033

Kinematics and workspace analysis of a modified portable exoskeleton arm

Xiao yongfei [1], Zhai xinpeng [1,2,a], Fu xiaochun [1,2], Ji xiangjing [1,2]

1Institute of Automation, Qilu University of Technology(Shandong Academic of Science),Shandong Provincial Key Laboratory of Robot and Manufacturing Automation Technology, Jinan,China

2 School of Electrical Engineering of Qilu University of Technology(Shandong Academy of Science) ,Jinan,China

zxp_ql@163.com

Abstract. In order to solve the problems such as small workspace, complicated mechanical structure and motion interferences of traditional exoskeleton arm, a modified exoskeleton mechanism is proposed by optimizing the angles between the shoulder joint axes. In this paper, the model of traditional exoskeleton arm and the modified one are both built. And the kinematic theory of the two models is analysed and discussed. Furthermore, the Monte Carlo method is used to simulate the workspace of the mechanism with the help of the Matlab software. The simulation results show that the modified exoskeleton arm has larger workspace than the traditional one. So it can better meet the needs of more application scenarios.

1. Introduction

The exoskeleton arm is a bio-mechatronic device that integrates electronic, mechanical, biology, advanced control and other technologies. It has broad application prospects in the fields of modern military, industry, disaster relief, medical rehabilitation and so et al. So it has important theoretical and application value. The human upper-limb, including the shoulder, elbow and wrist joint, has a complex structure. The exoskeleton arm can work well when the degrees of freedom (DOFs) of the exoskeleton arm are equal to or greater than the ones of the upper-limb. For example, pneumatic exoskeleton auxiliary clothing (PAL) was developed by Japan Kanagawa institute of technology[1]. Military exoskeleton XOS2 was designed by American Raytheon Company[2]. 7DOF exoskeleton arm CADEN-7 was developed by Washington University[3]. The complexity of mechanism is greatly increased. The cost of mechanism is expensive. It limit the applicable field. So many scholars have proposed using a simplified kinematic model in order to reduce the complexity of the design. For example, two serial parallelograms were used to simulate the shoulder joint by WU et al[4]. In other projects NEF et al. presented an optimization method by shifting a rotation axis in the ball-and-socket joint[5]. A circular track is formed to simulate the spherical motion produced by the glenohumeral joint (GH joint). Lee et al. proposed a simplified model based on the physiological structure of the upper limb[6]. It uses the spatial separation of the exoskeleton and the upper limb to compensate for the inaccuracy of the model. Simplified kinematics model cause the following problem: a) the

Content from this work may be used under the terms of the Creative Commons Attribution 3.0 licence. Any further distribution of this work must maintain attribution to the author(s) and the title of the work, journal citation and DOI.
Published under licence by IOP Publishing Ltd

IOP Conf. Series: Materials Science and Engineering **533** (2019) 012033 doi:10.1088/1757-899X/533/1/012033

simplified kinematics reduces the working space and limit the applicable field; b) Interference occurs between the branches and the arm.

The layout of this paper is organized as follows; The exoskeleton shoulder joint is analyzed and D-H model of traditional exoskeleton arm is established in section 2. In addition, the reasons for the small workspace of traditional exoskeleton arm are analyzed. So a modified portable exoskeleton arm is proposed by optimizing the angles between the shoulder joint axes. Compared with the traditional exoskeleton arm, the modified one has larger workspace. The forward kinematics of traditional exoskeleton arm and modified one are solved in section 3. The workspace of the two models is simulated in section 4. The simulation comparison and analysis of two models are also presented in this section. Finally the conclusion of this paper is shown in section 5.

2. Analysis of an exoskeleton shoulder joint

2.1. Analysis of a traditional exoskeleton shoulder joint

The most complicated part of the human upper limb is the shoulder joint. The traditional exoskeleton arms implement three mutually orthogonal series of joints which simulate the spherical motion produced by the GH joint. In order to analyze the traditional exoskeleton arm and modified one, a typical traditional exoskeleton D-H model and coordinate systems are established in this paper as shown in Figure 1.

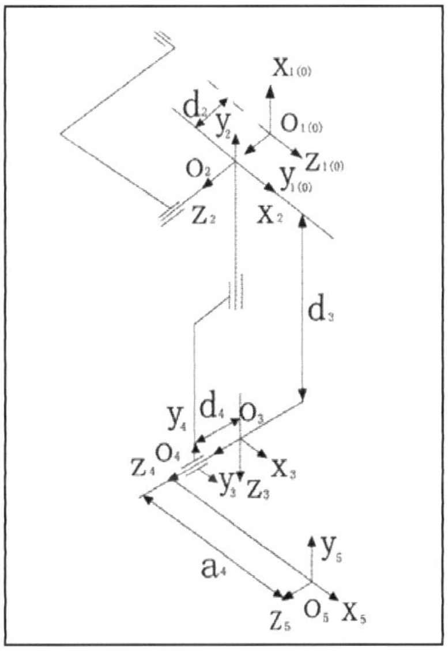

Figure 1. Traditional exoskeleton arm model

In order to meet the activities of daily living, the exoskeleton arm need achieve large working space. But it is difficult to reproduce the spherical motion in the traditional portable exoskeleton. The reason for this is that the design of the exoskeleton arm must be anthropomorphic. In addition, the three orthogonal joint axes will cause interference with the human arm or between the branches. Finally, the length of the arm should also be considered. Due to difference of individuals, everyone has different sizes of arms. That means the kinematic model must be able to adjust to suit different wearers. So the working space produced by the traditional portable exoskeleton arm can't assist people complete their most daily activities. Based on the previous work, a modified exoskeleton arm is proposed by optimizing the angles between the shoulder joint axes.

2.2. Design of a modified exoskeleton arm

As can be seen from the human anatomy, the shoulder joint is a ball-and-socket joint that can achieve abduction/adduction, flexion/extension, internal/external rotation. But three mutually perpendicular joint axes hard to perform such movements. To perform this motion, a serial manipulator requires three series of rotating joints that intersect each other. In order to maximize the theoretical space of the end effector (regardless of the limitations of the physical joint), the following two inequalities must be met between the joint axes.

$$\frac{\pi}{2}-\theta_3 \le \theta_2 \le \frac{\pi}{2}+\theta_3 \tag{1}$$

$$\pi-\theta_2-\theta_3 \le \theta_1 \le \theta_2+\theta_3 \tag{2}$$

The angles are all from 0 to π. The relationship between the shoulder joint axis and the end effector is shown in Figure 2.

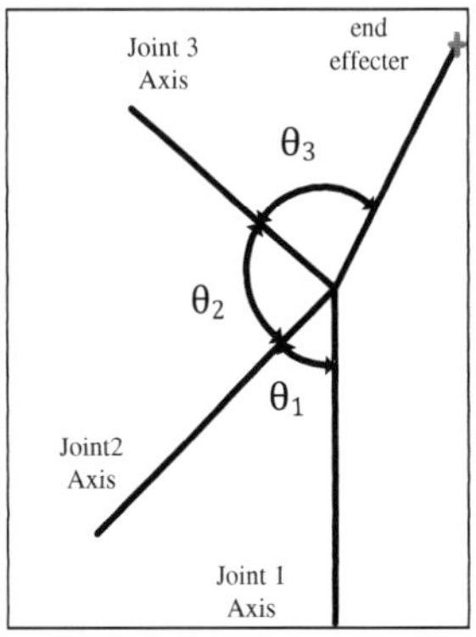

Figure 2. Relationship between shoulder joint axes and end effector[7]

To explain these two inequalities, the end effector and joint axis 3 are considered separately. The end effector is marked as a green dot, and the joint axis 3 is marked as a black line as shown in Fig 3(1). First rotating the end effector about the joint axis 3 will obtain a orange circle as shown in Fig 3(2).Then rotating the circle around the joint axis 2 produce an orange surface as shown in Fig 3(3).Rotating the orange surface around the joint axis 1 yields a complete sphere as shown in Fig 3(4). In order to ensure that workspace of the end effector is maximized, the orange surface must contain a complete large circle, and the joint axis 1 must intersect this circle. Inequality (1) describe the condition that the orange surface contains a complete large circle. Inequality (2) describe the condition under which the first joint axis intersects this circle.

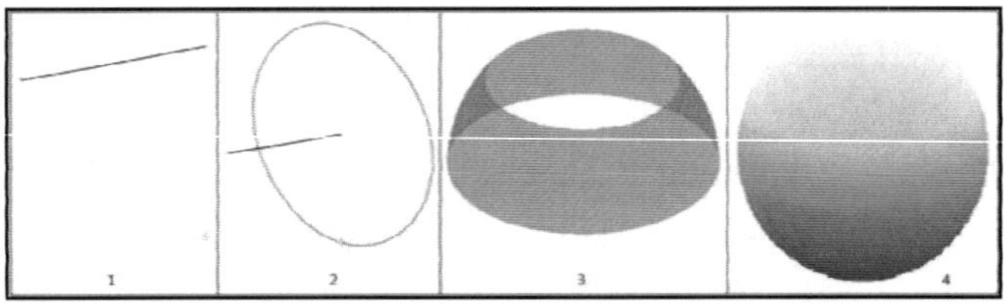

Figure 3. Step in determining workspace of three intersecting axis

The exact value of the angle selection between adjacent joint axes affects the position of the singular points in the workspace and the range of joint limits. Considering these factors, the modified exoskeleton arm is shown in Figure 4. Where the modified exoskeleton arm is worn on the 3D man model. The arm of the 3D model can swing freely like a normal person.

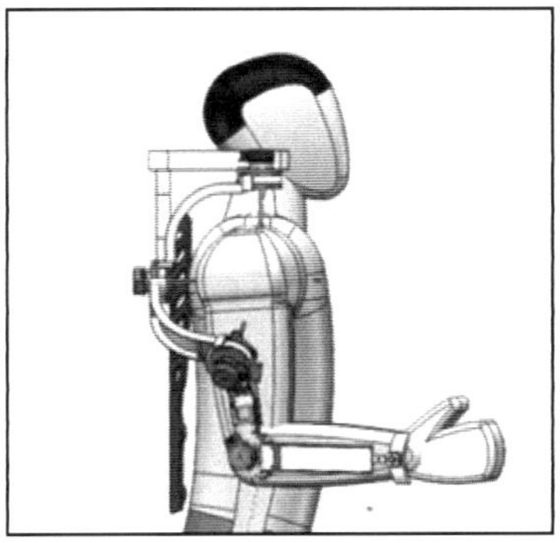

Figure 4. Modified exoskeleton arm of 3D model

3. Kinematics analysis of an exoskeleton arm

3.1. Forward kinematics analysis of a traditional exoskeleton arm

Due to the compact structure of the traditional exoskeleton arm and the small interval space between the branches and the arm, interferences may occur during the auxiliary human motion. It could cause serious injury for human arm or unable to reach the desire positions. So the joint angles should be restricted. The range limit of each joint is shown in Table 1

$$ {}_{i}^{i-1}T = \begin{bmatrix} c\theta_i & -s\theta_i & 0 & a_{i-1} \\ s\theta_i c\alpha_{i-1} & c\theta_i c\alpha_{i-1} & -s\alpha_{i-1} & -s\alpha_{i-1}d_i \\ s\theta_i s\alpha_{i-1} & c\theta_i s\alpha_{i-1} & c\alpha_{i-1} & c\alpha_{i-1}d_i \\ 0 & 0 & 0 & 1 \end{bmatrix} \quad (3) $$

Table 1. Joint limit of traditional exoskeleton arm

Joint Action	Joint Limit(degrees)	
	Min	Max
abduction/adduction	0	90
flexion/extension	-120	30
Internal/external rotation	45	90
flexion/extension of elbow	-90	30

The linkage coordinate systems were established on traditional exoskeleton model as shown in Figure 1. The D-H parameters are shown in Table 2

Table 2. D-H parameters of traditional exoskeleton arm

i	a_{i-1}	α_{i-1}	d_i	θ_i
1	0	0°	0	θ_1
2	0	90°	d_2	θ_2
3	0	90°	d_3	θ_3
4	0	-90°	d_4	θ_4
5	a_4	0°	0	θ_5

Substituting the parameters in the table 2 into the formula (3) can get the positive kinematics of the traditional exoskeleton arm.

$$ p_x' = (c_1c_2c_3c_4 - s_1s_3s_4 - c_1c_2s_4)a_4 - c_1c_2s_3d_4 - s_1c_3d_4 + c_1s_2d_3 - s_1d_2 \quad (4) $$

$$ p_y' = (s_1c_2c_3c_4 - s_1s_2s_4 + c_1s_3c_4)a_4 - s_1c_2s_3d_4 + c_1c_3d_4 + s_1s_2d_2 + c_1d_2 \quad (5) $$

$$ p_z' = (-s_2c_3c_4 - c_2s_4)a_4 + s_2s_3d_4 + c_2d_3 \quad (6) $$

Where $s_i = \sin(i), c_i = \cos(i)$

3.2. Forward kinematics analysis of a modified exoskeleton arm

The coordinate systems were established on the modlfied mechanism as shown in Figure 5. The amount of movement of the Y-axis is not represented in the D-H model, while the modified mechanism has a translational amount on the Y-axis. Therefore the translation and rotation transformation of coordinates should be used to solve the transformation matrix

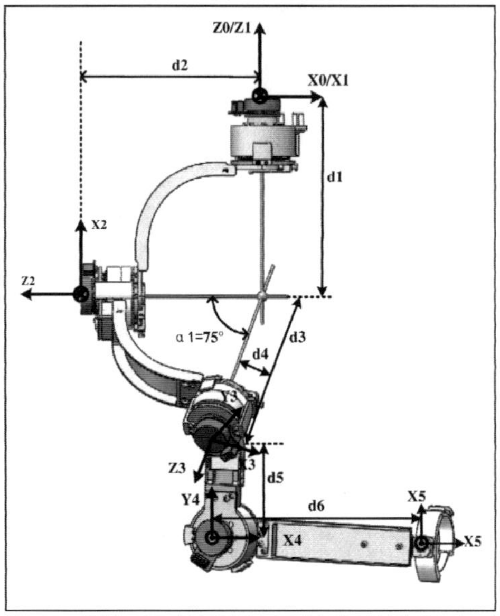

Figure 5. D-H linkage coordinate systems

$$ {}^0_1T = \begin{bmatrix} c_1 & -s_1 & 0 & 0 \\ s_1 & c_1 & 0 & 0 \\ 0 & 0 & 1 & 0 \\ 0 & 0 & 0 & 1 \end{bmatrix} \quad {}^1_2T = \begin{bmatrix} c_2 & -s_2 & 0 & 0 \\ 0 & 0 & -1 & -a_2 \\ s_2 & c_2 & 0 & -a_1 \\ 0 & 0 & 0 & 1 \end{bmatrix} \quad {}^2_3T = \begin{bmatrix} c_3 & -s_3 & 0 & 0 \\ a_9 s_3 & a_9 c_3 & -a_8 & -a_3 \\ a_8 s_3 & a_8 c_3 & a_9 & -a_4 \\ 0 & 0 & 0 & 1 \end{bmatrix} $$

$$ {}^3_4T = \begin{bmatrix} c_4 & -s_4 & 0 & 0 \\ a_{10} s_4 & a_{10} c_4 & a_{11} & -a_5 \\ -a_{11} s_4 & -a_{11} c_4 & a_{10} & a_6 \\ 0 & 0 & 0 & 1 \end{bmatrix} \quad {}^4_5T = \begin{bmatrix} 1 & 0 & 0 & 0 \\ 0 & 1 & 0 & -a_7 \\ 0 & 0 & 1 & 0 \\ 0 & 0 & 0 & 1 \end{bmatrix} $$

Where $a_1 = d_1, a_2 = d_2, a_3 = d_3 \sin 75^\circ, a_4 = d_2 - d_3 \cos 75^\circ \; a_5 = d_4 + d_5 \cos 40^\circ, a_6 = d_6 \sin 40^\circ, a_7 = d_6$

$a_8 = \sin 75^\circ, a_9 = \cos 75^\circ, a_{10} = \cos 45^\circ, a_{11} = \sin 45^\circ$

After the coordinate transformation calculation, the kinematic equation of the exoskeleton is obtained

$$ {}^0_1T\, {}^1_2T\, {}^2_3T\, {}^3_4T\, {}^4_5T = {}^0_5T = \begin{bmatrix} n_x & o_x & a_x & p_x \\ n_y & o_y & a_y & p_y \\ n_z & o_z & a_z & p_z \\ 0 & 0 & 0 & 1 \end{bmatrix} \tag{7} $$

$$ \begin{aligned} p_x = {} & a_2 s_1 - a_4 s_1 + a_5(c_1 c_2 c_3 - a_8 c_3 s_1) + a_7(s_4(c_1 c_2 c_3 + a_8 s_3 s_1 - a_9 c_1 s_3 s_2) + a_{11} c_4(a_9 s_1 + a_8 c_1 s_2) + a_{10} c_4(c_1 c_2 s_3 - a_8 c_3 s_1)) \\ & + a_6(a_9 s_1 + a_8 c_1 s_2) + a_3 c_1 s_2 \end{aligned} \tag{8} $$

$$p_y = a_4c_1 - a_2c_1 + a_5(a_8c_1c_3 + c_2s_1s_3 + a_9c_3s_1s_2) - a_7(s_4(a_8s_3c_1 - s_1c_2c_3 + a_9s_1s_3s_2) + a_{11}c_4(a_9c_1 - a_8s_1s_2) - a_{10}c_4(c_2s_1s_3 + a_8c_3c_1 + a_9c_3s_1s_2)) - a_6(a_9c_1 - a_8s_1s_2) + a_3s_1s_2 \tag{9}$$

$$p_z = a_7(s_4(c_3s_2 + a_9c_2s_3) + a_{10}c_4(s_2s_3 - a_9c_3c_2 - a_8a_{11}c_2c_4) - a_3c_2 - a_1 + a_5(s_2s_3 - a_9c_2c_3) - a_6a_8c_2 \tag{10}$$

4. Simulation and analysis of workspace

The workspace of the manipulator is an important kinematic indicator for measuring the performance of the robot[8]. It is also the basis for optimizing the structural and performance parameters. whether exoskeleton arm can accomplish the specific task can be judged by workspace. The workspace space of the exoskeleton refers to the set of points at which the reference point of the end of the exoskeleton can reach all discrete positions[9]. It commonly used methods including geometric drawing method, analytical method, numerical method, Monte Carlo method and so on[10]. Based on the analysis of kinematics, this paper simulates the working space of two exoskeleton models by Monte Carlo method. The two-dimensional and three-dimensional space maps obtained by Matlab simulation can visually observe the workspace of the mechanism.

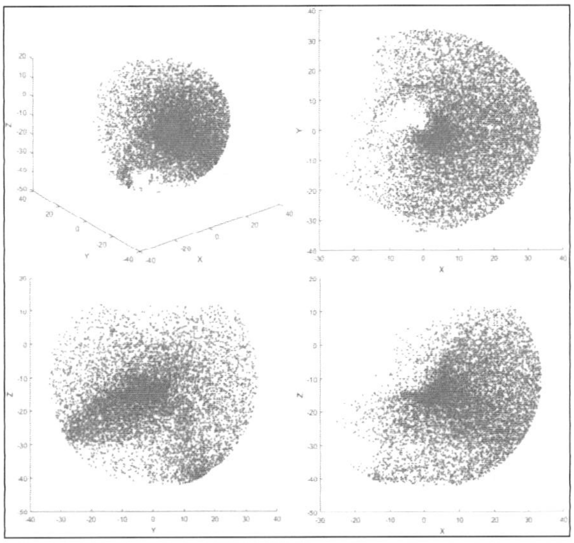

Figure 6. Workspace of a modified exoskeleton arm

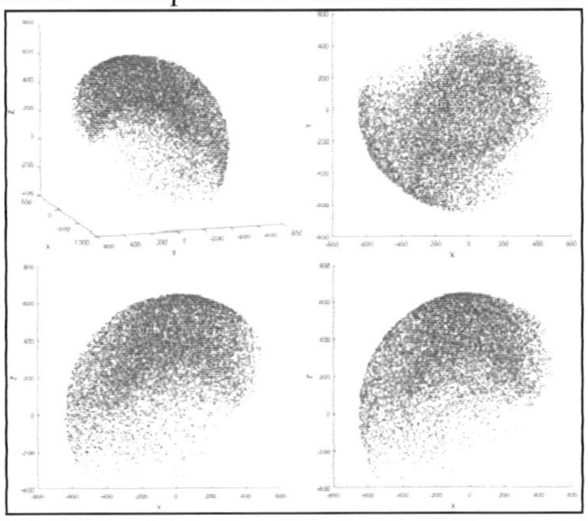

Figure 7. Workspace of a traditional exoskeleton arm

Based on theoretical analysis in section 2, the traditional exoskeleton has small workspace. The simulations of two models are shown in Figure 6 and 7. The working space of the modified exoskeleton arm is a notched sphere, while the workspace of the traditional one is a notched ellipsoid as shown in Figure 6 and 7. Compared with the ellipsoid, the sphere's workspace is obviously larger, which proves that the modified mechanism has a larger working space than the traditional one. In addition some of the extreme positions that modified exoskeleton arm can achieve compared with a traditional one is shown in Figure 8(a) and Figure 8(b). Traditional exoskeleton arm is difficult to reach these extreme positions due to compact structure or interferences. It also proves that the modified exoskeleton arm has larger working space than traditional one.

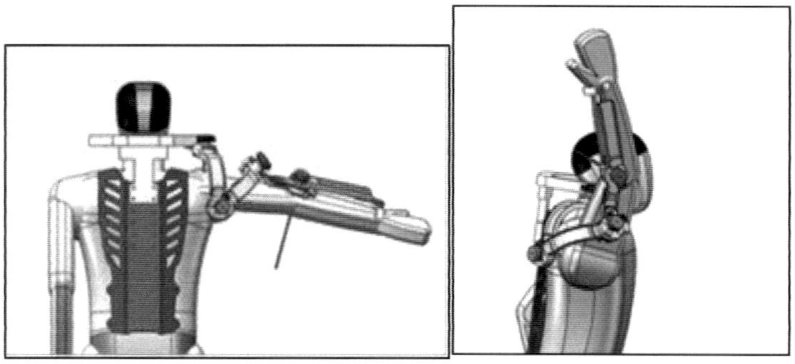

(a) (b)

Figure 8. Test of some extreme postures

5. Conclusion

Aiming at the shortcomings of the traditional exoskeleton arm, such as small working space and bulky structural, a modified portable exoskeleton mechanism is proposed. In addition, the kinematics of both exoskeleton models is analyzed in this paper. The Monte Carlo method was used to simulate the workspace of the mechanism with the help of the Matlab software. Compared with three mutually orthogonal series of joint, this configuration of exoskeleton arm has larger workspace. On the other hand, it can be seen from 3D model, the modified exoskeleton can reach some extreme positions. It also proves that the modified exoskeleton arm has larger workspace

Acknowledgements

This topic is supported by Shandong Province Key Research and Development Program (2017GGX30120).

Reference

[1] Park H S , Ren Y , Zhang L Q . IntelliArm: An Exoskeleton for Diagnosis and Treatment of Patients with Neurological Impairments[C]// IEEE Ras & Embs International Conference on Biomedical Robotics & Biomechatronics. IEEE, 2008.

[2] Zhiyong Y , Lihua G , Jing Z , et al. Research Status and Development Trend of Energy-assisted Exoskeleton Suit[J]. Journal of Shandong University of Science & Technology, 2012.

[3] Perry J C , Rosen J , Burns S . Upper-Limb Powered Exoskeleton Design[J]. IEEE/ASME Transactions on Mechatronics, 2007, 12(4):408-417.

[4] Wu T M , Wang S Y , Chen D Z . Design of an exoskeleton for strengthening the upper limb muscle for overextension injury prevention[J]. Mechanism & Machine Theory, 2011, 46(12):1825-1839.

[5] Nef T , Riener R . Shoulder actuation mechanisms for arm rehabilitation exoskeletons[C]// IEEE Ras & Embs International Conference on Biomedical Robotics & Biomechatronics. 2009.

[6] Lee H D , Lee B K , Kim W S , et al. Human–robot cooperation control based on a dynamic model of an upper limb exoskeleton for human power amplification[J]. Mechatronics, 2014, 24(2):168-176.

[7] Xiao Y , Zhu Y , Wang X , et al. Configuration optimization and kinematic analysis of a

wearable exoskeleton arm[C]// International Conference on Robotics & Automation Sciences. IEEE, 2017.
[8] Rastegar J . Manipulation workspace analysis using the Monte Carlo Method[J]. Mechanism & Machine Theory, 1990, 25(2):233-239.
[9] Da X U , Xiang X , Hua L I , et al. Analysis of Ammunition Auto-loading Robot Workspace Based on Graphical Method and Monte-Carlo Method[J]. Journal of Gun Launch & Control, 2014.
[10] Low K H, Yin Y. An integrated lower exoskeleton system towards design of a portable active orthotic device[J]. International Journal of Robotics & Automation, 2007, 22(1):32-43.

IOP Conf. Series: Materials Science and Engineering **533** (2019) 012034 doi:10.1088/1757-899X/533/1/012034

The Location of Electric Vehicle Charging Station Based on TLBO Optimization Algorithm

Hu Jinlei[1], Sun Yunlian[2], Yu Junwei[2], Lu Jue[2], Zou Qiwu[1], Xie Xinlin[1], Fu Bin[1]

1Qingyuan Power Supply Bureau of Guangdong Power Grid Co., Ltd., Qingyuan 511500, China
2Wuhan University, School of Electrical Engineering, Wuhan 430074, China

hujinlei@163.com, 3344147131@qq.com, yujunwei@whu.edu.cn

Abstract. With the large-scale application of electric vehicles, the construction of electric vehicle charging stations has become a prerequisite for urgent needs. In this paper, TLBO optimization algorithm is proposed. The optimal model of the location and the constant volume of the charging station is established based on the weighted distance and minimum distance from the user to the charging station. Based on the layout of the district network in the planning area, the construction plan of the electric vehicle charging station in the planning area is studied. Through a case study of a planning area, the results show that the TLBO optimization algorithm is applied to the location and constant volume of electric vehicle charging stations, with accurate calculation, rapid convergence, and good global optimization performance.

1. Introduction

As a zero-emission and green and energy-saving transportation vehicle, electric vehicles have received the attention of government departments and environmentalists since its introduction. However, due to its limited battery capacity, electric vehicles must be charged after traveling for a distance. As an energy supply system for electric vehicles, the charging station plays an extremely important role. Its rational layout in the city directly affects the efficiency of electric vehicle users.

At present, many experts and scholars at home and abroad put forward constructive methods and suggestions for the location planning of electric vehicle charging stations. The literature [1-3] focuses on the factors influencing the location planning of electric vehicle charging stations and the principles to be followed when constructing charging stations.

Literature [4,5] take the driving characteristics of the drivers, the public service infrastructure and the cost of the construction of charging stations into account.

Literature [6] considers the traffic flow of the traffic network, and takes the minimum of network loss and the minimum node voltage deviation as the constraint conditions to carry out the optimal planning of the electric vehicle charging station.

Literature [7] aims at minimizing the sum of the total cost and the network loss of charging stations in the planning area, and constructs a mathematical model for the optimal planning of the charging stations for electric vehicles.

Content from this work may be used under the terms of the Creative Commons Attribution 3.0 licence. Any further distribution of this work must maintain attribution to the author(s) and the title of the work, journal citation and DOI.
Published under licence by IOP Publishing Ltd

The above documents provide a method for site selection planning of electric vehicle charging stations from different perspectives. However, in the newly planned urban areas, the planning of electric vehicle charging stations shall be based on the premise of minimum construction costs and operating costs, and the convenience of user charging should be considered. The shortest distance to the charging station and the least amount of time spent are the factors.

From the user's point of view, this paper takes the shortest distance between the user and the charging station and the coverage of the charging station as the objective function. Taking into account the distribution center of the charging requirements of the electric vehicles in the planning area, the location and volume method of the charging station of the electric vehicle based on the teaching and learning optimization algorithm is proposed. , Actually calculated by a new city planning area, the method is suitable for the site selection of the charging station with good results.

2. The mathematical model

2.1. Algorithm introduction

TLBO (Teaching-Learning-Based Optimization) is a new swarm intelligence optimization algorithm proposed by Rao et al. [8,9] in 2010 for solving nonlinear global optimal problems. Simulation teacher teaching process and student self-learning process, students learn each other to improve the overall class performance. This method has many advantages such as high calculation accuracy, fast convergence rate, small computational load, good global convergence for nonlinear optimization, and no constraint on variation of algorithm parameters. The algorithm has no control parameters and only needs to set the number of groups and the number of cycles. You can solve the optimization problem.

The principle of using TLBO algorithm to deal with the optimization of n-dimensional space is as follows: The student representative group of a class, each individual is composed of n-dimensional variables, each represents n lessons of each student, and the number of students' knowledge acquired is measured by the fitness function. The current best solution in the group is taught as a teacher, and brings the score of the class to the level close to the teacher. In the "learning" phase, if a student is only richer, the rest of the students learn new knowledge from them.

The algorithm optimization process is shown in Figure 1.

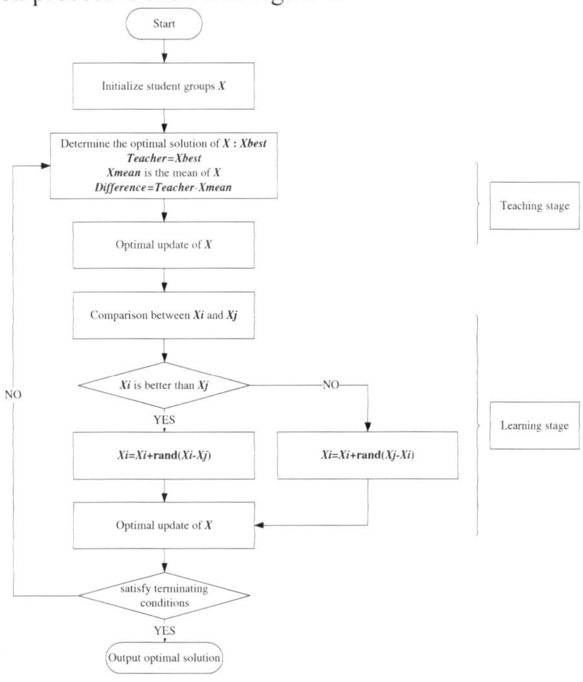

Figure 1. The flow chart of TLBO algorithm.

2.2. Model establishment

To divide the planning area into districts by division, assuming that there are N districts in a new town planning area, and the number of parking spaces in each district is $K=\{k_1,k_2...k_n\}$, The central point of each residential area is the charging demand location. The central coordinate point is shown in (1):

$$(X,Y)=\{(x_1,y_1),(x_2,y_2),...,(x_t,y_t),...,(x_n,y_n)\} \tag{1}$$

Considering that the electric vehicle charging stations in the planning area are mainly for the convenience of charging all users, they cannot be built outside the planning area. When a certain point i is selected as the charging station construction point in the planning area, the center point of each district is to the charging station. The distance is shown in (2).

$$D_i=\sum_{j=1}^{n}\sqrt{\left(x_j-\bar{x}_i\right)^2+\left(y_j-\bar{y}_i\right)^2} \tag{2}$$

According to the actual situation, the number of external demand of the charging points of the electric vehicles in each district is different. The location of the location of the charging station will shift to the partition with a large number of demand. The number of external charging piles of a certain district electric vehicle is used as the charging station. The weight coefficient of the address, its value is $\lambda=\{\lambda_1,\lambda_2,...,\lambda_n\}$,so the function is shown in (3).

$$F=\min\{\sum_i(\lambda_i*D_i)\} \tag{3}$$

2.3. Constrains

The sum of the number of charging piles to be built in each charging station shall meet the requirements for the charging load of all electric vehicles in the planning area, and the number of charging piles per charging station can meet the charging requirements of electric vehicles within its coverage area:

$$\sum P_j\ge\sum L_j \text{ and } P_j\ge L_j \quad (j=1,\ 2,\ ...,\ S) \tag{4}$$

Among them, S is the number of proposed charging stations; Pj is the number of charging piles to be built at the j-th charging station, and Lj is the number of charging piles for the electric vehicle charging needs for the coverage of the proposed j-th charging station.
The range of the charging station location.
The TLBO program search scope should be the planned area of the planning area.

$$(x_t)_{\min}\le\bar{x}_i\le(x_t)_{\max} \text{ and } (y_t)_{\min}\le\bar{y}_i\le(y_t)_{\max} \tag{5}$$

3. Charging station location planning process

3.1. Main ideas

Taking the coordinates of the center point of each zone as the load demand center point of the zone, and taking the weighted distance and minimum of the center point to the proposed charging station as the objective function, using the TLBO algorithm to select in the planning area under the constraint conditions. Address, calculate the number and location of the best charging stations, and use the output as the final result of the construction of the charging station.

3.2. Location and volume calculation process

According to the above analysis, the calculation process is as follows:
(1) Enter the coordinates of each partition;

(2) Enter the proposed number of electric vehicle charging stations;
(3) Enter the coordinate constraints of the proposed site;
(4) Input the external demand quantity of each district charging pile;
(5) Enter the number of iterations and the number of groups;
(6) Use TLBO optimization algorithm to calculate the coordinates of the proposed charging station;
(7) Calculation of the charging requirements for electric vehicles within the coverage and coverage of the proposed charging station;
(8) Calculate the number of electric vehicle charging piles within the coverage of the proposed charging station;
(9) Output the position coordinates, coverage of the proposed charging station for electric vehicles, and the number of charging piles to be built at each station.

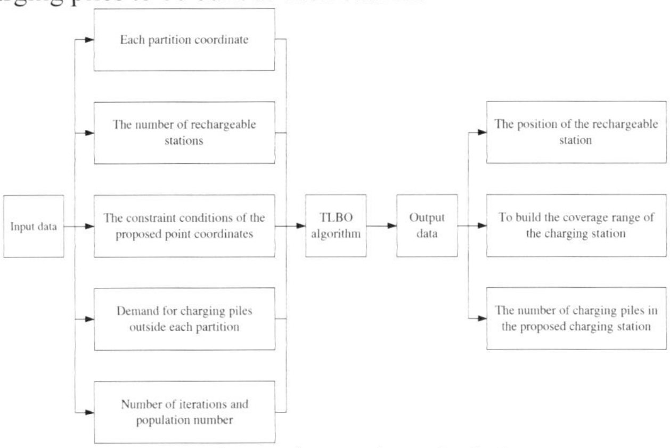

Figure 2. Location and capacity calculation process.

3.3. The location of charging piles based on TLBO algorithm
In this paper, within the entire planning area, the calculated location of the proposed charging station must meet two conditions: the distance from the charging station to the coverage area is the shortest, and the proposed location is close to the coverage area where the external charging piles have more demand.

4. Examples
Take a new town planning area as an example. The planning area covers an area of 5.7km2 and the total population is 115,000. The planning area is divided into 36 blocks according to the nature of the land. The coordinates of the 36 district centers are shown in Figure 3.

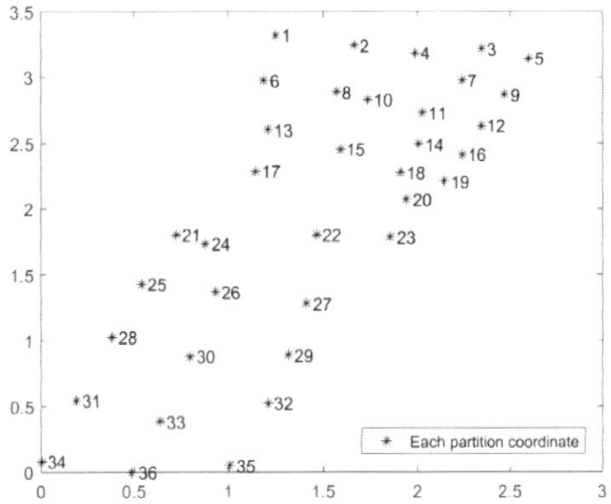

Figure 3. 36 partition locations.

In the figure, the horizontal and vertical coordinate units are km.

In each of the 36 districts, there is a parking lot for each district. Figure 4 shows the distribution of parking spaces in each district.

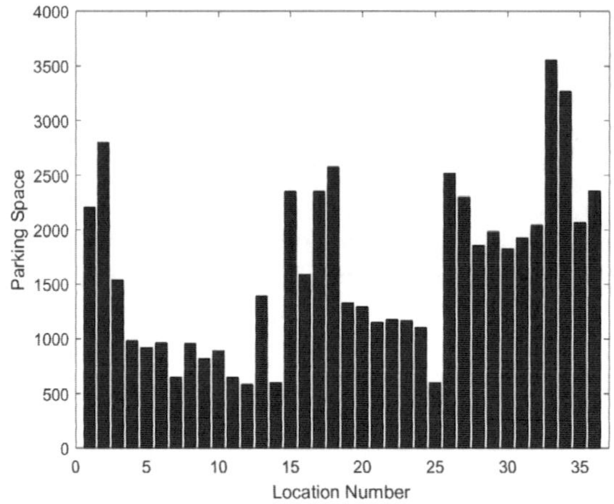

Figure 4. 36 partition locations.

Considering that the area of the planning area is only 5.7km2 and the number of charging stations is too large, construction costs and operating costs will be increased accordingly. However, for users, the time and distance to save are not many, so set at least one building. Do more to build 4 electric vehicle charging stations.

The proposed site must be within the planning area, that is, its abscissa should be in the range of (0, 3.32) and ordinate in the range of (0, 2.56). Exceeding this range indicates that the construction site is not desirable and should be rejected.

Assuming that the parking lot is the total number of cars in the planned area of the new city, the penetration rate of electric vehicles is calculated at 15%, and 15% of all electric vehicles in the subzone need to be charged at the charging station because the fast charging pile can be within 20-30 minutes. Complete charging of electric vehicles is completed, so 10 electric vehicles are set to share one fast charging pile. After calculation, the number of external charging piles in each subarea is shown in Figure 5.

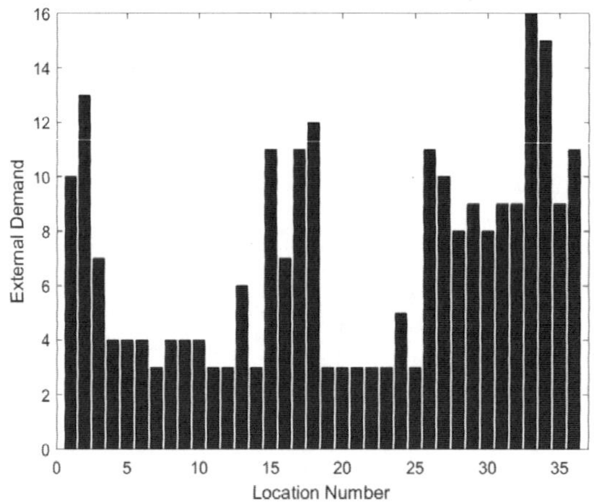

Figure 5. External demand of charged piles.

Set the population number to 500, the maximum number of iterations to 5000, and the overall variance to take 10-6 as the convergence criterion. Through the TLBO algorithm, the construction number and coordinates of charging stations, their respective coverage areas and the number of charging piles in the coverage area are obtained.

Table1. Construction index of charging station

Construcion number of charging station	Location	Coverage range	Demand numer of charing pile
1	(1.26,1.69)	1-36	128
2	(0.79,0.88)	21、24-36	67
	(1.72,2.58)	1-20、22-23	61
3	(0.60,0.27)	31-36	28
	(1.05,1.35)	21-30	47
	(1.74,2.71)	1-20	53
4	(0.48,0.19)	33-36	23
	(1.11,1.00)	25-30、32	35
	(1.44,1.80)	17、20-24	21
	(1.72,2.78)	1-16、18-19	49

According to the calculation results in Table 1, the proposed location coordinates of the charging station are marked on the partition location map, as shown in Figure 6.

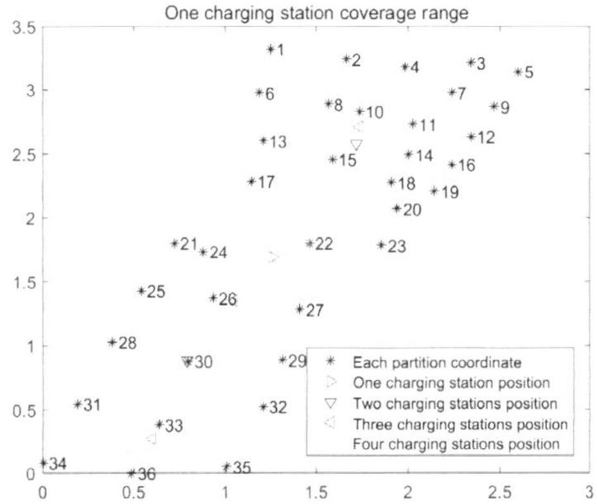

Figure 6. Construction position of charging station.

The coverage of the proposed charging station is marked on the map. Since the construction of a charging station covers 36 districts within the entire planning area, no sign is made for it. Two, three, and four charging stations are built and covered. The range is marked as shown in Figure 7.

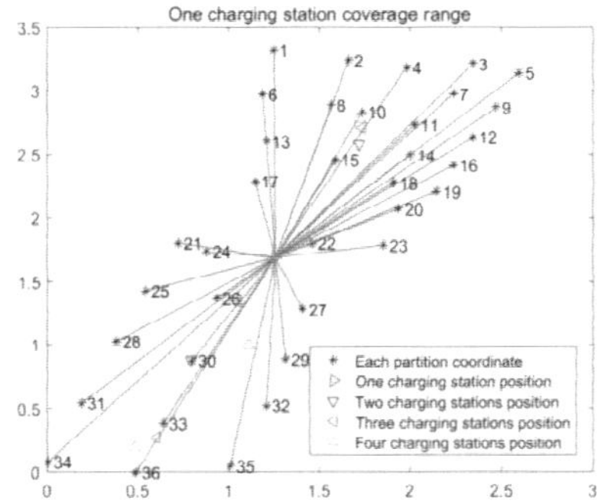

(a). One charging station coverage range.

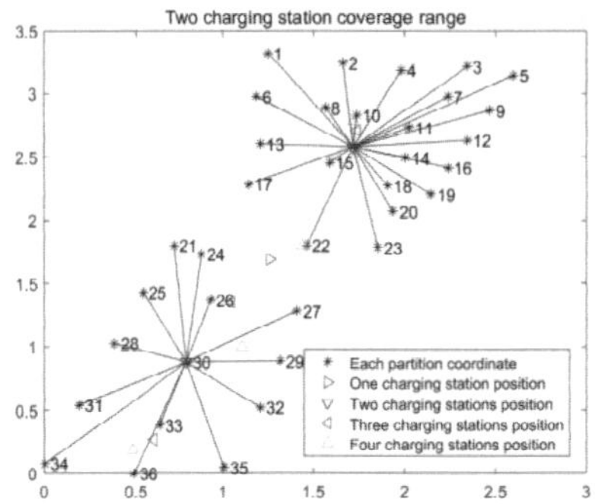

(b). Two charging station coverage range.

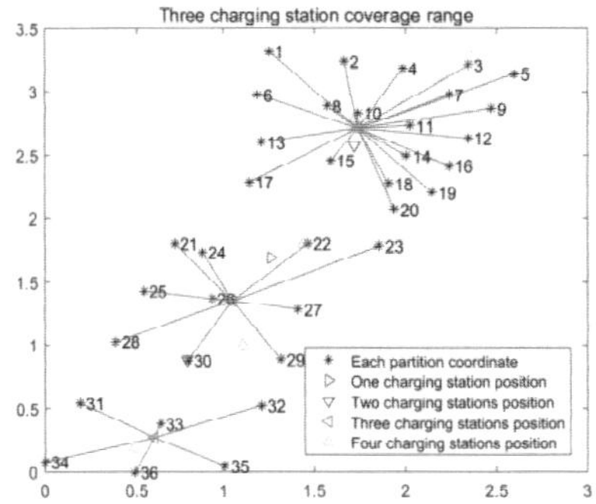

(c). Three charging station coverage range.

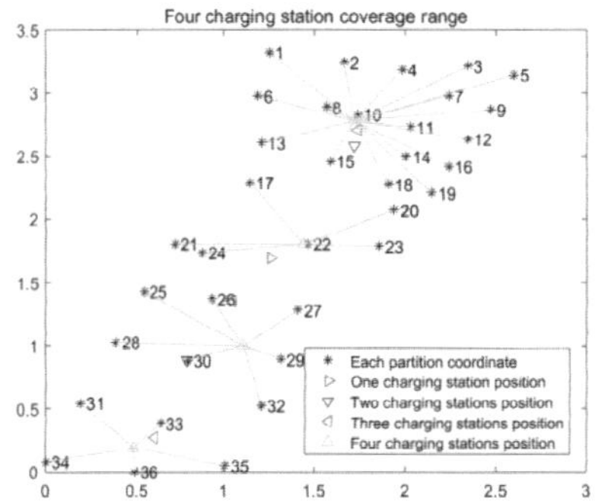

(d). Four charging station coverage range.

Figure 7. Coverage range of charging station.

5. Conclusion

Based on the construction of the planning area, this paper first divides the planning area before the site selection and capacity allocation, and regards each district center as the electric vehicle charging load center of the district, and the user's convenience of charging is shown by the shortest distance from the user to the charging station. The number of electric vehicle charging piles in the district is used as a weighting factor for the location of the charging station to comprehensively determine the construction location of the electric vehicle charging station. Using this as an objective function, the TLBO algorithm was used to find out the number of electric vehicle charging stations, the construction location, the coverage of each charging station, and the number of charging piles in the charging station. The optimal site selection and constant capacity of electric vehicle charging stations were achieved. In the end, the planning plan for the construction of the electric vehicle charging station in the entire planning area will be obtained.

The practical example shows that the TLBO optimization algorithm and the model proposed in this paper have certain feasibility. However, this article mainly considers the factors related to electric vehicles, such as the distance between the user and the charging station, external requirements of the district charging pile, and the structure of the planning area. The factors related to the distribution system, such as the distribution transformer capacity, are closely related to the construction of the charging station. The factors have not yet been considered. This will be further studied in the follow-up work.

References

[1] Ou S, Lin Z, He X, et al. Estimation of vehicle home parking availability in China and quantification of its potential impacts on plug-in electric vehicle ownership cost[J]. Transport Policy, 2018, 68: 107-117.
[2] Jordán J, Palanca J, del Val E, et al. A Multi-Agent System for the Dynamic Emplacement of Electric Vehicle Charging Stations[J]. Applied Sciences, 2018, 8(2): 313.
[3] Zhong W, Jing G, Peng Z, et al. Research on regional fast charging piles planning method of EV based on driving characteristics[C]//Energy Internet and Energy System Integration (EI2), 2017 IEEE Conference on. IEEE, 2017: 1-6.
[4] Zhang J, Zhou H, Li H, et al. Multi-objective planning of charging stations considering vehicle arrival hot map[C]//Energy Internet and Energy System Integration (EI2), 2017 IEEE Conference on. IEEE, 2017: 1-6.
[5] Su S, Zhao H, Zhang H, et al. Forecast of electric vehicle charging demand based on traffic flow model and optimal path planning[C]//Intelligent System Application to Power Systems (ISAP), 2017 19th International Conference on. IEEE, 2017: 1-6.
[6] Wang H, Wang G, Zhao J, et al. Optimal planning for electric vehicle charging stations considering traffic network flows[J]. Automation of Electric Power Systems, 2013, 37(13): 63-69.
[7] Moses M V A, Priya S, Ramys S, et al. Real-Time Forecasting of EV Charging Station Scheduling for Smart Energy Systems[J]. 2018.
[8] R.V. Rao, V.J. Savsani， D.P. Vakharia. Teaching－Learning-Based Optimization: An optimization method for continuous non-linear large scale problems［J］. Information Sciences. 2011,(1).
[9] R.V. Rao, V.J. Savsani， D.P. Vakharia. Teaching－learning-based optimization: A novel method for constrained mechanical design optimization problems［J］. Computer-Aided Design. 2011,(03).

IOP Conf. Series: Materials Science and Engineering **533** (2019) 012035 doi:10.1088/1757-899X/533/1/012035

A Novel Prediction Algorithm for the Cross Temperature Estimation of Blast Furnace

Yan Jin[1], Sen Zhang[1] and YiXin Yin[1]

[1]School of Automation and Electrical Engineering, University of Science and Technology Beijing, Beijing 100083, China
E-mail: Jiny_ustb@163.com, zhangsen@ustb.edu.cn, yyx@ies.ustb.edu.cn

Corresponding author: Sen Zhang (zhangsen@ustb.edu.cn)

Abstract. In order to predict the distribution of gas flow, we need to get the temperature of each point in blast furnace throat in advance. In this paper, firstly, two intelligent modeling methods are used to establish a multiple-input multiple-output prediction model, one is extreme learning machine (ELM) algorithm and the other is online sequential extreme learning machine (OS-ELM) algorithm. And the model is a single-step prediction model of temperature in blast furnace, single-step prediction means the prediction of temperature in the next moment. We use autocorrelation analysis to determine input vector and output vector of the model. The result of autocorrelation analysis indicates that the method of temperature sequence prediction has a higher prediction accuracy and better prediction stability than the method of single point prediction. Next, based on real industrial data, we make a comparison between the multiple-input multiple-output model and least squares support vector machines (LS-SVM) model used in common. The experiment results show that OS-ELM model has a better forecast effect than the ELM model and LS-SVM model.

1. Introduction

Facing the increasingly serious environmental pollution problems, the blast furnace iron-making should bear the responsibility of energy-saving emission reduction and green production in the process of iron and steel production. The process of blast furnace iron-making is highly complex. Therefore, reasonably operating the blast furnace according to the change of energy consumption are the key to achieve low consumption and high yield in blast furnace iron-making [1]. Blast furnace gas flow is the main source of thermal energy and chemical energy in the process of blast furnace smelting. Therefore, study on distribution of gas flow in blast furnace can help diagnose the furnace condition and optimize the burden distribution system [2]. Beyond that, it will also help achieve the purpose of stabilizing blast furnace operation and reducing coke ratio and raising smelting level [3].

The cross temperature measurement device is widely used to monitor the top temperature in blast furnace. And coincidentally, the top temperature in blast furnace can reflect the distribution of gas flow [4]. Wu Min, et al proposed a new extension evaluation model based on the distribution characteristics of blast furnace gas flow to evaluate the distribution of gas flow [2].This method provides qualitative analysis and quantitative indication for optimizing the burden distribution of blast furnaces. Using Bayesian techniques, particle swarm optimization algorithm, echo state network, expert knowledge methods, Zhang LiMin accurately predicted the change trend of blast furnace gas flow [3]. Zhou Ping, et al established the temperature estimation model of cross temperature

Content from this work may be used under the terms of the Creative Commons Attribution 3.0 licence. Any further distribution of this work must maintain attribution to the author(s) and the title of the work, journal citation and DOI.
Published under licence by IOP Publishing Ltd

measuring center of blast furnace based on the two intelligent modeling methods of multi-output support vector regression machine(M-SVR) and random vector functional-link networks (RVFLNs) [4].

All above provides a reference for the establishment of ELM prediction model [5-6] and OS-ELM [7-8] prediction model in this paper. The ELM model and OS-ELM model are used to predict the temperature of seventeen points in blast furnace, through these temperature we can know the distribution pattern of gas flow and adjust burden distribution operation in advance.

2. Correlation Analysis

For the prediction of temperature in blast furnace, the traditional prediction method is to consider the 17 temperature points as an isolated point, and each temperature point is predicted separately, which will ignore the cross correlation between points. Therefore, it is necessary to analyze the autocorrelation between 17 temperature points on time series.

Firstly, removing the noise from the original temperature data. Then taking out the data of any continuous 180 moments to calculate their correlation coefficient. The formula for calculating the correlation coefficient between the column vector X and Y can be expressed as:

$$\rho_{XY} = \frac{Cov(\mathrm{X},\mathrm{Y})}{\sigma_X \sigma_Y} = \frac{E(\mathrm{X}-\mu_X)(\mathrm{Y}-\mu_Y)}{\sigma_X \sigma_Y} \tag{1}$$

Where, is covariance, is mathematical expectation, is mean value, is standard deviation.

$$Cov(\mathrm{X},\mathrm{Y}) = E(\mathrm{X}-\mu_X)(\mathrm{Y}-\mu_Y) \tag{2}$$

For X, $\mu_X = E(\mathrm{X})$, ${\sigma_X}^2 = E(X^2) - E^2(X)$. Similarly, for Y. Therefore,

$$\rho_{XY} = \frac{E(\mathrm{XY}) - \mathrm{E(X)E(Y)}}{\sqrt{E(X^2) - E^2(X)}\sqrt{E(Y^2) - E^2(Y)}} \tag{3}$$

Correlation coefficient between temperature sequence for 180 moments is shown in Fig.1.

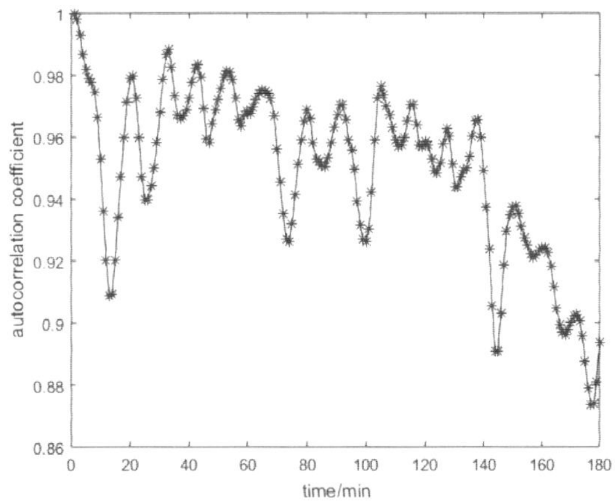

Figure 1. Correlation coefficient between temperature sequence.

When considering all 17 temperature points as a whole, the correlation coefficient between predicted moment and the current moment is 1. The correlation coefficient between the predicted time and the first six moments is over 0.97. And then correlation coefficient decreases gradually, but it always stays over 0.86. This indicates that when the 17 temperature points are considered as a whole, the correlation of temperature sequence has been decreasing as time goes away from the predicted moment. But it always maintains a great correlation.

The correlation coefficient of a single temperature point at different times is calculated as follows:

$$\hat{R}(m) = \frac{1}{N}\sum_{n=1}^{N-m} x(n)x(n+m), m = 0,1,\cdots(N-1) \tag{4}$$

$$\hat{\rho}(m) = \hat{R}(m) / \hat{R}(0) \tag{5}$$

Where, $\hat{R}(m)$ is autocorrelation function, $\hat{\rho}(m)$ is autocorrelation coefficient. The result of correlation coefficient for one single temperature point in 180 moments is shown in Fig.2.
When the temperature point is taken into separate consideration, the autocorrelation coefficient is shown in Fig.2. Correlation coefficient between the predicted time and the current time is 1. The autocorrelation coefficient in previous moment is about 0.9 and the forward two times is 0.7, subsequently, the autocorrelation coefficient decreases to less than 0.5.

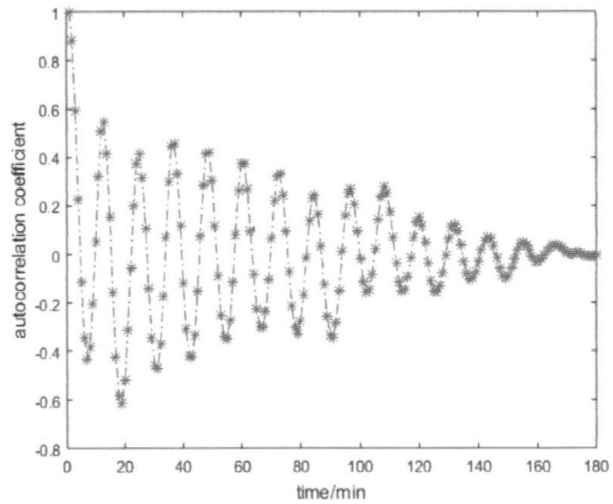

Figure 2. Correlation coefficient of single temperature point.

The comparison of the above two graphs can be seen that the correlation that considering the 17 temperature points as a whole is much greater than that of each temperature point considered alone.

3. Modeling of Temperature

In this chapter, the ELM algorithm and OS-ELM algorithm will be briefly introduced. And then the cross temperature prediction model of blast furnace based on OS-ELM algorithm will be introduced in detail.

3.1. ELM Theory

Extreme learning machine (ELM) is a single hidden layer feedforward neural network. The ELM algorithm is proposed by Huang [8-9] et al, which is a new simple and effective learning algorithm. The extreme learning machine algorithm does not need complex iteration in the training process of the model, and just need to set the number of neuron nodes in hidden layer. The weight matrix and the bias vector are randomly generated. After that, the output matrix of the hidden node is obtained. Through calculating the output matrix, we can gain the weight matrix in the hidden layer and the output layer, and fast calculate the global optimal solution.
For arbitrary sample set containing N distinct samples $(\mathrm{x}_i, \mathrm{t}_i)$, where $x_i = [x_{i1}, x_{i2}, \cdots, x_{in}]^T \in R^n$, is n-dimensional feature vector of the ith sample, $t_i = [t_{i1}, t_{i2}, \cdots, t_{in}]^T \in R^m$, is m-dimensional target vector of the ith sample(output vector). The mathematical model of a single hidden layer neural network containing L hidden nodes is expressed as:

$$\sum_{i=1}^{L}\beta_i g_i(x_k)=\sum_{i=1}^{L}\beta_i g(\omega_i\cdot x_k+b_i)=t_k,\quad k=1,2,\cdots,N \tag{6}$$

Where, $\omega_i=[\omega_{i1},\omega_{i2},\cdots,\omega_{in}]^T$ is the weight vector connecting the ith hidden neuron and all the input neurons. $\beta_i=[\beta_{i1},\beta_{i2},\cdots,\beta_{im}]^T$ is the weight vector connecting the ith hidden neuron and all the output neurons. b_i is the threshold of the i-th hidden neuron, $\omega_i\cdot x_k$ is the inner product of ω_i and x_k, $g(\omega_i\cdot x_k+b_i)$ is the output of the i-th hidden neuron.

The above N equations can be briefly expressed as:

$$\mathrm{H}\beta=\mathrm{T} \tag{7}$$

Where,

$$\mathrm{H}=\begin{bmatrix} g(\omega_1\cdot x_1+b_1) & \cdots & g(\omega_L\cdot x_1+b_L) \\ \vdots & \cdots & \vdots \\ g(\omega_1\cdot x_N+b_1) & \cdots & g(\omega_L\cdot x_N+b_L) \end{bmatrix}_{N\times L} \tag{8}$$

$$\beta=\begin{bmatrix} \beta_1^T \\ \vdots \\ \beta_L^T \end{bmatrix}_{L\times \mathrm{m}} \tag{9}$$

$$T=\begin{bmatrix} T_1^T \\ \vdots \\ T_N^T \end{bmatrix}_{N\times \mathrm{m}} \tag{10}$$

H is known as the hidden layer output matrix of the neural network. The ELM algorithm is transformed into the following optimization problem:

$$\min L_{ELM}=\frac{1}{2}\|\beta\|^2+C\frac{1}{2}\sum_{i=1}^{N}\|e_i\|^2 \qquad s.t.\ \ h(x_i)\beta=t_i-e_i \tag{11}$$

Where, e_i is the prediction error of the i-th sample, $h(x_i)$ is the i-th row of the hidden output matrix H. ELM algorithm uses the least square estimation theory, the output weight β is expressed as:

$$\beta=H^{\dagger}T=(H^TH)^{-1}H^TT \tag{12}$$

Where, $H^{\dagger}$ is the generalized inverse of H matrix, $H^{\dagger}=(H^TH)^{-1}H^T$.

3.2. OS-ELM Theory

The concept of OS-ELM algorithm is based on ELM algorithm and OS-ELM is developed from a single hidden layer feedforward neural network with additional hidden nodes and RBF hidden nodes. The OS-ELM can train sequential data one-by-one or block-by-block [10-11]. For N_0 arbitrarily distinct samples $(\mathrm{x}_i,\mathrm{t}_i)$, input vector is $\boldsymbol{x}_i=\left[x_{i1},x_{i2},\cdots,x_{in}\right]^T\in R^n$, output vector is $t_i=\left[t_{i1},\mathrm{t}_{i2},\cdots,\mathrm{t}_{in}\right]^T\in R^m, i=1,2,\cdots,N$. If a single hidden layer feedforward neural network has L hidden nodes, then the N_0 samples can be estimated by 0 error. It indicates that it exist β_i, $\mathbf{a}_i$ and b_i:

$$f_L(\mathbf{x}_j)=\sum_{i=1}^{L}\beta_i G(\mathbf{a}_i,\mathrm{b}_i,\mathbf{x}_j)=\mathbf{t}_j,\quad j=1,\cdots,N \tag{13}$$

$\mathbf{a}_i$ and b_i are the parameters of hidden nodes. β_i is output weight, $G(\mathbf{a}_i,\mathrm{b}_i,\mathbf{x}_j)$ is the output of the i-th hidden node with respect to input .When the hidden node is added:

$$G(\mathbf{a}_i,\mathrm{b}_i,\mathbf{x}_j)=g(\mathbf{a}_i\cdot\mathbf{x}_j+\mathrm{b}_i),\mathrm{b}_i\in R \tag{14}$$

Where, $\mathbf{a}_i$ is input weight vector, b_i is the bias of the i-th hidden node. When using RBF hidden node:

$$G(\mathbf{a}_i,\mathrm{b}_i,\mathbf{x}_j)=g(\mathrm{b}_i\ \|\ \mathbf{x}_j-\mathbf{a}_i\ \|),\mathrm{b}_i\in R^+ \tag{15}$$

$\mathbf{a}_i$ and b_i are the center and impact width of the i-th hidden node, R^+ is set of all real numbers. There is no need for complex iteration in model training. But it is necessary to set hidden layer neuron nodes to facilitate the generation of weight matrix and offset vector. Using the idea of the basic extreme learning algorithm, we get the initial hidden layer output matrix $\mathbf{H}_0$ and initial output weight that satisfy the equality $\|\mathbf{H}\beta\text{-}\mathbf{T}\|=\min_{\beta_0}\|\mathbf{H}\beta_0\text{-}\mathbf{T}\|$.

$$\mathbf{H}_0=\begin{bmatrix} g(a_1x_1+b_1) & \cdots & g(a_Lx_1+b_L) \\ g(a_1x_2+b_1) & \cdots & g(a_Lx_2+b_L) \\ \vdots & \ddots & \vdots \\ g(a_1x_{N_0}+b_1) & \cdots & g(a_Lx_{N_0}+b_L) \end{bmatrix}_{N_0\times L} \tag{16}$$

Where, $\beta_0=\mathbf{K}_0^{-1}\mathbf{H}_0^T\mathbf{T}_0$, $\mathbf{K}_0=\mathbf{H}_0^T\mathbf{H}_0$. When in the phase of online sequence update, data is in the learning process in the established model. When new data samples come into the model, the weight value will be updated according to the equation below:

$$\begin{aligned} &\mathbf{K}_{k+1}=\mathbf{K}_k+\mathbf{H}_{k+1}^T\mathbf{H}_{k+1} \\ &\beta^{(k+1)}=\beta^{(k)}+\mathbf{K}_{k+1}^{-1}\mathbf{H}_{k+1}^T(\mathbf{T}_{k+1}-\mathbf{H}_{k+1}\beta^{(k)}) \end{aligned} \tag{17}$$

Let $P_{k+1}=K_{k+1}^{-1}$, it can be introduced by the Woodbury formula:

$$\begin{aligned} &\mathbf{P}_{k+1}=\mathbf{P}_k-\mathbf{P}_k\mathbf{H}_{K+1}^T(\mathbf{I}+\mathbf{H}_{K+1}\mathbf{P}_k\mathbf{H}_{K+1}{}^T)^{-1}\mathbf{H}_{k+1}\mathbf{P}_K \\ &\beta^{(K+1)}=\beta^{(K)}+\mathbf{P}_{K+1}\mathbf{H}_{K+1}{}^T(\mathbf{T}_{K+1}-\mathbf{H}_{k+1}\beta^{(K)}) \end{aligned} \tag{18}$$

Let k=k+1, go back to the online learning stage.

4. Experimental Results

In this paper, we uses the operational production data of a blast furnace in Wuhan Iron and Steel Company. Data sampling interval is one minute. Selecting 1500 samples that are removed outliers and denoising the noise, where, nine hundred samples is the training set for the model, the other three hundred samples is the testing data for the model. Because the value of different temperature measurement points is largely different, it will cause a obvious fluctuation in the forecast results and affect the accuracy of the prediction model. Therefore, the data is normalized in this paper. Then, the ELM and OS-ELM models are used to predict the single temperature and 17 temperature values, after that we obtain the experimental results.

Fig.3 shows the error comparison results of ELM model for single point prediction and temperature sequence prediction of TE3036;Fig.4 is the error comparison results of OS-ELM model for single point prediction and temperature sequence prediction of cross temperature measurement point TE3036.

According to Fig.5 and Fig7, we can know that the prediction errors for temperature sequence prediction generally fall between [-0.05, 0.05],however, the prediction errors for single point prediction roughly fall between [-0.1,0.1].It can be found that the prediction error of OS-ELM algorithm is smaller than that of ELM algorithm. It shows that the OS-ELM prediction model has higher prediction accuracy.

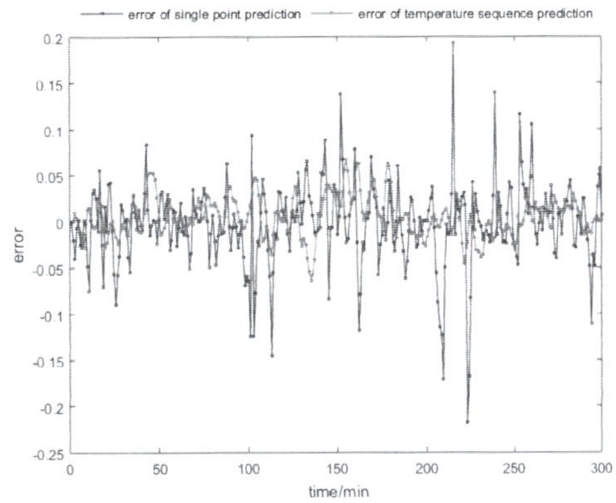

Figure 3. Error of ELM model for single point prediction and temperature sequence prediction.

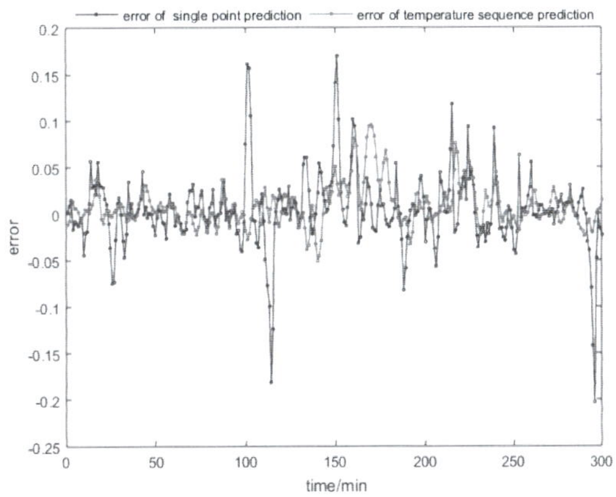

Figure 4. Error of OS-ELM model for single point prediction and temperature sequence prediction.

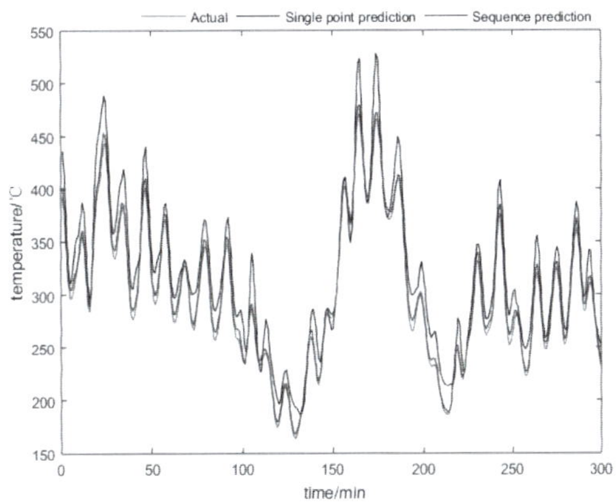

Figure 5. Comparison between the prediction value using the ELM prediction model and the real value.

Fig.5 shows the comparison between the prediction value using the ELM prediction model and the real value of TE3036; Fig.6 shows the comparison between the prediction value using the OS-ELM prediction model and the real value of TE3036.When the two different prediction models are used to predict the temperature, the method of temperature sequence prediction can better track the real value than that of single point prediction. And 0S-ELM model has a higher prediction accuracy.

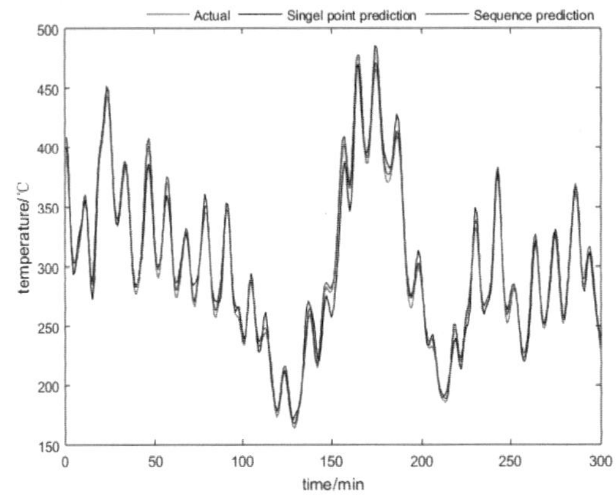

Figure 6. Comparison between the prediction value using the OS-ELM prediction model and the real value.

In order to evaluate the performance of the established model more comprehensively, in this paper, three evaluation formula are given below:

1) Root mean square error (RMSE):

$$RMSE=\sqrt{\frac{1}{n}\sum_{i=1}^{n}(y(k)-\hat{y}(k))} \tag{19}$$

2) Mean absolute percentage error (MAPE):

$$MAPE=\frac{1}{n}\sum_{i=1}^{n}\frac{|y(k)-\hat{y}(k)|}{y(k)}*100 \tag{20}$$

3) Correlation coefficient between real target value series and prediction value series (CC):

$$CC=\frac{\frac{1}{n}\sum_{i=1}^{n}(y(k)-\bar{y})(\hat{y}(k)-\bar{\hat{y}})}{\sigma(\mathrm{y})\sigma(\hat{\mathrm{y}})} \tag{21}$$

Where, $\sigma(\mathrm{y})$ is the standard deviation of the actual target value sequence, $\sigma(\hat{y})$ is the standard deviation of the predicted value sequence. $\bar{y}$ is the mean value of the real values, $\bar{\hat{y}}$ is the mean value of the predicted values.

Table1. A slightly more complex table with a narrow caption.

model	RMSE		MAPE		CC	
	SPP	TSP	SPP	TSP	SPP	TSP
ELM	0.0827	0.0593	4.5184	3.9350	0.7593	0.8024
OS-ELM	0.0359	0.0297	3.2857	2.0503	0.8819	0.9254

In the above table, The SPP refers to Single Point Prediction. The TSP refers to Temperature Sequence Prediction. Tab.1 is the comparison of prediction results of ELM forecasting model and OS-ELM forecasting model for single point prediction and temperature sequence prediction. It can be found that the OS-ELM model can follow the tracks of unstable real values. And the root mean square error obtained by the method of temperature sequence prediction is minimum value, and the mean absolute percentage error is also minimum value, and correlation coefficient between real target value series and prediction value series is maximum value.

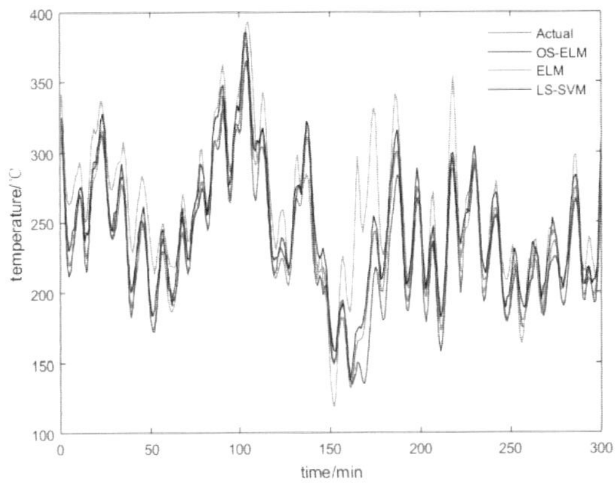

Figure 7. Comparison of prediction results for TE3039

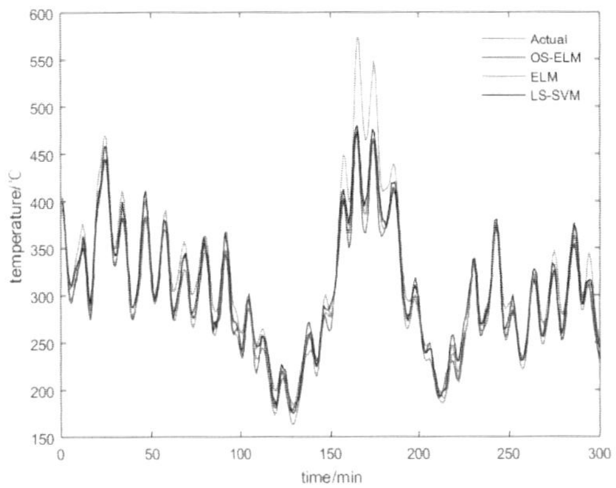

Figure 8. Comparison of prediction results for TE3036

Fig.7 shows the prediction results of the three prediction models of ELM, OS-ELM and LS-SVM for TE3039. Fig.8 shows the three prediction results of the prediction models of ELM, OS-ELM and LS-SVM for TE3036. From Fig.7 and Fig.8, we can see that the ELM model, OS-ELM model and LS-SVM model can track the real value very well. However, the prediction effect of OS-ELM model is obviously better than that of ELM model and LS-SVM model, and the prediction effect of OS-ELM model is more stable.

5. Conclusions

In this paper, we analyse the correlation between temperature at different moments. The results show that thinking of all 17 temperature points as a whole has a greater relevance than thinking of each temperature point separately. Therefore, the traditional method of single point prediction is abandoned in the paper. We establish a multiple-input multiple-output prediction model to predict the temperature of 17 points in blast furnace. The ELM algorithm and OS-ELM algorithm are used in the model. Compared with the existing single point prediction model, the multiple-input multiple-output prediction model has higher prediction accuracy and better prediction stability.

Acknowledgments

This work is supported by National Natural Science Foundation (NNSF) of China under Grant No. 61673056 and 61673055.

References

[1] Lin Shi, WenLei Liu, JiangPeng Li, et al."Prediction model of gas utiliza-tion rate based on blast furnace throat gas distribution" Journal of Inner Mongolia University of Science and Technology, 2016, 35(2): 121-125.

[2] WU M, NIE Q, XU Y, et al."Method of gas flow distribution extension evaluation in a bell-less blast furnace and its application" Journal of Central South University (Science and Technology), 2010, 3: 033.

IOP Conf. Series: Materials Science and Engineering **533** (2019) 012035 doi:10.1088/1757-899X/533/1/012035

[3] LinMIN Zhang. "Research of Blast Furnace Flow Prediction Algorithm and its Application in Furnace Condition Analysis"YanShan University, 2015.
[4] Ping Zhou, Lei You, JiPing Liu, et al."Centre Temperature Estimation of Blast Furnace Cross Temperature Measuring Based on M-SVR and RVFLNs"Journal of Northeastern University (Natural Science), 2017, 38(5): 614-619.
[5] G. B. Huang, Q. Y. Zhu, C. K. Siew. "Extreme learning machine: a new learning scheme of feedforward neural networks" IEEE International Joint Conference on, 2004, 2: 985-990.
[6] W. Zong, G. B. Huang, Y. Chen, " Weighted extreme learning machine for imbalance learning," Neurocomputing, 2013, 101(3):229-242.
[7] H. Zhang, S. Zhang, Y. Yin, "Online Sequential ELM Algorithm with Forgetting Factor for Real Applications," Neurocomputing, 2017, 261.
[8] Y. Lan, Y. C. Soh, G. B. Huang. "Ensemble of online sequential extreme learning machine"Neurocomputing, 2009, 72(13-15): 3391-3395.
[9] W. Xiao, J. Zhang, Y. Li, et al, "Class-specific cost regulation extreme learning machine for imbalanced classification,"Neurocomputing, 2017,261.
[10] H. Zhang, S. Zhang, Y. Yin, "Online Sequential ELM Algorithm with Forgetting Factor for Real Applications," Neurocomputing, 2017, 261.
[11] N. Y. Liang, G. B. Huang, P. Saratchandran, et al, "A Fast and Accurate Online Sequential Learning Algorithm for Feedforward Networks," IEEE Transactions on Neural Networks, 2006, 17(6):1411-1423.

IOP Conf. Series: Materials Science and Engineering 533 (2019) 012036 doi:10.1088/1757-899X/533/1/012036

Design of Digital Control Phase-Shifting Full-Bridge Switching Power Supply

Hucheng He, Yanzhao Liu, Chao Wang, Chenyang Zhang
School of Electrical and Control Engineering, Xi' an University of Science and Technology, Xi'an, China

liuyanzh_544671227@qq.com

Abstract. With the development of switching power supply towards miniaturization and high efficiency, digital and soft switching technology has become the research focus of switching power supply. Phase-shifted full-bridge DC/DC converter is commonly used in medium and high power applications due to its advantages of low switching loss, high efficiency and low output current ripple. Based on the existing switching power supply and digital control technology, a digital control phase-shifting full-bridge switching power supply is designed. Its output voltage ranges from 0 to 12 V.The output current ranges from 0 to 50 A. The voltage and current level can be adjusted according to the user's power demand, and the output mode of steady voltage or steady current can be set. The parallel operation of multiple power supply is realized by RS485 communication protocol.

1. Introduction

Switching power supply has experienced the development of circuit topology transformation, control technology and power device. The traditional analog control switching power supply can not meet all the application needs, its shortcomings include high output noise and ripple, low reliability and flexibility, etc[1]. The whole machine system is in urgent need of switching power supply which can realize intelligent functions such as quick and accurate response, monitoring and so on. The digital embedded technology is introduced into the field of power electronic switching power supply, and the digital sampling, operation, control output, system monitoring and human-machine interface function are realized[2].

Soft-switching technology is the core technology to solve the reliability problem of digital control switching power supply. The use of resonance principle to achieve zero voltage and zero current conduction or turn off is the basic characteristics of PWM soft-switching[3]. The main circuit of medium and high-power switching power supply is basically a Full-Bridge Converter (FBC) structure, and its soft-switching has three ways of working accordingly, namely ZVS, ZCS and ZVZCS[4]. This paper designs a digitally controlled phase-shifted full-bridge switching power supply. It plays the characteristics of high precision and high speed of microcontroller. By combining traditional power electronics with modern embedded system, the output precision, intelligence, integration and system stability of switching power supply are improved.

2. Design of system scheme

A microcontroller STM32F103RC is used as the main control unit of the digital control power supply designed in this paper. The main circuit is a full-bridge inverter topology, which adopts phase-shifting

Content from this work may be used under the terms of the Creative Commons Attribution 3.0 licence. Any further distribution of this work must maintain attribution to the author(s) and the title of the work, journal citation and DOI.
Published under licence by IOP Publishing Ltd

full-bridge control mode, and can realize the function of ZVS/ZCS soft switch. In the design, the touch screen is utilized to display voltage and current, and realize the function of the human-machine interface. An isolated optocoupler called HCPL7840 is utilized to sample the voltage and current. The PWM waveform output by the main control unit is driven by a chip called M57962L to control the opening or closing of the IGBT. Thus realizing the purpose of controlling the output voltage and output current.

The hardware circuits of the whole project include the main control circuit based on STM32, voltage sampling and current sampling circuit, rectifier and filter circuit, full-bridge inverter circuit, auxiliary power supply circuit, host computer and panel communication circuit and so on. The overall block diagram of the system is shown in figure 1.

The main technical parameters of the system are as follows.

(1) the input rated voltage is 220VAC($\pm$15%).

(2) the output voltage range is 0 to 12 VDC.

(3) the output current range is 0 to 50A.

(4) the maximum output power is 600W.

(5) the switching frequency is 100 kHz.

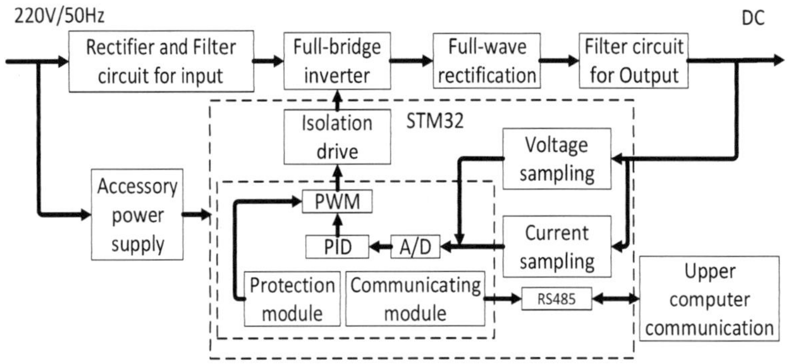

Figure 1. Block diagrams of digital controlled switching power supply system.

3. Design of system hardware

3.1. Selection of main topology of power supply

Full-bridge DC/DC converter is a kind of power conversion circuit suitable for medium and high power switching power supply. It is composed of full-control single-phase full-bridge inverter, high-frequency transformer and output rectifier filter circuit. Its main features are simple circuit structure, easy to improve circuit frequency, high transformer utilization, fast system dynamic response, high system stability, strong anti-high-frequency interference capability, etc[5].

In recent years, the DC/DC converter with single-phase full-bridge inverter as the main circuit has been widely studied, the control technology has been further improved and mature, and its application is very common. In this design, phase-shifted full-bridge is chosen as the topology of DC/DC converter, as shown in figure 2.

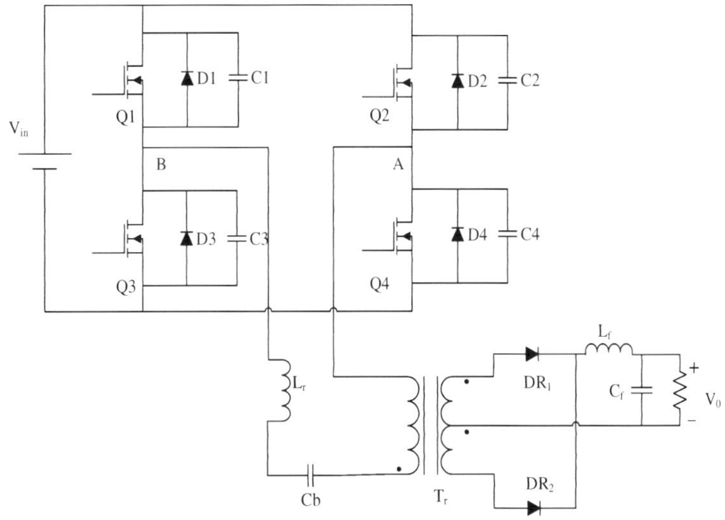

Figure 2. Topology of phase-shifted full-bridge DC-DC converter.

3.2. Design of control circuit

The control circuit is designed based on STM32 microcontroller. The microcontroller can operate at a frequency of up to 72 MHz, on-chip integrated high-speed memory (256 KB of memory and 48KB of SRAM). It has abundant multi-function I/O interface and analog-to-digital converters and other peripherals, greatly simplifying the difficulty of engineering design. The figure 3 gives a key part of the circuit.

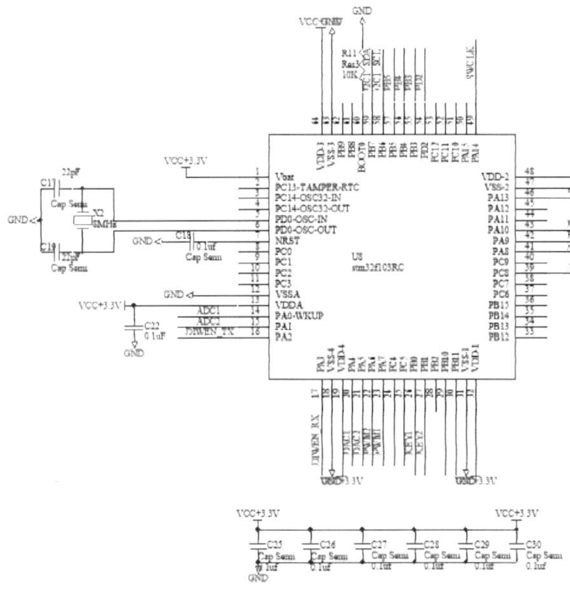

Figure 3. The control circuit.

3.3. Design of voltage sampling and current sampling circuits

In order to realize the function of digital closed-loop control and protection, the system must collect the output voltage, output current and other signals of power supply. The required level signals are obtained by isolating amplifiers, and the power circuit is electrically isolated from the STM32 control board.

In order to realize the function of electrical isolation between main circuit and control circuit of power supply, an isolated optocoupler called HCPL7840 is used to detect output current and output voltage. The output voltage sampling circuit is shown in figure 4. In the output current sampling circuit, the shunt is used to convert the current signal into the voltage signal for easy sampling. The schematic diagram is similar to the output voltage sampling schematic diagram.

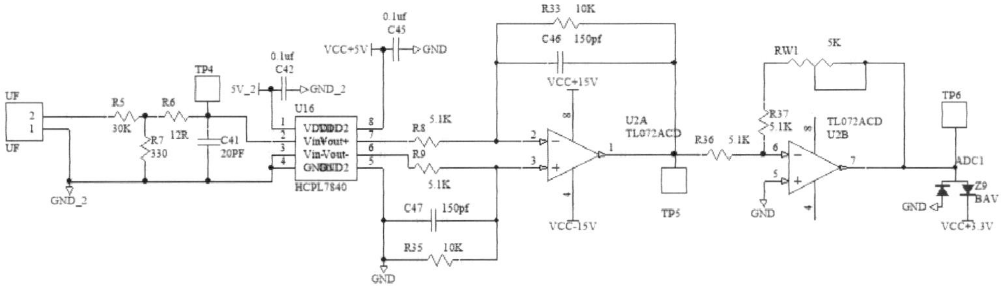

Figure 4. The output voltage sampling circuit.

3.4. Design of main device parameters

3.4.1. Selection of single-phase uncontrolled rectifier bridge. The single-phase alternating current generates a direct current after uncontrolled rectification, and the direct current is supplied to the inverter through capacitive filtering. In this paper, the rectifier bridge is used for uncontrolled rectification. According to the technical specifications of switching power supply, the maximum input power can be obtained as

$$P_{\mathrm{im}} = P_{\mathrm{om}} / \eta \tag{1}$$

The maximum input current can be obtained as

$$I_{\mathrm{im}} = P_{\mathrm{im}} / V_{\mathrm{min}} \tag{2}$$

The average current flowing through diode can be obtained as

$$I_{\mathrm{D}} = I_{\mathrm{im}} / 2 \tag{3}$$

The maximum reverse voltage for diodes can be obtained as

$$V_{\mathrm{RM}} = \sqrt{2} * V_{\mathrm{max}} \tag{4}$$

According to the two main parameters V_{RM} and I_D, and considering that enough margin should be left in the actual work, the single-phase uncontrollable rectifier bridge suitable for this subject can be selected.

3.4.2. Capacity calculation of input filter capacitor. It can be seen from the previous section that the single-phase alternating current is subjected to uncontrolled rectification to obtain a relatively large pulsating DC voltage. The function of the input filter capacitor is to smooth the DC voltage. Therefore, the selection of the input filter capacitor is more important. If it is too small, the DC voltage ripple is too large. Conversely, the charging current pulse width is narrowed and the amplitude is increased, resulting in a decrease in the input power factor and an increase in EMI.

In general, the pulsation of DC voltage after rectifying and filtering is 15% of the peak value of the lowest input AC voltage. Set the range of known AC input voltages to V_{inmin} to V_{inmax}, follow these steps to calculate the capacity of C_{in}

$$V_{inmin} = \sqrt{2}V_{line} * (1 - 15\%) \tag{5}$$

$$V_{inmax} = \sqrt{2}V_{line} * (1 + 15\%) \tag{6}$$

The pulsation value of DC voltage after rectifying and filtering can be obtained as

$$V_{pp} = \sqrt{2}V_{line} * 30\% \tag{7}$$

The value range of DC voltage after rectifying and filtering is $\left(\sqrt{2}V_{line(min)} - V_{pp}\right)$ to $\sqrt{2}V_{line(max)}$. When the input DC voltage is the minimum, the energy W_{in} stored in capacitor C_{in} in each cycle is as follows

$$W_{in} = P_{im}/f_{min} \tag{8}$$

Where f_{min} is the frequency and P_{im} is the maximum input power. The input filter capacitance can be obtained as

$$C_{in} = \frac{W_{in}}{[(\sqrt{2}V_{line(min)})^2 - (\sqrt{2}V_{line(min)} - V_{pp})^2]} \tag{9}$$

3.4.3. Selection of spacer capacitance. In order to reduce the loss of the capacitor and improve the efficiency of the converter, the capacitors should not be too small. At the same time, in order to reduce the volume and save the cost, the capacitors should not be too large. In summary, the following conditions should be satisfied for the selection of the capacitors C_b. It is obtained as

$$C_{bmin} \geq \frac{2CV_i}{10\%V_i - \frac{L_r V_o}{nL_f}} \tag{10}$$

Where C is the shunt capacitor of the switch.

3.4.4. Design of primary side series resonant inductor. When selecting resonant inductor, it should be ensured that the energy stored by the inductor can meet the conditions of realizing soft switching. It is obtained as

$$\frac{1}{2}L_r I_p^2 > CV_i^2 + \frac{1}{2}C_{TR}V_i^2 \tag{11}$$

Where I_p is the current passing through the resonant inductance, C is the shunt capacitor of the switch, and C_{TR} is the parasitic capacitance of the primary winding of the transformer.

4. Design of system software

This system adopts C language to design the software and realizes the functions of digital control power supply. The software includes main control program, A/D conversion subroutine, PI subroutine, host computer communication subroutine, subroutine to realize the function of over-voltage and over-current protection and so on. The main control program includes the initialization of each function of the system, configuration enabling interrupt, waiting interrupt, etc. The main control program block diagram is shown in figure 5.

IOP Conf. Series: Materials Science and Engineering **533** (2019) 012036 doi:10.1088/1757-899X/533/1/012036

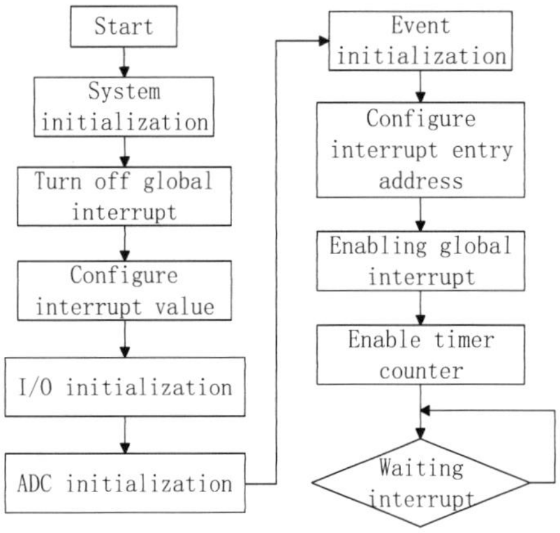

Figure 5. Main control program block diagram.

Switching converter is a time-varying, nonlinear system, which cannot establish an accurate mathematical model. The advantage of the fuzzy PID algorithm used in this paper is that it does not need accurate modeling and is very suitable for the DC/DC converter system.The flow chart of the algorithm is shown in figure 6. Set the current reference value to $r(k)$, the sampling value is $j(k)$. Set the fuzzy rule according to the error value $e(k)$ and the change rate $ec(k)$ of the deviation, then the adjustment amount of the PID parameter can be adjusted.

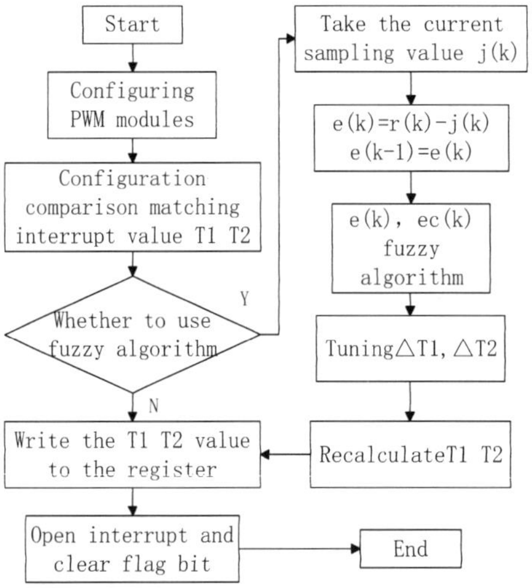

Figure 6. PID algorithm flow chart.

5. Debugging and testing

In this paper, a digital control phase-shifting full-bridge switching power supply is built and tested in the laboratory to verify its integrity. The experimental platform is shown in figure 7. High voltage differential voltage probe P2200 and AC/DC current probe A622 are used to detect voltage and current signals, which are fed into digital storage oscilloscope DS1074Z-S.

IOP Conf. Series: Materials Science and Engineering **533** (2019) 012036 doi:10.1088/1757-899X/533/1/012036

Figure 7. The power prototype.

Figure 8 is the IGBT driver waveforms at 90 degrees phase shift , and figure 9 shows the output voltage waveform. Through the analysis of the experimental results, it can be seen that the digital control phase-shifted full-bridge switching power supply system designed in this paper can achieve the required voltage stabilization effect in the voltage stabilization stage, and the output voltage ripple of the system is also controlled within the design scope. The results show that the digital control phase-shifting full-bridge switching power supply system meets the design requirements.

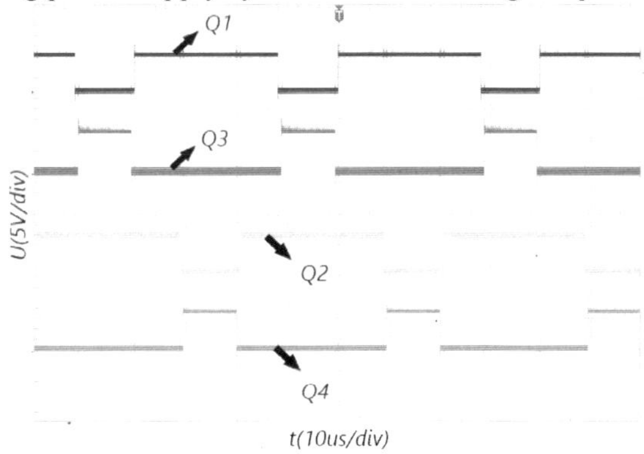

Figure 8. The IGBT driver waveforms at 90 degrees phase shift.

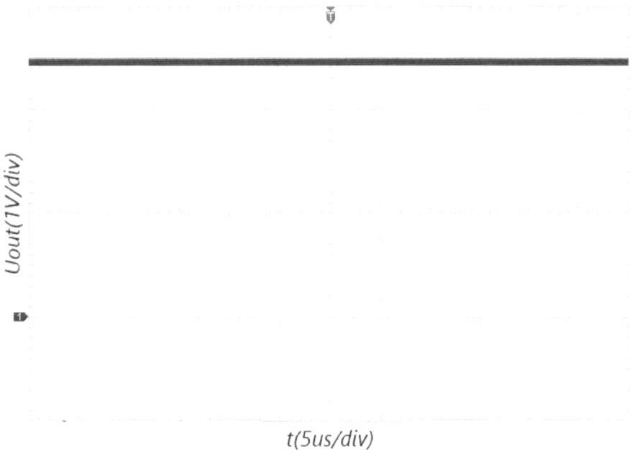

Figure 9. The output voltage waveform.

6. Conclusions

The digitally controlled DC power supply designed by the paper improves the power supply from low ripple, high precision, high efficiency and strong human-machine interaction. The range of output voltage is 0 to 12V, the range of output current is 0 to 50A. The touch screen realizes human-machine interaction and real-time monitoring of voltage and current. It has the advantages of convenient operation, high control precision, high voltage stability and strong practicability. If the digital control power supply has deeper development in the field of soft switching technology, and further reduce the noise and reduce the switching loss while further improving the operating frequency and output power of the converter, it will be further popularized and applied in the future.

Acknowledgments

The Project Supported by Natural Science Basic Research Plan in Shanxi Province of China(Program No. 2018JZ5014).

References

[1] F. Sobrino-Manzanares,A. Garrigós.A generic FPGA-based PWM generator with automatic device fault recovery for fuel cell,interleaved,multi-phase and multi-switch DC/DC boost converters. International Journal of Hydrogen Energy,2017,13876-13888.

[2] Shaowu Li,Honghua Liao,Hailing Yuan,Qing Ai,Kunyi Chen. A MPPT strategy with variable weather parameters through analyzing the effect of the DC/DC converter to the MPP of PV system.Solar Energy,2017,175-184.

[3] Shady A. Mohammed,Hakan Dogan,Mehmet Tamer Ozgun. An 85%-efficiency reconfigurable multiphase switched capacitor DC-DC converter utilizing frequency,switch size,and interleaving scaling techniques. Microelectronics Journal,2017,155-161.

[4] Mon, J.,Gonzalez,D.,Gautier,C. et al. Coupled Interleaved Multicellular Parallel converters operated under switching frequency modulation. Power Electronics and Applications (EPE'14-ECCE Europe), 2014 16th European Conference on, 2014,1-7.

[5] S. Saravanan,N. Ramesh Babu. Analysis and implementation of high step-up DC-DC converter for PV based grid application. Applied Energy,2017,64-72.

IOP Conf. Series: Materials Science and Engineering 533 (2019) 012037 doi:10.1088/1757-899X/533/1/012037

A Camera Tracking System Based on Closed-loop Kernelized Correlation Filters

Muzi Li, Bo Cheng, Shuai Zhao, and Junliang Chen

State Key laboratory of Networking and Switching Technology, Beijing University of Posts and Telecommunications, Beijing, China

Corresponding author: cedricLi@bupt.edu.cn

Abstract. Camera tracking is an important application in the computer vision. With the progress of the tracking-by-detection algorithm, industrial cameras can track targets autonomously. However, during the long-time tracking, it's prone to miss target owing to heavy object occlusion, environment changing and objects appearance variations. Our camera tracking system is based on the kernelized correlation filter to track object, and we add a self-verification module to judge if the current tracking results are reliable. This can be useful in solving target missing when object meets occlusion or variations, and avoid model drift during long-time detection. Last but not the least, we keep the targets in the corner of the image by controlling our camera, avoiding object's moving out of view. The kernel correlation filter bounded with self-verification and re-detection modules boost the performance. Extensive experiments on the OTB-2013 benchmark show that it performs better than state-of-the-art methods.

1. Introduction

Visual object tracking is a very popular topic in computer vision for its numerous applications, such as intelligent vehicles, human-machine interaction, and surveillance[1]. Even though the significant progress has been made in object tracking, there still remains a lot of challenge due to object occlusion and significant appearance changes.

In recent years, there are many object trackers, The tracking mechanism may be similar, a tracker identify targets after a detector inspect it, and then update the tracker according to object moving and its variations[2]. Bolme et al. [3] introduced correlation filters to visual tracking for this first time by minimizing the output sum of squared error. And then Correlation Filter based trackers have been very popular for its high efficiency and robustness among those object trackers. There are two reasons for Correlation Filter based trackers' success. One thing is that this kind of trackers can produce thousands of training samples through image patch shifting. The lack of training data avoids other trackers' progressing. And the other thing is its fast training and detecting via circular correlation via Fast Fourier Transform(FFT)[4].

Even though Correlation Filter based trackers are so popular, there still remains a lot to be improved. In a long- time detection, the model update is easy to drift and leads to a wrong result under the situation of object occlusion and appearance variations. Most Correlation Filter trackers consider tracking process as an open-loop process. They neither check the reliability of their results, nor generate any feedback to reshape their filter. So it's not hard to know that the ignored abnormity and

Content from this work may be used under the terms of the Creative Commons Attribution 3.0 licence. Any further distribution of this work must maintain attribution to the author(s) and the title of the work, journal citation and DOI.
Published under licence by IOP Publishing Ltd

localization error in a few frames will cause the bad results. There are two import things to fix the problem. The first one is to identify whether the current tracking result is reliable. And the other thing is to add re-detection module to the filter.

In order to solve the existing problems in long-time tracking, we designed a camera tracking system based on closed-loop kernelized filter tracker. We add a self-verification module after the tracking output to judge whether we need to run our re-detection module. What's more, a re-detection module is also added to compensate the localization error and reduce model drifting. Through a feedback loop, the re-detection model can retain the model's discriminative ability in long time tracking. We send control commands to our camera to turn a certain angle when the target deviates from the center of the image beyond the pre-set threshold..

2. Architecture

Our camera tracking system contains two parts in the figure 1. The first part is about our camera which provides the function of real-time video streaming and camera control. And the second part is our object tracker consists of kenerlized correlation filter, self-verification and re-detection. This part helps track the target in a long time.

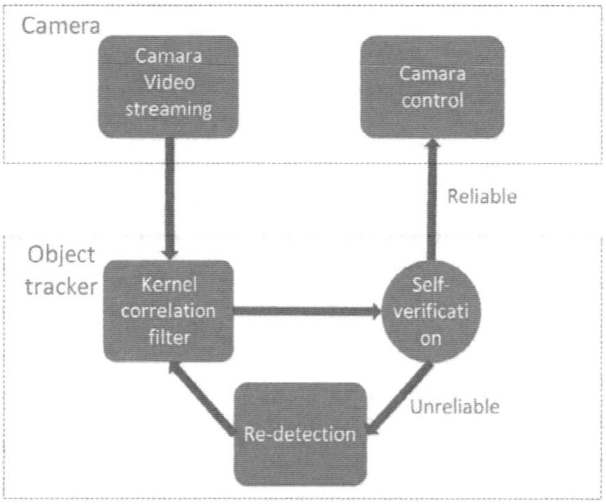

Figure 1. Architecture of the Camera Tracking System

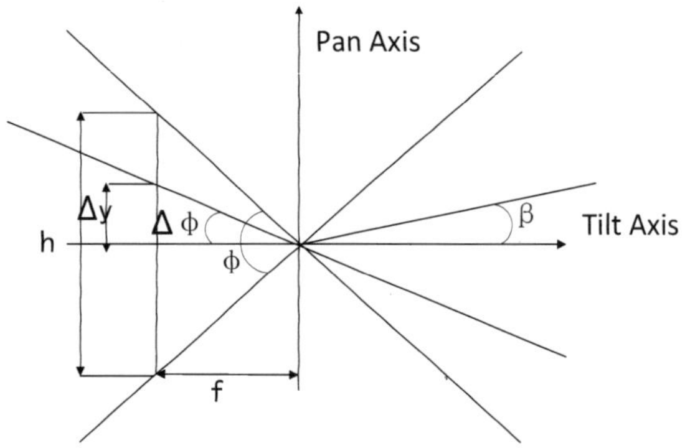

Figure 2. Relationship between Offset and Rotation Angle

2.1. Camera Function

The camera's real-time video stream is streamed in RTSP format. We use openCV tools to receive the video stream, and then extract the first frame as the input of our subsequent model initialization. The target position is also marked by the detection algorithm.

2.1.1. Real Time Video Streaming. The camera's real-time video stream is streamed in RTSP format. We use openCV tools to receive the video stream, and then extract the first frame as the input of our subsequent model initialization. The target position is also marked by the detection algorithm.

2.1.2. Camera Control. For a camera tracking system, target's being out of view should be avoided. It can be hard for a object tracker to achieve the goal. The target is always moving, we should control the moving of camera to keep the target in proper position. The way we do is to send control commands to our camera to turn a certain angle when the target deviates from the center of the image beyond the pre-set threshold.

Our experimental camera is a kind of PTZ camera. In the PTZ camera, there are three kinds of rotation axes including horizontal rotation axis of the turntable (Pan axis), the vertical axis of rotation (Tilt axis) and the optical axis of the camera. Every two of them are perpendicular to each other and intersect at point O, and the three meet the Cartesian coordinate system [5]. The X and Y axes of the camera coordinate system coincide with the Tilt and Pan axes, respectively, the Z axis is the optical axis of the camera.

When the target is at the edge of the image, it is necessary to control the motion of the camera to rotate around the Pan axis and the Tile axis, so that the coordinate position of the target is at a reasonable position of the image. During the control process, we should consider the coordinates' (Δx ,Δy) changes and the angle of rotation($\Delta\theta$, $\Delta\varphi$) as the most important parameters. And then we will discussed about the relationship between $\Delta\theta$, $\Delta\varphi$ and Δx , Δy.

We suppose that the horizontal and vertical angels of the field captured by the camera are l and ψ. The camera focus length is f when imaging. The horizontal length of the image is L pixels and the vertical direction is h pixels. Among these parameters, only f can't be measured. So we take the coordinate(x, y) as an example.

As the figure 2 shows, the Pan axis intersects the Tilt axis and they are perpendicular to each other. and then we get the $\Delta\varphi$,

$$\mathrm{f} \times \tan(\varphi/2) = \mathrm{h}/2 \tag{1}$$

$$\mathrm{f} \times \tan \Delta\varphi = \Delta y \tag{2}$$

$$\Delta\varphi = arctan\frac{2\Delta y \tan(\varphi/2)}{h} \tag{3}$$

In the precise camera model, the Tilt axis is project to the plane of x and y plane and there exits a error angle is β.

$$\Delta\varphi = \frac{\Delta y \tan(\varphi/2)}{h \tan(\varphi/2+\beta)}\varphi \tag{4}$$

We can also get the relationship between $\Delta\theta$ and Δx. In this equation, Δy equals to the amount of the target's change in the y axis.

2.2. Kernelized Correlation Filter

Kernelized Correlation Filter tracker is a kind of tracking-by-detection algorithm. The core function is to cyclically shift the tracking target area to mark the target object as a positive sample and the rest of the surrounding environment is a negative sample[6]. It constructs a training discriminant classifier on the basis that the positive sample is 1, and the negative sample is 0. The sample labelling method is well reflected by the weight of each negative sample, and is assigned [0,1] according to the distance between the target and the center of the sample. The closer the distance is to the target, the closer the value is to 1, and it is closer to 0 in the contrast. The experiment can prove that the sample labelling works well in the discriminant classifier training. The Kernelized Correlation Filter tracker constructs the training samples of the classifier via the matrix cyclic offset, and it selects the candidate region with the largest similarity as the new tracking target, so that the data matrix becomes a circular matrix. Then, based on the characteristics of the cyclic matrix, the solution of the un-linear kernel problem is transformed into the discrete Fourier transform domain in order to avoid the process of inverting the marix. This way can reduce both storage and computation of the algorithm by several orders of magnitude and greatly improve the processing speed.

2.2.1. Ridge Regression. Ridge regression can perform as well as those sophisticated methods. The goal is to minimize the squared error over samples x_i and regression targets.

$$\min_w \sum_i (f(x_i) - y_i)^2 + \lambda||w||^2 \tag{5}$$

Since ridge regression admits a simpler closed-form solution, w is given by:

$$w = (X^T X + \lambda I)^{-1} X^T y \tag{6}$$

In this eq, per row x_i and each regression trarget element of y are in the data matrix. To minimize the squared error w, there exists a large number of linear equations. And in the next section, he computational complexity will be greatly reduced With the introduction of the circulant matrix.

2.2.2. Circulant Matrices. Focusing on only single-channel signals can help reduce our computation complexity. We refer to an $n \times 1$ vector representing a sample patch as the base sample which is denoted as x. And then we can model the permutation matrix by a cyclic shift operator. Shifting x by one element, we model a small translation. A larger translation can be made by the permutation matrix. By this way we can both get half of the shifts set as positive samples, and the other half as negative samples. And the translation lead us to the circulant matrix.
The fact that all circulant matrices are made diagonal by the Discrete Fourier Transform(DFT) , regardless of the generating vector x [7]. And the expression is:

$$X = F\,\mathrm{diag}(X^{\wedge})\,F^H \tag{7}$$

F is a DFT matrix and X^ is the DFT of the generating vector. x^= F(x). And DFT is a kind of linear operation.

2.2.3. Kernel Trick. This section will map the inputs of a linear problem to a complicated non-linear spatial feature space with the kernel trick. The map function is φ(x) , the solution w can be expressed as:

$$w = \sum_i \alpha_i \varphi(x_i) \tag{8}$$

In this eq, α ={α 1 ,α 2 ,...,αi ,... }, αi is the coefficient of the responding training sample x_i. Using the kernel function κ, we can get the expression

$$\varphi^T(x)\varphi(x`) = \kappa(x, x`) \tag{9}$$

$$K_{ij} = \kappa(x_i, x_j) \tag{10}$$

And the final regression function is eq11, the kernel trick bring the model to a high-dimensional feature space, however the complexity will grow larger as the number of samples grow.

$$f(z) = \omega^T z = \sum_{i=1}^{n} \alpha_i \kappa(z, x_j) \tag{11}$$

2.2.4. Fast Training and Detection. With the addition of kernel trick, the solution of the ridge regression is:

$$\alpha = (K + \lambda I)^{-1} y \tag{12}$$

Since K is circulant matrix formed by cyclic shifting of the data sets, we can diagonalize Eq aas fast as linear case. There are several kinds of kernel that meets the demand of circulant matrix, such as Radial Basic Function Kernels, Dot-product kernels, Addictive kernels and Exponentiated addictive kernels.

$$\alpha^{\wedge} = \frac{y^{\wedge}}{k^{\wedge xx} + \lambda} \tag{13}$$

Hat ^ is the DFT of a vector and the k^(^xx) is the first row of the kernel matrix.
When KCF calculates the regression function for a newly input image z, the calculation is of great computation complexity which will cause much time consuming. We can reconstruct the image to get the test samples and training samples of the kernel function, and the established core function matrix is

$$K^z = C(k^{xz}) \tag{14}$$

In the eq, k^xz is the vector of the first row. Then we can calculate the full detection response of the cyclic shift structure based on the testing sample z

$$f^z(z) = k^{xz} \odot \alpha^{\wedge} \tag{15}$$

2.3. *Self-verification*

location errors in a few frames can lead to the model's drifting and tracking failure. The self-verification module is to check the reliability of the detection results carefully. And in this section, we put up an easy way to tell a better result from a bad result.We utilize the responses map of HOG feature as the criterion for the reliability of the tracking results. The peak to sidelobe ratio (PSR) is regarded as the most import parameter. It can help quantify the sharpness of the correlation peak. When the correlation between current frame and previous frames is high, the PSR value is also higher. And in the contrast, it will be lower, if the correlation is down. So we propose the tracking results' score as:

$$S_h^{(i)} = \frac{\max\left(f_h^{(i)}\right) - \mu_i}{\sigma^i} \tag{16}$$

In the eq, f_h^((i)) is the i-th response map of HOG-based correlation filter, and σ^i are the mean and standard deviation of the response. Minimizing the discrepancy between the tracking results and expected response can help us find better threshold. Under the assumption that the sub-image catered at the ground-truth position has the minimum discrepancy to the expected response. The discrepancy function can be written as:

$$\arg\min_{X_p} D(R(X_p, h_t), N) \tag{17}$$

the sub-image X_p is the filter response and the h_t is the filter trained from previous frames. And D is a function to calculate the discrepancy between the actual response map and the expected response map N.

$$D\left((X_p, h_t), N\right) = W_p\left(P_{expected} - P(X_p, h_t)\right) + W_s(S_{expected} - S(X_p, h_t)) \tag{18}$$

In the eq, P_expected is the peak value of the expected fileter response and P(X_p,h_t) is the real peak value of the tracking filter response. S(X_p,h_t) is the PSR response map and S_expected is the expected one. W_p and W_s are different weights.

2.4. *Self-detection*

Under the assumption that the object target is prone to appear in the area near the last reliable frame for a little interval, we only search from our predefined sub-image patches according to our G(X_p,h_t) score.

$$\arg\max_{X_p} G(X_p, h_t) = w_p\, P(X_p, h_t) + w_s\, S(X_p, h_t) \tag{19}$$

The problem comes to select the best one which can maximum the G score. A step-wise method can be used in this proble:

$$x_{i,j} = x_0 + i \times \Delta T \cos(j \times \Delta\theta + \frac{(-1)^i + 1}{4} \times \Delta\theta) \tag{20}$$

IOP Conf. Series: Materials Science and Engineering **533** (2019) 012037 doi:10.1088/1757-899X/533/1/012037

$$y_{i,j} = y_0 + i \times \Delta T \sin(j \times \Delta\theta + \frac{(-1)^i+1}{4} \times \Delta\theta) \quad (21)$$

In the eq, ΔT and $\Delta\theta$ stands for every distance step and angle step. R_d is the radius of the re-detection area circle, determined by the target size and the PSR value. And m= R_d/ΔT, n= $2\pi/\Delta\theta$, m x n candidate sub-images are extracted.

Our Kernelized Filter tracker can work well in most scenarios. In order to keep the computation efficiency, the re-detection module only works under the situation that the self-verification consider the tracking results as unreliable results.

3. Experiment Analysis

We implemented the proposed tracking algorithm in MATLAB and conducted all experiments on a computer with an Intel I7-4700HQ, 2.40GHZ CPU, 8GB RAM and conducted extensive experiments on OTB-2013 datasets. Evaluation Metrics: we use the success rate as our main evaluation metrics.

3.1. Evaluation of Our Proposed Filter Tracker

We evaluate the proposed algorithm on OTB-2013. It can be seen in figure 3 that, the precision rate of our kernelized filter tracker has been increased from 0.514 to 0.677. Among the compared tracking methods, out tracker achieves the second result with a success rate better than the threshold of 0.600. And in the figure 4, our tracker gets a better score than the ECO algorithm when target gets out of view.

3.2. Actual Scenario

In the actual scenario of our camera tracking system, the red frame area is the area where the target object is located, and the yellow one contains more area used by the re-detection module. We can see from the figure 4, the tracking results are considered as unreliable results in our self-verification module when the white woman passes the pole. And then, larger detection area is added in the re-detection part. The result can be seen that the woman can be detected right and our tracking system continues following the woman, which proves that our tracking system works well under the situation of object occlusion.

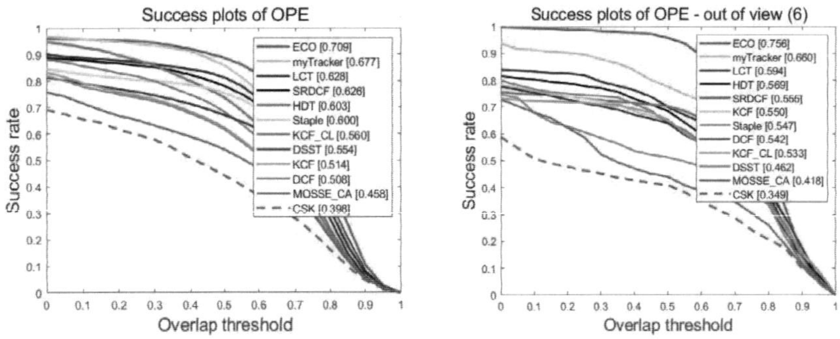

Figure 3. Experiment evaluation on OTB-2013 benchmark.

Figure 4. Actual senario of object occlusion.

4. Conclusion

The camera tracking system is of great significance in the era of rapid development of computer vision applications. The target tracking algorithm is applied to the camera. Based on the algorithm of the kernelized filter tracker, the system can keep a real-time object tracking and keep the model work well in a long term. The self-verification module help the system to judge the reliability of the tracking results. The self-detection module run under the unreliable situation, and help search more participants patches to localize target. What's more, we also keep the object target in the center of the screen via camera control. In all, Our camera tracking system effectively provides long-term target tracking and the correctness of the occlusion object to continue tracking.

In our future work, we hope to add color features and deep learning features to the tracker to improve the precision, and add scale feature to deal with the size variation of the tracking object due to distance changes.

Acknowledgments

This work is supported by National Natural Science Foundation of China (Grant No. 61501048), Beijing Natural Science Foundation (Grant No. 4182042), Fundamental Research Funds for the Central Universities (No. 2017RC12), and National Natural Science Foundation of China (U1536111, U1536112).

References

[1] Arnold WM Smeulders, Dung M Chu, Rita Cucchiara, Simone Calderara, Afshin Dehghan, and Mubarak Shah. Visual tracking: An experimental survey. IEEE Transactions on Pattern Analysis and Machine Intelligence, 36(7):1442–1468, 2014.

[2] Sayanan Sivaraman and Mohan Manubhai Trivedi. Looking at vehicles on the road: A survey of vision-based vehicle detection, tracking, and behavior analysis.IEEE Transactions on Intelligent Transportation Systems, 14(4):1773–1795, 2013.

[3] David S Bolme, J Ross Beveridge, Bruce A Draper, and Yui Man Lui. Visual object tracking using adaptive correlation filters. In IEEE Conference on Computer Vision and Pattern Recognition, pages 2544–2550, 2010.

[4] R. Patnaik and D. Casasent, "Fast FFT-based distortion-invariant kernel filters for general object recognition," in Proceedings of SPIE, vol. 7252, 2009, p. 1.

[5] A. Maksai, X. Wang, and P. Fua, "What players do with the ball: A physically constrained interaction modeling," in Proceedings of IEEE Conference on Computer Vision and Pattern Recognition, 2016.

[6] João F Henriques, Rui Caseiro, Pedro Martins, and Jorge Batista. High-speed tracking with kernelized correlation filters. IEEE Transactions on Pattern Analysis and Machine Intelligence, 37(3):583–596, 2015.

[7] Z. Hong, Z. Chen, C. Wang, X. Mei, D. Prokhorov, and D. Tao, "Multistore tracker (muster): A cognitive psychology inspired approach to object tracking," in Proceedings of IEEE Conference on Computer Visionand Pattern Recognition, 2015, pp. 749–758.

IOP Conf. Series: Materials Science and Engineering **533** (2019) 012038 doi:10.1088/1757-899X/533/1/012038

Analysis of Misalignment Effect on Wireless Charging System for Inspection Robots

Han Liu[1,2], Linlin Tan[1,2] , Xueliang Huang[1,2] and Dariusz Czarkowski[3]

[1]School of Electrical Engineering, Southeast University, Nanjing, 210096, China
[2]Jiangsu Provincial Key Lab of Smart Grid Technology and Equipment, Nanjing, 210096, China
[3]Department of Electrical and Computer Engineering, Tandon School of Engineering, New York University, New York, 11218, USA

corresponding author's e-mail address: xlhuang@seu.edu.cn

Abstract. To solve the issues including the inflexibility, connector wear, and electric shock risk in contact charging system for inspection robots, the wireless charging system for inspection robots is introduced in this paper. The circuit model of this paper is developed based on the mutual inductance theory. The expressions of the receiving power and the efficiency are derived. The coordinate calculation method for mutual inductance between non-coaxial rectangular spiral coils according to the movement characteristics of inspection robot is presented. According to the calculation method, the characteristics of the receiving power and the efficiency varying with the position change of the receiving coil are investigated. The location of the positioning device for starting the wireless charging system can be selected based on the system characteristics. The theoretical analysis is verified by experiments.

1. Introduction

The inspection robots are wildly applied in smart factories and smart substations to replace the manual inspection tasks. At present, the inspection robots are mainly charged with contact through the manual or mechanical plug-in. The manual contact charging method is inflexible and not suitable for the smart unmanned development. Another contact charging method is realized through the automatic mechanical structure, which also causes the connector wear and electric shock risk.

In recent years, the wireless power transfer technology is developed and applied in the fields including electronic devices, medical implants [1-2], electric vehicles [3-7], rail transit, and the drones. In [8], the optimizing compensation topologies for inductive power transfer at different mutual inductances are proposed. The optimizing three-phase three-layer coil array for omnidirectional wireless power transfer is investigated in [9]. The multiphase inductively coupled resonant converter is proposed for wireless electric vehicle charger in [10]. A two-loop control method in wireless power transfer system is studied in [11].

In this paper the wireless power transfer technology is applied in the charging system for inspection robots to solve the issues mentioned before. Firstly the structure of the wireless charging system for the inspection robot is introduced. Secondly the circuit model of this wireless charging system is developed according to the circuit theory of mutual inductance. Then the coordinate calculation method for mutual inductance between non-coaxial rectangular spiral coils with alignment according

Content from this work may be used under the terms of the Creative Commons Attribution 3.0 licence. Any further distribution of this work must maintain attribution to the author(s) and the title of the work, journal citation and DOI.
Published under licence by IOP Publishing Ltd

to the movement characteristics of inspection robot is investigated. Based on the calculation method, the characteristics of the receiving power and the efficiency varying with the position change of the receiving coil are studied. The location of the positioning device for starting the wireless charging system can be selected according to the obtained system characteristics. The theoretical analysis is verified by experiments and the conclusion is drawn finally.

2. System description

The sketch of the wireless charging system for inspection robot proposed in this paper is shown in Figure 1. The transmitting coil is installed underground. The receiving coil is installed under the chassis of the inspection robot. The robot moves along the fixed line trajectory. Hence the lateral misalignment is ignored in this paper. The positioning device is set to detecting the access of the robot. In order to improve the flexibility of the wireless charging system and reduce the requirement of the stopping location, the location of the positioning device is selected to determine the wider effective charging area where the receiving power and the efficiency are acceptable. During the access process of the robot, the wireless charging system starts when the robot reaches the location of the positioning device.

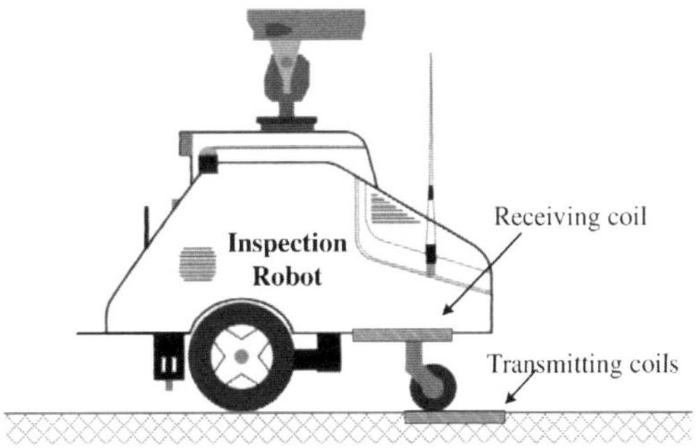

Figure 1. The sketch of the wireless charging system for inspection robot.

The structure of the wireless charging system is shown in Figure 2. The transmitting side includes the high-frequency power source, the transmitting coil, and the LCC compensation circuit. The receiving side includes the receiving coil, the LC compensation circuit, the rectifier, the DC converter and the battery. The energy is transferred wirelessly through the magnetic coupling between the transmitting coil and the receiving coil.

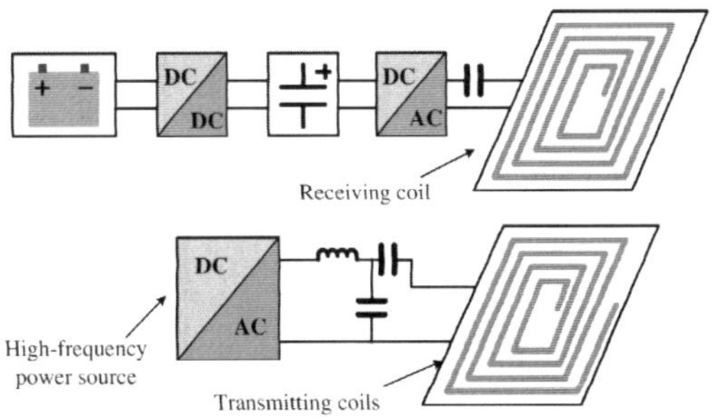

Figure 2. The structure of the wireless charging system for inspection robot.

3. Circuit model of system

The circuit model of the wireless charging system for the inspection robot is developed in this paper. The equivalent circuit of this system is shown in Figure 3. U_s represents the high-frequency power source. L_1, C_1, and C_t are the inductance and capacitors of the compensation circuit in transmitting side. L_t and R_t are the inductance and the resistance of the transmitting coil respectively. L_r and R_r denote the inductance and resistance of the receiving coil respectively. C_r is the compensation capacitor in receiving side. M is the mutual inductance between the transmitting and receiving coils. R_L represents the equivalent AC resistance in the receiving side.

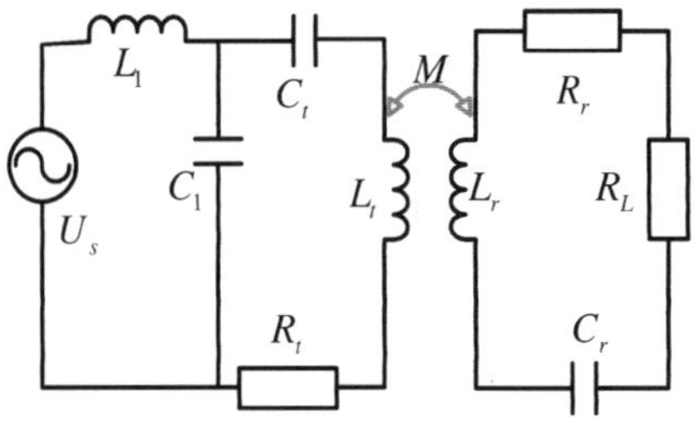

Figure 3. The equivalent circuit of the wireless charging system for the inspection robot.

To maintain the stability of the current through the transmitting coil, the components of compensation circuit in transmitting side should satisfy that $C_1L_1=\frac{1}{\omega^2}$ and $C_t(L_t-L_1)=\frac{1}{\omega^2}$, where $\omega=2\pi f$, and f is the operating frequency.

Based on the circuit theory, the current through the transmitting coil is deduced as

$$I_t=\frac{U_s}{j\omega L_1} \tag{1}$$

The receiving coil is compensated with the capacitor, which satisfies that $C_rL_r=\frac{1}{\omega^2}$.

According to the theory of the mutual inductance, the current in receiving coil can be calculated as

$$I_r=\frac{MU_s}{L_1(R_L+R_r)} \tag{2}$$

The receiving power is deduced as

$$P=\frac{M^2U_s^{\,2}R_L}{L_1^2(R_L+R_r)^2} \tag{3}$$

The efficiency can be expressed as

$$\eta=\frac{\omega^2M^2R_L}{R_t(R_L+R_r)^2+\omega^2M^2(R_L+R_r)} \tag{4}$$

According to (3) and (4), the receiving power and the efficiency are influenced by the mutual inductance between the transmitting coil and the receiving coil. During the moving process of the robot, the changing relevant position between the transmitting and receiving coils causes the variation of the mutual inductance. The stopping position of the robot will determine the characteristics of the charging system while the system operation after the robot access is detected by the positioning device. Hence the system characteristics during the moving process of the receiving coil are investigated in this paper. The system characteristics can provide a basis for the selection of the location of the position device to detect the robot access.

4. Mutual inductance calculation and system analysis

The transmitting and receiving coils in wireless charging system for inspection robots are rectangular spiral coils. The schematic diagram of non-coaxial rectangular spiral coils is shown in Figure 4. In the transmitting coil, the maximum length is L while the maximum width is W. The number of turns of transmitting coil is N_1. In the receiving coil, the maximum length is l while the maximum width is w. The number of turns of receiving coil is N_2. The distance between the adjacent turns of every coil is d. In the Cartesian coordinate system for a three-dimensional space, as shown in Figure 4, coordinate origin O denotes the center of the transmitting coil and the point O' is the center of the receiving coil. A_i is the start point of the ith turn in the transmitting coil. B_i, C_i, D_i, and E_i are the next points of the ith turn in transmitting coil. A_{i+1} is the end point of the ith turn and the start point of the $(i+1)^{th}$ turn. A_j' is the start point of the jth turn in receiving coil. B_j', C_j', D_j', and E_j' are the next points of the jth turn in receiving coil. A_{j+1}' is the end point of the jth turn and the start point of the $(j+1)^{th}$ turn.

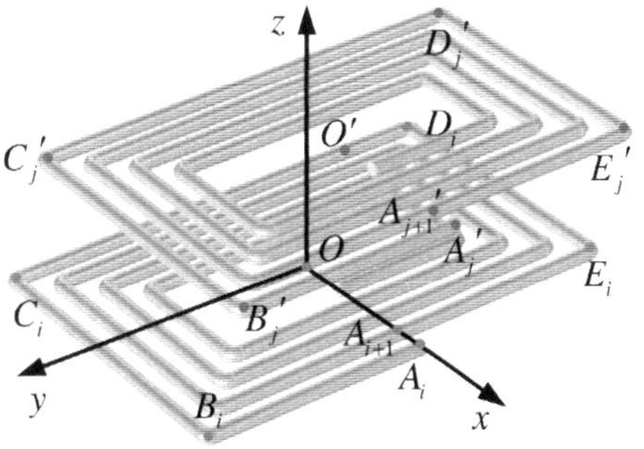

Figure 4. The schematic diagram of non-coaxial rectangular spiral coils.

The coordinate calculation method for mutual inductance between non-coaxial rectangular spiral coils is studied in this paper. In the Cartesian coordinate system for a three-dimensional space, O is denoted as $O(0,0,0)$ and O' is represented by $O'(O_x,O_y,O_z)$. Hence, the points in transmitting coil are expressed by

$$\begin{aligned}
&A_i\left(\frac{W}{2}-d(i-1),0,0\right), i=1,2,\cdots,N_1\\
&B_i\left(\frac{W}{2}-d(i-1),\frac{L}{2}-d(i-1),0\right)\\
&C_i\left(-\frac{W}{2}+d(i-1),\frac{L}{2}-d(i-1),0\right)\\
&D_i\left(-\frac{W}{2}+d(i-1),-\frac{L}{2}+d(i-1),0\right)\\
&E_i\left(\frac{W}{2}-d\bullet i,-\frac{L}{2}+d(i-1),0\right)\\
&A_{i+1}\left(\frac{W}{2}-d\bullet i,0,0\right)
\end{aligned} \quad . \tag{5}$$

And the points in receiving coil are expressed by

$$\begin{aligned}
&A_j'\left(\frac{w}{2}-d(j-1)+O_x,O_y,O_z\right), j=1,2,\cdots,N_2\\
&B_j'\left(\frac{w}{2}-d(j-1)+O_x,\frac{l}{2}-d(j-1)+O_y,O_z\right)\\
&C_j'\left(-\frac{w}{2}+d(j-1)+O_x,\frac{l}{2}-d(j-1)+O_y,O_z\right)\\
&D_j'\left(-\frac{w}{2}+d(j-1)+O_x,-\frac{l}{2}+d(j-1)+O_y,O_z\right)\\
&E_j'\left(\frac{w}{2}-d\bullet j+O_x,-\frac{l}{2}+d(j-1)+O_y,O_z\right)\\
&A_{j+1}'\left(\frac{w}{2}-d\bullet j+O_x,O_y,O_z\right)
\end{aligned} \tag{6}$$

In this coordinate system, according to (5) and (6), the ith loop in transmitting coil is written as

$$\boldsymbol{l}_{p(i)}=\overrightarrow{A_iB_i}+\overrightarrow{B_iC_i}+\overrightarrow{C_iD_i}+\overrightarrow{D_iE_i}+\overrightarrow{E_iA_{i+1}} \,. \tag{7}$$

The jth loop in receiving coil is

$$\boldsymbol{l}_{s(j)} = \overrightarrow{A_j'B_j'} + \overrightarrow{B_j'C_j'} + \overrightarrow{C_j'D_j'} + \overrightarrow{D_j'E_j'} + \overrightarrow{E_j'A_{j+1}'} \ . \tag{8}$$

Hence, the mutual inductance between the ith loop in transmitting coil and the jth loop in receiving coil can be expressed as

$$M_{ij} = \frac{\mu_r \mu_0}{4\pi} \int_{l_{p(i)}} \int_{l_{s(j)}} \frac{\mathrm{d}\boldsymbol{l}_{p(i)} \bullet \mathrm{d}\boldsymbol{l}_{s(j)}}{R_{ij}} \tag{9}$$

where μ_r is the relative magnetic permeability ($\mu_r \approx 1$, in air), μ_0 is the magnetic permeability of vacuum ($\mu_0 = 4\pi \times 10^{-7}$ H/m), and R_{ij} represents the distance between current microelements of the ith loop in primary coil and of the jth loop in secondary coil.

The total mutual inductance between non-coaxial rectangular spiral coils can be calculated according to

$$M = \left| \sum_{i=1}^{N_1} \sum_{j=1}^{N_2} M_{ij} \right| = \left| \sum_{i=1}^{N_1} \sum_{j=1}^{N_2} \frac{\mu_r \mu_0}{4\pi} \int_{l_{p(i)}} \int_{l_{s(j)}} \frac{\mathrm{d}\boldsymbol{l}_{p(i)} \bullet \mathrm{d}\boldsymbol{l}_{s(j)}}{R_{ij}} \right| . \tag{10}$$

According to the coordinate calculation method, the characteristics of the mutual inductance between the transmitting and receiving coils varying with the receiving coil position can be obtained. Combining (10) with (3) and (4), the receiving power and the efficiency of the system can be calculated when the receiving coil is at different positions. The characteristics of the receiving power and the efficiency varying with the position of the receiving coil are shown in Figure 5 and Figure 6.

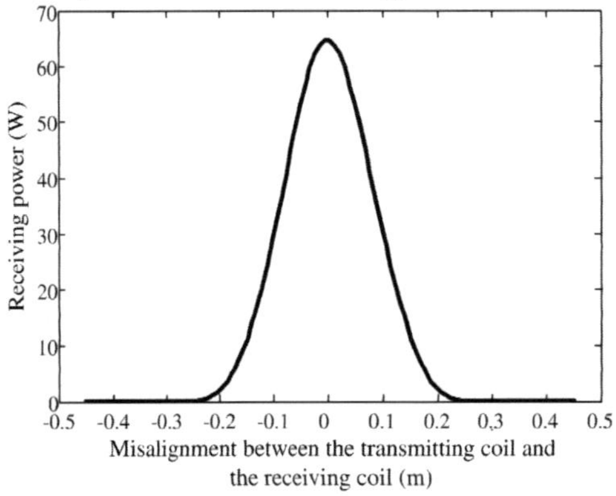

Figure 5. The characteristics of the receiving power varying with the position of the receiving coil.

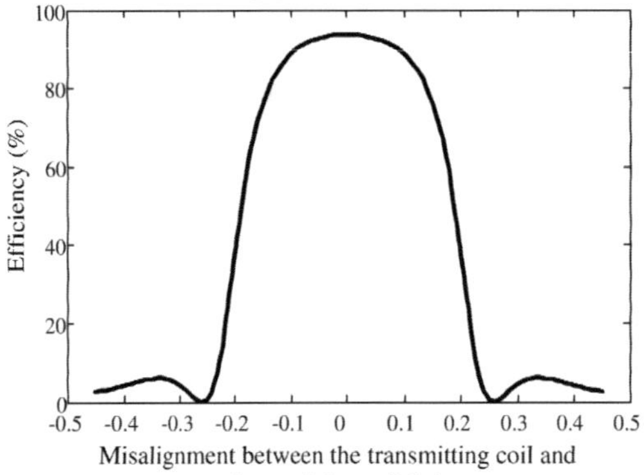

Figure 6. The characteristics of the efficiency varying with the position of the receiving coil.

According to the characteristics shown in Figure 5 and Figure 6, the receiving power and the efficiency change with the misalignment between the transmitting coil and the receiving coil. The receiving power and efficiency increase when the misalignment decreases and reach the maximum when the misalignment is zero. Hence, the location of the positioning device for detecting the robot access and starting the wireless charging system can be selected according to the characteristics of the receiving power and the efficiency varying with the position of the receiving coil.

5. Experimental verification

To verify the characteristics in the theoretical analysis part, the experiment is carried out in this part. The prototype of the proposed wireless power transfer system is shown in Figure 7. The transmitting side includes the high-frequency inverter, the LCC compensation circuit, and the transmitting coil. The receiving side includes the receiving coil, the LC compensation circuit, and the load. The output voltage of the high-frequency inverter is 20 V. The operation frequency of this system is 85 kHz. The transmitting coil and the receiving coil are identical rectangular spiral coils. The length is 0.30 m. The width is 0.30 m. The number of turns is 7. The vertical height between the primary and secondary coils is 0.10 m. The distance between the adjacent turns of every coil is 0.01 m. L_1=5.0 μH. C_1 is 700.5 nF (the calculated value is 701.2 nF) and C_p is 231.7 nF (the calculated value is 232.2 nF). C_s is 173.6 nF (the calculated value is 174.4 nF). The load resistance is 4.2 Ω.

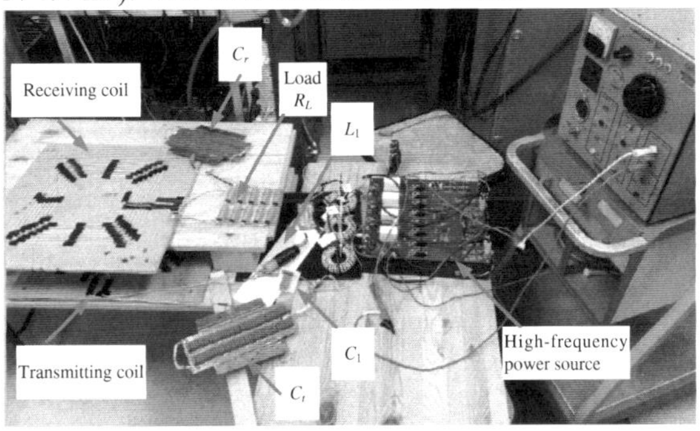

Figure 7. The prototype of the proposed wireless power transfer system.

Changing the position of the receiving coil, the voltage of the load is measured. The experimental characteristic of the receiving power varying with the position of the receiving coil is obtained as shown in Figure 8.

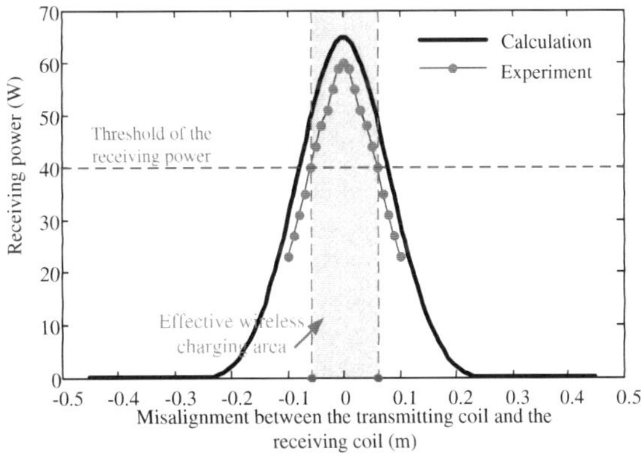

Figure 8. The experimental characteristic of the receiving power varying with the position of the receiving coil.

According to the curve of the receiving power varying with the position of the receiving coil, the receiving power and efficiency increase when the misalignment decreases and reach the maximum when the misalignment is zero. The receiving power is low when the positioning device is set too far away from the central position of the transmitting coil and reaches the maximum when the positioning device is set at the central position of the transmitting coil. If the threshold of the receiving power is set as 40 W, the effective wireless charging area can be determined as shown in Figure 8. The waveforms of the output voltage of the inverter and the voltage of the load are shown in Figure 9 and

Figure 10 comparatively when the receiving coil is at the edge position and the central position of the effective wireless charging area. The receiving power is 40.2 W and the efficiency is 81.3 % at the edge position. The receiving power is 60.2 W and the efficiency is 84.6 % at the central position.

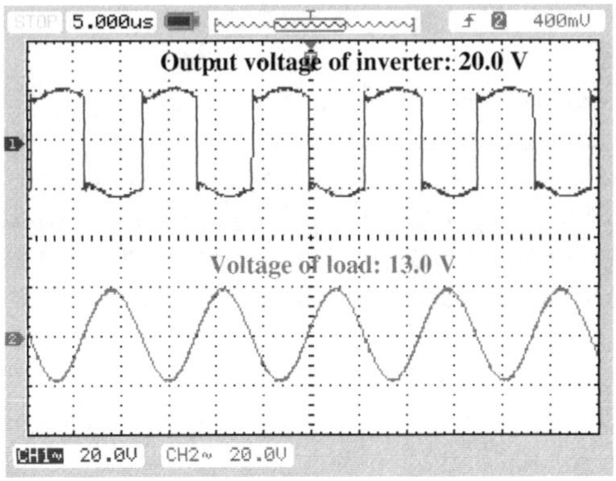

Figure 9. The waveforms of the output voltage of the inverter and the voltage of the load at the edge position.

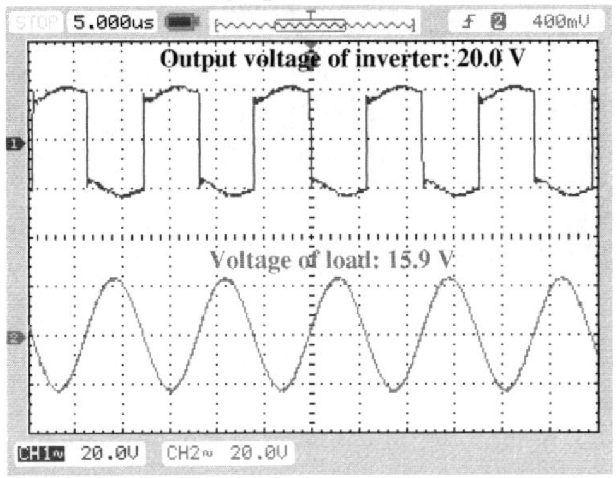

Figure 10. The waveforms of the output voltage of the inverter and the voltage of the load at the central position.

6. Conclusions

The wireless charging system for inspection robots is introduced in this paper to solve the issues in contact charging system for inspection robots. The circuit model of this paper is developed and the expressions of the receiving power and the efficiency are derived. The coordinate calculation method for mutual inductance between non-coaxial rectangular spiral coils according to the movement characteristics of inspection robot is studied. According to the calculation method of the mutual inductance, the characteristics of the receiving power and the efficiency varying with the position change of the receiving coil are investigated. The receiving power and efficiency increase when the misalignment decreases and reach the maximum when the misalignment is zero. The location of the positioning device for starting the wireless charging system can be selected according to the system characteristics.

Acknowledgement

This work was supported in part by National key R & D project (NO.2018YFB0106300), in part by the National Natural Science Foundation of China (51877036), in part by the Nature Science Youth Foundation of Jiangsu Province (BK20150617), in part by the Scientific Research Foundation of Graduate School of Southeast University (YBJJ1728), and in part by the Postgraduate Research & Practice Innovation Program of Jiangsu Province (KYCX17-0088).

References

[1] B. Wang, A. P. Hu and D. Budgett. *Maintaining middle zero voltage switching operation of parallel-parallel tuned wireless power transfer system under bifurcation [J]*. IET Power Electronics, **7**, 1, (2014): 78-84.

[2] D. Ahn and S. Hong. *Wireless Power Transmission With Self-Regulated Output Voltage for Biomedical Implant [J]*. IEEE Transactions on Industrial Electronics **61**, 5, (2014) : 2225-2235.

[3] D. Kosmanos et al. *Route Optimization of Electric Vehicles Based on Dynamic Wireless Charging [J]*. IEEE Access, **6**, (2018): 42551-42565.

[4] L. Tan et al. *A Segmented Power-Efficiency Coordinated Control Strategy for Bidirectional Wireless Power Transmission Systems With Variable Structural Parameters [J]*. IEEE Access, **6**, (2018): 40289-40301.

[5] I. Karakitsios, F. Palaiogiannis, A. Markou and N. D. Hatziargyriou. *Optimizing the Energy Transfer, With a High System Efficiency in Dynamic Inductive Charging of EVs [J]*. IEEE Transactions on Vehicular Technology, **67**, 6, (2018): 4728-4742.

[6] A. A. S. Mohamed, S. An and O. *Mohammed. Coil Design Optimization of Power Pad in IPT System for Electric Vehicle Applications [J]*. IEEE Transactions on Magnetics, **54**, 4, (2018): 1-5.

[7] R. Ruffo, V. Cirimele, M. Diana, M. Khalilian, A. L. Ganga and P. Guglielmi. *Sensorless Control of the Charging Process of a Dynamic Inductive Power Transfer System With an Interleaved Nine-Phase Boost Converter [J]*. IEEE Transactions on Industrial Electronics, **65**, 10, (2018): 7630-7639.

[8] V. T. Nguyen, S. Yu, S. Yim and K. Park. *Optimizing compensation topologies for inductive power transfer at different mutual inductances [C]*. 2017 IEEE PELS Workshop on Emerging Technologies: Wireless Power Transfer (WoW). (2017): 153-156.

[9] Y. Cheng, G. Chen, G. Qian, M. S. E. Sendi, M. Ghovanloo and G. Wang. *Optimizing three-phase three-layer coil array for omnidirectional wireless power transfer [C]*. 2017 IEEE Wireless Power Transfer Conference (WPTC). (2017): 1-3.

[10] M. Bojarski, E. Asa, K. Colak and D. Czarkowski. *Analysis and Control of Multiphase Inductively Coupled Resonant Converter for Wireless Electric Vehicle Charger Applications [J]*. IEEE Transactions on Transportation Electrification. **3**, 2, (2017): 312-320.

[11] S. Samanta and A. K. Rathore. *Small-Signal Modeling and Closed-Loop Control of a Parallel–Series/Series Resonant Converter for Wireless Inductive Power Transfer [J]*. IEEE Transactions on Industrial Electronics. **66**, 1, (2019): 172-182.

Research on the PFC Mechanism of Forward-Flyback Converter

Liu shulin[1], Hu chuanyi[1], Cao jian[1],Yan jizhi[1] ,Wu hao[1]

[1] College of Electrical and Control Engineering, Xi'an University of Science and Technology, Xi'an, China

e-mail:416877367@qq.com

Abstract. The forward-flyback converter is used in power factor correction to improve the performance of conventional forward PFC and flyback PFC converters. By analyzing the working principle of a single-transistor single-transformer forward-backward combined PFC converter in one power frequency cycle, it is pointed out that there are two working modes—flyback mode and forward flyback mode; through its PFC mechanism research. The relationship between the duty cycle of each operating mode when achieving unity power factor is obtained, and the duty cycle when operating in flyback mode is the maximum duty cycle within a power frequency cycle. Finally, the open-loop control model of the forward-flyback PFC converter is established by using MATLAB. The duty cycle is changed according to the specified rules and the PFC is implemented, which verifies the correctness of the theoretical analysis.

1. Introduction

Single-phase active power factor correction (PFC) technology can greatly improve the power factor of the uncontrolled rectifier circuit, effectively reduce the harmonic content of the grid-side current, and it is an important means to improve power quality [1].

Flyback PFC converters have the advantages of input and output isolation, simple structure and low cost, it have become one of the most commonly used single-stage PFC converters [2-5]. However, due to transformer leakage inductance and power device stress and other factors, it is mainly used for applications below 100W [6].

The forward PFC converter has the characteristics of isolate the input and buck output and it has a high energy transmission efficiency [7]. However, due to its inherent structure, it is necessary to add a magnetic reset circuit. When the instantaneous input voltage is low, the converter cannot achieve energy transfer, a dead zone with zero input current occurs, and the dead time is long at low voltage input, causing its cannot satisfy the harmonic requirements of IEC61000-3-2 [8].

The forward-flyback converter can work in both forward and flyback modes and significantly improve the circuit performance. It has been widely studied and concerned by people in the industry[9-11]. It works in flyback mode when the input voltage instantaneous value is low in PFC, the unity power factor can be obtained [12] and the input current has no dead zone. The transformer core not only resets naturally, but also transfers the magnetizing energy to the load with the high energy transmission efficiency [13].The documents [6-7] studied the forward-flyback PFC converters with dual secondary windings in resonant control, the performance of the forward and flyback parts of the converter are compared in critical mode (CRM), respectively. It was pointed out that when the flyback

Content from this work may be used under the terms of the Creative Commons Attribution 3.0 licence. Any further distribution of this work must maintain attribution to the author(s) and the title of the work, journal citation and DOI.
Published under licence by IOP Publishing Ltd

part works in CRM, it can obtain higher energy transmission efficiency and power factor, and its parameter design methods were summarized. Reference [12] compares the input current distortion degree and power factor of the forward-flyback combined PFC converter with constant duty cycle and variable duty cycle control, and points out that the former can obtain the best working performance. At present, a variety of forward-back converter converter configurations have been proposed [14-16]. Among which the single-transistor single-transformer configuration [16] is the simplest, and it has higher research value with the advantages of dual-transformer structure and double-secondary-side winding structure.

Therefore, the forward-flyback application in PFC was studied, the working mode of the converter in a power frequency cycle and the mechanism of unity power factor were analyzed. Finally, the simulation of an open-loop control model verified the correctness of theoretical analysis.

2. Work mode analysis

The main circuit topology of a single-tube single-transformer forward-flyback combined PFC converter is shown in Figure 1. Its forward part can only work in current discontinuous conduction mode (DCM). The main working waveform in a power frequency cycle is shown in Figure 2, where: v_g and i_g are the input voltage and input current after rectification and filtering; $i_{o,for}$ and $i_{o,fly}$ are the output currents of the forward and flyback sections, respectively.

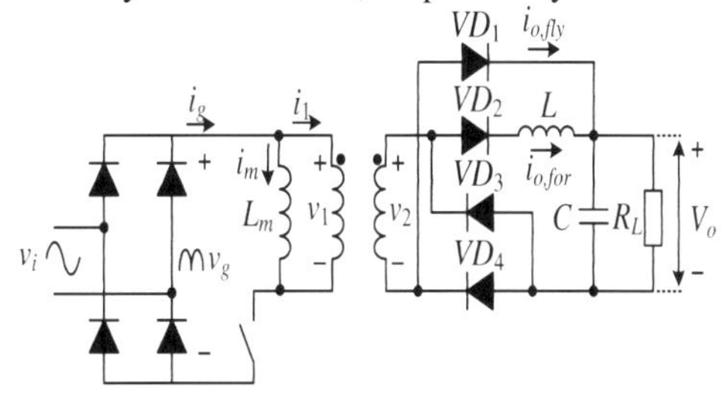

Figure 1. Forward-flyback PFC converter

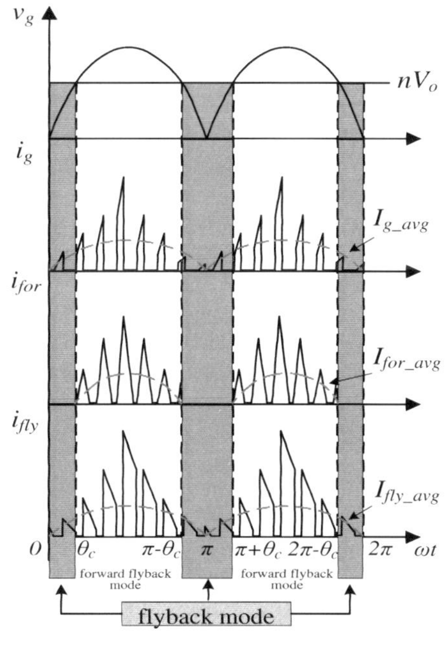

Figure 2. Waveform of forward-flyback PFC converter

It can be seen that the converter has two operating modes: flyback mode and forward-flyback mode.

Flyback mode: When the input voltage v_g is lower than nVo, the diodes VD2 and VD4 are reversed during the switch on period, and the forward part cannot work. At this point, only the flyback part powers the load, and the converter shows the operating characteristics of the flyback PFC converter.
Forward-flyback mode: When the input voltage v_g is higher than nVo, both the forward and flyback parts can provide energy for the load.

3. Power factor correction

The mechanism of power factor correction implemented by the forward-flyback converter is analyzed following. To simplify the analysis, the following assumptions is made first.
(1) All devices are ideal, regardless of the effects of parasitic parameters.
(2) The switching frequency is much larger than the power frequency. In a switching cycle, the input voltage can be regarded as a constant value approximately.
(3) The output filter capacitor is large enough that the output voltage remains constant during one power cycle during steady-state operation.
In order to ensure the output voltage stability and obtain the unity power factor, the following relationship should be satisfied.

$$\begin{cases} v_i(\theta)=\sqrt{2}V_{in\_rms}\sin\theta \\ i_i(\theta)=\sqrt{2}I_{in\_rms}\sin\theta \\ v_o(\theta)=V_o \end{cases} \quad (1)$$

Where $v_i(\theta)$ is the AC input voltage; $i_i(\theta)$is the AC input current; $v_o(\theta)$is the output voltage; V_{in_rms} is the AC input voltage rms value; I_{in_rms} is the AC input current rms value.
According to instantaneous power conservation, the following equation can be obtained:

$$\begin{aligned} P_o(\theta)&=v_o(\theta)i_o(\theta)=V_o i_o(\theta) \\ &=P_{in}(\theta)=2V_{in\_rms}I_{in\_rms}\sin^2\theta \end{aligned} \quad (2)$$

Among them,$p_o(\theta)$ is the output average power in one switching period; $p_{in}(\theta)$ is the instantaneous input power; $i_o(\theta)$is the average output current in one switching period.
Therefore, the following equation can be obtained according to equation (2).

$$i_o(\theta)=2\frac{V_{in\_rms}}{V_o}I_{in\_rms}\sin^2\theta=i_{o,for}(\theta)+i_{o,fly}(\theta) \quad (3)$$

A. Flyback mode power factor correction mechanism
The input power frequency sinusoidal AC voltage rectified by the diode rectifier bridge can be expressed as:

$$v_g(\theta)=\sqrt{2}V_{in\_rms}|\sin\theta| \quad (4)$$

Let $vg(\theta)=nVo$, and the critical phase angle θ_c of the converter operating in positive-flyback mode is:

$$\theta_c=\arcsin\frac{nV_o}{\sqrt{2}V_{in\_rms}} \quad (5)$$

In the formula, n is the transformer turns ratio; Vo is the direct current output voltage.
When $(n-1)\pi+\theta_c<\theta<n\pi-\theta_c$, the converter operates in the forward-flyback mode; otherwise it operates in the flyback mode.
When the converter works in flyback mode, the flyback part first works in DCM as $v_g(\theta)$ rises from zero. According to the relationship between the input voltage and output voltage, the following equation can be obtained:

$$i_o(\theta)=i_{o,fly}(\theta)=\frac{v_g(\theta)^2 d(\theta)^2}{2L_m V_o f} \quad (6)$$

where, $d(\theta)$ is the switch conduction ratio (duty cycle); Lm is the transformer magnetizing inductance; f is the switching frequency.
The equation (3) can be brought into the above formula. To achieve unity power factor, the duty cycle $d(\theta)$ should be

$$d_1 = \sqrt{2fL_m \frac{I_{in\_rms}}{V_{in\_rms}}} \tag{7}$$

It can be seen that the duty cycle is constant as the instantaneous value of the input voltage increases. With the increase of $v_g(\theta)$, if the flyback part enters continuous current mode (CCM), according to the relationship between the input voltage and output voltage, the duty cycle $d(\theta)$ should be changed according to the following formula.

$$d_2(\theta) = \frac{nV_o}{V_g(\theta) + nV_o} \tag{8}$$

The expression of the output current of the converter can be expressed by the formula (3) at this point, and the unity power factor also can be achieved.
B. Forward-flyback mode power factor correction mechanism
The converter operates in the forward-flyback mode when $v_g(\theta) > nVo$. During the on-time of the switch, the current through the inductor L rises linearly, and the peak current is:

$$I_{LP}(\theta) = \frac{v_g(\theta) - nV_o}{nLf} d(\theta) \tag{9}$$

During the off-time of the switch, the time required for the inductor current to drop to zero is $t_d(\theta)$.

$$t_d(\theta) = \frac{I_{LP}(\theta)}{V_o / L} = \frac{v_g(\theta) - nV_o}{nfV_o} d(\theta) \tag{10}$$

Therefore, the average output current of the forward part of a switching cycle is:

$$\begin{aligned} i_{o,for}(\theta) &= \frac{1}{2T}\left[t_d(\theta) + d(\theta)T\right] I_{LP}(\theta) \\ &= \frac{d^2(\theta) V_{in\_rms}^2 \sin^2\theta}{Lf}\left(\frac{1}{n^2}\cdot\frac{1}{V_o} - \frac{1}{n}\cdot\frac{1}{\sqrt{2}V_{in\_rms}\sin\theta}\right) \end{aligned} \tag{11}$$

If the forward part and the flyback part both work in DCM at the same time, the average output current of the converter in one switching cycle is:

$$i_o(\theta) = i_{o,fly}(\theta) + i_{o,for}(\theta) \tag{12}$$

Equations (6) and (11) can be brought into the above formula to obtain formula (13).

$$i_o(\theta) = \frac{d^2(\theta) V_{in\_rms}^2 \sin^2\theta}{L_m f}\left[\frac{1}{V_o}\left(1 + \frac{L_m}{n^2 L}\right) - \frac{L_m}{nL}\cdot\frac{1}{\sqrt{2}V_{in\_rms}\sin\theta}\right] \tag{13}$$

Conclusions are drawn combine the formulas (3) and (13).when the forward part and the flyback part both work in the DCM at the same time, the duty ratio of the unity power factor should be changed by the following relationship.

$$d_3(\theta) = \frac{d_1}{\sqrt{1 + \frac{1}{n^2}\cdot\frac{L_m}{L} - \frac{1}{n}\cdot\frac{L_m}{L}\cdot\frac{V_o}{\sqrt{2}V_{in\_rms}\sin\theta}}} \tag{14}$$

It can be seen that when $\theta = \theta_c, d_3(\theta_c) = d_1$,and when $\pi - \theta_c > \theta > \theta_c$, $d_\theta < d_{FB}$. Therefore, d_1 is the maximum duty cycle in this mode.

When the flyback enters the CCM as $v_g(\theta)$ rises, the relationship between the input voltage and the output voltage is:

$$V_o = \frac{d(\theta)}{1-d(\theta)} \cdot \frac{\sqrt{2}V_{in\_rms} \sin \theta}{n} \tag{15}$$

and its duty cycle changes according to the following rules at this time:

$$d_4(\theta) = \frac{nV_o}{v_g(\theta)+nV_o} \tag{16}$$

The expression of the output current is shown in equation (3).It also enables unity power factor. Synchronous formulas (14) and (16) yield the following formula.

$$\frac{nV_o}{\sqrt{2}V_{in\_rms} \sin \theta_d + nV_o} = \sqrt{\frac{2fL_1I_i}{V_{in\_rms}\left[1+\frac{L_1}{n^2L}-\frac{V_oL_1}{\sqrt{2}nV_{in\_rms}L \sin \theta_d}\right]}} \tag{17}$$

According to the above formula, the critical angle θ_d of the flyback part entering the CCM can be obtained.

4. Example and simulation analysis

An example circuit of 24V/100kHz is designed to verify the theoretical analysis aforementioned and use MATLAB performs simulation analysis. The circuit parameters are shown in Table 1.

Table 1. Parameters table

Parameter	***Setting value***
Input voltage(v_i)	$220V_{ac}$
Output voltage(V_o)	$24V_{dc}$
Output current (I_o)	3A
Switch frequency (f)	100kHz
Inductance (L)	40uH
Secondary inductance (L_2)	200uH
Transformer ratio (n)	2

For simplicity, the AC input voltage and the full-bridge rectification part are directly operated by using a sine signal source, and the output is connected to a voltage control source to provide power for the component model. To make the simulation closer to the ideal situation, use an ideal switch instead of a power switch and set the diode's turn-on voltage drop to zero, and replace the output filter capacitor with an ideal voltage source in order to ignore the effects of twice power-frequency ripple. The simulation circuit is shown in Figure 3.

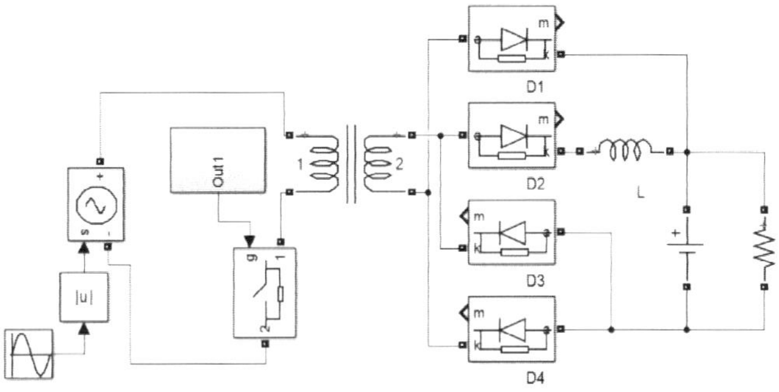

Figure 3. Circuit of simulation

According to the formulae (5) and (17), the critical phase angles of the converter operating in the forward- flyback mode and the flyback part entering the CCM are: θc=8.875° and θd=31.143°. In order to verify the PFC mechanism of the converter, open-loop control was used for simulation analysis. The open-loop control model is shown in Figure 4.

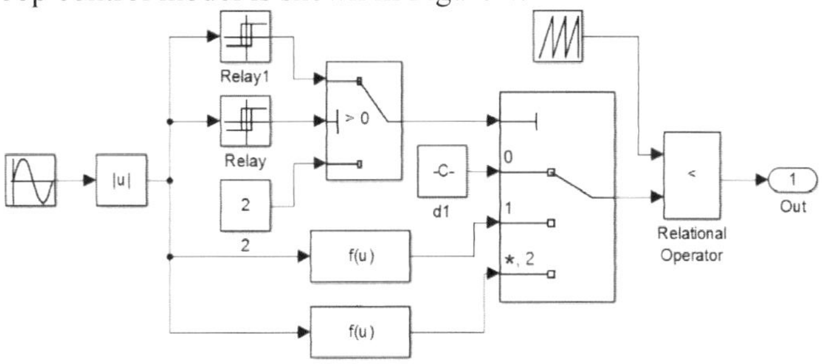

Figure 4. open-loop control model

According to the above control method, the duty cycle of the inverter operating in the flyback mode is constant and its expression is as shown in Equation (7); when the converter is operating in the forward-flyback mode, the duty cycle of the flyback part operates in the DCM and CCM is shown as the formulas (14) and (16), respectively. The simulation results are shown in Figure 5.

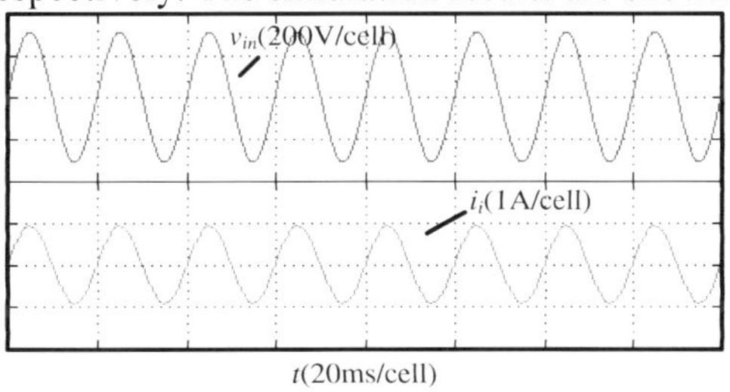

(a) Input voltage and current waveform

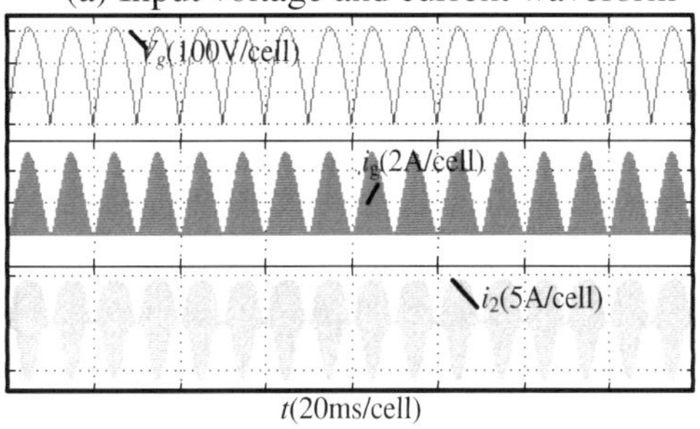

(b) Converter input voltage/current waveform and transformer secondary current waveform

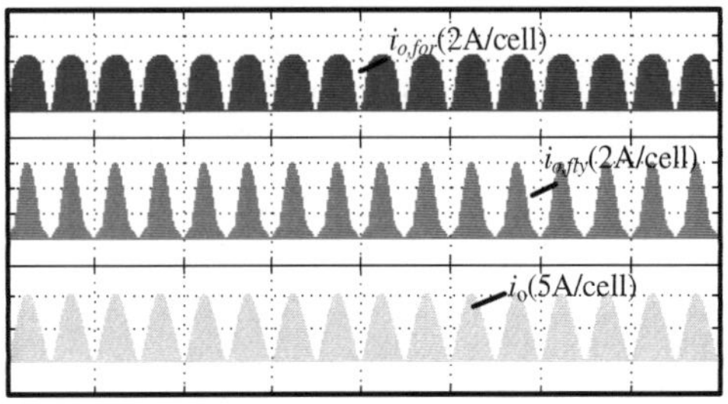

(c) Output current of the forward and flyback sections and the output current waveform of the converter

Figure 5. Waveform of simulation

The simulation results are consistent with the theoretical analysis. The duty cycle changes according to the specified rule in each mode of the converter, and the input current follows the input voltage, and the unity power factor can be achieved.

5. Conclusion

The forward-flyback converter combines the characteristics of the forward and flyback converters. There are two working modes of the converter during one power frequency cycle when applied to PFC: the flyback mode and the forward-flyback mode. And unity power factor is realized as long as the duty cycle changed according to a specified law in each mode of the converter. Maximum duty cycle is obtained when the converter is operating in DCM of the flyback part in one power frequency cycle, and then the duty cycle gradually decreases as the input voltage rises.

Acknowledgment

Project Supported by National Natural Science Foundation of China (51777167， 51604217). Shaanxi province technical innovation guidance special plan project(2017CGZH-XNGJ-09).

References

[1] Zhang Yi, Lin Zhengqi, Zhang Junming et al. A Primary Control Strategy for Boost-Flyback Single-Stage PFC Converter[J]. Transactions of China Electrotechnical Society,2015,30(04):155-160+178.

[2] Li Y C, Chen C L. A novel primary-side regulation scheme for single-stage high-power-factor AC－DC LED driving circuit[J]. IEEE Transactions on Industrial Electronics, 2013, 60(11): 4978-4986.

[3] Yang Y, Ruan X, Zhang L, et al. Feed-forward scheme for an electrolytic capacitor-less AC/DC LED driver to reduce output current ripple[J]. IEEE Transactions on Power Electronics, 2014, 29(10): 5508-5517.

[4] Yan Tiesheng, Xu Jianping, Zhang Fei, et al. Analysis of output voltage ripple for flyback PFC converter[J]. Advanced Technology of Electrical Engineering and Energy, 2013,33(09):41-46(in Chinese).

[5] Yan Tiesheng, Xu Jianping, Zhang fei, et al. Variable On-time Controlled Critical Conduction Mode Flyback PFC Converter[J]. Proceedings of the CSEE, 2013,33(27):60-68+10(in Chinese).

[6] Li J, Xie X, Peng K, et al. Optimum design on the single-stage Forward-flyback PFC converter with two different QR controls[C]//Applied Power Electronics Conference and Exposition (APEC), 2015 IEEE. IEEE, 2015: 1869-1874.

[7] Xie X, Li J, Peng K, et al. Study on the single-stage forward-flyback PFC converter with QR

control[J]. IEEE Transactions on Power Electronics, 2016, 31(1): 430-442.
[8] Wang Nan, Xu Jianping, Liu Xueshan. A Single-Switch Buck-Flyback Power Factor Correction Converter[J]. Proceedings of the CSEE, 2015,35(03):679-687(in Chinese).
[9] Chen F, Hu H, Shen J, et al. Design and analysis for ZVS forward-flyback DC-DC converter[C]//Energy Conversion Congress and Exposition (ECCE), 2011 IEEE. IEEE, 2011: 116-121.
[10] Xie Xiaogao, Zhao Chen, Zheng Lingwei, Qian Zhaoming. A Third-winding and Current-driven Hybrid Driving Scheme for SRs of Active-clamped Forward-flyback Converters[J]. Proceedings of the CSEE，2012,32(33):31-36+172(in Chinese).
[11] Chen L, Hu H, Zhang Q, etal. A boundary-mode forward-flyback converter with an efficient active LC snubber circuit[J]. IEEE Transactions on Power Electronics, 2014, 29(6):2944-2958.
[12] Chen F, Amirahmadi A, Batarseh I. Zero voltage switching Forward-Flyback Converter with efficient active LC snubber circuit[C]//Applied Power Electronics Conference and Exposition (APEC), 2014 Twenty-Ninth Annual IEEE. IEEE, 2014: 2041-2047.
[13] Wang Guoli, Jin Xinmin. Research on a Flyback-Forward Converter[J]. Power Electronics, 2001, 35(2):4-6(in Chinese).
[14] Shi Jianjiang, Hong Feng, Yan Yangguang. Study of Two-Transistor Forward-Flyback Combined Converter[J]. Transactions of China Electrotechnical Society, 2003, 18(5):87-90(in Chinese).
[15] Chen F, Amirahmadi A, Batarseh I. Zero voltage switching Forward-Flyback Converter with efficient active LC snubber circuit[C]//Applied Power Electronics Conference and Exposition (APEC), 2014 Twenty-Ninth Annual IEEE. IEEE, 2014: 2041-2047.
[16] Himmelstoss F A, Votzi H L. Comparison of a new combined four-diode forward-flyback converter with the classical forward converter[C]//Signals, Circuits and Systems, 2009. ISSCS 2009. International Symposium on IEEE, 2009:1-4.

Multi-sensor cross cueing technology based on grey wolf algorithm

WEI Daozhi[1],XIE Jiahao[2,a]
[1] Air and Missile Defense College, Air Force Engineering University, Xi'an 710051, China
[2] Graduate College, Air Force Engineering University, Xi'an 710051, China

[a]Corresponding author:18222517021@163.com

Abstract. In order to solve the problem of poor convergence and low precision of the current Multi-sensor Detection Network algorithm, a multi-sensor cross cueing technology based on Grey Wolf algorithm is proposed. By endowing the grey wolf with autonomous decision-making ability, the algorithm can jump out of the local optimal solution as soon as possible, and the convergence degree of the algorithm is improved. The simulation results show that the algorithm can find the optimal solution in a short time and maintain good stability, and the optimization ability is further strengthened. By comparing with other algorithms, it is more effective to prove that the algorithm has better solution quality.

1. Introduction

The formation of multi-sensor network belongs to the multi-objective assignment problem in multi-sensor management. The concept of multi-sensor cross cueing technology is systematically proposed in document [1].The application of this technology in target detection improves the reliability and timeliness of detection. In document [2], the discrete particle swarm optimization (DPSO) algorithm is proposed in multi-sensor management. In the application of multi-objective allocation, the algorithm has good stability; literature [3] proposes an algorithm based on colored Petri nets, which proves the superiority of the algorithm; literature [4] considers the asynchronism of communication, proposes a DTC algorithm for distributed tasks, and solves the sensor problem considering asynchronous communication of sensors. Resource management problem; document [5-6] mainly uses contract network and improved contract network algorithm to achieve multi-sensor multi-objective allocation. Although the convergence speed of the algorithm is faster, its solving ability needs to be further enhanced; document [7] proposes an energy-efficient distributed sensor management algorithm with good convergence characteristics; document [8] based on the theory of artificial bee colony algorithm, an improved artificial bee colony algorithm with stronger searching ability is proposed, and the convergence and robustness of the algorithm are validated effectively.

From the above literature, it can be concluded that in the current multi-sensor network formation algorithm, although the convergence and stability of the algorithm is good, the quality of the algorithm needs to be further strengthened, and the optimization ability of the algorithm is weak, the real-time performance and effectiveness are poor. In view of the above problems, a multi sensor network scheme based on improved grey wolf algorithm is proposed.

Content from this work may be used under the terms of the Creative Commons Attribution 3.0 licence. Any further distribution of this work must maintain attribution to the author(s) and the title of the work, journal citation and DOI.
Published under licence by IOP Publishing Ltd

2. The establishment of the model

Suppose that the Multi-sensor Network contains m sensors, and there are n targets at a certain time. Among them, at the moment k, the sensor network set up for the target t is recorded as. $S_t = \{s_1, s_2, ..., s_m\}$ Let the detection network C be a $m \times n$ one-order0-1 matrix, then we have:

$$c_{ij} = \begin{cases} 0, & \text{Sensors are not added to the dynamic network of the target} \\ 1, & \text{Sensors are added to the dynamic network of the target} \end{cases}$$

The detection accuracy A of the target by the sensor is a matrix of $m \times n$ order, which a_{ij} is the detection accuracy of the target t_j in sensor s_i. The accuracy of the sensor on the target is calculated from equation (1).

$$a_{ij} = \frac{n_{ij} t_{ij}}{NT} \tag{1}$$

In the formula, n_{ij} indicates the number of targets detected at a certain time, t_{ij} indicates the effective time at which the target is detected at a certain time, N indicates the total number of probe targets required by the network, and T indicates the total time of the target motion.

When using a sensor s_i to detect a target, the energy consumption is:

$$\cos t(s_i) = -\exp(-\frac{l(s_i)}{\alpha_1 \times r(s_i) + \beta_1}) \tag{2}$$

Where α_1 and β_1 are constants, take $\alpha_1 = 0.01$, $\beta_1 = 0.1$, $l(s_i)$ is the number of targets that can be detected at the same time, and $r(s_i)$ is the farthest detection distance.

When establishing the objective function of the multi-sensor detection network model, it should be made to meet the total detection accuracy A of the multi-sensor detection network C, and the maximum total energy consumption $COST$ of the network at the same time is the minimum. Therefore, the following objective function can be obtained:

$$\max A = \max \left\{ \sum_{i=1}^{m} \sum_{j=1}^{n} a_{ij} \right\} \tag{3}$$

$$\min COST = \min \left\{ \sum_{i=1}^{m} \sum_{j=1}^{n} \cos t_{ij} \right\} \tag{4}$$

The constraints are:

$$\sum_{j=1}^{n} c_{ij} \le M_i \tag{5}$$

$$\sum_{i=1}^{m} c_{ij} \ge 1 \tag{6}$$

The evaluation index of the multi-sensor detection network refers to the value of the optimal solution of the multi-sensor detection network, that is, the fitness value. The fitness value should be proportional to the total target detection accuracy A and inversely proportional to the maximum total energy consumption $COST$, so it is used to represent the most of the detection network. The value of the optimal solution has the following formula:

$$\Phi(X) = \frac{\sum_{i=1}^{m} \sum_{j=1}^{n} a_{ij}}{\sum_{i=1}^{m} \sum_{j=1}^{n} \cos t_{ij}} \tag{7}$$

3. Grey wolf algorithm

The hunting behaviour of grey wolf population mainly includes following, pursuing and approaching the prey; hunting and encircling the prey until the prey no longer moves. The mathematical models surrounding prey are as follows:

$$D = \left| C g X_P^t - X^t \right| \tag{8}$$

$$X^{t+1} = X_P^t - A g D \tag{9}$$

Where D indicates the distance between the prey and the wolf, t indicates the current iteration, X_P^t indicates the position vector of the prey, X_P indicates the position vector of the wolf, A and C are the reference vectors. $A = 2agr_1 - a$, $C=2gr_2$, where a decreases from 2 linear to 0 in the iteration process, and r_1, r_2 is a random vector on [0,1].

Grey wolf population can use α、β、δ wolf position to determine the location of prey. The formula for updating its position is as follows:

$$D_i = \left| C_j g X_i^t - X^t \right| \left(i = \alpha, \beta, \delta; j = 1, 2, 3 \right) \tag{10}$$

$$X_j = X_i^t - A_j g D_i \left(i = \alpha, \beta, \delta; j = 1, 2, 3 \right) \tag{11}$$

$$X^{t+1} = \frac{X_1 + X_2 + X_3}{3} \tag{12}$$

To illustrate the above formula, figure 1 gives a graphical representation of the location of the grey wolf population.

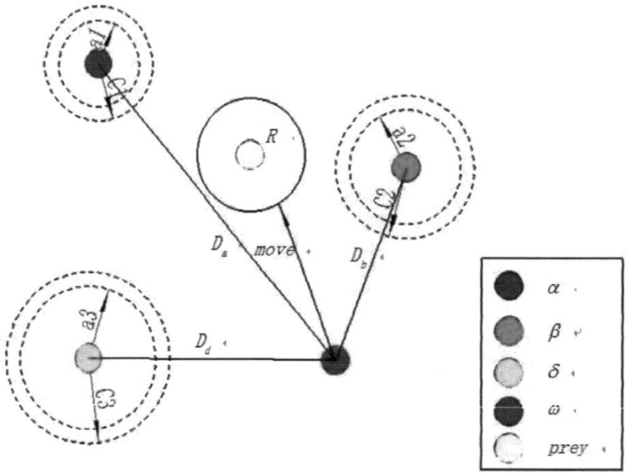

Figure 1. The process of own position updating

The realization of basic Grey Wolf algorithm mainly depends on changing the value to realize the change of parameters. In the process of decreasing the equivalent value from 2 to 0, the corresponding value also obtains any value in [-2a, 2a]. When $|A| \le 1$, the wolves besieged the prey. On the contrary, when $|A| > 1$, the wolves dispersed and continued to search for the optimal solution, and then they would be trapped away from the optimal solution. In the case of searching for local optimal solution, the location of α, β, δ wolves plays an important role in the grey wolf population, which may lead to the existence of wolves in a rush and scatter situation.

In view of the situation that the basic Grey Wolf algorithm proposed above is easy to make the optimization process fall into local optimal solution, a new improved grey Wolf algorithm is proposed. In grey wolf population, each wolf can follow the previous method to find the optimal solution, but at the same time each wolf can follow the changes of the external environment. Have the ability of independent decision-making, through the analysis of the position of wolves higher than their own level, real-time decision-making,change their position to achieve the search process of the optimal solution.

According to the classification of wolves based on the basic wolf swarm algorithm, wolves are divided into four levels. Assuming that the decision-making factor is D , the total number of wolves is recorded as R , the level L_i at which the first wolf is at the moment t , the decision-making model of each level is given as follows:

$$D_i = \frac{R - L_i}{R - 1} \tag{13}$$

The procedure of the algorithm is shown below.

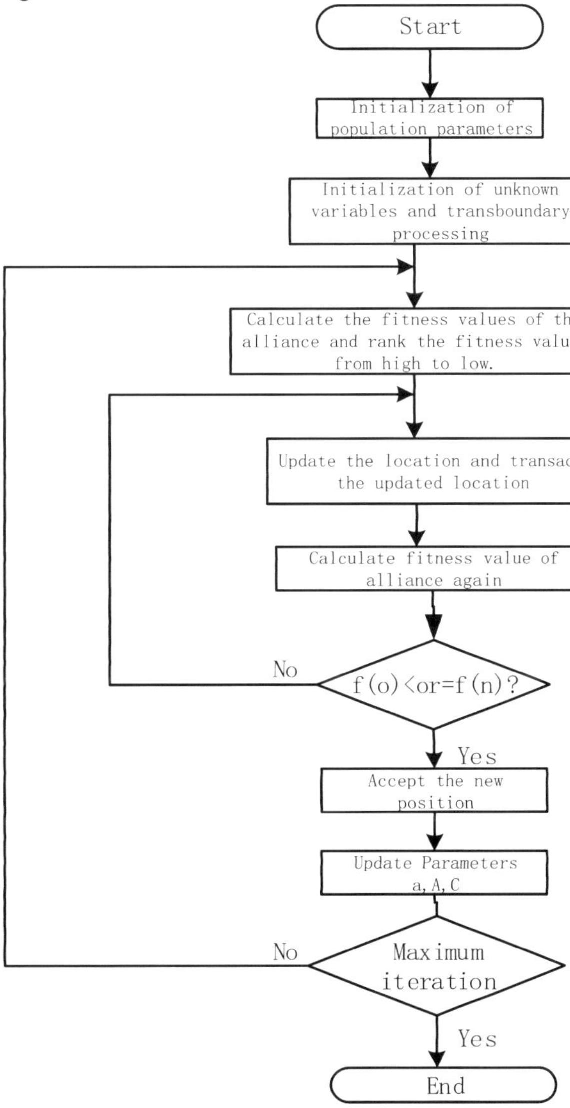

Figure 2. The procedure of the algorithm

4. Simulation

Assuming that there are 10 sensors in a sensor network and eight enemy targets are detected at the same time, the detection capability of each sensor is given below, as shown in Table 1.

Table 1. Detecting precision of sensors to targets

	T1	T2	T3	T4	T5	T6	T7	T8	Level	Ability	Cost
S1	0.981	0.031	0.413	0.386	0.987	0.659	0.678	0.614	0.05	8	0.03
S2	0.439	0.861	0.543	0.471	0.259	0.546	0.757	0.605	0.11	6	0.03
S3	0.171	0.196	0.969	0.986	0.071	0.800	0.743	0.127	0.08	5	0.02
S4	0.008	0.231	0.074	0.163	0.015	0.035	0.392	0.713	0.21	8	0.04
S5	0.277	0.093	0.047	0.944	0.499	0.649	0.655	0.632	0.15	10	0.02
S6	0.442	0.915	0.275	0.462	0.212	0.234	0.171	0.097	0.12	8	0.03
S7	0.040	0.162	0.617	0.942	0.802	0.970	0.600	0.757	0.07	12	0.04
S8	0.051	0.353	0.397	0.330	0.569	0.915	0.147	0.764	0.06	12	0.02
S9	0.043	0.081	0.706	0.888	0.329	0.934	0.042	0.157	0.07	8	0.03
S10	0.655	0.596	0.387	0.031	0.516	0.171	0.715	0.670	0.08	8	0.02

4.1. Comparison of optimal solutions of coalition schemes before and after algorithm improvement

The grey wolf algorithm and the improved grey wolf algorithm are used to calculate the optimal solution[9-10] of the multi-sensor network, and 100 Monte Carlo experiments are carried out. The results are shown in Figure3 and the alliance scheme is shown in Table 2.

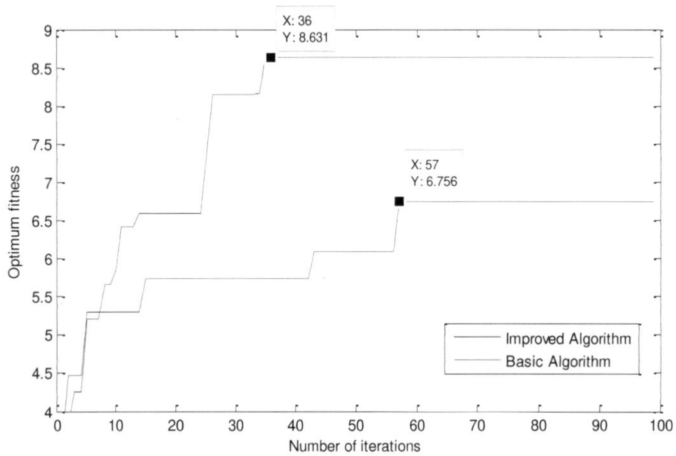

Figure 3. The contrast of the former and the latter

Table 2. Multi-sensor detective coalition cases

	Basic Algorithm			Improved Algorithm		
Target	Sensor	Accuracy	Cost	Sensor	Accuracy	Cost
1	1、2、4、5	0.985	0.15	1、5	0.858	0.05
2	2、4	0.893	0.07	3、6	0.920	0.05
3	1、3	0.863	0.05	3、6	0.886	0.05
4	3	0.655	0.04	3	0.655	0.04
5	3、5	0.911	0.04	2	0.757	0.02
6	2、3、5	0.743	0.02	5	0.800	0.04
7	2、5	0.632	0.03	3	0.655	0.03
8	2、4、6	0.785	0.02	4、6	0.756	0.03

From Figure 3, it can be seen that the improved grey wolf algorithm and grey wolf algorithm can effectively solve the optimal solution of multi-sensor network. The improved grey wolf algorithm achieves stability in 36 iterations. The optimal solution of the alliance is 8.631. The grey wolf algorithm achieves stability after 57 iterations and the optimal fitness of the alliance is 6.756. The simulation results show that the improved grey wolf algorithm has better convergence than the improved grey wolf algorithm and detects the alliance. The coalition fitness is higher, which shows that the improved grey wolf algorithm has improved the search ability.

4.2. Comparison of different algorithms

The improved wolf swarm algorithm proposed in this paper is compared with artificial bee swarm algorithm and particle swarm optimization algorithm. Fifty Monte Carlo experiments are carried out to study the influence of different algorithms on the optimal solution of network scheme. The experimental results are shown in Figure 4 below.

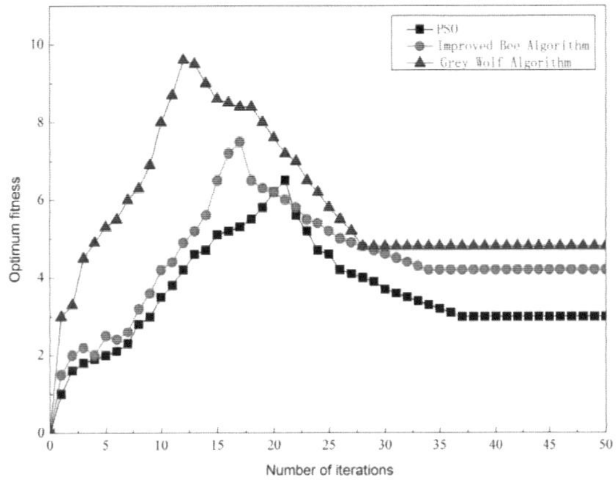

Figure 4. The comparison of three algorithms

The simulation results of figure 4 shows that the convergence of the three algorithms is very good. The algorithm in this paper achieves the optimal coalition solution after 13 iterations, with the fitness value of 9.633, and the running time of the algorithm is the shortest; the improved bee colony algorithm achieves the optimal coalition solution after 18 iterations, with the fitness value of 7.515; and the particle swarm optimization algorithm runs the longest. After 21 iterations, the optimal solution is obtained, but the coalition fitness is the lowest at this time. This shows that the algorithm has short running time, fast convergence speed and stability, which can make the detection coalition maintain a better fitness value.

5. Conclusion

This paper proposes a multi sensor network scheme based on Improved Grey Wolf algorithm. Firstly, the target observation model is established. Secondly, the multi-sensor target detection model is established and the objective function and evaluation function of the alliance are put forward. Through the improvement and optimization of the basic grey wolf algorithm, the grey wolf has the right of autonomous decision-making to jump out of the local optimal solution and enhance the optimization degree of the algorithm to the optimal solution of the alliance. The Improved Grey Wolf algorithm has the advantages of simple design idea, high practicability, high solution quality and good convergence.

Acknowledgments

This work was supported by the National Natural Science Foundation of China [61273275,61603410] and Young Talent Fund of University Association for Science and Technology in Shannxi, China[20170107] .

References

[1] D D, R C A. Proceedings of the 2007 American Control Conference. 2007: 5406-5411.
[2] DP, CP. Service orientation in holonic and multi-agent manufacturing and robotics studies in

computational intelligence, 2014, 5 (44): 81-95.
[3] A, G. International Journal on Software Tools for Technology Transfer, 2016, (17): 1-22.
[4] Z. Li, C. Liu, F. Meng, K. Zhou, Proc. Inst. Mech. Eng. C 227, 1481 (2013)
[5] S. Li, Z. Liu, IEEE Trans. Ind. Electron. 56, 3050 (2009)
[6] H. Wu, J. Hu, Y. Xie, Characteristic model-based intelligent adaptive control (China Science and Technology Press, Beijing, (2009)
[7] S. Li, H. Du, X. Yu, IEEE Trans. Auto. Ctrl. 59, 546 (2014)
[8] M S, M S M, L A. Advances in Engineering Software, 2014, 69(3) : 46-61.
[9] K G M, K V. Journal of Computational Science, 2015, 8: 109-120.
[10] SS, M S Z, MS M. Neural Computing and Applications, 2015, 26(5): 1256-1263.

A New Principle of Breaker Failure Protection Based on Station Information Sharing

Yong Pan, Wei Han, Jinfeng Geng, Yao Liu

The Electrical Power Research Institute of Henan Electric Power Company, Zhengzhou, 450052, China

The corresponding author's e-mail address: 540327446@qq.com

Abstract. In the case of terminal fault of long line, the failure protection of traditional circuit breaker has the problem of refusing to operate due to the insufficient sensitivity of voltage locking elements. A new circuit breaker failure protection scheme based on station area information sharing is proposed in this paper. The new scheme will replace the protection export signal with the actual protection criterion. In addition, it sets up two current transformers to detect whether the breaker is cut off after the circuit breaker is cut off, and checks the redundant current of the bus, which can improve the accuracy of the current detection. Besides, it uses reclosing checkout voltage device signal to supplement the original compound blocking signal. Theoretical analysis and simulation results show that the scheme can effectively cover the full length of the transmission line and has high reliability.

1. Introduction

Circuit breaker failure protection, as a backup protection after protection action, it cannot remove faults due to various reasons[1-4]. At present, line breaker failure protection consists of voltage blocking elements, starting circuit composed of protection action and current discrimination, time element and trip outlet circuit. In order to avoid the protection outlet contacts being jammed and not returned, and to start the failure protection by mis-touching or mis-energizing the terminals, the voltage element is usually used as the locking element in the action logic of the failure protection[5-8]. However, with the growing strength of the power system, more and more substations are connected with the system. When the metal short circuit fault occurs at the end of the long line of the substation, the bus voltage is also difficult to decrease, which will lead to the failure protection being refused to operate due to the inadequate sensitivity of the compound voltage blocking element.

At present, there is no effective and thorough solution to the problem of insufficient sensitivity of compound voltage blocking for circuit breaker failure protection[9-10].Therefore, exploring a new and reliable criterion of circuit breaker failure protection, getting rid of the dependence of using single electrical quantity of voltage or current as blocking element, and thoroughly resolving the risk of rejection of current failure protection due to sensitivity of blocking element, will be important for protecting electrical equipment and maintaining stable operation of the system.

With the development of smart substation, station protection is becoming more and more mature, and the acquisition of various electrical quantities in the station becomes easier[11-15]. On this basis, a new circuit breaker failure protection scheme based on station and information sharing is proposed in this paper. The new scheme will replace the protective export signal in the criterion with the actual

Content from this work may be used under the terms of the Creative Commons Attribution 3.0 licence. Any further distribution of this work must maintain attribution to the author(s) and the title of the work, journal citation and DOI.
Published under licence by IOP Publishing Ltd

protection criterion. In addition, two current transformers are set up to detect whether the breaker is cut off, and the redundant current of the bus is used to check. It uses reclosing checkout voltage device sigal to supplement the original compound blocking signal.

2. Analysis of failure scenario of breaker failure protection

The malfunction of the breaker failure protection will have serious consequences. To prevent it from malfunctioning, the voltage component should be used as the blocking component in the action logic of the failure protection. Compound voltage blocking component ports consisting of phase low voltage, zero-sequence voltage and negative-sequence voltage or logic are generally used. This method is proved to be simple and reliable by practice and can effectively prevent maloperation of failure protection. However, with the increasingly strong power system, more and more substations are strongly connected with the system. When a metal short circuit fault occurs at the end of the long line of the station, the bus voltage is also difficult to drop, which will cause the failure protection to refuse to operate because of the inadequate sensitivity of the voltage locking elements. The substation with this problem is usually a hub substation or an important substation. Once the protection fails, it will lead to serious power grid accidents.

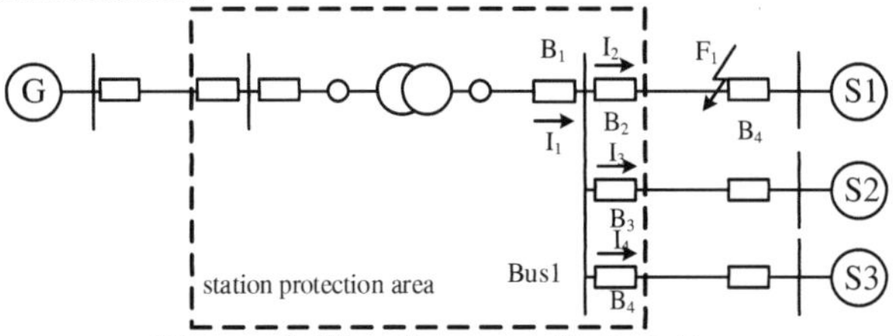

Figure 1. The diagram of substation

Case: As shown in Fig 1, when a fault occurs at point F1 at the end of a long outlet line in a substation, the pilot protection removes the circuit breakers B2 and B4. If the circuit breaker B2 fails, the circuit breaker B2 failure protection is started.

The current circuit breaker failure protection needs to be blocked by Bus-1 compound voltage value. When the terminal of the long line fails, the Bus-1 voltage changes little, so the failure protection cannot be opened. At this time, it needs to wait for the upper level zero-sequence current III or IV protection to be removed. The removal time is longer, which has a greater impact on equipment safety. In order to solve the problem of insufficient sensitivity of compound voltage blocking, the current countermeasures are mainly carried out from two aspects: A and B.

A. Reduce the setting value of compound voltage blocking elements

According to the regulation, the setting value of the compound voltage locking element should synthetically ensure that there is enough sensitivity when any terminal of the line connected by the busbar fails. The negative sequence voltage and the zero sequence voltage should reliably avoid the unbalanced voltage under normal conditions, so the space of reducing the voltage setting value is limited and the problem cannot be solved completely. In addition, the low value setting will reduce the reliability of failure protection and cause the possibility of misoperation.

B. Increase the function of reopening and locking

For lines with insufficient voltage sensitivity, voltage locking can be relieved by adding protective action contacts. The essence of this method is to cancel the voltage lock in a short time, which can effectively guarantee the failure protection to be open. However, this method cannot solve the problem of non-six-unit failure protection, for which the additional secondary circuit will also leave operating hidden trouble with the change of system structure.

Therefore, this paper will explore a new and reliable circuit breaker failure protection blocking principle, get rid of the dependence of the single electrical quantity using voltage as a blocking component, and completely solve the current failure protection risk due to the sensitivity problem of the blocking component.

IOP Conf. Series: Materials Science and Engineering 533 (2019) 012041 doi:10.1088/1757-899X/533/1/012041

3. New scheme for breaker failure protection

With the development of smart substation, the station area protection is becoming more and more mature, and the acquisition of electrical quantities in the substation becomes easier. Fig 2. shows the schematic diagram of station-area information sharing in 220 kV substation. The transformer is configured in a dual way, and the checkout voltage signal of reclosing check is uploaded to the station control layer via intra-station Ethernet.

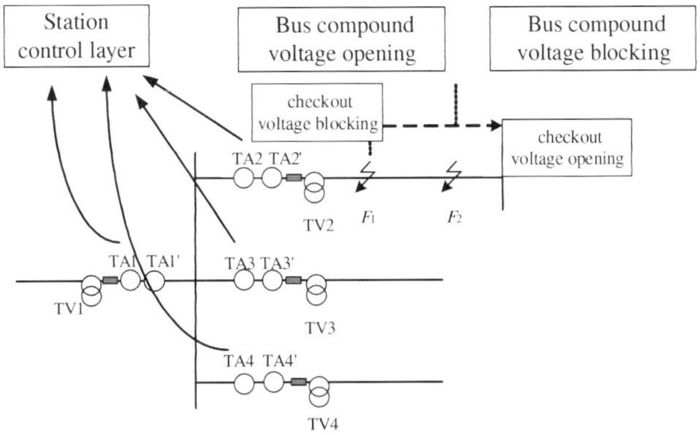

Figure 2. The diagram of substation area information sharing

This scheme considers replacing the protection exit signal in the failure protection starting criterion with the actual protection criterion judgment signal. Two current transformers are set up to detect the current or no-current of the circuit breaker. At the same time, based on the Kirchhoff current law, the bus redundant current is used for verification, which can improve the accuracy of the current and no-current detection. The checkout voltage of reclosing non-voltage device is used to supplement the original double-voltage locking signal, which greatly improves the reliability of failure protection. Fig 3. shows the logic of the new scheme for breaker failure protection, which consists of three parts: A~C.

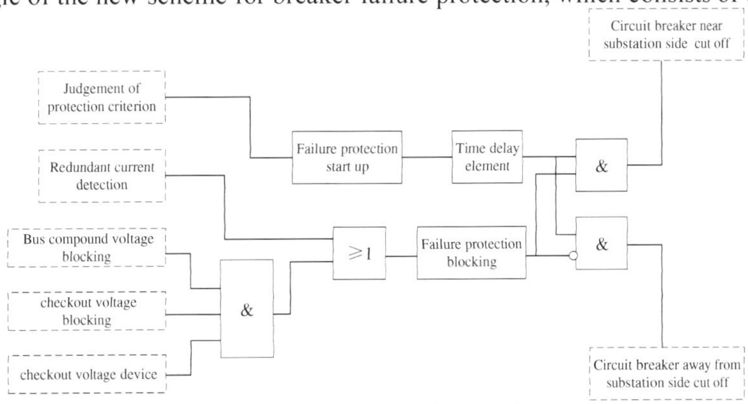

Figure 3. The logic of new schem

A. Protection criteria judgment section

In the station control layer, the local protection criterion is completely copied, and the same protection principle, protection logic and setting value are adopted. Based on the collected local information, the local protection criterion is calculated to generate the action signal of the local protection outlet. The scheme of local protection outlet calculated by software can effectively avoid the collision of hardware devices and abnormal signal of protection outlet caused by misoperation.

B. Redundant detection section with current

The scheme adopts the local protection criterion and delays Δt after startup, and then carries out detection whether the breaker is cut off. The setting of Δt should consider both the calculation time of the local protection criterion and the breaking time of the circuit breaker. In theory, if the circuit breaker is cut off, the current is 0. If the circuit breaker is not cut off, the current is a value greater than 0. By setting reasonable current threshold to detect whether the circuit breaker is cut off. In order to enhance the accuracy and reliability of the detection, two current transformers are arranged around each circuit breaker for current/no current detection, and the redundant current of busbar is used for checking. As shown in Figure 4, there are three redundant current information: I2, I2'and bus redundancy current (I1-I3-I4) for the current transformer TA2. Setting check error e is:

$$\begin{aligned} e &= I_2 - (I_1 - I_3 - I_4) \\ e' &= I_2' - (I_1 - I_3 - I_4) \end{aligned} \tag{1}$$

The current corresponding to the smaller value in the formula 1 is the result of the judgement of whether there is a current or not. If the protection criterion detects a current when it lasts for Δt after the protection criterion starts, it is judged that the circuit breaker is not cut off, and then opens failure protection; if there is no current, it is judged that the circuit breaker is cut off, and then blocks failure protection.

C. checkout voltage signal section of Reclosing

When the line reclosing is to be performed, it is first necessary to detect whether the line voltage is 0. Generally, a voltage transformer is installed on the single phase to detect whether there is voltage. Based on information sharing in station area, the checkout voltage detection signal can be reliably utilized to complement the bus voltage locking logic without additional economic cost.

As shown in Figure 2, assume that fault location F1 is the critical fault point for detecting checkout voltage signal and F2 is the critical point for the busbar recompression lockout. Adopting "and" blocking logic, the new scheme can reliably block and open circuit breaker failure protection after failure. When the phase failures occur without checkout voltage devices, the voltage locking criterion is not added, and the scheme of detecting redundant current is still reliable. Voltage blocking can be used for the phase with checkout voltage device, which can further enhance the reliability of failure protection without additional economic cost.

4. Simulation studies

In PSCAD / EMTDC, the 220kV system model as shown in Fig 4. is built. The length of each line are L1=60km, L2=20km, L3=50km. The positive sequence parameters of the circuit are $r_1 = 0.012\Omega/km$, $x_1 = 0.1045\Omega/km$, $c_1 = 0.01272uF/km$. The zero sequence parameters of the circuit are $r_0 = 0.0948\Omega/km$, $x_0 = 0.2894\Omega/km$, $c_0 = 0.009uF/km$. The threshold of current or no current is set to 0.5 A. Bus voltage closed phase locked voltage threshold is set to 50 V. No-voltage threshold is set to 20 V. The criterion of line differential protection action is: $|\dot{I}_2 + \dot{I}_5| > 0.5*(|\dot{I}_2| + |\dot{I}_5|)$

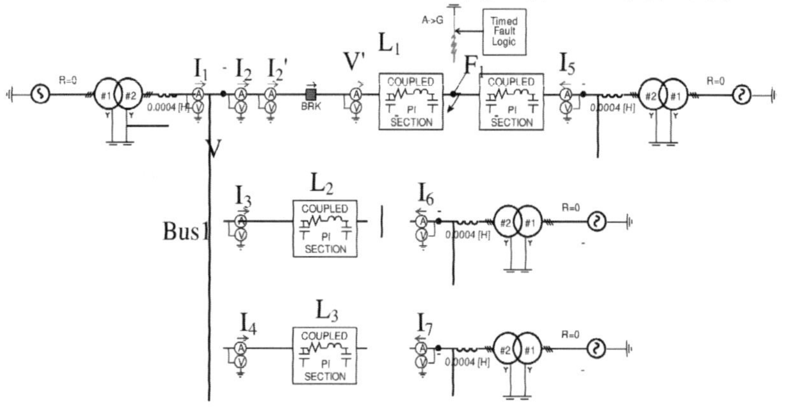

Figure 4. The simulation model of PSCAD

The single phase grounding fault is located at different locations of the line to verify the effectiveness of the proposed scheme.

CASE1. The fault location is at the outlet of the circuit breaker (4km).

The fault time is 0.125s, and the fault is removed at 0.25s time. Simulation shows that the current and voltage waveforms are shown in Fig 5 and 6.

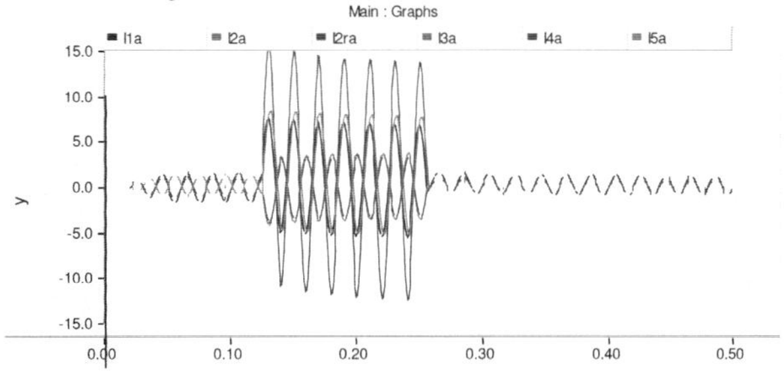

Figure 5. Current waveform in case 1

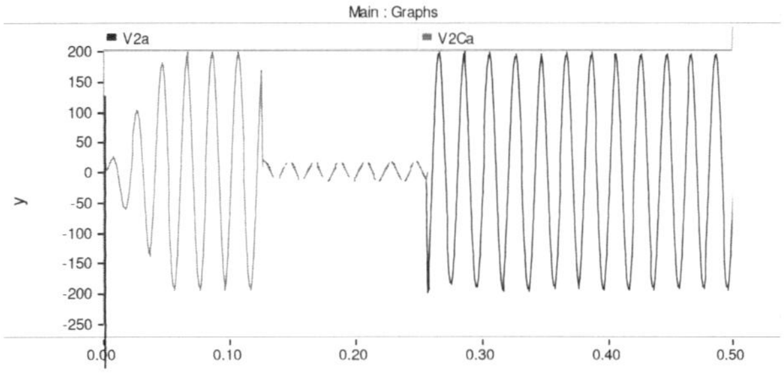

Figure 6. Voltage waveform in case 1

The simulation results show that the differential current is 19.59 A, greater than the braking current 9.89 A, and so the circuit breaker failure protection starts. Table 1 shows the current and voltage signals before and after the circuit breaker cut off.

Table 1. The current and voltage signals before and after the circuit breaker disconnected in case 1

		circuit breaker state	
		not disconnected	disconnected
Current/A	Busbar redundancy	13.1455	0.0004
	Line redundancy 1	13.1438	0.0002
	Line redundancy 2	13.1826	0.5382
	Error 1	0.0019	0.0006
	Error 2	0.4561	0.5382
Voltage/V	Bus phase low voltage	17.1154	192919
	Check no-voltage device	16.6368	0.0005

Table 1 shows that the error 1 of circuit breaker before and after interruption is small, then the line redundant current 1 is selected to judge whether the breaker is cut off. When the circuit breaker is not cut off, the redundant current 1 is 13.1438 A, which is larger than the threshold value. The phase low voltage of bus compound voltage blocking is 17.1154 V, and the voltage of checkout voltage device is 16.6386 V. Therefore, it opens circuit breaker failure protection. Redundant current of circuit breaker after disconnection is 0.0002 A, phase low voltage of line compound voltage blocking is 192919 V, and voltage of checkout voltage device is 0.0005 V. Therefore, the failure protection is blocked.

CASE2. The fault location is at the middle of line L1(30km).

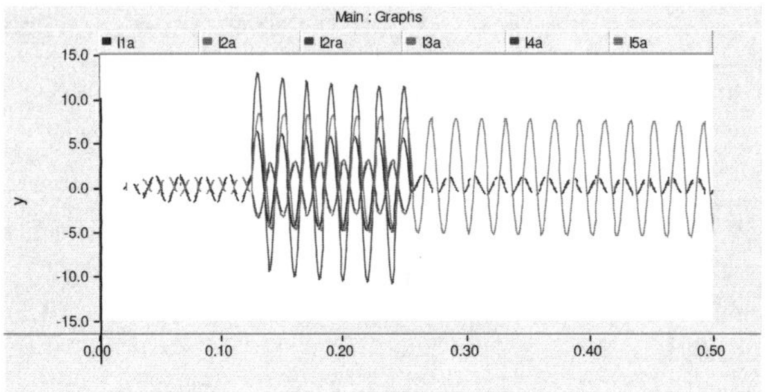

Figure 7. Current waveform in case 2

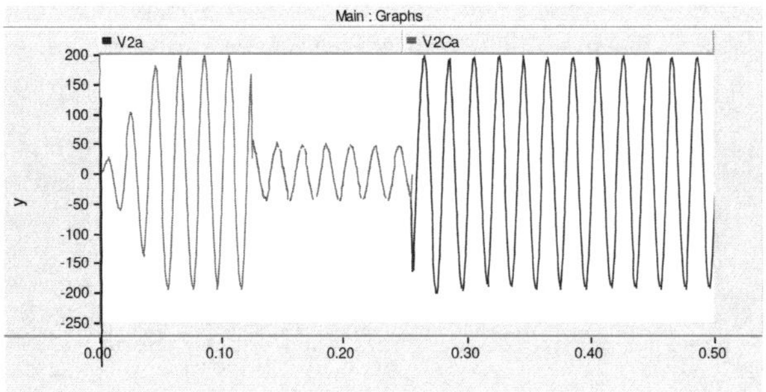

Figure 8. Voltage waveform in case 2

Other conditions remain unchanged. The fault location is located in the middle 30 km of line L1. Simulation shows that the current and voltage waveforms are shown in Fig 7 and 8. It show that the differential current is 17.53A, greater than the braking current 8.86A, and so the circuit breaker failure protection starts. Table 2 shows the current and voltage signals before and after the circuit breaker disconnected.

Table 2. The current and voltage signals before and after the circuit breaker disconnected in case 2

		circuit breaker state	
		not disconnected	disconnected
Current/A	Busbar redundancy	11.0814	0.0001
	Line redundancy 1	11.0793	0.0002
	Line redundancy 2	11.1293	0.5382
	Error 1	0.0022	0.0003
	Error 2	0.4862	0.5382
Voltage/V	Bus phase low voltage	45.7035	199085.4
	Check no-voltage device	45.4133	0.0028

Table 2 shows that the error 1 of circuit breaker before and after interruption is small, then the line redundant current 1 is selected to judge whether the breaker is cut off. When the circuit breaker is not cut off, the redundant current 1 is 11.0793 A, which is larger than the threshold value. The phase low voltage of bus compound voltage blocking is 45.703.5 V, and the voltage of checkout voltage device is 45.413.3 V. Therefore, it opens circuit breaker failure protection. Redundant current of circuit breaker after disconnection is 0.0002 A, phase low voltage of line compound voltage blocking is

199085.4 V, and voltage of checkout voltage device is 0.003 V. Therefore, the failure protection is blocked.

CASE3. The fault location is at the end of line L1(59km).

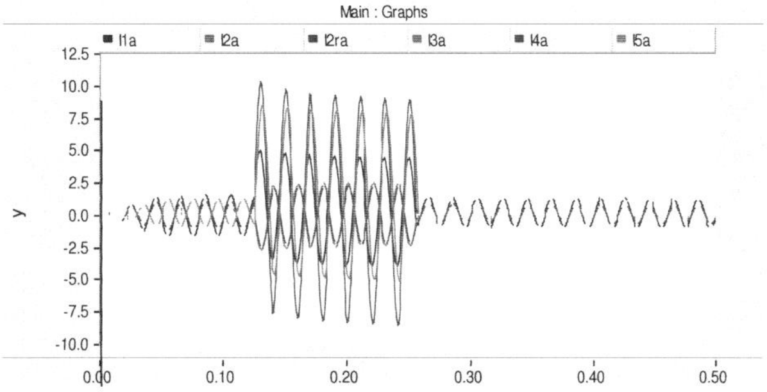

Figure 9. Current waveform in case 3

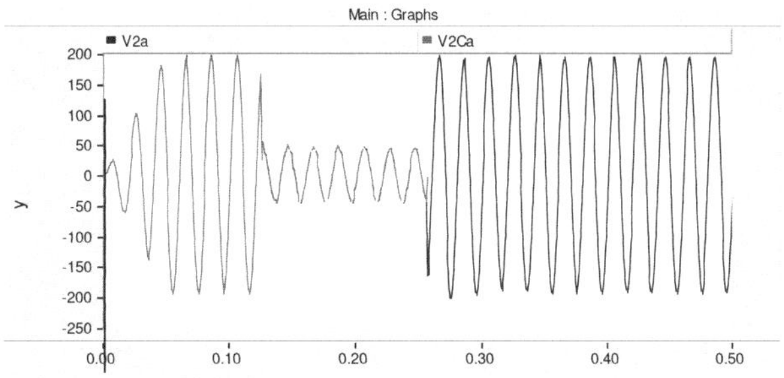

Figure 10. Voltage waveform in case 3

The fault location is located in the end 59 km of line L1. Simulation shows that the current and voltage waveforms are shown in Fig 9 and 10. It show that the differential current is 15.23A, greater than the braking current 7.71A, and so the circuit breaker failure protection starts. Table 3 shows the current and voltage signals before and after the circuit breaker disconnected.

Table 3. The current and voltage signals before and after the circuit breaker disconnected in case 3

		circuit breaker state	
		not disconnected	disconnected
Current/A	Busbar redundancy	8.7677	0.0004
	Line redundancy 1	8.7682	0.0002
	Line redundancy 2	8.8318	0.5382
	Error 1	0.0012	0.0006
	Error 2	0.5008	0.5382
Voltage/V	Bus phase low voltage	75.0536	199630.5
	Check no-voltage device	74.9214	0.0068

Table 3 shows that the error 1 of circuit breaker before and after interruption is small, then the line redundant current 1 is selected to judge whether there is current or no current. When the circuit breaker is not disconnected, the redundant current 1 is 8.7682 A, which is larger than the threshold value. The phase low voltage of bus compound voltage blocking is 75.0536 V, and the voltage of

IOP Conf. Series: Materials Science and Engineering **533** (2019) 012041 doi:10.1088/1757-899X/533/1/012041

checkout voltage device is 74.9214 V. Therefore, it opens circuit breaker failure protection. Redundant current of circuit breaker after disconnection is 0.0002 A, phase low voltage of line compound voltage blocking is 191630.5 V, and voltage of checkout voltage device is 0.0068 V. Therefore, the failure protection is blocked.

5. Conclusion

A new circuit breaker failure protection scheme based on station area information sharing is proposed in this paper. High-precision current/no-current detection is realized by using sampling information of dual-configuration current transformer and current information of back bus. At the same time, for the phase equipped with checkout voltage device, the checkout voltage signal is used to supplement the original compound voltage blocking signal. The theoretical analysis and simulation results show that the scheme can effectively cover the whole length of the line and has high reliability. The new scheme makes full use of the advantages of information sharing in the station domain and has a high practical prospect.

References

[1] P. Wu, T. Wang, W. Han and C. Chen, "Analysis on the false action and operation criteria for the generator circuit breaker failure protection," 2011 International Conference on Advanced Power System Automation and Protection, Beijing, 2011, pp. 215-218.

[2] N. T. Stringer and D. Waser, "An innovative method of providing total breaker failure protection," in IEEE Transactions on Industry Applications, vol. 32, no. 5, pp. 1011-1016, Sept.-Oct. 1996.

[3] B. Kasztenny, V. Muthukrishnan and T. S. Sidhu, "Enhanced Numerical Breaker Failure Protection," in IEEE Transactions on Power Delivery, vol. 23, no. 4, pp. 1838-1845, Oct. 2008.

[4] M. N. Noran, M. I. Ridwan, M. S. M. Shokri, H. R. Sabtu and Z. Shukri, "IEC 61850 based breaker failure protection scheme design for TNB double busbar substation," 2014 IEEE Innovative Smart Grid Technologies - Asia (ISGT ASIA), Kuala Lumpur, 2014, pp. 697-702.

[5] Y. Yue, G. Yarong, L. Yanjun and Z. Rongrong, "The delay optimization of circuit breaker failure protection based on the CT current tail truncation algorithm," 2016 IEEE International Conference on Power System Technology (POWERCON), Wollongong, NSW, 2016, pp. 1-7.

[6] M. N. Noran and Z. Shukri, "Adaptive breaker failure protection scheme for double busbar substation using IEC 61850 GOOSE message communication," 2015 International Symposium on Technology Management and Emerging Technologies (ISTMET), Langkawai Island, 2015, pp. 207-212.

[7] M. Wang, W. Leterme, J. Beerten and D. Van Hertem, "Robustness evaluation of fast breaker failure backup protection in bipolar HVDC grids," 13th IET International Conference on AC and DC Power Transmission (ACDC 2017), Manchester, 2017, pp. 1-7.

[8] Z. Darabi, B. Falahati, M. J. Mousavi and M. Ferdowsi, "On circuit breaker failure protection in 61850-based substations," 2012 IEEE Power and Energy Society General Meeting, San Diego, CA, 2012, pp. 1-6.

[9] N. T. Stringer and D. Waser, "An innovative method of providing total breaker failure protection," IAS '95. Conference Record of the 1995 IEEE Industry Applications Conference Thirtieth IAS Annual Meeting, Orlando, FL, USA, 1995, pp. 1165-1169 vol.2.

[10] Y. Xue, M. Thakhar, J. C. Theron and D. P. Erwin, "Review of the Breaker Failure Protection practices in Utilities," 2012 65th Annual Conference for Protective Relay Engineers, College Station, TX, 2012, pp. 260-268.

[11] B. Zhao, C. Yang, Y. Wang and H. Wang, "Functions and implementation of substation-area backup protection in smart substation," 2017 IEEE 2nd Information Technology, Networking, Electronic and Automation Control Conference (ITNEC), Chengdu, 2017, pp. 218-222.

[12] Y. Liu, H. Gao, W. Gao and F. Peng, "Development of a Substation-Area Backup Protective Relay for Smart Substation," in IEEE Transactions on Smart Grid, vol. 8, no. 6, pp. 2544-2553,

Nov. 2017.
[13] F. Peng, H. Gao and Y. Liu, "Transformer Sympathetic Inrush Characteristics and Identification Based on Substation-Area Information," in IEEE Transactions on Power Delivery, vol. 33, no. 1, pp. 218-228, Feb. 2018.
[14] Q. Li, Z. Zhou, D. Du, Z. Li, W. Li and X. Wang, "A novel substation area backup protection for smart substation," 2013 IEEE PES Asia-Pacific Power and Energy Engineering Conference (APPEEC), Kowloon, 2013, pp. 1-4.
[15] H. Gao, Y. Liu, G. Zou, D. Cui, M. Liu and X. Li, "Principle and implementation of substation-area backup protection for digital substation," 12th IET International Conference on Developments in Power System Protection (DPSP 2014), Copenhagen, 2014, pp. 1-5.

IOP Conf. Series: Materials Science and Engineering **533** (2019) 012042 doi:10.1088/1757-899X/533/1/012042

Research on Optimal placement for GIC mitigation with Blocking Device

Xiaoliang Ning, Qing Liu, Xikai Cui, Chendong Wang, Yinglan Wang

College of Electrical and Control Engineering, Xi'an University of Science & Technology, Xi'an, Shaanxi 710054 China

Corresponding author: 1065180397@qq.com

Abstract. Geomagnetically induced current (GIC) caused by geomagnetic storms flows between the power grid and the earth would threaten the stable operation of the power system. Capacitor blocking device connected in series at transformer neutral point can directly block the DC path. However, due to the complexity of the DC path in the power system, before the installation of the blocking device, if global considerations are not taken into account, the adverse effects of GIC will not be reduced. Therefore, to study the position optimization of the capacitor blocking device is of great value. This optimization of placement of capacitor blocking device is proposed in this paper. Applying the idea of genetic algorithm, the installation quantity and installation position of the blocking device were calculated and compared with the traditional GIC non-optimal method, which verified the superiority of the genetic algorithm optimization solution.

1. Introduction

The geomagnetically induced current(GIC) caused by the geomagnetic storm flows between the power grid and the earth and poses a threat to the safe operation of the power system [1]. With the development of the national economy, power grid in China continues to develop toward high voltage, long distance and large capacity. These factors will increase the risk of damage to the power grid. At the same time, modern society has higher and higher requirements for power supply reliability. Therefore, the suppression of GIC in the power grid has attracted extensive attention [2][3][4].

The capacitor blocking device connected at the neutral point of the transformer can directly block the DC path. Therefore, installing capacitor blocking devices to reduce the adverse effects of DC bias on the power grid has commonly used in the power grid. However, due to the complexity of the DC path in the power system, when the DC blocking device is installed, without global consideration, it may not reduce the adverse effects of GIC and will also cause huge waste of resources.

Based on Xinjiang Planning Power Grid, this paper presents an optimization algorithm called multi-population genetic algorithm(MPGA) for the placement of capacitor blocking devices: under the premise that the effective GIC of all transformers in the system is less than the limit current, minimize the number of installations of blocking devices with the optimization of genetic algorithm. The numerical results indicate that by adopting the MPGA, noticeably less number of the capacitive blocking devices are required compared with the traditional placement.

Content from this work may be used under the terms of the Creative Commons Attribution 3.0 licence. Any further distribution of this work must maintain attribution to the author(s) and the title of the work, journal citation and DOI.
Published under licence by IOP Publishing Ltd

2. GIC modeling

Due to the GICs of substations or transmission lines between different voltage levels may be mutually impact, the grid consisting of different voltage levels is considered in the modeling in this paper.
The frequency component of GIC is mainly between 0.0001 Hz and 0.1 Hz. It can be regarded as a DC component compared with the grid frequency of 50 Hz, and it can only flow through the DC path which consists of transmission lines, substation transformers and the earth [4].

2.1. GIC calculation

Compared with single-voltage-level GIC calculations, multi-voltage-level GIC calculations differ only in transformer node and branch processing, and the calculation principle is basically the same [5][5].
Assuming that the geoelectric field E caused by the geomagnetic storm is uniform and constant, the induced voltage of the line in the power grid can be superimposed from the induced voltage of the line in the north and east direction, that is [4]:

$$U = E_E L_E + E_N L_N \tag{1}$$

where E_E, E_N are the magnitude of the induced geoelectric field in the north and east directions, and L_N and L_E are the effective distances of the transmission lines in the north and east directions, are given by

$$L_N = [111.133 - 0.56\cos(2\varphi)] \cdot \Delta Lat \tag{2}$$

$$L_E = [111.5056 - 0.01872\cos(2\varphi)] \cdot \cos(\varphi) \cdot \Delta Long \tag{3}$$

where ΔLat and $\Delta Long$ are the latitude difference and longitude difference of the substation at both ends of the transmission line respectively, and φ is the average of the latitude of the substation at both ends.

In order to make the final placement suitable for mitigating the adverse effects of the induced electric field in all possible directions on the power system, to calculate GIC for every possible angle in the range of 0°-360° is of great value [7]. Since the magnitude of the induced electric field can be linearly superimposed, the direction can also be orthogonal decomposed in both the east and the north directions. Therefore, after the value of the induced electric field Amplitude is given, the electric field is rotated in the range of 0°-360°, suppose the angle between the direction of the induced electric field and the east direction is θ, so that the E_E and E_N are given by

$$E_N = Amplitude \cdot \sin\theta \tag{4}$$

$$E_E = Amplitude \cdot \cos\theta \tag{5}$$

For any node i in the grid, there is

$$J_i = u_i y_i + \sum_{k=1}^{N} u_i y_{ki} - \sum_{k=1}^{N} u_k y_{ki} \tag{6}$$

where J_i is the sum of the branch current sources connected to the node i which can be calculated by the induced electric field. y_i is the grounding conductance of node i, y_{ki} is the branch conductance between nodes k and i; after calculating the node voltages u_i and u_k, the GIC can be obtained.
Its matrix form is

$$J = YV \tag{7}$$

2.2. Effective GIC

Since the autotransformer is generally used in China's UHV power grids, when the neutral point of the transformer is installed with a block device, the current flowing through the neutral point is zero, however there is still existing bias current in the series winding [8][8]. Therefore, we use the effective bias current to evaluate the influence of the GIC on the transformer as defined by

$$I_{eff} = I_H + I_L / k_t \tag{8}$$

where k_t is the transformer turn ratio, and I_H and I_L are GICs flowing into the autotransformer winding through the series winding and the common series of the transformer.

The maximum of the maximum effective GIC and the average of the maximum effective GIC are used to evaluate the effect of the installing of blocking device on the grid are given by

$$I_{eff}\_\max = \max\left(\max\left(I_{ij}, j \in [0, 2\pi]\right)\right), i \in [1, N] \tag{9}$$

$$I_{eff}\_ave = \frac{1}{N}\left(\sum_{i=1}^{N} \max\left(I_{ij}, j \in [0, 2\pi]\right)\right), i \in [1, N] \tag{10}$$

In (9) and (10), i is the substation number, j is the angle between the direction of the induced electric field and the east direction and N is the number of substations.

3. Optimization approach for capacitor blocking device placement

3.1. GIC equivalent model of Xinjiang planning power grid

In order to implement the global energy interconnection strategy, serve Xinjiang, and promote social stability and long-term stability in Xinjiang, Xinjiang State Grid Corporation plans to build a main grid of 750kV and above in Xinjiang. Xinjiang Planning Power Grid includes six 1000kV substations and thirty-seven 750kV substations, forty 750kV AC transmission lines, and five 1000kV AC transmission lines shown in Fig 1.

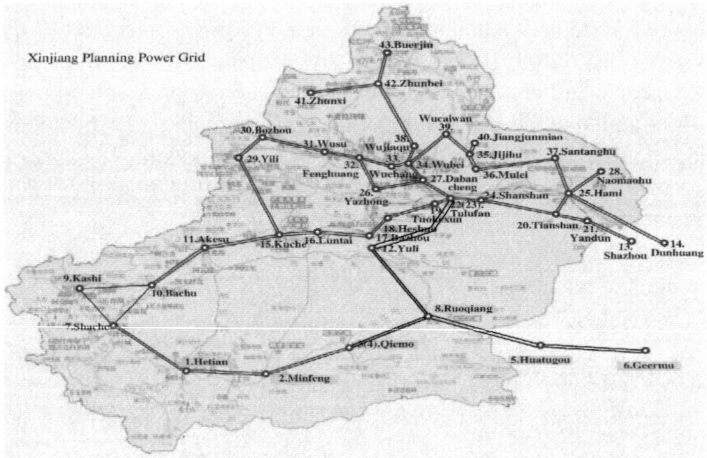

Figure 1. Xinjiang Planning Power Grid geographical connection diagram

The GIC equivalent model of the Xinjiang planning grid is established by the GIC calculation modeling method described above. Each substation in the grid has two autotransformers running in parallel. The transformer winding branch is equivalent to the series winding resistance and the common winding resistance. The substation to the ground branch is equivalent to the grounding resistance, and the transmission line is equivalent to the series connection of the induced voltage source and DC resistance. Assuming that the GIC of each phase flowing through the line is equal, only the single-phase GIC will be considered in the calculation, the grounding resistance is 0.3Ω, and the line resistance is 0.01205Ω per unit length.

It can be seen from [10] that the DC current allowed by the transformer is related to the heating limit of the transformer winding and the iron core, the local overheating limitation of the transformer structural parts, and the operational stability of the power system. When the DC current injected into the transformer I_{dc} < 10 A, The transformer works well and there is basically no obvious fault. However, when the 10 < I_{dc} < 50 A, the temperature rise of the winding of the transformer increases, and the additional loss caused by the magnetic flux leakage increases rapidly, which cause the transformer to overheat and have hidden troubles. Therefore, we set 15A as the limit current for the transformer stable operation.

The single-phase GIC of each transformer of each substation node is calculated which is shown in Fig 2. if the limit current for single-phase GIC of the transformer is 15A, the single-phase GIC of most substations of the grid under the influence of the electric field of 1V/km is larger than the limit current. Thus, to take effective measures to minimize the negative impacts of GIC on power system is of great necessity.

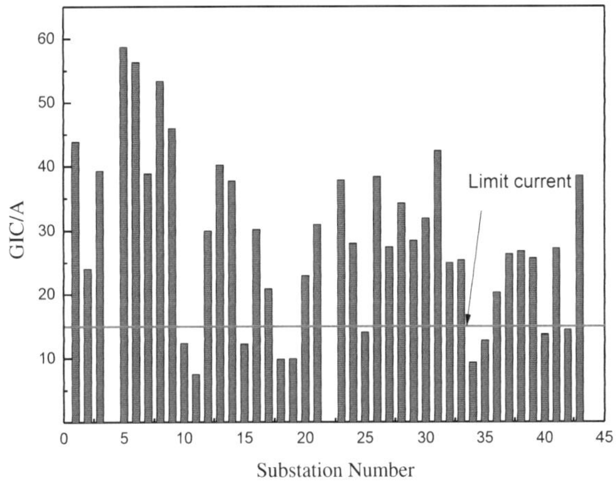

Figure 2. The single-phase maximum GIC of each transformer of each substation.

3.2. GIC suppression effect of traditional placement method

Installing the capacitor blocking device at the neutral point of the transformer can directly block the flow path of the GIC, but the installation of the blocking device will have a certain impact on the transformer in the neighbouring area, and this impact cannot be quantified. In addition, the cost of capacitor blocking device is high, it is unrealistic to install blocking device at the neutral point of all the transformers.

The traditional placement method is to install the blocking device according the GIC of each transformer, that is to say, calculate the maximum GIC of each transformer under the condition that the amplitude of the induced electric field is 1V/km, the magnitude of the induced electric field is in the range of 0°-360°, and then sort the transformers by maximum GIC of each transformer in descending order, finally the capacitor blocking device is preferentially installed at the transformers with larger GIC, until the maximum effective GIC of all transformers of the grid satisfies the limit current.

It can be seen from the Fig 3. that when the placement is not considered to be optimized, the maximum effective GIC of the grid is not monotonously decreasing as the number of blocking device installations increases, even when the number of installations is 35, the maximum effective GIC of the grid is much larger than when the blocking devices are not installed. The maximum effective GIC of the grid is significantly decreased and satisfies the limit current until the number of installations of the blocking device reaches 40. As shown in Fig 4., the single-phase maximum effective GIC of each transformer of substation No.11 and No. 34 dose not monotonically diminish with the number of blocking device installations either, especially the GIC flowing through the transformer of No. 34 substation whose GIC was smaller before the installation of the blocking device will passively increase and even will be much larger than the limit current because of the installation on other transformers in the grid.

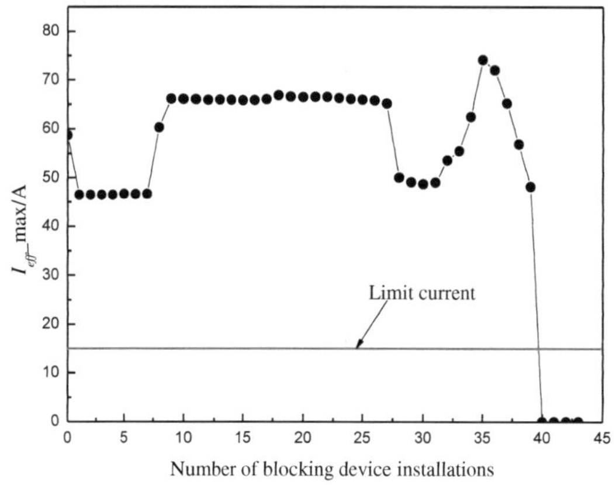

Figure 3. The I_{eff_max} of the grid while the transformers with larger GIC are blocked one by one by using the traditional placement.

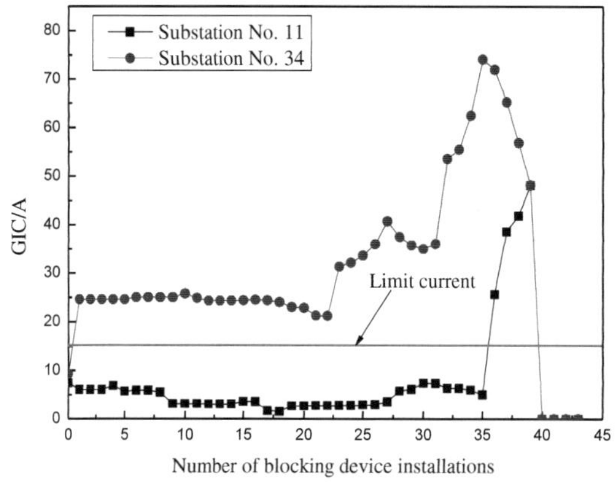

Figure 4. The single-phase maximum effective GIC of each transformer of substation No.11 and No.34 while the transformers with larger GIC are blocked one by one by using the traditional placement.

In summary, in the case of global considerations, installing blocking device on the substations with larger maximum effective GIC cannot reduce the impact of the geomagnetic storm on the power grid even if the GIC of the installed transformer can be reduced.

3.3. Algorithm principle for optimizing the blocking device placement

For the grids with a small number of substations, all the candidate solutions to the problem of the placement of the capacitor blocking devices can be enumerated by the exhaustive method. Calculate the maximum effective GIC of each solution, and find the optimal solution under the set limit current. However, as the number of substations in the grid increases, it will be much more difficult to solve the problem and may not even be solved.

It is a discrete problem whether or not the blocker is installed, and the multiple population genetic algorithm (MPGA) does not depend on the gradient in the optimization process. It has strong robustness and global search ability and is very suitable for solving such problems.

The MPGA begins to evolve from several sets of solutions(individuals) that represent solutions to the problem. Each individual in the population corresponds to a gene code to represent a solution to the optimization problem. For the optimization problem of the blocking device, An N-bit binary code is used to represent the solution to the problem. Each bit of the code has two encoding modes, 0 or 1, which indicates whether the corresponding substation is equipped with a blocking device or not. Each individual is evaluated its superiority and inferiority according to the objective function and obtains its corresponding fitness value. Since the evolutionary principle of genetic algorithm is the process of increasing the fitness value, when construct the objective function, if the maximum effective GIC of the system satisfies the limit by adopting a solution placement, then the objective function of this solution is the number of "1" in its binary code, and its assigned fitness value is larger, while the fitness value is smaller of the one does not satisfy the limit. Individuals in the initial population compete for excellent individuals according to the fitness value, and generate new individuals after a series of genetic operations. The new individuals continue to evolve according to the same objective function, and assign fitness values. After several generations, the algorithm converges to one or several global optimal solutions.

3.4. Optimization result of the blocking device placement

The global optimal solution is obtained using the MPGA, with the following parameters: the transformer single-phase maximum effective GIC limit current is 15A, the amplitude of the induced electric field is 0.5V/km, chromosome length is 43, population number is 10, initial population size is 80, the crossover probability and mutation probability of each population are random numbers in [0.9, 0.95] and [0.01, 0.1], the retention algebra for the optimal individual is 20. Under the constraint that the maximum effective GIC of all transformers of the grid is less than 15A.

After running the MPGA optimization program for several times, the objective function can always converge to 29, which means the impact on the Xinjiang grid can be reduced to an acceptable level with only 29 blocking devices installed when the grid is affected by the magnitude of the induced electric field of 0.5V/km. The maximum effective GIC of all transformers of the grid is 14.8999A with the placement of the installation position: 1-3, 5-21, 23, 24, 27, 29, 37-39, 42 and 43 substation.

Under the same constraint that the maximum effective GIC of all transformers of the grid is less than 15A, the number of blocking devices of the MPGA optimal placement is much less compared with the traditional placement.

The single-phase maximum effective GIC of all substations in the grid with the optimal placement obtained by MPGA is shown in Fig. 5, from which we can intuitively observe the inhibitory effect of GIC.

In order to further elucidate the superiority of MPGA in solving the problem of blocking devices under different conditions, the problem of blocking devices under the influence of different induced electric field in Xinjiang planning power grid was solved using MPGA.

After running the optimization program, the optimization results and the GIC inhibition effect with the MPGA optimal placement shown in Table 1. can be obtained.

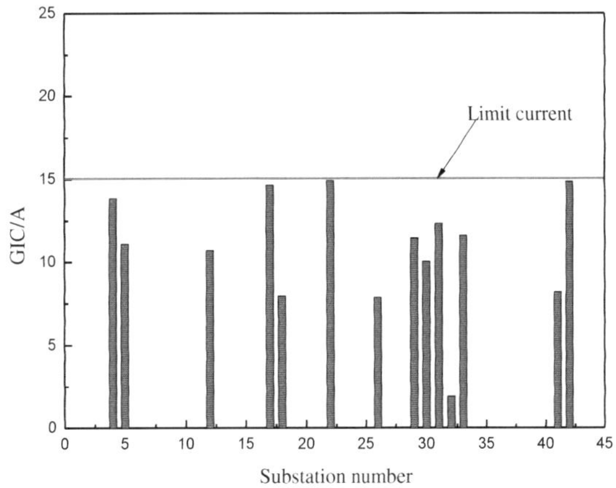

Figure 5. The single-phase maximum effective GIC of each substation with the MPGA optimal placement.

Table 1. The optimization results for different induced electric field

Induced geoelectric field amplitude(V/km)	0.5	0.75	1
Blocking device installation position	1-3、5-21、23、24、27、29、37-39、42、43	1-3、5-21、23-25、27、28、33-40、42、43	1-3、5-21、23-29、32-40、42、43
Number of blocking devices	29	35	38
I_{eff_max}(A)	14.8999	14.7178	14.8343
I_{eff_ave}(A)	10.8112	11.9525	11.0065

Fig. 6 shows the number of blocking devices required for the grid under the influence of different induced geoelectric fields for two cases: 1) solve the problem using traditional placement method and 2) solve the problem using MPGA. In this figure, when the limit current is 15A, the number of blocking device of the MPGA optimal placement for different induced electric field is smaller compared with the traditional placement method.

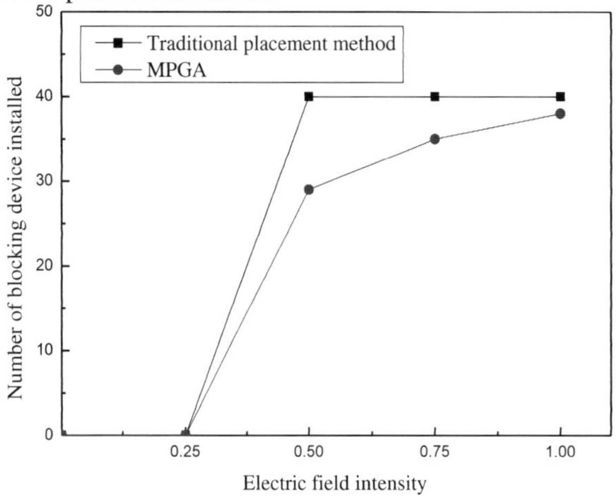

Figure 6. Comparison of the number of blocking devices between the traditional placement method and the MPGA

The above figures and tables show that MPGA has the irreplaceable superiority in solving the problem of the placement of the blocking device compared to the traditional placement method.

4. Conclusion

1) Installing the blocking device only at several substations with larger effective GIC can reduce the maximum effective GIC of the installed substations, however the maximum effective GIC of other substations can be passively increased and even exceed the limit current. Therefore, it is of great research significance to optimize the installation placement of the blocking device.

2) In the optimization calculation of the GIC blocking device placement for mitigating the effects of GIC of induced electric field magnitudes ranging from 0.5V/km-1V/km, in order to limit the maximum effective GIC of all transformers of the power grid to 15A or less, the number of blocking device of optimal placement obtained by MPGA is significantly smaller than the traditional placement method.

3) After adopting the optimized placement obtained by the MPGA optimization algorithm, the maximum and average values of the maximum effective GIC of the transformer are significantly reduced, which means that the adverse effects of GIC is effectively suppressed.

4) The future research plan for this paper includes multi-objective optimization calculation with more constraints, such as transformer reactive power loss caused by GIC, power flow constraints of grid operation, and finding more efficient algorithms to solve the problem.

Acknowledgments
I would like to express my heartfelt gratitude for the support in part by the National Key R&D Program of China (No. 2016YFC0800100).

References

[1] E. Mulasalihovic', S. Traxler, H. Yamaguchi. Effects of geomagnetically induced currents on the magnetic performance of transformer cores Journal of Magnetism and Magnetic Materials 320 (2008) e920 – e924
[2] L. Trichtchenko and D. H. Boteler, " Effects of recent geomagnetic storms on power systems, " in Proc. Int. Symp. Electromagn. Compat. Conf., 2007, pp. 265-268.
[3] W. A. Radasky and J.G. Kappenman, "Impacts of geomagnetic storms on EHV and UHV power grids, "in Proc.Asia-PacificInt.Symp.Electromagn. Compat. Conf., 2010, pp. 695-698.
[4] ZHENG Kuan, LIU Lianguang, BOTELER D H,et al.Modeling geomagnetically induced currents in multiple voltage levels of a power system illustrated using the GIC-Benchmark case[J], Proceeding of the CSEE,2013(16):179-186+3.
[5] R. Pirjola, "Study of effects of changes of earth resistances on geomagnetically induced currents in an electric power transmission system, "Radio Sci., vol. 43, 2008.
[6] R. Horton, D. Boteler, T. J. Overbye, R. Pirjola, and R. C. Dugan,"A test case for the calculation of geomagnetically induced currents," IEEE Trans. Power Del., vol.27, no.4, pp.2368 – 2373, Oct.2012.
[7] REZAEI-ZARE A. Optimal Placement of GIC Blocking Devices for Geomagnetic Disturbance Mitigation[J]. IEEE Transactions on Power Systems, 2014, 29(6): 2753-2762.
[8] Zhu H, Overbye T J. Blocking device placement for mitigating the effects of geomagnetically induced currents[J]' IEEE Transactions on Power Systems, 2015, 30(4):2081-2089ETEMADI A H.
[9] REZAEI-ZARE A. Reactive Power Loss Versus GIC Characteristic of Single-Phase Transformers[J]. IEEE Transactions on Power Delivery, 2015, 30(3): 1639-1640.
[10] YAO Yingying, "Research on DC Biasing of Large Power Transformers[J]," Shenyang University of Technology, 2000,11

IOP Conf. Series: Materials Science and Engineering **533** (2019) 012043 doi:10.1088/1757-899X/533/1/012043

Research on the forward distance detection algorithm based on the camera switching

MengYuan Chen[1 2 3], Dehong Tian[1],Xiao Xiao[1],ChengYuan Bian[1]

1 Anhui Polytechnic University, Anhui Key Laboratory of Electric Drive and Control, AnhuiWuhu,241000

2 Univ Sci & Technol China, Dept Precis Machinery & Precis Instrumentat,AnhuiHefei,230027

3 Manufacturing Informatics Centre, SATM, Cranfield University

mychen@ahpu.edu.cn

Abstract. In order to promote vehicle active safety and to decrease collision accident, In this paper, an algorithm for detection of vehicle distance in front based on single and double camera switching is presented. This algorithm has done the corresponding difference processing to the day and night two different situations. By switching between single and double cameras, the image of the car bottom shadow, the characteristics of the vertical direction of the vehicle, the characteristics of the car taillight and other features was taken, and the position of the vehicle in front was locked. Finally, the feature information of the vehicle is converted into the distance information. Experiments show that, this system designed in the paper, achieved the anticipated function, can accurately detect multi-lane and short distance vehicle distance, and the efficiency of the system also achieved the real-time requirement.

1. Introduction

In the field of ADAS (Advanced Driving Assistance System) research, vehicle ranging ahead and reflecting road condition information is a key technology in collision warning and fatigue driving warning. In several safety aid systems commonly used today, millimeter scale radar ranging is far away. But it is so expensive that cannot be widely promoted. Ultrasonic ranging is not widely used because of its physical characteristics. Due to the advantages of abundant information data, stable performance and price advantage, machine-vision-based detection has become a hot spot in current research[1].

At present, the advanced vehicle detection of machine vision is mainly based on the image prior feature knowledge of vehicle tail. Vehicle feature extraction is a key part of vehicle distance detection. In the previous studies, the method of hybrid Gaussian function was used to describe the multi-level background to achieve the purpose of separating vehicle features[2][3]; In literature [4], vehicle shadow and edge are used as the main features of detection. Variable model is adopted to accurately locate the vehicle ahead, and radar data is integrated to improve the accuracy of the system.; The vehicle distance detection is also realized by the vehicle modular method, and the vehicle components are extracted by using the main components of the vehicle feature in [5][6]. And improve the stability of the system by grouping the adjacent areas of the feature. In the literature[7][8], the Gabor filter is used to obtain the direction angle and the lane line, and the vector division is used to complete the

Content from this work may be used under the terms of the Creative Commons Attribution 3.0 licence. Any further distribution of this work must maintain attribution to the author(s) and the title of the work, journal citation and DOI.
Published under licence by IOP Publishing Ltd

community division. However, these methods are computationally intensive and require too much CPU to operate in an embedded system.

For the binocular ranging camera system, the key step is feature point matching. Due to the diversity of physical shapes, since the beginning of stereo vision research, the stereo matching of objects is the most difficult, difficult to overcome, and extremely important thing[9][10]. On the one hand, due to the geometry and physical characteristics of the target object, noise interference and distortion, and the physical optical characteristics of the camera itself, it will be integrated into a single image brightness value problem in the camera, that is the problem of gray value. On the other hand, the determination of depth is also one of the difficulties. After obtaining the parallax image through stereo matching, the correlation algorithm can be used to calculate and determine the depth image, and restore the 3D information restored to the original scene[11][12][13]. Among the many factors that affect the accuracy of distance measurement by visual ranging, the main interference is digital quantization (light intensity will be quantized into an integer in the camera), camera calibration error, the use of feature detection algorithms, and the accuracy of matching positioning[14][15].

2. Vehicle feature detection

This paper proposes a robust solution that distinguishes the detection environment into day and night. In the daytime environment, the gray level is used to analyze the threshold to capture the shadow of the car bottom, and in the case of capturing the shadow of the bottom of the car, the characteristics of the vertical direction of the vehicle are analyzed, and finally the vehicle is locked; in the night environment, first in the YUV The color space filters and normalizes the taillights of the vehicle, and finally completes the vehicle detection under the taillight judgment mechanism. The flow chart is shown in Figure 1.

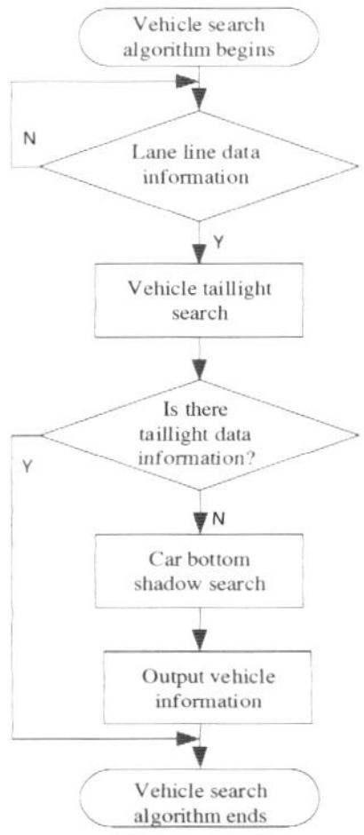

Figure 1. Vehicle search algorithm flow chart

For the position where the camera is erected, it can limit the detection range of the vehicle during daytime and nighttime road conditions, which can effectively reduce the calculation amount of the system. Therefore, the detection range is limited for two different situations during the day and night, day and night. The scene ROI range is shown in Figure 2.

(a) Range of ROI detected during the day (b) Night detection ROI range

Figure 2. Day and night detection ROI range diagram

IOP Conf. Series: Materials Science and Engineering **533** (2019) 012043 doi:10.1088/1757-899X/533/1/012043

2.1Daytime vehicle feature detection

In this paper, the Jiugongge template is used to analyze the binarized shadow candidate points (the binarized shadow of the vehicle bottom shadow candidate is 255, and the pixel point of the candidate point is not set to 0. Figure 3 shows the Jiugongge template. The center point E of the Jiugongge is the first undercarriage shadow candidate point searched vertically in the ROI range of the daytime scene. The eight connected directions at this point are searched. The whole grouping process is shown in Figure 4. The red square part represents the pixel point with the brightness value of 255, that is, the pixel point which may be the shadow of the bottom of the car. If there is any direction in the eight connected directions of E in Figure 4(a) 255 pixels, this pixel is assigned as the vehicle shadow candidate point, and marked as unavailable, the next round of search is shown in Figure 4 (b), then E is Figure 4 (b) F, and because there is a point of 255 in the D direction at this time, but it is marked as unavailable in the previous round, so this search only has 255 points in the F direction, so loop until the center point E All eight connected directions are 0, and then exit the group. These searched points are represented as points of the same group. If there is no 255 point in the eight directions of E when starting grouping, then the grouping is directly exited, as shown in Figure 4(c), that is, only the pixel corresponding to E in the community at this time.

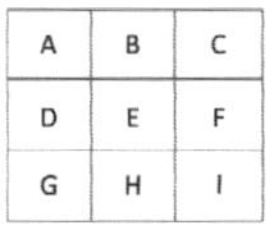

Figure 3. Jiugongge grouping algorithm illustration

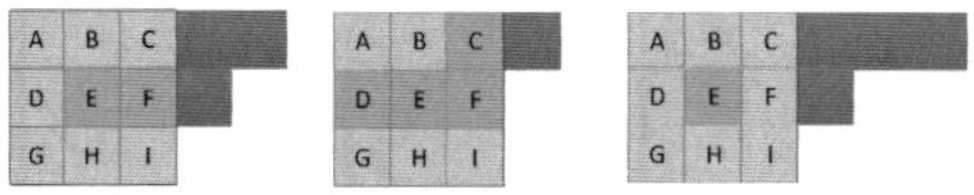

(a) There are 255 points in the eight connected directions - the end of the group

(b) There are 255 points in the eight connected directions - the end of the group

(c) There is no point of 255 in the eight connected directions - the end of the group

Figure 4. Cluster process diagram

The concept of close distance and medium distance is introduced in the article. The definition of close distance is within 15 meters. As shown in Figure 5, the red line in Figure 5 is the boundary line between medium and long distance and close distance.

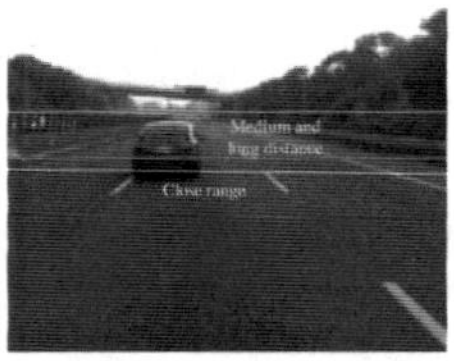

Figure 5. Close-up and long-distance conceptual illustration

For the analysis of the bottom shadow of the car, the main feature information is captured by the difference between the bottom shadow and the road surface brightness value. For the close-up bottom shadow, the bottom shadow value is shown in Figure 6, and can be analyzed from Figure 6. In the range shown in the figure, the information of the bottom shadow is sufficient and the brightness value is lower than 35. Considering the climatic factors in the actual scene, the close-up bottom shadow brightness search threshold is set to 40.

IOP Conf. Series: Materials Science and Engineering **533** (2019) 012043 doi:10.1088/1757-899X/533/1/012043

(a) Close-up bottom shadow of the car (b) luminance value data of 1 in Figure a

Figure 6. Close-up vehicle shadow data

When searching for the shadow of the bottom of the car, the order of search in the ROI range is from bottom to top, and the feature points that match the brightness threshold of the bottom shadow are searched column by column, as shown in Figure 7. In the car bottom shadow search ROI range, the pixels whose brightness threshold is lower than 40 are searched, and the cluster width of these pixels is greater than the quarter lane width, and these points are used as the vehicle bottom shadow candidate points, as shown in Figure 8.

Figure 7. Shaded shadow search Schematic

Figure 8. Schematic diagram of shadow candidate points at the bottom of the vehicle

(1) Short-distance vehicle shadow candidate points

After searching for the candidate points of the vehicle bottom shadow, it is necessary to further confirm whether these candidate points are the shadow of the bottom of the vehicle. The method adopted in this paper is to judge the road surface and the shadow of the vehicle bottom by the position of the bottom shadow candidate point position down one-half of the lane width. Whether there is a luminance difference in the candidate points, and whether the point cluster is greater than a quarter of the lane width, and if the condition is established, it is determined to be a valid vehicle bottom shadow. The selection of the candidate points for the short-distance bottom shadow is shown in Figure 9.

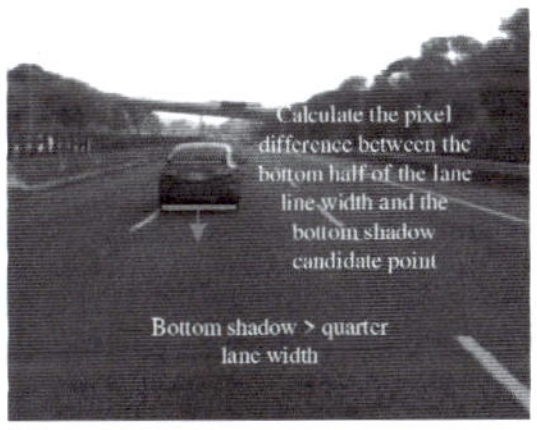

Figure 9. Close-up shaded pixels at the bottom of the car

(2) Medium and long distance vehicle shadow candidate points

The search for the mid-distance bottom shadow candidate points is the same as the close distance. First, the pixels with the brightness difference and the group width greater than the quarter lane width are searched. If this condition is true, the pixels are saved first; The center of the pixel points upward to search for the feature information of the vehicle edge point in the horizontal direction about the width of the lane width. If there is edge feature information, and the number of points accounts for more than 70% of the total number of detection points, it is effective for the shadow of the vehicle bottom. Information data, as shown in Figure 10.

(a) Medium and long distance search is greater than a quarter lane width

(b) Medium and long distance search for vehicle edge feature points greater than the threshold

Figure 10. Medium and long - distance vehicle feature detection

Affiliations of authors should be typed in italic 8-point Arial. They should be preceded by a numerical superscript corresponding to the same superscript after the name of the author concerned. Please ensure that affiliations are as full and complete as possible and include the country.

2.2Night vehicle feature detection

(1) Analysis of color space characteristics of taillights

At night, it is difficult to capture the shadow of the car and the feature points on the edge of the vehicle, which makes it impossible to detect the vehicle at night using the same algorithm during the day. However, at night, the taillight signal of the vehicle is obvious, and the vehicle taillight can be captured to capture the vehicle at night. In order to ensure the real-time performance of the system and improve the computing efficiency, the YUV format image is not converted into other color spaces. The numerical analysis of the vehicle taillights in the YUV color space is shown in Figure 11.

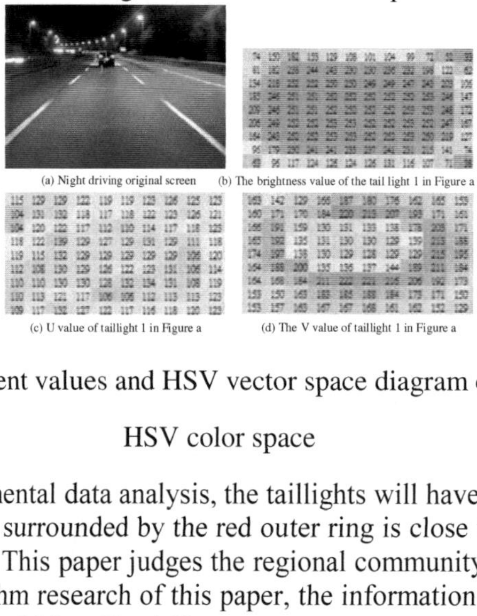

(a) Night driving original screen (b) The brightness value of the tail light 1 in Figure a

(c) U value of taillight 1 in Figure a (d) The V value of taillight 1 in Figure a

Figure 11. The component values and HSV vector space diagram of the taillights in the HSV color space

After a large amount of experimental data analysis, the taillights will have a halo of red outer ring, and the brightness value of the area surrounded by the red outer ring is close to the saturation value of the brightness value (value is 255). This paper judges the regional community with such characteristics as Vehicle taillights. In the algorithm research of this paper, the information of capturing taillights of the vehicle is finally defined as: $Y > 200$, $V > 160$.

(2) Description of the taillight capture process

After capturing the characteristics of the vehicle taillights in the YUV space, some of the binarized white spots are retained, and the information of the accurate taillights needs to be searched from these spots. Interference noise will also be captured due to threshold filtering in the YUV space. In order to eliminate these very similar noise disturbances, each spot needs to be grouped. In this paper, the Jiugongge grouping algorithm is used, and then the normalization of each community is carried out, and the size of the community after normalization is compared. The two communities are similar in size and within the limited value, it is considered as the taillight of the vehicle.

In the ROI detection range, the preliminary taillight candidate information captured by the threshold set by the YUV space is searched in the order from left to right. When the first information that matches the characteristics of the taillight is found, the current record is recorded. The location and area size information, as shown in Figure 12, then continue to search for information that matches the characteristics of the taillights. If it is found, it is judged whether the areas of the two are similar. The similarity is the characteristic information of the effective taillights. If the camera is assumed When the horizontal correction is not made, the tail lights in the picture are not horizontally paired. In this case, it is necessary to search within the entire ROI range.

Figure12. Search for rear taillights candidate points

If the characteristic area of the taillights is not similar, the taillight candidate point information is updated, and the rightward search is continued until the boundary of the ROI range. If the symmetrical taillight information is still not found after continuously searching for the ROI boundary, it is determined that the noise is cleared. Previously recorded, continue to search up column by column. As shown in Figure 13, if both 1 and 2 meet the characteristics of the taillights, but the areas are not similar, the candidate points are updated to 2, and the symmetrical taillights are searched from 2 to the right.

Figure 13. Rear lights are not similar to the schematic diagram

(3) Taillight feature point normalization comparison

As can be seen from the previous description, an essential step in locking the taillights is to normalize the preliminary information of the taillights and compare the size of the area. In the normalization calculation, a standard size is first determined to standardize the target to be compared. The target refers to the taillight information in this paper. Define the normalized size, that is, the normalization size is normalization size, and the community length and width of the taillight candidate community are recorded as grouplength and groupwidth, and the normalized community length and width are grouplength non and groupwidth non, then the community is normalized. The expression is:

```
If (Candidate taillight community pixel)
   While (Traverse all community pixels)
         groupwidth go
      grouplength go
         groupwidth_non = groupwidth/normalization_size
         grouplength_non=grouplength/ normalization_size
```

If the normalized community length and width calculated by the two communities are similar, it is considered to be the taillight of the vehicle taillight, otherwise it is considered to be interference noise.

3. Distance conversion and measure-ment

3.1 Lane line correction mechanism

In order to enhance the stability and robustness of the system, the lane line correction mechanism has been added to ensure that lane line detection can be made under the more vulnerable to the interference of environment conditions such as light and shade. The algorithm flow chart of lane correction is shown in Figure 14, and the test results of the algorithm in the actual experiment are shown in figure.

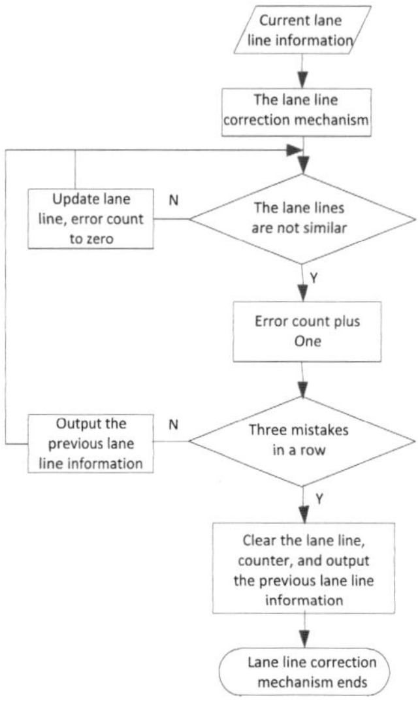

Figure 14. lane line correction flow chart

Figure 15. lane line correction results

In Figure 15, figure (a) is the original map of the driving image. As the lane line here is completely damaged, the driver can no longer distinguish the main lane visually. Lane in figure (b), using the modified algorithm, the algorithm for this kind of situation to make the lane line, as shown in figure can see, in the lane line cannot be in a short time detection, algorithm to ride in this short period of time according to the pictures before a couple of reserved the lane line information on the lane line information make general judgment, and for display.

3.2. Monocular distance model

(1) Establishment of the distance model

When the vehicle is running, the camera shakes and the shooting angle will change, which causes the camera's multiple built-in input parameters to change. From this aspect, its applicability is greatly limited. At the same time, the amount of computation in the actual embedded system is large, which is directly related to whether the system can meet the requirements of real-time processing. Based on the above considerations, this paper adopts a simple and practical monocular ranging method, as shown in Figure 16, which is based on the lane width information. When the vehicle characteristic information is found, a fixed position lane width information is displayed in the screen. For the substrate, the ratio of the lane widths of A and B is calculated, corresponding to the distance conversion table that realizes the measurement, thereby obtaining the actual distance information. The base of the lane line information is generally set as the base where the lowest position of the detected vehicle is detected, and the calculation formula of the lane ratio mentioned above is expressed as $\frac{B}{A}$

Figure 16. Substrate selection and lane width ratio diagram

In the monocular ranging method proposed above, the distance conversion table is important data measured in actual experiments. In the monocular ranging strategy of this paper, the data displayed on the screen is based on this data table, and different The road width corresponds to different data tables, and the specific data is shown in Table 1.

Table 1. 3 m and 4 m width lane distance search form

	actual measured lane width – 3 m		actual measured lane width – 4 m	
actual distance (m)	the pixel width occupied by the screen width	lane width ratio	the pixel width occupied by the screen width	lane width ratio
6	412	1	59	1
9	278	0.66903	342	0.678439
12	209	0.50942	255	0.507041
15	167	0.39951	206	0.410384
18	139	0.33463	173	0.33333
21	118	0.28431	147	0.286032
24	105	0.25245	130	0.251393
27	94	0.22354	118	0.224852
30	85	0.20343	106	0.203626
33	77	0.18382	96	0.183252
36	71	0.16872	88	0.167893
39	65	0.15482	81	0.155632
42	63	0.14751	79	0.147801

Figure 17. Experimental results of single - head distance measurement

(2) Analysis of experimental detection results

The measured results show that the system has extremely stable performance in the daytime, night and road conditions when the distance is more than 10 meters. The experimental results in different situations are shown in Figure 15.

In Figure17(a), when the distance between the vehicles is too far, the bottom shadow can not be searched, the taillight information is used to calculate the distance. In Figure 17(e), it can be seen that the system can also adopt the previous case when the lane line information is lost. The lane line correction mechanism finds the lane line and calculates the shadow position of the vehicle bottom, and performs the calculation of the distance conversion; when there are complicated road signs in Figure

17(f), the system can still work stably. In Figure 17(c), because the light interference is too strong, the vehicle correction mechanism cannot be correctly started, resulting in an error in the capture of the vehicle, resulting in a large error between the actual measurement and the actual distance. All the types of this situation are used in this paper. The environment rarely appears, and for the most part, the overall system work presented in this paper is stable.

3.3This paper binocular distance model distance detection strategy and evaluation

In the binocular vision ranging strategy of this paper, the parallax of the vehicle in the left and right pictures is first obtained. In order to achieve this goal, the results obtained by the vehicle search algorithm described above need to be obtained, and the left view picture can be obtained through the vehicle search algorithm. Figure 18 (a) The coordinate point (x1, y1) of the vehicle, searching for the position of the same vehicle on the right-view screen, and obtaining the coordinate point (x2, y2) of the vehicle of Figure. 18(b), the disparity value is |x2-x1| And the parallax calculation in day and night is the same, but the way to capture coordinate points is different.

(a) Left perspective picture (b) Right perspective picture

Figure 18. Left and right perspective coordinate points for schematic drawing

After obtaining the coordinate points and calculating the disparity value, the next important step is to convert the real distance. This paper uses the geometric relationship to find the relationship between the parallax and the distance, thus indirectly calculating the distance of the current vehicle. As shown in Figure 19, some geometric relationships between the parallax and the actual distance are calibrated, and according to these relationships, the true distance can be derived, where Object represents an object in a three-dimensional space, and d represents a target object in the three-dimensional physical space and the true state of the camera. distance C_x, C_y is the optical center of the camera lens, w1 and w2 represent the optical center spacing of the two cameras, f represents the focal length of the camera, and O_x and O_y represent the projected position of the object in the three-dimensional physical space in the two-dimensional image.

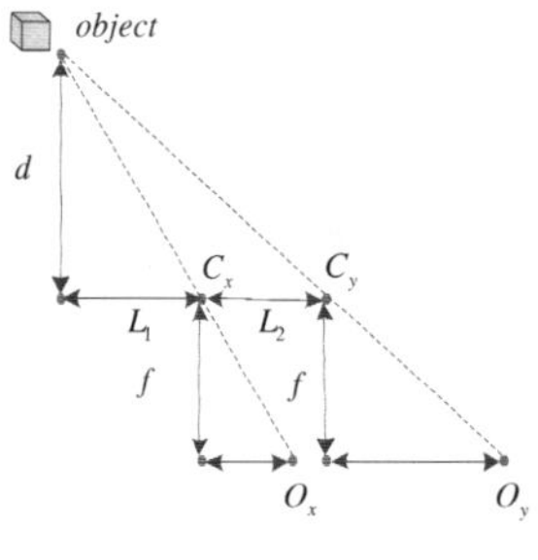

Figure 19. The Geometric Relationship between the Three - dimensional Physical Physical Position of Target and the Projection Position in Two - dimensional Image

From the description of the above geometric relationship, the geometric relations (1) and (2) can be obtained:

$$d : f = L1 : O_x \tag{1}$$

$$d : f = (L1 + L2) : O_y \tag{2}$$

After finishing, take the dimension to get the formula (3):

$$d = \frac{f(m) * L2(m)}{O_x - O_y(pixel)} * \frac{1}{m_p(\frac{m}{piexl})} \tag{3}$$

In order to obtain the error between the distance and the actual distance obtained by the parallax conversion, in the actual experiment, the corresponding table of the actual distance and the parallax measurement when the camera spacing is 20 cm is first measured in advance. The distance between the known parameters of the two cameras is 0.22 meters, the focal length is 0.00367 meters, and the experimentally measured data is substituted into equation (3), and the error value of the theoretical distance between the actual distance and the parallax calculation is observed. The specific data is shown in Table 2.

It's observed from table2 that after the distance between the two-lens parallax distance conversion exceeds 10 meters, the distance converted from the formula and the actual distance error become more and more larger, which seriously affects the accuracy of the system and poses a danger to the driver. In addition, it can be seen from table 2 that after the distance of the two-lens parallax exceeds 20 meters, there is almost no difference in the parallax value, and the distance cannot be resolved by the parallax value. The single-lens ranging method proposed above cannot accurately measure the distance for too close distance, and the double-lens parallax distance conversion cannot accurately measure the distance at a long distance. Therefore, in order to make the system accurate distance measurement in the far, medium and close distance, the advantages of combining single and double lens distance conversion are integrated. In the long range, the single lens is used for distance conversion, while the close distance is used for double lens parallax distance conversion.

Table 2. Comparison of Theoretical and Actual Distance of Parallax Ranging

camera spacing - 20 cm			
actual distance (m)	parallax (pIxel)	theoretical distance (m)	error rate(%)
6	59	6.00	
9	42	8.37	7.00%
12	37	9.65	19.58%
15	32	11.46	23.60%
18	29	12.15	32.50%
21	27	13.05	37.85%
24	26	13.72	42.83%
27	25	14.27	47.14%
30	24	14.75	50.83%
33	24	14.75	55.30%
36	24	14.75	59.03%
39	23	15.39	60.54%
42	22	15.99	60.02%

4. Experimental results and analysis

Figure 20 and Figure 21 show the results of the system's distance detection in the single-and-double-lens switching mode, including day and night. As shown in Figure 18, even if only part of the characteristics of the vehicle can be detected within the range of ROI set before the article, the system can accurately detect the distance of the vehicle; as shown in Figure 21, there is a street lamp at night. In the case of reflective road surface, the system can still achieve stable functions.

(a) Frame- 185 (b) Frame- 186

Figure 20. Daytime distance detection results

(c) Frame- 47 (d) Frame- 48

Figure 21. Night distance detection results

The strategy of single- and double-lens switching detection distance proposed in the paper is analyzed, and the characteristics of system distance detection are summarized. The results of the analysis by a large number of experimental data analysis systems are shown in Table 3.

Table 3. System distance detection analysis table

testing scenarios	total number of pictures	correct detection number	error detection number	detection rate
daytime scene 1	4200	4125	75	98.21%
daytime scene 2	6050	5952	98	98.38%
daytime scene 3	4545	4505	40	99.11%
night scene 1	3200	3125	75	97.65%
night scene 2	6000	5455	545	90.91%
all scenes	23995	23162	833	96.52%

When detecting a vehicle, the vehicle is detected by the characteristics of the bottom shadow, the vertical direction of the vehicle, and the taillights. These methods are not only computationally intensive, but also have high accuracy. Based on this, a method based on single and double lens switching is proposed to detect the distance between vehicles. This strategy makes clever use of the geometric relationship between vehicle features and distance, which ensures the accuracy of detection. This greatly reduces the amount of computation of the system.

5. Acknowledgments

This work was supported by 2018 Natural Science Foundation of Anhui, China (1808085QF215), 2018 Foundation for Distinguished Young Talents in Higher Education of Anhui, China (gxyqZD2018050) and Anhui Key Research and Development Programs (Foreign Scientific and Technological Cooperation, 1804b06020375).

6. References

[1] Wei Kai,Sheng Jian-ping,Xiong Kai.Study on forward vehicle detection and distan-ce measurement based on shadow features[J].Machinery Design &Manufacture,2013(2) :131-133.

[2] Cheng Wen Dong,Yuan Wei,Zhang Ming Fang,Vehicle detection and longitudinal distance measurement based on fusion characteristics of rear edge.[J]. Machinery Design &Manufacture,2017(7) :152-156.

[3] Qi Yong Long.Wen Bo.Design of a wireless vehicle detection sensor [J]. Foreign Electronic Measurement Technology, 2014, 33 (1): 50-53.

[4] Guo Lei,Li Ke-qiang,Wang Jian-qiang.A feature-based vehicle detection method[J].Automotive Engineering,2006,28(11):1031-1035

[5] Chao Kui,Feng Yu Cai.Color Image Representation and Retrieval Technology Based on Color and Shape Features [J], Journal of Computer-Aided Design & Computer Graphics, 2011, 3（10）: 906-91.

[6] Shen Huan,Li Yu.Ming.A method of monocular vision detection for lane lines in str-uctured

roads [J] .Journal of Instrumentation and Instruments,2010,31(2):397-403.
[7] Zhu Zhou,Lu Xiao Bo.Algorithm of Vehicle Occlusion Based on Ellipse Fitting [J] .Chinese Journal of Scientific Instrument,2015,36(1):209-214.
[8] Sun Shu Cheng,Lang Lang,Wang Hu.Design and Implement- ation of Tracing Intelligent Vehicle Based on Path Recognition [J]. Journal of Anhui University of Technology,2012,27(1):60-63.
[9] Wang Ning,Cai Shun Yan.An Automated Vision Measurement Method Based on Uncertain Angle [J]. Science Technology and Engineering,2016,16(4):226-230.
[10] Qi Yue. An Embedded System Based on Video Sensor is a Distance Test System [J]. Journal of Southwest China Normal University (Natural Science Edition),2012,37(1):104-107.
[11] Zhang Feng-jing,Zhou Jiang,Xia Wei-jie.Research of vehicle security spacing measurements based on binocular stereovision [J].CAAI Transactions on Intelligent Systems,2011,6(1):79-84
[12] Jianzhuang,You ping Chen,Jingming Xie,Haiping Lin.Model-based Lane Detection and Lane Following for Intelligent Vehicles[C].Intermnation Conference on Intelligent Human-Machine Systems and Cybernetics,2010:170-175.
[13] Gopalan,R.;TsaiHong;Shneier,M.Chellappa,R.ALearningApproach Towards Detection and Tracking of Lane Marking[J].Intelligent Transportation Systems,2012,13(3):1088-1098.
[14] AS Huang,S Teller,Lane boundary and curb estimation with lateral uncertainties, The 2009 International Conference on Autonomic and Trusted Computing,2012:330-336.
[15] CY Chang, CH Lin,"An Efficient Method for Lane-Mark Extraction in Complex Conditions,"in 9th International Conference on Autonomic and Trusted Computing,2012:330-336.

Cloth Recommender System Based on Item Matching

Yan Lei[1], **Long Chen**[1] **and Ziyu Guan**[1,a]

[1]Northwest university, Xi'an, China

Corresponding author: ziyuguan@nwu.edu.cn

Abstract. Recommender systems have become an essential part in our life. Most social Websites use recommender systems to enhance user experience. In online shopping Websites such as Amazon, Clothing is one of the most popular domains, therefore a recommender is of great significance. Previous recommender systems were often focused on retrieving items based on user preference, i.e. similarity to the previous items purchased by the user. However, in Clothing domain, the matching relationships between candidate items and the previously purchased is also important for recommendation. For example, a user may want to buy a new jean rather than suit pants if he/she has just purchased a shirt. This kind of matching relationships also frequently occurs in other life contexts. In this paper, we aim to recommend new clothes that can better match the clothes purchased by a user. This new recommendation strategy would work better in the Clothing domain and complement the current recommendation literature. Experiment results show that our method can lead to better recommendation performance in the Clothing domain.

1. Introduction

Today's society is a network explosion world. Various network applications are endlessly emerging. In these network applications, recommendation systems are a key component. In the online shopping context, recommendation systems build a user preference model based on user purchase records or rating records, then recommend suitable products for target users according to this preference model. The research of recommendation systems not only personalizes user's experience, but also brings a great potential value and potential customers to many companies.

As a specific application of the recommendation system, clothing recommendation is important. Because the clothing is a popular domain in online shopping. Previous recommender systems often focus on retrieving items based on user preference. For example, when we want to purchase the jean, the recommender system will suggest a series of pants based on our preference. This suggestion may bring inconvenience to the user.

Many researchers believe that the common ground between purchasing the previous cloth and purchasing the new clothe is to meet the user preference or requirement. But this is different from our actual situation. When we buy clothes, we will carefully consider the quality or price of clothes. One of the key but overlooked aspect is that we also consider whether the item can match the purchased. If we have purchased one cloth A and this cloth cannot match the previously purchased, we will not wear it. Especially, If we purchase a new cloth which can match the previously purchased better, the cloth A will be forgotten. Therefore, matching relationship between the candidate and the purchased is also important for recommendation systems. The matching is related to the cloth color, texture, shape, catalog, description and so on.

Content from this work may be used under the terms of the Creative Commons Attribution 3.0 licence. Any further distribution of this work must maintain attribution to the author(s) and the title of the work, journal citation and DOI.
Published under licence by IOP Publishing Ltd

In this paper, we aim to recommend new clothes that can match clothes purchased better. Of course, we think this cloth also needs to meet the target user preference or requirement. When users buy the cloth, they will not buy the cloth hated or disliked. So, in our paper, we also consider the user preference. We take a series of experiments to test our design for recommendation systems. And final results show that our method can lead to a better recommendation performance in the clothing recommendation domain.

In summary, the contribution of this paper can be summarized as following:

(1) We find that matching strength between new items and previous items purchased can affect users' purchases and this is always ignored by current recommender systems.
(2) We fuse user preference and cloth matching into a new model.
(3) We realize and evaluate our model on real datasets.

In section 2, we formally define our primary problem. In section 3, we discuss recommender system related works. In the section 4, we demonstrate our models and our algorithms for the problem mentioned above. In the section 5, we compare our method with other baselines on Amazon dataset. In the section 6, we summarize our paper and provide the future research direction.

2. Problem Description

We first give the notations in our work. Items (clothes) and users are denoted as I= {I1, I2, ...In} and U= {U1, U2, ...Um} respectively, where n is number of items, m is the number of the users. Context information include two parts: non-visual content I^N and visual context I^v. Non-visual context involves some clothes attributes, such as brand, text, suitable age, category. Visual context are visual features of clothes such as shape, line, outline. The definition of our problem is given as following.

Problem: given a user record U_i. Item attributes I^N and image I^v are employed as context information. The target is to predicting the score of item Ii on user U_i.

3. Relate Work

Recommendation is a long-term research problem which has attracted much attention. The collaborative filtering (CF) [1] style algorithms are most popular. Further, methods of matrix decomposition are slowly emerging [2][3]. For example, A.T. Hofmann et al proposed a matrix factorization SVD [4]. This Matrix factorization approach is proposed to capture the implicit feature in documents. It specializes in mining the statistical information of documents but ignoring the time-varying user preferences which is a key factor for recommendation. To fill this gap [5], takes dynamics temporal into consideration and fuses time attribute into the matrix factorization of SVD. Since the user preferences are changing over the time. This time SVD++ approach achieves a good performance in the condition of a large scale of training data. However, it is a data-hungry approach. When trained on a small dataset, the performance will decline. Besides, it cannot well capture the user preferences because this model is poor at detecting strong associations among a small set of closely related items. We use a variant time SVD++ model to capture user preference.

Clothes recommendation systems get more and more attention since the Vartak M proposes a combination based recommendation system [6]. This paper recommends user one cloth combination according user's occasion description. But this paper just provides a good idea and a simply realize method. Further, some researchers propose scenario-oriented recommendation [7] [8] [9]. These methods all consider the cloth combination effect, but they ignore the cloth matching strength between the candidate and the user's previous item purchased. Our work exactly considers this relationship and measures the candidates' matching score.

4. Method

The key idea of our method is to fuse matching strength and user preference into a variant time SVD++ model. The final prediction relies on two important parts: user preference and matching strength. We design two corresponding models to capture them. In the following, we first describe the user preference model which is design for user interests mining. Then a cluster method is employed to compute the matching strength between new items and purchased ones. Last, we design a prediction function to combine above two scores.

4.1. The Preference Model of the User

When choosing a piece of clothing, different persons have different interests. These interests range from colors to styles. For instance, some customers like black and some others prefer white. The user preference is a key factor to affect the purchasing behavior. Hence, we propose a preference model to mining user interests. Most matrix factorization (MF) approaches are used to discover the user preferences. But preferences change over the time, e.g. a fashion style in last year maybe not popular in this year. These MF methods do not take this factor into account except the time SVD++. Time SVD++ fuses the time into the model to better capture the tendency of user preference. However, one fatal flaw of time SVD++ is that the ability measuring item relationship is weak. Because this model uses an indirect way (user's single vector) to measure the relationship of two items and just only gets a weak association. Unfortunately, in the clothing recommendation, a weak clothing association will cause a bad recommend.

On the contrary, neighborhood approach evaluates the preference of a user for an item based on ratings of similar items by the same user. Using the neighborhood approach can get a strong association between two items. But the neighborhood approach cannot capture user preference well. Because this method ignores the rating matrix's implied information. After considering the two methods pro and con, in this paper, we fuse this neighborhood into time SVD++. This guarantees our method can capture the strong association between the clothes and user's feature.

Our model as following:

$$r_{ui}(t) = \mu + b_i(t) + b_u(t) + \gamma q^T p_u(t) + (1-\gamma)\frac{\sum_{j\in N(j)} sim(i,j) * r_{vj}}{\sum_{j\in N(j)} sim(i,j)} \tag{1}$$

We add the neighbor effect to detect strong associations among a small set of closely related clothes. The r_u , i represents the predicting score of cloth Ii for user u at the time t. The mu represents the average rating of the user u, and the bi(t) represents the average rating of Ii at the time t. The $b_u(t)$ represents the average rating of user u at the time t. The back part represents the Matrix factorize. The Pu(t) represents the implied characteristics of user u and the qi represents the implied characteristics of the item. The N(i) denotes the target neighbor set of item Ii. This equation uses Matrix factorized to get the implied characteristics of user u or item Ii based on the time. So it can capture user implied characteristics and item's implied characteristics. This equation combines the advantage of the time SVD++ and the neighborhood model, so this model can mine the user's cloth preference better.

In practical, we can find that people purchased some clothes on the same time. If the purchased time is closer, the clothes are more relevant. So we take the purchased time into consideration. The equation as following:

$$simt(i,j) = \frac{\sum_{v\in u}(r_{vi} - r_i^*)(r_{vj} - r_j^*)z(v_i, v)}{\sqrt{\sum_{v\in u}(r_{vi} - r_i^*)z(v_i, v)}\sqrt{\sum_{v\in u}(r_{vi} - r_j^*)z(v_j, v)}} \tag{2}$$

$$z(v_i, v_j) = e^{-|t_{vi} - t_{vj}|} \tag{3}$$

We add the exponential decay z (v_i, v_j) to control the time effect. We can use equation (1) to predict the user preference and get score about the preference.

4.2. The Matching Strength

In daily life, in clothing domain, the matching relationship between candidate items and the previously purchased is also important for recommendation. Further, this kind of matching relationship also frequently occurs in other life contexts. Bad matching will greatly reduce the user purchase intention. In this paper, we aim to recommend new clothes that can match the purchased better. Further, we use the matching strength between the new cloth and the previous cloth combination purchased as the recommend standard. When without considering the calculated consumption, using a loop compute can get the whole matching score, called the naive method in this paper. However, in practice, it is necessary to reduce calculate amount when considering the actual matching scale will reach M*2N. The M represents candidates amount and the N represents the amount of clothes purchased. In this paper, we propose the cluster method to reduce computation. In the next, we will describe the two kinds of methods in detail.

Naive Method: naive method gets matching strength score by computing cloth combination similar strength. Because a good cloth combination always has large similar parts. These similar parts usually include two aspects. The one aspect is the cloth's no-visual context information such as the brand, the texture, the suitable crowd. The other aspect is the visual features [10] [11] [12] include texture, outline, line and so on. So, in order to get an accurate matching strength score, we must consider these two aspects.

In this paper, the naive method uses the cosine [13] to measure the similar strength among clothes. After considering these two aspects, the matching strength can be inferred as following:

$$M(i,j) = \partial sim(I_i^N, I_j^N) + (1-\partial) sim(I_i^V, I_j^V) \quad (4)$$

$$sim(i,j) = \frac{I_i . I_j}{|I_i| * |I_j|} \quad (5)$$

We set parameter ∂ to control the weight difference of no-visual feature and visual feature. When the visual feature is more important than the no-visual feature, the ∂ will be smaller. Of course, when the no-visual feature is more important than the visual feature, the ∂ will be greater. In a word, this method is very simple. When the matching scale is small, the method can quickly get the matching strength score. But this will be invalid when the matching scale is huge.

Improved method: in order to reduce calculated consumption, we adapt cluster method to classify clothes purchased into several big sets. This is inspired by our real world, e.g.: we always cluster clothes purchased into different combinations. When we consider new clothes, we need to compare the matching score between this cloth and each combination. If the matching scores are not good, we will not consider to buy this cloth. Therefore, we only need to compute matching strength between candidate and these several big sets after clustering. The scale will be reduced to M*N by using this improved method. Our improved method relies on two parts. The first part, we cluster clothes purchased into several big sets. The second part, we compute matching strength between the new cloth and each cluster. We will describe these two parts in detail.

We cluster these items by the hierarchical clustering AGES. We use C to represent the cluster in which the element can match other. All elements in the same cluster have a similar taste, or be used into the same occasion. We also use a feature vector to describe the profile of the cluster. The profile can be described as following:

$$P(C_i) = [f_1, f_2, ..., f_{n+1}, f_{n+m+1}] \quad (6)$$

We use n+m+1 dimensions to describe the cluster profile, the n dimension represents cluster's no-visual feature; the m dimension represents cluster visual feature; and last 1 dimension represents cluster last update time.

We use a matrix Ma to record the suitability (matching strength) among cluster, the matrix element Ma(i, j) represents the suitability between cluster C_i and C_j. We set parameter min_{dis} as the minimum suitability threshold in cluster. When the suitability is larger than the mindis, we think corresponding clusters can match each other well. When the suitability is less than the min_{dis}, we think corresponding clusters cannot match each other well. The algorithm is as algorithm 1 as follow Table1. Firstly, we initialize clusters and the suitability matrix, and select the maximum suitability which needs to meet the requirement of min_{dis}. Then we merge and update the two corresponding clusters, and repeat the process until without the suitability. During merging two clusters into one cluster, we need to update the profile of this cluster. When we are updating cluster, we select maximum features per dimension in two clusters. This cluster's profile can describe margin level features of the cluster. If one candidate can match this cluster, the suitability between then will ranges over (d,1), and d represents suitability of candidate and the cluster. The updating profile of clusters as following:

$$U(C_i,C_j)=\sum_{f_k\in p(C_i)}\sum_{f_t\in p(I_j)}\max(f_k,f_t) \tag{7}$$

This improved method has a small computation, and it will quickly find the candidate with higher score.

Table 1. Matching Algorithm

Item matching cluster Algorithm 1	
Input: item set S, the item no-visual context feature I^N, Item visual feature I^C	
(1)	For each j in [1, m] do
(2)	Init-cluster(C_i)
(3)	End for
(4)	For each i in [1, m] do
(5)	Ma (i, j) = M (C_i, C_j) Reference Formula (6)
(6)	End for
(7)	While(True):
(8)	C_i,C_j,s = Findmaxsuit(Ma)
(9)	If s<min_{dis}:
(10)	Break
(11)	C_i=Merging (C_i, C_j)
(12)	Renew-number ()
(13)	Update(Ma)
(14)	End for
Output: C=C_1,C_2,.....Ck	

4.3. The Final Composition

This improved method has a small computation, and it will quickly find the candidate with higher score.

$$r_{ui}^*=r_{ui}*\frac{\sum_{C_j\in N_{I_i}} M(C_j,I_i)*|c_j|}{\sum_j M(C_j,I_i)*|c_j|} \tag{8}$$

r_{ui}^* denotes the final score of candidates I_i. r_{ui} denotes the preference score of I_i .N_{I_i} represents the cluster set which can match with the I_i harmoniously. In this paper, harmonious matching when the matching strength beyond the misdis. The right part of equation represents the harmonious matching set proportion based items purchased. This proportion can reflect this candidate matching strength. This combination means that we not only consider user preference, but also consider the matching.

5. Experiment

The experiments are designed to prove the effectiveness of our method. And we conduct comparisons of our method with other baseline method on Amazon dataset.

5.1. Datasets
Amazon dataset: the dataset is sourced from Amazon.com. The dataset includes reviews (ratings, text, helpfulness, votes), product metadata (descriptions, category, price, brand, asion, tittle) and product visual feature. This visual feature has employed a pre-trained convolution neural network, and obtain visual features. We use the visual feature to conduct experiments. In our experiments, we just use one catalog Clothing, Shoes and Jewellery. The statics information about the Amazon datasets as the table (2).

Table 2. Statics about the Amazon datasets (before) during Mar.2003-Jul.2014

Dataset	users	Items	Total
reviews	199,748	331,173	5,748,920
metadata	-	1,503,384	1,503,384
Visual features	-	1,503,384	3.2G

5.2. Preprocessing
Most user items in Amazon dataset just have few purchase records which has insufficient information for clothes recommendation. Therefore, we remove these users' records. We choose 500 users and corresponding items' metadata information to conduct the experiment. After preprocessing, we get the user record, including (user-ID, user-name, item-ID, rating, time), the item information record includes (item-ID, item-title, price, asin, brand, description, catalog, visual feature).

5.3. Baselines
We compare our model MPS with two baselines: SVD++, time SVD++.
SVD++: this baseline first uses matrix decomposition to process the user rating matrix, then obtains two singular vectors, and finally uses these two singular vectors to describe the potential characteristics of users and products.
time SVD++: This model considers the user's dynamic preference. This model takes the user's purchase time as part of the user rating, then use matrix decomposition to process the user rating matrix.

5.4. Metric
In our experiment, we use RMSE metric and RMSE D-value to evaluate our model.
RMSE: the quality of the results is measured by their root mean squared error.

$$RMSE = \sqrt{\frac{\sum_{(u,i)\in Testset} (r_{ui} - r_{ui}^*)^2}{|Testset|}} \tag{9}$$

This measure puts more emphasis on large errors compared with the alternative of mean absolute error. Nonetheless, there is evidences that small improvements in RMSE terms can have a significant impact on the quality of the top few presented recommendations.
RMSE D-values: the RMSE difference value.

5.5. Performance
We conduct some experiments in this dataset. First, we conduct our method in different user's feature f and bins b. We set ∂=0.5, γ=0.5, min_{dis}=0.5. Then we show our method MPS training performance in the validation set. We run the dataset for twenty times repeatedly and get the average value of RMSE. The result as the figure 1.
In the figure 1, we intuitively discover that features affect the time the global minimum is found. When the f equals 20 and b equals 20, the global minimum shows at the iteration number equals 20. The other's global minimum shows at the iteration number nearly equals 10. More features lead to

more redundant information, and increasing the time overhead of matrix decomposition. So when the f equals 10, we find the global minimum later. Then, the user's feature also affects the model shape. When the feature equal to 10, the model as a whole is linear. And when the feature equal to 20, the model as a whole presents a curve with a slower rate of decline. When the user feature increasing, the step size of each updating will become smaller during the parameter training process of the model. User feature increasing will directly increase the parameter training amount. Finally, users' bins affect the model training speed. As our bias increasing, the average squared error of the model increases rapidly after the model reaching a minimum.

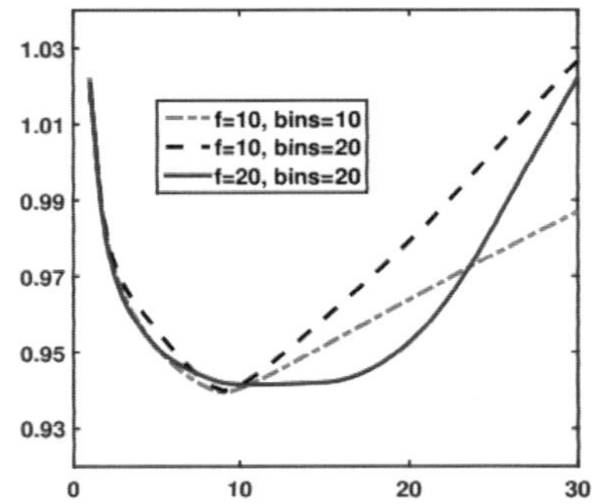

Figure 1. The model performance on the different features and bins in validation set

Secondly, we compare with baseline methods. We set ∂=0.5, γ=0.5, min_{dis}=0.5. And we show the RMSE difference value in the different feature and different bins, comparing with our method. The result is shown in the figure 2. The figure 2 describes the RMSE D-value.

From Figure 2, we can see that our overall model performs better than other methods. In the Amazon dataset, the association between the cloth and cloth is more strong than the movie dataset. Matrix decomposition models maybe poor at detecting strong associations among a small set of closely related items, such as SVD++, time SVD++. Our method combines the neighbour effect, so we can capture the associations and have a good performance. Our method MPS considers the cloth matching, so the performance is better than cft SVD++ (not consider cloth matching). Then, the SVD++ performs better than the time SVD++. Users' preference in clothing are not as personal as movies and music. The preference of clothing is largely influenced by the public aesthetics, so the change in clothing preferences usually needs a long time span. So the SVD++ has a good performance than the time SVD++.

From our experiment, we can find that this new strategy would work better in the clothing domain and complement the current recommendation literature.

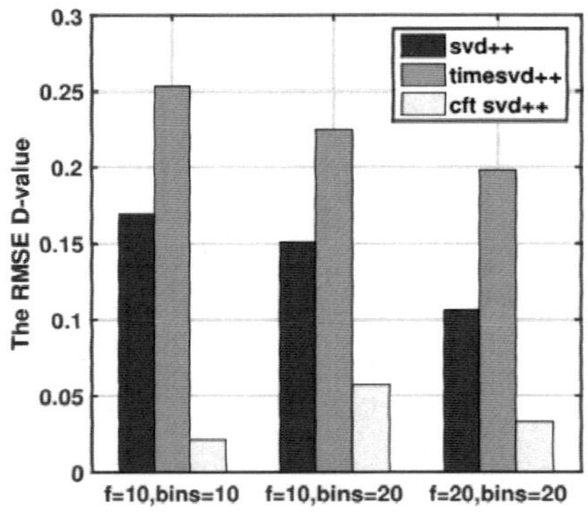

Figure 2. The comparison performance based on our Method MPS in the different feature and bins

6. Conclusion

Now, we will summary this paper. First of all, one of the key but overlooked aspect is the matching strength between the candidate item and the previously purchased. This matching relationship frequently occurs in our daily life. In this paper, we recommend new cloth that can match the previous clothes purchased by a user. We fuse user preference and matching strength into one new model. We conduct and evaluate our model, and experiments proved that our method is more effective than other methods. However, besides the matching with the cloth purchased and the user preference, the cloth recommend is affected by many other factors, such as regional. In the further, we will further consider more factors to produce a more accurate clothing recommend.

Acknowledgments

Authors wishing to acknowledge assistance or encouragement from colleagues, special work by technical staff or financial support from organizations should do so in an unnumbered Acknowledgments section immediately following the last numbered section of the paper.

References

[1] David Goldberg, David Nichols, Brian M Oki, and Douglas Terry 1992. Using collaborative filtering to weave an information tapestry ACM 35,12 (1992), 61–70

[2] David M Blei, Andrew Y Ng, and Michael I Jordan. 2003 Latent dirichlet allocation. J Machine Learning Research Archive 3 (2003), 993–1022.

[3] Andrew Fano and Scott W Kurth 2003.Personal choice point: helping users visualize what it means to buy a BMW. In International Conference on Intelligent User Interfaces. 46–52.

[4] Thomas Hofmann. 2013. Latent semantic models for collaborative filtering ACM.

[5] Yehuda Koren. 2010. Collaborative Filtering with Temporal Dynamics Commun ACM.

[6] Manasi Vartak and Samuel Madden. 2013. CHIC: a combination-based recommendation system. SIGMOD.

[7] Edward Shen, Henry Lieberman, and Francis Lam. 2007.What am I gonna wear: scenario-oriented recommendation. In International Conference on Intelligent User Interfaces

[8] Bo Wang, Jichang Guo, and Yan Zhang. 2015. An Analysis and Application of Fast Nonnegative Orthogonal Matching Pursuit for Image Categorization in Deep Networks. Mathematical Problems in Engineering

[9] Domonkos Tikk 2007 Major components of the gravity recommendation system SigKDD Explorati ons.

[10] Ruining He and Julian Mcauley. 2016. Ups and Downs: Modelling the Visual Evolution of Fashion Trends with One-Class Collaborative Filtering. In International Conference on World Wide Web. 507–517.

[11] Julian Mcauley, Christopher Targett, Qinfeng Shi, and Anton Van Den Hengel 2015. Image-Based Recommendations on Styles and Substitutes. SIGIR

[12] Sida Gu, Xiaoqiang Liu, Lizhi Cai, and Jie Shen. 2017. Fashion Coordinates Recommendation Based on User Behavior and Visual Clothing Style. In Proceedings of the 3rd International Conference on Communication.

[13] David C Anastasiu and George Karypis 2015. PL2AP: Fast Parallel Cosine Similarity Search. In Proceedings of the 5th Workshop on Irregular Applications: Architectures and Algorithms (IA3 '15). ACM, New York

Design of Lightweight Convolutional Neural Network Based on Dimensionality Reduction Module

Yue Zhou[1,a], Yanyan Feng[1], Shangyou Zeng[1] and Bing Pan[1]

[1]College of Electronic Engineering, Guangxi Normal University, 541000 Guilin, China

Abstract. The traditional CNN extracted image features is insufficient, so the classification accuracy is not ideal. It also brought overmuch model parameters and calculation. Based on these problems, this paper proposes a dimension reduction residual module. First, reducing the dimension of the output feature map. Then, feature extraction used two different sets of convolution kernels. It can get more sufficient and different characteristic information. Two sets constitute the cascaded layer. Finally, the cascaded layer act as the input of the next layer. This module can reduce parameters. Meanwhile, it increases the depth of the network and enriches the diversity of feature acquisition. A new convolution neural network is build through this module. The performance of new network and other recognition algorithms is compared on GTSRB and 101_food datasets. The new network model is reduced to about 6.2MB, and the classification accuracy can reach 98.2% on GTSRB, 72.3% on 101_food. The experimental results show that this module can effectively improve network performance and control model size.

1. Introduction

Since Geoffrey Hinton proposed the concept of deep learning[1] in 2006, deep learning has become the hottest research field of artificial intelligence. It has made breakthroughs and solved the long-term stagnation in the field of artificial intelligence. The situation has set off an innovative revolution. In terms of recognition or natural language processing, deep learning algorithms outperform other algorithms. So far, deep learning algorithms have become the top algorithms in the field of artificial intelligence. The artificial intelligence technology represented by deep learning has begun to slowly change people's daily life. Researchers have made major breakthroughs in many aspects, such as automatic driving and real-time voice processing. It has entered a new stage.

CNN[2] is an important part of the deep learning algorithm. Up to now, various excellent models have been proposed, such as VGG[3], GoogLeNet[4], ResNet[5]. The CNN network model has been pursuing the recognition accuracy. With the development of the network, the recognition accuracy is continuously improved, and the parameter quantity is also increasing. However, the development of computer hardware cannot keep up with artificial intelligence and big data, so in this era of massive data, we need to streamline the parameters of the network, thereby reducing the requirements on computer hardware. To ease the pressure on the computer and pursue the better recognition accuracy, this paper proposes a lightweight convolutional neural network named ReduceNet, which can run on a limited-capacity hardware possibly.

In order to design a lightweight CNN with high recognition rate, we mainly made the following contributions: (1) lightweight convolution sampling module Reduce module is proposed. In the case of the same input and output, the module has fewer parameters than the standard convolution. (2) Based on the basic Reduce module,

Content from this work may be used under the terms of the Creative Commons Attribution 3.0 licence. Any further distribution of this work must maintain attribution to the author(s) and the title of the work, journal citation and DOI.
Published under licence by IOP Publishing Ltd

we propose an improved module ReduceV2 module. The module has stronger expressive ability than the Reduce module. (3) Based on the proposed two modules, two lightweight and efficient convolutional neural network architectures are designed and compared, ReduceNet and ReduceV2Net.

2. Convolution Neural Network

The CNN training process is divided into forward and backward propagation. Forward propagation is used for extracting feature and obtaining image features through a series of transformations such as convolution, down sampling, and so on. Backpropagation uses the traditional BP mechanism to propagate the error forward by layer and use the chain derivation to update the convolution kernel (ie, weight and bias). In the application of machine learning, the convolutional input is usually a multi-dimensional array of data. The parameters of the convolution kernel are the multi-dimensional optimized arrays.

The excitation layer of CNN is mainly used to perform nonlinear features. The excitation function has been changing. They include the early saturated linear functions such as sigmoid, tanh and Relu[6]. The Relu solves the phenomenon of gradient dispersion to some extent. It is now widely used. In the forward calculation and back propagation calculation, it makes the network avoid the sigmoid or tanh complex exponential calculation and function derivation. Due to simple calculation, the CNN saves the computational memory and reduces the training time.

However, the work done by the incentive layer is far from enough to move the network to the hardware. We still need to do more. The related work we have done is shown below.

3. Convolution Neural Network Model Based on Dual Channels and Cascading Modules

3.1. Model Principle

The traditional convolutional neural network adopts a alternate channel between convolutional layer and downsampling layer. In the process of feature extraction, model parameters and computational complexity are large. Due to the unity of the channel, the features acquisition are not sufficient. Figure 1 shows the standard convolution structure.

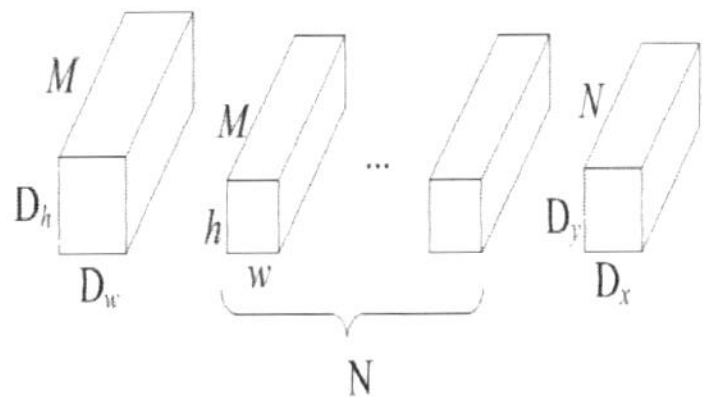

Figure 1. Tradictional 3*3 convolution kernels convolution structure

The Reduce module is the core building block of the ReduceNet model. Divide a convolutional layer into a dimension reduction layer and a sampling layer, each with a ReLU activation layer. The dimension reduction layer uses the 1x1 convolution kernel to reduce the number of the input feature map. The sampling layer uses two c (G1 and G2) to extract features. G1 uses a layer of 3*3 convolution kernels. G2 uses two layers of 3*3 convolutions. The Reduce module is shown in Figure 2.

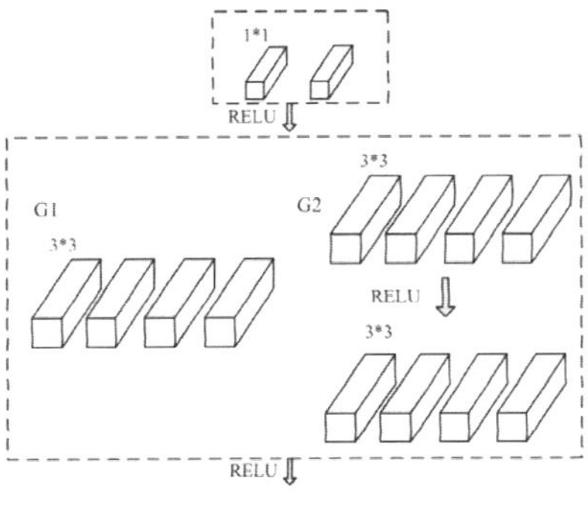

Figure 2. Reduce model

The Reduce module adopts two important strategies: (1) Using a 1×1 compressed convolution layer, reduces the number of input, which can greatly reduce the convolution computational complexity and parameters. (2) Feature extraction use two different scales of convolutions. The traditional network is basically a stack of convolution layers. VGG used large number of 3×3 convolution layers. Although its accuracy is improved, there are more parameters. In fact, the feature maps can use multiple and different sizes convolution kernels to obtain different scale features. Then combine these features to get better features[7]. By applying these two strategies, ReduceNet reduces the model size while ensuring network accuracy.

The Reduce module is explained by 128 feature map inputs. Firstly, 128 input are reduced to 64 by 64 1*1 convolutional. Then, the layer uses dual channel extraction feature. One channel uses 64 3*3 convolution kernels. The other channel of two 16 3*3 convolutional layers. Lastly, the output is the layer after the cascade of two channels. The output is the input of the next layer. When the input and output are 128 feature maps, the traditional 3*3 convolution parameter is 128*3*3*128=147456. The Reduce module parameters are 128*1*1*64+64*3*3*64+64*3*3*64+64*3*3*64=118784. Compared with the parameters, the Reduce module is less.

In the traditional CNN, the areas of the Pooling are not overlapping. Non-overlapping ignores the neighboring pixels influence on the features, which will cause the network accuracy to decrease. Overlapping Pooling can alleviate the over-fitting to some extent. However, overlapping structures may introduced noise. If Max Pooling is too much overlap, there is a greater possibility of amplifying the noise. If adding the Avg Pooling, it can reduce the noise introduced. Therefore, ReduceNet uses the overlapping Max+Max+Max+Avg in the Pooling layer.

The structure of ReduceNet is the alternate stack of the Reduce module and the downsampling layer, inserting BatchNorm layer[8] and ReLU active layer between each layer, as shown in Figure 3. Most of the parameters come from the fully connected layer. It results in over-fitting. In order to prevent over-fitting and achieve lightweight, ReduceNet uses the global average pooling instead of the fully-connected layer [9].

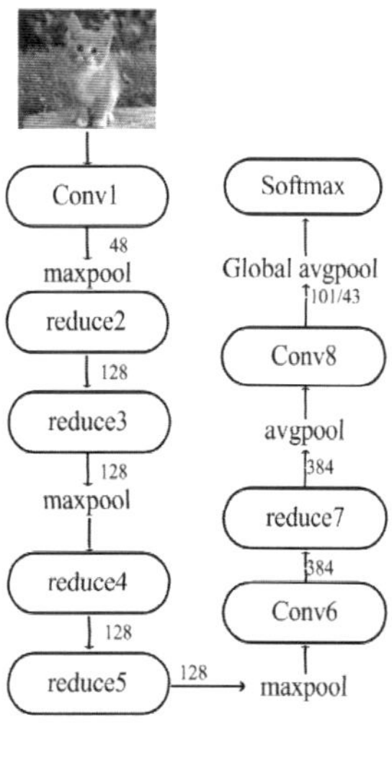

Figure 3. ReduceNet

3.2. Deformation Model

As shown in Figure 4, Suppose the input of the layer is x, the output is F(x), and the activation values of the upper and lower layer are added as H(x), ie H(x) = F(x) + x. The output H(x) is then input to the next layer. Add an identity mission, convert the original function H(x) to F(x)+x. This connection structure is called shortcut connection. It decompose a problem into multiple scale direct residual problems. It can optimize the training effect. Furthermore, this simple addition does not add extra parameters and calculations, but it can greatly increase the training speed of the model and improve the training effect.

The traditional CNN increases the depth of the network by stacking convolutional layers, thereby improving the claasification accuracy. However, excessive convolution layers can cause gradient dispersion problems. It is impossible to effectively update the gradient to the previous layer, so that parameters cannot be updated.

When the number of layers is deepened, the shortcut connection can solve the degradation problem well. It can better spread the gradient to a shallower level in the backpropagation process. Thus, it solves the gradient dispersion problem of traditional CNN. The shortcut connection is added between the input and output. When the Reduce module gets shortcut connection, it becomes the deformed structure named ReduceV2 module, as shown in Figure 5. Then, the ReduceV2 network is designed with the ReduceV2 module.

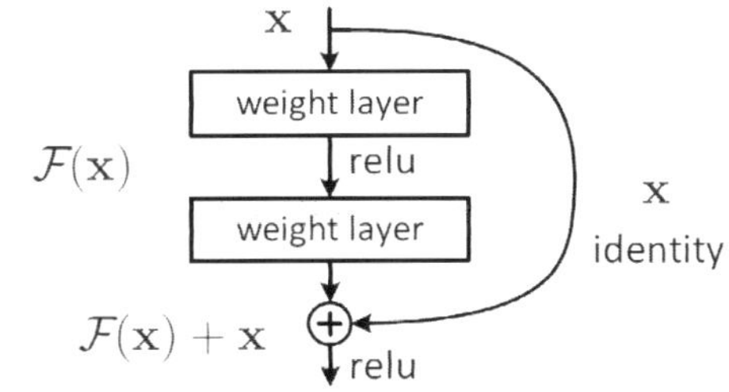

Figure 4. Shortcut connection

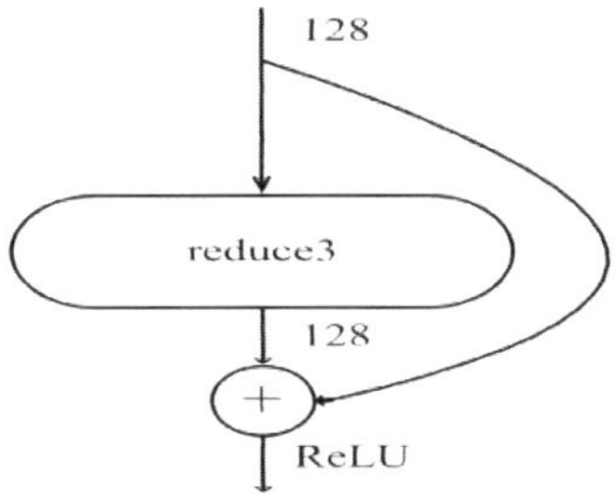

Figure 5. ReduceV2 module

4. Experiment

4.1. Experiment Setup

This experiment was conducted using the caffe[10] of deep learning framework. The computer was configured as an i7-6700K quad-core CPU, Ubuntu 14.04 operating system, 32 GB memory, and NVIDIA-GTX1070 GPU. Experimental images come from the currently widely used 101_food dataset and GTSRB dataset.

The 101_food dataset has a total of 101,000 sheets, 101 food images, and 1000 images for each type of food. According to 3:1, it is divided into training set and test set. As shown in Figure 6.

The GTSRB dataset is a traffic sign dataset with a total of 51,831, 43 traffic signs. The training set contains 39,209 images and 12,432 test sets. All images were taken in a natural environment. As shown in Figure 7.

Figure 6. 101_food partial sample

Figure 7. GTSRB partial sample

When networks train on the GTSRB dataset, the solver file is configured with an initial learning rate of 0.005, a learning rate change of multistep, a gamma of 0.1, a stepvalue of 24000 and 48000, a maximum number of iterations of 60000, and a test of 500 times per training. When networks train on the 101_food dataset, the solver file was configured with an initial learning rate of 0.005, a learning rate change of multistep, a gamma of 0.1, a stepvalue of 40000, 80000, and 120000, a maximum number of iterations of 150000, and a test of 500 times per training.

Data enhancement, normalization, and de-averaging of all images prior to training[11]. All images are converted to 256*256 specifications, and the average value of the training images is subtracted for each pixel of the image. All images are randomly cropped to a size of 227*227 in five orientations (upper left, upper right, lower left and lower right and middle) and horizontally flipped to increase the amount of data.

4.2. Experimental Results and Analysis

This experimental evaluation uses the ReduceNet, ReduceV2Net, the traditional CNN (TraNet) and others published identification methods to identify the performance on the GTSRB and 101_food datasets. Through the experimental data, Table 1 and Table 2 are obtained, which are the accuracy and model size obtained by each network.

Table 1 Performance of each model on 101_food

Network model	Accuracy(%)	Model size(MB)
TraNet	53.1	78.9
AlexNet	56.7	234.8
VGG[12]	59.3	553.6
ResNet[12]	67.4	95.1
DCNN[13]	68.4	—
FOOD-DCNN[13]	70.4	—
Squeezenet[13]	52.3	3.2
ReduceNet	71.0	6.4
ReduceV2Net	72.3	6.4

Table 2 Performance of each model on GTSRB

Network model	Accuracy(%)	Model size(MB)
LeNet-5	87	—
SqueeNet	94.2	3.0
AlexNet	95.9	233.8
TraNet	96.6	77.9
GoogLeNet	96.5	41.8
ReduceNet	98.2	6.3
ReduceV2Net	98.2	6.3

Figure 8 shows the variation of the test accuracy of the two networks during 101_food training. When the maximum number of iterations set to 15000 is reached, the classification accuracy of ReduceNet and ReduceV2Net is still significantly higher than other methods. Compared with the SqueezeNet model, the model size has increased, but the accuracy has increased by about 19%.

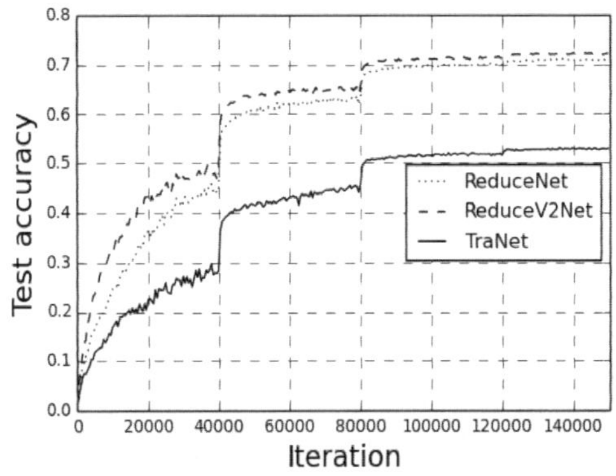

Figure 8. The model on the 101 _food training curve

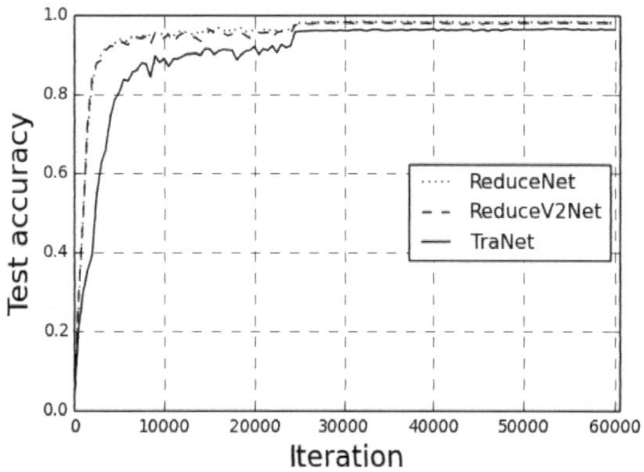

Figure 9. The model on the GTSRB training curve

Figure 9 shows the variation of the test accuracy of the two networks during GTSRB training. It can be seen that when the maximum number of iterations is reached, the classification accuracy of two network are better than other method. The classification accuracy of two network can reach 98.2%.

In terms of accuracy, the ReduceNet and ReduceV2Net has better performance than the others methods. In terms of model size, the ReduceNet and ReduceV2Net is more lightweight. Compared with the SqueezeNet model, although the model size has increased, the accuracy of model classification has been greatly improved. It can be seen from the experimental results that compared with other network models, the classification performance of ReduceNet and ReduceV2Net is better and lightweight. So they are more convenient for network hardware application and bring better performance than others published identification methods. Combining two comparative experiments can lead to the below conclusions:

(1) This module can effectively improve network performance and control model size.
(2) Dual-channel feature extraction contributes to the overall recognition rate
(3) The dimensional reduction of the 1*1 convolutional can effectively control the network parameters
(4) Using global pooling instead of the traditional fully connected layer can effectively reduce network parameters.

5. Conclusion

Based on the Reduce module, we proposed an more optimized ReduceV2 module. The module is used to design the lightweight ReduceV2Net. The experiment found that the ReduceV2 module has stronger performance in classification. To verify the effectiveness of two networks, we conducted experiments. Experimental data proves that the ReduceNet and ReduceV2Net have a good effect in reducing the model size and improving the classification accuracy. They are convolutional neural network model with less memory requirements and convenient portability. The latter work is to continue to improve on this module, further test the performance of the module in other fields, such as target detection, image segmentation, etc. Test and optimize on larger dataset to compare the performance of the network.

References

[1] Hinton G E, Salakhutdinov R R. Reducing the Dimensionality of Data with Neural Networks[J]. Science, 2006, 313(5786):504.
[2] LECUNY,BOTTOU L,BENGIO Y,et al.Geadient-based learning applied to document recognition[J].Pr Oceeding of the IEEE,1998,86(11);2278-2324.
[3] Simonyan K, Zisserman A. Very Deep Convolutional Networks for Large-Scale Image Recognition[J]. Computer Science, 2014.
[4] Szegedy C, Liu W, Jia Y, et al. Going deeper with convolutions[C]// IEEE Conference on Computer Vision and Pattern Recognition. IEEE Computer Society, 2015:1-9.
[5] He K, Zhang x, Ren S, et al. Deep Residual Learning for Image Recognition[C]// IEEE Conference on Computer Vision and Pattern Recognition. IEEE Computer Society, 2016:770-778.
[6] V. Nair, G. E. Hinton. Rectified linear units improve restricted boltzmann machines[C].In Proc. 27th International Conference on Ma-chine Learning, 2010.
[7] Szegedy C, Vanhoucke V, Ioffe S, et al. Rethinking the inception architecture for computer vision[C] //Proc of the IEEE Conference on Computer Vision and Pattern Recognition. 2016: 2818-2826.
[8] Ioffe S, Szegedy C. Batch Normalization: Accelerating Deep Network Training by Reducing Internal Covariate Shift[J]. Computer Science, 2015.
[9] Lin M, Chen Q, Yan S. Network In Network[J]. Computer Science, 2013.
[10] Jia Y, Shelhamer E, Donahue J, et al. Caffe: Convolutional Architecture for Fast Feature Embedding[C]// ACM International Conference on Multimedia. ACM, 2014:675-678.
[11] Krizhevsky A, Sutskever I, Hinton G E. Imagenet classification with deep convolutional neural networks[C]//Advances in Neural Information Processing Systems. 2012: 1097-1105.
[12] Heravi E J, Aghdam H H, Puig D. An optimized convolutional neural network with bottleneck and spatial pyramid pooling layers for classification of foods[J]. Pattern Recognition Letters, 2017.
[13] Yanai K, Kawano Y. Food image recognition using deep convolutional network with pre-training and fine-tuning[C]// IEEE International Conference on Multimedia & Expo Workshops. IEEE, 2015:1-6.
[14] Iandola F N, Moskewicz M W, Ashraf K, et al. SqueezeNet: AlexNet-level accuracy with 50x fewer parameters and <1MB model size[J]. 2016.

IOP Conf. Series: Materials Science and Engineering **533** (2019) 012046 doi:10.1088/1757-899X/533/1/012046

Short Text Classification Improved by Feature Space Extension

Yanxuan Li

School of Computer Science, Beijing University of Posts and Telecommunications, Beijing, China

liyanxuanandy@163.com

Abstract. With the explosive development of mobile Internet, short text has been applied extensively. The difference between classifying short text and long documents is that short text is of shortness and sparsity. Thus, it is challenging to deal with short text classification owing to its less semantic information. In this paper, we propose a novel topic-based convolutional neural network (TB-CNN) based on Latent Dirichlet Allocation (LDA) model and convolutional neural network. Comparing to traditional CNN methods, TB-CNN generates topic words with LDA model to reduce the sparseness and combines the embedding vectors of topic words and input words to extend feature space of short text. The validation results on IMDB movie review dataset show the improvement and effectiveness of TB-CNN.

1. Introduction

As the development of mobile Internet, a large amount of meaningful short text data is generated in many ways, such as community comments, instant message applications, personal blogs, etc [1]. Therefore, how to deal with short text classification has become a problem that needs to be solved urgently. For instance, an effective classification method to distinguish malicious community comments and non-malicious ones can help create a nice user experience. However, short text has a series of features, such as shortness, sparsity, lack of semantic and contextual information [1-2]. It brings challenges for traditional methods to achieve good performance. As for reason, traditional methods of word representation, such as bag-of-word (BOW), has difficulty to express contextual information well [3]. In addition, statistical learning methods used for long documents classification need adequate word co-occurrence, such as Naive Bayes and Support Vector Machine (SVM) [1]. Moreover, it is exactly what short text lacks. Thus, these traditional methods have encountered their limitations in terms of short texts.

In related works, researches on short text mainly focus on the extension of feature space and the collection of contextual information. One way is to use external resources to enrich the feature of text. Mehran Sahami [4] and Danushka Bollegala [5] used the results returned by web search engine to measure the semantic similarity of short text. Evgeniy Gabrilovich [6] and Xiaohua Hu [7] utilized Wikipedia as an external knowledge source to compute semantic relatedness. However, with the help of external resources, these methods are quite time consuming and greatly depend on the quality of resources. Another way is to apply model-based methods, especially deep learning methods [8]. As is mentioned, word representation methods like bag-of-word does not have the ability to obtain contextual information. So Mikolov [9-11] proposed continuous bag-of-words (CBOW) model and

Content from this work may be used under the terms of the Creative Commons Attribution 3.0 licence. Any further distribution of this work must maintain attribution to the author(s) and the title of the work, journal citation and DOI.
Published under licence by IOP Publishing Ltd

continuous skip-gram model as a high-quality word embedding method to make up for the defects of bag-of-word model. As for classification models, Sida Wang [12] took the log-count ratio of Naive Bayes as the eigenvalue of SVM and got excellent performance in many sentiment classification datasets. Peng Wang [3] proposed a novel convolutional architecture for short text classification. Yoon Kim [13] presented a simple convolutional neural network with one-dimensional convolution layer and pre-trained word vectors, which got improvement on sentiment analysis and question classification tasks [14].

Based on the related work, we are motivated that to deal with short text classification, extending the feature space of text is an effective method. In this paper, we propose a novel topic-based convolutional neural network model (TB-CNN). The approach is an extension on convolutional neural network by extending feature space with topic words generated by LDA. We train the model on IMDB movie review dataset and compare the result with the state-of-the-art models. The experimental result shows that the proposed model obtains better performance than baseline models.

The rest of this paper is organized as follows: Section 2 describes the framework of the proposed method and how each part works. We present the evaluation result on dataset and compare the proposed model with baseline models in Section 3. At last, we summarize our work and draw conclusions in Section 4. In addition, we introduce future work in this section.

2. The Proposed Method

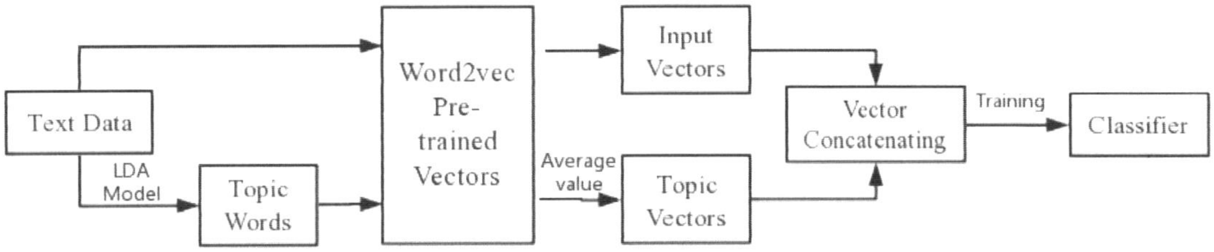

Figure 1. Framework of TB-CNN

The framework of the proposed approach is presented in this section, as shown in Figure 1. Firstly, we generate topic words with LDA model [15]. Next, we use pre-trained word vectors to transform input words and topic words into embedding vectors. Then, we calculate the mean of topic vectors for each topic and regard the result as the vector representation of this topic. With the input and topic vectors, we concatenate them to enrich the feature space. As for the classifier, we employ convolutional neural network and train it on IMDB dataset.

2.1. Word2Vec

Word2Vec is a neural network probabilistic language model proposed by Google in 2013 [9]. It embeds words with one-hot representation into a dense feature space with low dimension. So compared with traditional methods, Word2Vec model can solve the natural language processing problems with high dimensional sparse feature space [16].

The main viewpoint of Word2Vec is that words with similar contexts have similar semantics. So through training, word vector will carry contextual information. Another breakthrough achievement of Word2Vec is to provide a method to measure the similarity between words. Because the similarity between word vectors can represent their similarity in semantics, and the former can be measured by word distance such as cosine similarity.

In this paper, we use the pre-trained vector published by Google, which is trained with Google News dataset. It contains 300-dimensional vectors for 3 million words and phrases. Since the pre-trained word vector may not be suitable for specific tasks, they will be fine-tuned for our task during training.

2.2. Latent Dirichlet Allocation

Latent Dirichlet Allocation (LDA) is a topic generation model, also known as a three-layer Bayesian probability model [17]. It contains three-layer structure of words, topics and documents The model considers that each word in an article is obtained through a process of choosing a topic with a certain probability and then selecting a word from the topic with a certain probability. In addition, the relation between document and topic as well as the relation between topic and word obey a polynomial distribution. We employ LDA model here to identify the topic information hidden in the text. All mathematical symbols, abbreviations and their interpretations are integrated in Table 1.
In LDA model, the process of generating a document is as follows:

- Determine an article's multinomial distribution θ based on the Dirichlet distribution with parameter α.
- Sample from the polynomial distribution θ of the topic and generate the topic z of each word of the document d.
- Sample from the Dirichlet distribution with parameter β and generate the word distribution φ corresponding to the topic z.

Sample from the polynomial distribution φ of the word and finally generating the word w.

Table 1. Mathematical Symbols, Parameters and Their Interpretations of LDA Model.

Name Of Symbols and Parameters	Explanation
D	Corpus used to train the model
M	Number of documents in the corpus
V	Number of words in the corpus
N	Number of words in a document
w	A single word, expressed as a one-dimensional vector
w	A document, $\mathbf{w} = (w_1, w_2, \ldots, w_N)$, w_n means the represents the nth word
ε	Parameter of Poisson distribution
α	Parameter of Dirichlet distribution
β	Parameter of Dirichlet distribution
θ	The topic distribution of a document, θ~Dirichlet(α)
z	Topic of a document, z~Multinomial(θ)
k	Number of topics

The aim of LDA model is assigning latent topics to every document to estimate the document-topic distribution matrix and topic-word distribution matrix. The probability formula for the entire corpus is shown below.
The probability distribution formula for joint distribution of θ, w and z is formula (1). p(θ | α) is Dirichlet distribution with parameter α. $p(z_n|\theta)$ is a multinomial distribution with parameter θ and $p(w_n|z_n,\beta)$ is a multinomial distribution with parameter z_n and β.

$$p(\theta, z, w \mid \alpha, \beta) = p(\theta \mid \alpha) * \prod_{n=1}^{N} p(z_n \mid \theta) * p(w_n \mid z_n, \beta) \tag{1}$$

Because θ obeys a continuous distribution and z obeys a discrete distribution, for formula (1), we can integrate θ, sum z, and then multiply the generated probability formulas of all the documents in the training corpus as a joint distribution relationship.

With LDA model, we can obtain the document-topic distribution matrix, from which we can know the topic of each document. Moreover, we can find the latent keywords of each topic from the topic-word distribution matrix by ranking the probability of belonging to the topic.

2.3. Topic-based Convolutional Neural Network (TB-CNN).

2.3.1. Topic Vector Concatenating. We implement topic vector concatenating before network training. From the distribution matrixes generated by LDA model, we know the correspondence between documents and topics and the latent keywords of each topic. With pre-trained vectors of Word2Vec, we can obtain the vector representation of these keywords. For each topic, we calculate the mean of keyword vectors and regard this mean vector as the vector representation of this topic. And for each input word, we know the topic to which its document belongs. Thus, we concatenate each input word vector and its corresponding topic vector, as shown in Figure 2. The process of concatenating vectors can be summarized as shown below:

- Get the input document matrix of which the dimension is x*y. x is the number of input words of the input document, and y is the dimension of word vector.
- Find the topic words of input document with document-topic distribution matrix and transform them into word vectors.
- Calculate the mean of these word vectors as topic vector and broadcast the y-dimensional vector to an x*y topic matrix.
- Concatenate the input matrix and the topic matrix in the 1-axis dimension and obtain an x*(2y) matrix.

Vector concatenating is an important part of our approach. It represents the exploration and expansion of the short text feature space and provides a major contribution to better model performance.

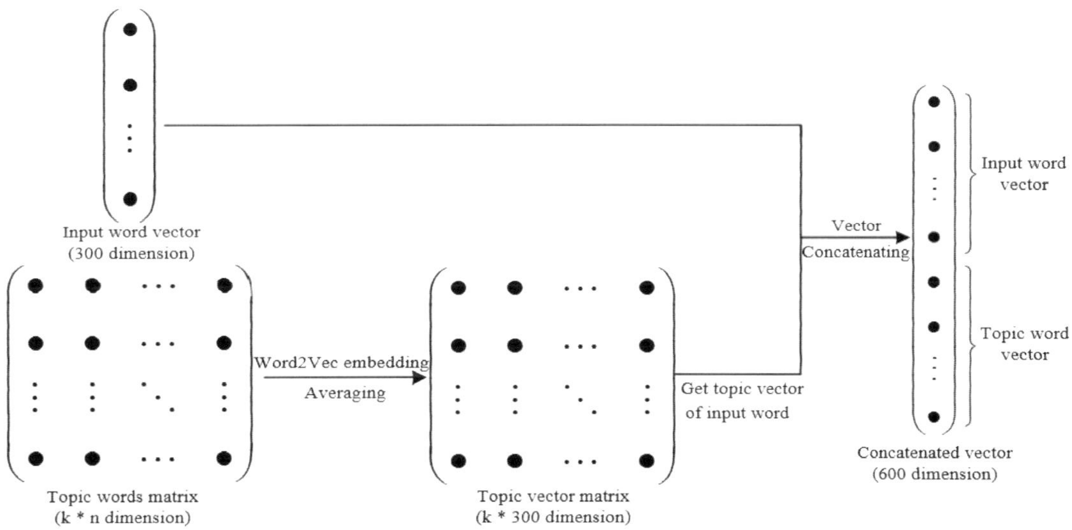

Figure 2. Process of topic vector concatenating

2.3.2. Architecture of Convolutional Neural Network. Convolutional neural networks are widely used in computer vision as part of deep learning. However, based on its characteristics, it can also be applied to the field of natural language processing. Firstly, the input layer receives a sentence of length n with necessary padding as formula (2):

$$X = \begin{bmatrix} X_1 \\ X_2 \\ \vdots \\ X_n \end{bmatrix} \tag{2}$$

We modify the embedding layer with vector concatenating mentioned in section 2.2.2, based on pre-trained vector of Word2Vec. As for convolutional layer, we perform the classification with the basic CNN architecture described by Yoon Kim [13]. We use one convolutional layer with one-dimensional convolutional kernel and multichannel. The process of a convolution operation can generate a feature map from the window of words of every filter, as shown in formula (3) and (4): (w is the parameter of convolutional layer, x represent a window of word, h is the size of window and b is a bias term)

$$c_i = f(w \cdot x_{i:i+h-1} + b) \tag{3}$$

$$c = [c_1, c_2, \cdots, c_{n-h+1}] \tag{4}$$

Then, a max-pooling layer is applied. Finally, we employ a dense layer and use softmax function as activation function to transform the score vector into a probability distribution:

$$\phi(x_i, W_z) = W_z \hat{f} \tag{5}$$

$$p(c_j \mid x_i, W_z) = \frac{\exp(\phi(x_i, W_z))}{\sum_{j=1}^{|C|} \exp(\phi_j(x_i, W_z))} \tag{6}$$

3. Experiments

3.1. Dataset

To validate the effectiveness of the proposed method, we conducted experiments respectively on IMDB movie review dataset. The description of dataset is shown in Table 2.
It is worth mentioning that truncation and zero-padding are applied to dataset so that the length of all input data will be 200.

Table 2. Description of IMDB Dataset.

Description	Values
Size of training and testing size	25000/25000
Average number of words	220
Maximum number of words	2361
Minimum number of words	5
Classes	Positive or negative

3.2. LDA Model

The main parameter that needs to be adjusted in the LDA model is the number of topics. To measure the quality of the LDA model, we use perplexity as index. The meaning of perplexity is the degree of uncertainty when a LDA model judges whether a document belongs to a topic. Therefore, a lower perplexity means a better LDA model. The perplexity of LDA model with different number of topics is shown in Table 3. As a result, we take the number of topics as 16 with lowest perplexity.

Table 3. Perplexity of LDA model with different number of topics.

Number of topic	Perplexity
10	2370.438
11	2374.400
12	2372.564
13	2373.258
14	2371.939
15	2369.386
16	**2354.322**
17	2372.139
18	2364.307
19	2360.964

3.3. Pre-trained Word Vector
In our experiments, we initialized the embedding layer with Word2Vec pre-trained word embeddings, which is publicly available. Its training corpus is Google News with every vector of 300 dimension. The size of vocabulary is 3 million, but we only load the vectors of the words found in our vocabulary and save some RAM.

3.4. Hyper-parameters of CNN
The design of the convolutional layer structure is the most important step in the task of convolutional neural networks. In our method, the number of convolutional layers and the dimension of the convolution kernels are already determined. Therefore, the hyper-parameter to be adjusted is the filter region size. Table 4 shows the effect of different multiple region sizes. The description of our network is shown in Table 5.

Table 4. Model Accuracy of CNN with Different Region Size.

Multiple region size	Accuracy (%)
(2,3,4)	91.39
(3,4,5)	91.31
(4,5,6)	**91.40**
(5,6,7)	91.19
(4,4,4)	90.86
(5,5,5)	91.01
(6,6,6)	90.80

Table 5. Description of TB-CNN.

Description	Values
Pre-trained word vector	Word2Vec
Filter region size	(4,5,6)
Activation function	ReLU
Pooling	1-max pooling
Dropout rate	0.5

3.5. Result
Compared with the state-of-the-art methods, we introduce four famous methods as baselines. The comparison between our method and the baselines is demonstrated in Table 6 and Table 7. From the result, the proposed TB-CNN achieves the best performance with topic information added in feature space. Since we extend the feature space with our method, time for model generation and testing slightly increased.

Table 6. Result for Methods above. **MNB**: Multinomial Naive Bayes [12].**Bow+SVM**: Linear SVM on Bag of Words Features. **NBSVM**: SVM with Naive Bayes Features [12]. **Textcnn**: Convolutional Neural Network for Sentence Classification [13].

Methods	Accuracy	*Precision*	Recall	*F1-score*
MNB	86.59	-	-	-
BoW+SVM	87.80	-	-	-
NBSVM	91.22	-	-	-
TextCNN	88.64	87.32	90.20	88.58
TB-CNN	**91.40**	**92.65**	**91.91**	**91.37**

Table 7. Complexity of TextCNN and TB-CNN.

Methods	Time for model generation and testing (s)
TextCNN	164
TB-CNN	498

4. Conclusion
In the field of natural language processing, short text classification is an emerging and foundational task. This paper proposes a novel convolutional neural network named TB-CNN for better performance in short text classification. LDA model is used to generate topic words of each document and enrich the feature space of short text. Pre-trained word vectors are applied to generate a better representation of text with more contextual and semantic information. As a result, TB-CNN gains better performance on IMDB movie review dataset against the state-of-the-art models.
In the future, we will focus on the exploration of short text feature space. For more effective features, we will do research in how to generate topic vector with higher quality and how to add topic information into the feature space, aiming at a further improvement in short text classification.

Acknowledgments
My deepest gratitude goes foremost to Professor Jiali Bian, my supervisor, for her constant encouragement and guidance. She has walked me through all the stages of the writing of this paper. Without her consistent and illuminating instruction, this paper could not have reached its present form. Last my thanks would go to all those who have helped me, including my friends and my fellow classmates who gave me their help and time in listening to me and helping me work out my problems during the difficult course of the paper.

References
[1] G Song, Y Ye, X Du, et al. Journal of Multimedia, 9, 635. (2014)
[2] M Chen, X Jin, D Shen. IJCAI, 1776-1781. (2011)
[3] P Wang, J Xu, B Xu, et al. WI-IAT, 1, 75-78. (2015)
[4] M Sahami, TD Heilman. Proceedings of the 15th international conference on World Wide Web, 377-386. (2006)
[5] D Bollegala, Y Matsuo, M Ishizuka. Proceedings of the 16th international conference on World Wide Web, 7, 757-766. (2007)

[6] E Gabrilovich, S Markovitch. IJCAI, 7, 1606-1611. (2007)
[7] X Hu, X Zhang, C Lu, et al. KDD, 389-396. (2009)
[8] A Hassan, A Mahmood. ICCAR, 705-710. (2017)
[9] T Mikolov, K Chen, G Corrado, et al. arXiv, 1301.3781. (2013)
[10] Q Le, T Mikolov. International Conference on Machine Learning, 1188-1196. (2014)
[11] T Mikolov, W Yih, G Zweig. NCAAL, 746-751. (2013)
[12] S Wang, CD Manning. ACL, 90-94. (2012)
[13] Y Kim. arXiv, 1408.5882. (2014)
[14] DM Blei, AY Ng, MI Jordan. JMLR, 3: 993-1022. (2003)
[15] Y Zhang, B Wallace. arXiv, 1510.03820. (2015)
[16] F Sun, H Chen. ICIEA, 1189-1194. (2018)
[17] Q Chen, L Yao, J Yang. ICALIP, 749-753. (2016)

A New Improved Boosting for Imbalanced Data Classification

Zongtang Zhang, JiaXing Qiu and Weiguo Dai

Navy Submarine Academy, Qingdao, China

qtxy_robin@126.com, Qiujx8701@126.com, dwg1968@163.com

Abstract. As one of the most important component of artificial intelligence, machine learning is getting more and more attention. AdaBoost, a classic machine learning algorithm, is widely used. However, when faced with imbalanced data classification, AdaBoost's recognition rate of minority samples is low due to ignoring class imbalance. In many cases, minority samples are of high value. For this shortage, combining the theory of margin and cost-sensitive idea, a new Boosting algorithm called CMBoost is proposed based on cost-sensitive margin statistical characteristics, which is firstly through optimizing margin statistical characteristics to improve formal algorithm and then extended by cost-sensitive. Experimental results on the UCI dataset show that the CMBoost algorithm is superior to AdaBoost for imbalanced data classification problem.

1.Introduction

In recent years, imbalanced data classification has become a hot topic in the field of data mining and machine learning. Imbalanced dataset refer to the number of samples of one class in the dataset far less than the number of other classes, in which the majority of classes are called negative classes, while the minority is called positive class. The problem of classification of imbalanced dataset exists in people's real life and industrial production. For example, E-mail filtering[1], fraud detection[2], medical diagnosis[3], DNA microarray data analysis[4], software defect prediction[5], etc. In these applications, the classification accuracy of positive class is often more important. Therefore, to improve the classification accuracy of positive class has become the focus of research on imbalanced dataset.

AdaBoost[6] is a kind of strong learning algorithm to improve the weak learning algorithm performance, which has been applied to various fields of pattern recognition successfully[7,8,9]. The original AdaBoost algorithm treats both positive and negative samples equally and is not suitable for imbalanced data classification. Cost-sensitive AdaBoost introduces classification cost into AdaBoost algorithm, which can be used for imbalanced data classification. One of them is weighting method, which modifies the weight of Boosting update steps and is simple implementation, but lacks of theoretical support. Another one is the loss function method, which directly minimizes the expectation of cost sensitive loss. The algorithm is highly theoretical, but the realization is complex, which requires numerical method to solve.

In this paper, a new Boosting algorithm called CMBoost is proposed based on cost-sensitive margin statistical characteristics. Firstly, on the basis of generalization error bound based on the margin distribution, it is proposed to reduce the generalization error by optimizing the statistical characteristics of the margin, and then extend it with cost sensitivity, so that the algorithm can adapt to the imbalanced data. Experimental results on the UCI dataset show that the CMBoost algorithm is superior to the AdaBoost algorithm.

Content from this work may be used under the terms of the Creative Commons Attribution 3.0 licence. Any further distribution of this work must maintain attribution to the author(s) and the title of the work, journal citation and DOI.
Published under licence by IOP Publishing Ltd

2. AdaBoost

The AdaBoost[10] algorithm was firstly proposed for bi-class classification. The algorithm takes as input a training set $\{(x_1,y_1),\cdots,(x_N,y_N)\}$ where with the sample x_i is an attribute value vector as a realization of the attribute set $X=\{X_1,X_2\cdots X_N\}$, and class label y_i assumes a value in Y, with two classes, assuming that $Y=\{-1,1\}$. AdaBoost calls a given base learning algorithm repeatedly in a series of rounds $t=1,2,\cdots,M$. The weight of the ith training sample on the iteration t is denoted by $D^t(i)$. Initially, all weights are set equally.

The base learner's task is to come up with a base classifier $h_t: X \to Y$ based on the distribution D^t to minimize the classification error. Once the base classifier h_t has been trained, AdaBoost chooses a parameter $\alpha_t \in R$ which measures the performance of the classification h_t. The data distribution D^t is then updated. The final classification criterion F is a weighted majority vote of the M base classifiers where α_t is the weight assigned to h_t.

3. CMBoost algorithm

We use $\Pr_D[\cdot]$ to refer as the probability with respect to D and $E[\cdot]$ to denote the expected values. Let H be a base classifier space and $C(H)$ denote the convex hull of H. A voting classifier $f \in C(H)$ is of the following form

$$f=\sum \alpha_i h_i \ \ with \sum \alpha_i = 1 \ \ and \ \alpha_i \geq 0 \tag{1}$$

For an example (x,y), the margin with respect to the voting classifier $f=\sum \alpha_i h_i$ is defined as $yf(x)$ or ρ, in other words

$$yf(x)=\rho=\sum_{i:y=h_i(x)} \alpha_i - \sum_{i:y\neq h_i(x)} \alpha_i \tag{2}$$

Schapire[11]derived the generalization error bounds based on the margin distribution, as shown in Theorem 1

Theorem 1: D_{tr} is made up of N training samples which are selected randomly from a distribution $Dist$ on the training set $X \times Y$. Assuming that the base classifier space H is limited and $\delta > 0$, for all $\theta > 0$, every voting classifier $f \in C(H)$ satisfies the following bound with probability at least $1-\delta$

$$\begin{aligned}\Pr_{Dist}[yf(x)<0] &\leq \Pr_{D_{tr}}[yf(x)\leq\theta] \\ &+O\left(\frac{1}{\sqrt{N}}\left(\frac{\ln N \ln|H|}{\theta^2}+\ln\frac{1}{\delta}\right)^{1/2}\right)\end{aligned} \tag{3}$$

As we can see, the generalization error bounds relate to the margin distribution, but how do we get a good margin distribution? We know that statistical characteristics are good tools for describing distributions. Therefore, on the basis of bounds of margin in the past, the generalization error bounds derived from literature[12] reveal the relationship between them and the statistical characteristics of margin, as shown in Theorem 2:

Theorem 2: D_{tr} is made up of $N(N>5)$ training samples which are selected randomly from a distribution $Dist$ on the training set $X \times Y$. For all $\delta > 0$, every voting classifier $f \in C(H)$ satisfies the following bound with probability at least $1-\delta$:

$$\begin{aligned}\Pr_{Dist}[yf(x)<0] \leq \frac{1}{N^{50}} &+ \inf_{\theta\in(0,1]}\{\Pr_{Dtr}[yf(x)<\theta] \\ &+ N^{-2/(1-E^2_{Dtr}[yf(x)]+\theta/9)} \\ &+ \frac{3\sqrt{\mu}}{N^{3/2}}+\frac{7\mu}{3N}+\sqrt{\frac{3\mu}{N}\Gamma(\theta)}\}\end{aligned} \tag{4}$$

where

$$\mu = 144 \ln m \ln(2|H|)/\theta^2 + \ln(2|H|/\delta)$$
$$\Gamma(\theta) = \Pr_{D_{tr}}[yf(x) < \theta] \Pr_{D_{tr}}[yf(x) \geq 2\theta/3] \tag{5}$$

In the theorem, $E[yf(x)]$ is the margin mean and $\Gamma(\theta)$ reflects the size of the margin variance. It can be seen that the margin statistical characteristics of training sample set directly affect the size of generalization error. When the size of the training sample set and the complexity of the base classifier are fixed, if the margin mean gets larger and the margin variance gets smaller at the same time, the generalization error will get smaller.

Therefore, the algorithm performance is improved by maximizing the margin mean and minimizing the margin variance. If mg_{ave} is the margin mean, mg_{var} is the margin variance and $\lambda_1 > 0, \lambda_2 > 0$ are weighing coefficient, we can construct the margin statistical characteristics

$$mg_{sc} = \lambda_1 mg_{var} - \lambda_2 mg_{ave} \tag{6}$$

Then the improved AdaBoost algorithm is represented by the following formula:

$$\min_{\alpha} \ mg_{sc}$$
$$s.t. \ \ \alpha \geq 0, 1^T \alpha = D \tag{7}$$

where

$$mg_{ave} = \frac{1}{N} \sum_{i=1}^{N} \rho_i \tag{8}$$

$$mg_{var} = \frac{1}{N-1} \sum_{i>j} (\rho_i - \rho_j)^2 \tag{9}$$

We have

$$\begin{aligned} mg_{sc} &= \lambda_1 mg_{var} - \lambda_2 mg_{ave} \\ &= \lambda_1 \frac{1}{N-1} \sum_{i>j} (\rho_i - \rho_j)^2 - \lambda_2 \frac{1}{N} \sum_{i=1}^{N} \rho_i \\ &= 2\lambda_1 [\frac{1}{2(N-1)} \sum_{i>j} (\rho_i - \rho_j)^2 - \frac{\lambda_2}{2N\lambda_1} \sum_{i=1}^{N} \rho_i] \end{aligned} \tag{10}$$

Defining

$$A = \begin{bmatrix} 1 & c & \cdots & c \\ c & 1 & \cdots & c \\ \vdots & \vdots & \ddots & \vdots \\ c & c & \cdots & 1 \end{bmatrix}_{N \times N}, c = \frac{-1}{N-1} \tag{11}$$

$$\lambda = -\frac{\lambda_2}{2N\lambda_1} I_{N \times 1} \tag{12}$$

$$\rho = [\rho_1; \rho_2; \cdots; \rho_N] \tag{13}$$

We get

$$mg_{sc} = 2\lambda_1 (\frac{1}{2} \rho^T A \rho + \lambda^T \rho) \tag{14}$$

Then we can rewrite (7) as

$$\min_{\alpha, \rho} \frac{1}{2} \rho^T A \rho + \lambda^T \rho$$
$$s.t. \ \ \alpha \geq 0, 1^T \alpha = D \tag{15}$$
$$\rho_i = y_i \boldsymbol{h}_i^T \alpha, \forall i = 1, 2, \cdots, N.$$

where

$$\boldsymbol{h}_i = [h_1(x_i); h_2(x_i); \cdots; h_M(x_i)] \tag{16}$$

$$\alpha = [\alpha_1; \alpha_2; \cdots; \alpha_M] \tag{17}$$

As $\boldsymbol{x}^T A\boldsymbol{x} \geq \boldsymbol{0}$ is always holds for any real non-zero vector $\boldsymbol{x}$, so A is positive semidefinite matrix.

In the training sample set, the size of the positive sample set L_+ is n_+ and the size of the negative sample set L_- is n_-, then the imbalance degree of the sample set is $r_u = n_- / n_+$. Suppose the weight of each sample when calculating the margin mean is d_i, the weight when calculating the margin variance is k_i and the cost sensitive parameter is c, then

$$d_i = \begin{cases} 1 + cr_u \left|\ln r_u\right|, i \in L_+ \\ 1 + c\dfrac{1}{r_u}\left|\ln \dfrac{1}{r_u}\right|, i \in L_- \end{cases} \tag{18}$$

$$k_i = \begin{cases} 1 + c\dfrac{1}{r_u}\left|\ln \dfrac{1}{r_u}\right|, i \in L_+ \\ 1 + cr_u \left|\ln r_u\right|, i \in L_- \end{cases} \tag{19}$$

We have cost-sensitive margin mean

$$cmg_{ave} = \frac{1}{N}\sum_{i=1}^{N} d_i \rho_i \tag{20}$$

and cost-sensitive margin variance

$$cmg_{var} = \frac{1}{N-1}\sum_{i>j} (k_i \rho_i - k_j \rho_j)^2 \tag{21}$$

Then we can construct the cost-sensitive margin statistical characteristics:

$$\begin{aligned} cmg_{sc} &= \lambda_1 cmg_{var} - \lambda_2 cmg_{ave} \\ &= 2\lambda_1 [\frac{1}{2(N-1)}\sum_{i>j} (k_i \rho_i - k_j \rho_j)^2 \\ &\quad - \frac{\lambda_2}{2N\lambda_1}\sum_{i=1}^{N} d_i \rho_i] \end{aligned} \tag{22}$$

Defining

$$K = diag(k_1, k_2, \cdots, k_N) \tag{23}$$

$$B = K^T A K \tag{24}$$

$$\boldsymbol{d} = [d_1, d_2, \cdots, d_N] \tag{25}$$

$$\lambda_g = -\frac{\lambda_2}{2N\lambda_1}\boldsymbol{d} \tag{26}$$

Then

$$cmg_{sc} = 2\lambda_1 (\frac{1}{2}\rho^T B \rho + \lambda_g^T \rho) \tag{27}$$

Then the CMBoost algorithm is represented by the following formula:

$$\begin{aligned} &\min_{\alpha, \rho} \frac{1}{2}\rho^T B \rho + \lambda_g^T \rho \\ &s.t. \quad \alpha \geq \mathbf{0}, \mathbf{1}^T \alpha = D \\ &\qquad \rho_i = y_i \boldsymbol{h}_i^T \alpha, \forall i = 1, 2, \cdots, N. \end{aligned} \tag{28}$$

As $x^T Bx \geq 0$ is always holds for $x^T Ax \geq 0$, then B is positive semidefinite matrix. So (28) is a convex quadratic problem.
Because the H is unknown, we can't solve the problem by usual QP solvers. As in LPBoost[13], column generation can be used to attack this problem. Then we apply column generation to this problem.
The Lagrangian of (28) is

$$\begin{aligned} L(\alpha,\rho,\boldsymbol{u},r,\boldsymbol{q}) &= \frac{1}{2}\rho^T B\rho + \lambda_g{}^T\rho + r(1^T\alpha - D) \\ &\quad -\boldsymbol{q}^T\alpha + \sum_{i=1}^{N} u_i[\rho_i - y_i\boldsymbol{h}_i^T\alpha] \end{aligned} \tag{29}$$

The infimum of L is

$$\begin{aligned} \inf_{\rho,\alpha} L &= \inf_{\rho}[\frac{1}{2}\rho^T B\rho + (\boldsymbol{u}+\lambda_g)^T\rho] \\ &\quad + \inf_{\alpha}[(r1^T - \boldsymbol{q}^T - \sum_{i=1}^{N} u_i y_i \boldsymbol{h}_i^T)\alpha] - Dr \end{aligned} \tag{30}$$

For finite infimum, $r1^T - \boldsymbol{q}^T - \sum_{i=1}^{N} u_i y_i \boldsymbol{h}_i^T = \boldsymbol{0}$ must hold. So we have

$$\sum_{i=1}^{N} u_i y_i \boldsymbol{h}_i^T \leq r1^T \tag{31}$$

For the first term in L

$$\frac{\partial[\frac{1}{2}\rho^T B\rho + (\boldsymbol{u}+\lambda_g)^T\rho]}{\partial\rho_i} = 0, \forall i \tag{32}$$

This results in $\rho = -B^{-1}(\boldsymbol{u}+\lambda_g)$, and the infimum is $-\frac{1}{2}(\boldsymbol{u}+\lambda_g)^T B^{-1}(\boldsymbol{u}+\lambda_g)$.
Through putting the results together, the dual is

$$\begin{aligned} &\max_{r,\boldsymbol{u}} -\frac{1}{2}(\boldsymbol{u}+\lambda_g)^T B^{-1}(\boldsymbol{u}+\lambda_g) - Dr \\ &s.t. \sum_{i=1}^{N} u_i y_i \boldsymbol{h}_i^T \leq r1^T \end{aligned} \tag{33}$$

We can reformulate (28) as

$$\begin{aligned} &\min_{r,\boldsymbol{u}} \frac{1}{2D}(\boldsymbol{u}+\lambda_g)^T B^{-1}(\boldsymbol{u}+\lambda_g) + r, \\ &s.t. \sum_{i=1}^{N} u_i y_i \boldsymbol{h}_i^T \leq r1^T \end{aligned} \tag{34}$$

To speed up the convergence, we add the most violated constraint by solving the following problem:

$$h^{'}(\cdot) = \arg\max_{h(\cdot)} \sum_{i=1}^{N} u_i y_i h(x_i) \tag{35}$$

4. Evaluation method

In traditional classification learning, classification accuracy is generally used as the evaluation criterion. However, for imbalanced dataset, it is unreasonable to use classification accuracy to evaluate the performance of classifiers. Therefore, for imbalanced data, many scholars propose evaluation methods such as F-measure[14] and G-mean[15], most of which are based on the confusion matrix (see Table 1). In Table 1, TP and TN respectively represent the number of positive and negative samples of the correct classification, while FP and FN respectively represent the number of positive and negative samples of the wrong classification.

Table 1. Confusion matrix for bi-class problem

	Be classified into positive class	Be classified into negative class
True positive class	TP	FN
True negative class	FP	TN

F-measure is an evaluation criterion for imbalanced data classification problem, which is defined as follows

$$F-measure=\frac{(1+\beta^2)\times Recall\times Precision}{\beta^2\times Recall+Precision} \tag{36}$$

$$Recall=\frac{TP}{TP+FN},\ Precision=\frac{TP}{TP+FP} \tag{37}$$

Only when the recall and precision are both large, F-measure will be large. Therefore, it is reasonable to evaluate the classification performance of classifiers for positive class.
G-mean is an evaluation criterion for the overall classification performance , which is defined as follows:

$$G-mean=\sqrt{Positive\ Accurary\times Negative\ Accurary} \tag{38}$$

$$Positive\ Accurary=Recall=\frac{TP}{TP+FN} \tag{39}$$

$$Negative\ Accurary=\frac{TN}{TN+FP} \tag{40}$$

G-mean is to maximize the accuracy of two classes while maintaining the balance of the classification accuracy of positive and negative class. In this article, F-measure is used to evaluated the classification performance of the positive class, while G-mean is used to evaluated the overall classification performance.

5. Experiment and analysis

To test the effectiveness of the proposed method, several UCI dataset are selected. For data containing multiple classes, one of them is taken as positive and the rest as negative. The information of UCI dataset is given in Table 2. The dataset is divided into two halves. One half is training sample set and the other is test sample set. F-measure and G-mean are used to evaluate the performance of AdaBoost and CMBoost which is shown in Table 3.

Table 2. Information of UCI dataset

Dataset	Total size	Positive class size	Negative class size	Imbalance degree
heart	303	139	164	1.18
sonar	208	97	111	1.14
vehicle	846	212	634	2.99
wine	178	59	119	2.02
wpbc	198	47	151	3.21
segment	2310	330	1980	6.00
vote	435	168	267	1.59

Table 3. Algorithm performance comparison

Dataset	Algorithm	F-measure	G-mean
heart	AdaBoost	0.78	0.81
	CMBoost	0.82	0.83
sonar	AdaBoost	0.74	0.76
	CMBoost	0.78	0.82
vehicle	AdaBoost	0.26	0.41
	CMBoost	0.42	0.68
wine	AdaBoost	0.94	0.96
	CMBoost	0.94	0.95
wpbc	AdaBoost	0.77	0.44
	CMBoost	0.86	0.59
segment	AdaBoost	0.43	0.74
	CMBoost	0.58	0.86
vote	AdaBoost	0.93	0.95
	CMBoost	0.94	0.95

From Table 3, first of all, we can see that the mean F-measure of CMBoost is 7% larger than that of AdaBoost , while the mean G-mean of CMBoost is 8.8% larger than that of AdaBoost. Besides, for almost balanced dataset like heart, sonar, vote, AdaBoost performs very well. While for imbalanced dataset like vehicle, segment, the performance of AdaBoost is bad. The CMBoost has more enhancements on the imbalanced dataset than the balanced dataset.

6. Conclusion

Many domains are affected by the problem of class imbalance. It can be challenging to construct a classifier that effectively identifies the examples of the positive class. Several techniques have been proposed for dealing with the problem of class imbalance. In this paper, we have proposed a new algorithm, called CMBoost, for alleviating the problem of class imbalance. We compare the performance of CMBoost with AdaBoost from UCI dataset, finding that CMBoost performs better when compared to AdaBoost, especially for imbalanced dataset.

References

[1]Dai HL. Class imbalance learning via a fuuzy total margin based support vector machine. Applied Soft Computing, 31: 172-184, 2015
[2]Deng X, Tian X. Nonlinear process fault pattern recognition using statistics kernel PCA similarity factor. Neurocomputing, 121(18):298-308, 2013
[3]Ozcift A, Gulten A. Classifer ensemble construction with rotation forest to improve medical diagnosis performance of machine learning algorithms. Computer Methods Programs Biomedicine, 104(3):443-451, 2011
[4]Yu H, Ni J, Zhao J. ACOSampling: An ant colony optimization-based undersampling method for classifying imbalanced DNA microarray data. Neurocomputing, 101:309-318, 2013
[5]Wang S, Yao X. Using class imbalance learning for software defect prediction. IEEE Trans on Reliability, 62(2):434-443, 2013
[6]Freund Y, Schapire R E. A decision-theoretic generalization of on-line learning and an application to Boosting[J]. Journal of Computer and System Sciences, 55(1): 119-139, 1997
[7]Peng X J, Setlur S, Govindaraju V, Ramachandrula S. Using a boosted tree classifer for text segmentation in handannotated documents. Pattern Recognition Letters, 33(7): 943-950, 2012

[8]Ren J F, Jiang X D, Yuan J S. A complete and fully automated face verification system on mobile devices. Pattern Recognition, 46(1): 45-56, 2013

[9]Ren S K, Hou Y X, Zhang P, Liang X R. Importance weighted AdaRank. Proceedings of the 7th International Conference on Advanced Intelligent Computing. Berlin, Heidelberg: Springer-Verlag, 6838: 448-455, 2012

[10]Sun Y M, Kamel M S, Wong A K C, Wang Y. Cost-sensitive Boosting for classification of imbalanced data. Pattern Recognition, 40(12): 3358-3378, 2007

[11]RE Schapire, Y Freund, P Bartlett, WS Lee. Boosting the margin: a new explanation for the effectiveness of voting methods. Annals of Statistics, , 26(5): 1651-1686 1998

[12]Wei Gao, Zhihua Zhou. On the doubt about margin explanation of boosting. Artificial Intelligence, 203:1-18, 2013

[13]A. Demiriz, K.P. Bennett, and J. Shawe-Taylor. Linear programming boosting via column generation. Mach. Learn., 46225–254, 2002

[14]Han Hui, Wang Wenyuan, Mao Binghuan. Borderline-SMOTE：A New Over-Sampling Method in imbalanced Data Sets Learning. Pine of the International Conference on Intelligent Computing: 878—887, 2005

[15]SU C T, Chen hongsheng, Yih Y.Knowledge Acquisition through Information Granulation for Imbalanced Data．Expert Systems Tools and Applications, 31(3):531—541, 2006

IOP Conf. Series: Materials Science and Engineering **533** (2019) 012048 doi:10.1088/1757-899X/533/1/012048

Influence Maximization: A Time-Space Efficient Algorithm

Ganming Xia

Beijing University of Posts and Telecommunications, Beijing 100876, China

Gnmingxia@gmail.com

Abstract. Influence maximization is the problem of finding a small subset of nodes (seed nodes) in a social network that could maximize the spread the influence. The algorithm problem of computing influence estimation and influence maximization have been extensively studied for decades. In this paper, we studied the reverse influence sampling method proposed by Byogs et al. in 2014, and some other improved algorithms. We proposed a new algorithm TESA (Time-Space Efficient Algorithm). On large social networks, TESA saves more space than existing algorithms and runs faster. We used data sets tested in the literature to evaluate TESA experimentally and showed that it was superior to the most advanced solutions in terms of running time and memory footprint.

1. Introduction

Recently, many large social networking sites, such as Facebook, Friendster, etc., have achieved great success, mainly because they are able to connect people, and the results are very good. They also link small, separate social networks. At the same time, they are becoming a huge communication and marketing platform, where a lot of information can be spread to a very wide range in a very short time. In this paper, we present our work to address one of the challenges of effectively identifying influential individuals in large social networks.

Generally speaking, social networks are relatively large, so operation time and operation space are two main factors to be considered in solving the problem of maximum influence. Kempe et al. showed that the influence maximization problem is NP-hard [1], and Chen et al. showed for the influence estimation problem that computing the exact influence of a single seed is #P-hard [2]. Moreover, as Feige proved in 1998, the problem is hard to approximate to anything better than $1 - (1 - 1/s)^s$ of the optimum for a seed set of size s [3].

In order to capture the propagation of influence spread, Kempe et al. [1] introduced the Independent Cascade (IC) model [4]: an independent random variable is assigned to each directed edge (u, v); the variable reflects the level of influence from u to v. Starting from a vertex, in each step, information spreads to vertex neighbors with the probability equal to level of influence of the vertex over each neighbors. Kempe et al. Showed that IM on the IC model encodes the classic maximum coverage problem and therefore is NP-hard [5]. Chen et al. [6] shwoed for the IE problem that computing the exact influence of a single seed is #P-hard. Kempe et al. [1] showed that IM on the IC model is monotone and submodular, and therefore, a Greedy algorithm produces provably-good quality solutions. More precisely, the influence of the approximate Greedy solution with given number of seeds is $(1 - 1/e - \epsilon)$ of the optimal solution, for any $\epsilon > 0$ [7]. IC becomes a standard model of influence spread, and we are using it for our algorithm.

Content from this work may be used under the terms of the Creative Commons Attribution 3.0 licence. Any further distribution of this work must maintain attribution to the author(s) and the title of the work, journal citation and DOI.
Published under licence by IOP Publishing Ltd

According to the research results of Kempe et al., other improved algorithms have been published successively in this theory, such as [2, 8, 9, 10, 11, 12]. However, there are still many problems in dealing with large-scale social networks.

In 2014, Brogs et al. [13] proposed a different approach: Reverse Influence Sampling (RIS) method. The main idea of RIS is to select a random node from the existing nodes and represent it with v, and then determine how many nodes can affect v nodes. If a vertex u usually affects different randomly selected vertices, then u is a good candidate for the most influential vertices. This can be done by simulating the influence process using the IC model in the graph, and the edge direction in the graph is reversed. RIS is a fast algorithm for IM, obtaining the near-optimal approximation factor of $(1 - 1/e - \epsilon)$, for any $\epsilon > 0$, in time $O((m + n)k\epsilon^{-2} \log(n))$, where n is the number of vertices, m is the number of edges, and k is the number of seeds. Borgs et al. showed RIS runtime to be optimal (up to a logarithmic factor) [5]. Although this algorithm solves the problem of maximizing influence well, there are still some shortcomings. In the common IM algorithm and its improved algorithm, there are two main problems: one is not running fast enough, and the other is a large memory footprint.

In this paper, we deeply analyzed the space saving algorithm proposed by Popova et al. [5]. The algorithm has significant effect on space saving, but the running time is long and many useless subgraphs are generated. We propose a TESA (time-space efficient algorithm) based on this algorithm, which allows to scale-up computing of IM and IE to massive graphs with billions of edges and faster. We compared the time space performance of TESA with state-of-the-art algorithms. TESA takes less time and less space for the same graph.

2. Preliminaries

We describe in this section some notations and definition.

2.1. Notations

Let $G = (V, E, p)$ be a directed graph, and vertex is set as $|V| = n$, edge Set $|E| = m$, and p: $E \rightarrow [0,1]$ is the existence of the probability of the edge. S is the seed set.$\sigma(S)$ is the influence spread of a seed set S under the Independent Cascade (IC) model [14].

2.2. IM and IE problems

Influence Estimation Problem (IE). Given a graph $G = (V, E, p)$ and a seed set $S \subseteq V$, compute the influence spread $\sigma(S)$ of S.

Influence Maximization Problem (IM). Given a graph $G = (V, E, p)$ and an integer k, find a seed set $S \subseteq V$ of size k that maximizes $\sigma(S)$.

2.3. Social networks

A social network represents a social structure made up of individuals or organizations and their relations (friendship, co-authorship of scientific papers, co-appearance in a movie, and following-followers et al.). Social networks are described busing a Graph $G(V, E)$ where V is the set of nodes representing the individuals or organizations and E is the set of edges representing the social ties.

3. Proposed Data Structures

Although the existing greedy algorithm has a good result for solving IM problems, it still needs a great cost for large networks. We proposed a new algorithm on the existing RIS algorithm and its improvement algorithm, which can well solve the problem of running space and time.

3.1. RIS algorithm

RIS method is the generation of subgraphs by arbitrary selected nodes in the graph, which is called the influencing set. If a vertex often appears in the influencing set, then the vertex is the candidate for the optimal influence in the graph. Moreover, RIS is shown in [13] to be runtime-optimal (up to a logarithmic) with respect to network size.

The weight of the hypergraph is defined as the number of graph edges "touched" by RIS [5]. RIS randomly selects the edges to follow with a given edge probability p. During the search, each side that hits the node is counted as a "touch" and helps calculate the weight. Note that the edge is considered "touched" whether the search selects it or not. An approximation of how large the weight should be to ensure an optimal solution is defined in [13].

3.2. The Time Space Efficient Algorithm

In this part, we mainly introduce the algorithm, TESA. Based on the existing RIS algorithm and its improved algorithm, we proposed an improved algorithm to make the algorithm more efficient in time and space. We show the pseudo-code of the algorithm in algorithm 1, followed by a detailed description of the main features of the algorithm.The main idea of TESA algorithm is that we classify all the nodes in the graph by using the principle of 80/20. In the real world, the most influential people are often the minority. 20 percent have more influence than 80 percent. A node with a larger degree indicates that the node has a greater influence in the graph. We count all the nodes in the graph, select 20 percent of the nodes before the degree to form a candidate set. Then, RIS method was used to select any node in the candidate set to generate the subgraph. Meanwhile, we adopted

Algorithm 3 TESA

Input: directed graph G, precision $\epsilon \in (0,1)$, number of seeds k

Output: seeds set $S \subseteq V$ of size k, spread $\sigma(S)$

```
c ← 4(1 + ϵ)(1 + 1/k)
R ← cmkϵ^-2 log(n)
H ← BuildHypergraph(R)
Average ← AverageDegree(G)
return GetSeeds(H)
procedure BuildHypergraph(R)
sk_degree ← 0
sk_num ← 0
averageDegree ← 0
    while H_weight < R do
        v ← random vertex of 20% maximum degree of G^T
        sk ← DFS in G^T starting from v
        sk_num = sk_num + 1
        for each u ∈ sk do
            count (u) ← count (u) + 1
            sk_degree ← sk_degree + G^T.outdegree (u)
            averageDegree ← averageDegree + G^T .outdegree(u)
        if sk_cardinality > 2 and averageDegree > average then
            append sk to hypergraph H
        H_weight ← H_weight + sk_degree
    return H
procedure GetSeeds(H)
    S ← ∅
    σ (S) ← 0
    for i = 1,. . . ,k do
        vi ← argmax_v {count (v)}
        S.insert (v_i)
        σ (S) ← σ (S) + count (v_i ) * n/sk_num
        scan H
        if vi ∈ sk_j then
            for each u ∈ sk_j do
                count (u) ← count (u) - 1
```

IOP Conf. Series: Materials Science and Engineering 533 (2019) 012040 doi:10.1088/1757-899X/533/1/012040

```
            count (v_i) ← 0
        output S, σ (S)
podedure AverageDegree(G)
        total ← 0
        degree ← G.outdegrees()
        i ← 0
        while degree.hasNext() do
            total ← total + degree.next()
            i ← i + 1
            if i == n – 1 then
                break
        return total / n
```

the DFS (Depth First Search) algorithm in the process of generating the subgraph. But we don't see all the nodes as the same. To consider a general situation in real world networks, every node is different. For a specific spreading probability p, and p is usually very small in real world networks. When p is small, the multi-hop neighbors of v can be ignored, and only the nearest neighbors are counted toward the degree [7]. Therefore, when we use DFS to calculate the subgraph, we only calculate the candidate node three-hop neighbors. This is more in line with the actual situation, and can save time and improve the calculation efficiency when calculating the subgraph.

The major differences of TESA algorithm comparing to RIS and other RIS-based algorithms.

(1) We use webgraph, a highly efficient, and actively maintained graph compression framework [15].

(2) We use the Pareto principle, where influential people are often only 20 percent of the total population. So we used Pareto principleto constructed a candidate set.

(3) We use DFS algorithm to calculate the influence spread of the seed node, and we only calculate three-hop neighbors, which greatly accelerates the calculation speed and saves space.

4. Experimental Results

We have conducted extensive experiments on several real-world graphics and have tested our IM and IE solutions. At the same time, for the sake of simplicity, we only include the most interesting and problem-solving results in IM and its analysis. Due to the space constraints, we only present results for three real world graphs. The datasets are available from the Laboratory for Web Algorithmics [16, 17] (http://law.di.unimi.it/datasets.php).

Although the size of the network we consider is in the medium range, each node in the network can be sampled ultiple times, and the number of edges used by all our algorithms is huge.

Table 1. Datasets ordered by m.

Dataset	N	M
UK100K	100,000	3,050,615
CNR2000	325,557	3,216,152

Equipment.The experiments were conducted on a desktop computer with processor 4.0 GHz Intel Core i7 (8-core), RAM 16GB 2400MHz DDR4.

Parameters.The parameters we used in the experiment are as follows. k is the seed size. p is the probability of edge existence. In these tests, k takes values of 5, 10, 25, and p takes values of 0.1, 0.01, 0.001.

Arora et al. [18] comprehensively evaluated and tested the most advanced IM algorithms. Arora et al. tested several well-known IM algorithms, including CELF++, TIM, IRIE, and PMC, and compared their performance, runtime, and memory consumption. The algorithms tested by Arora et al. were released before May 2016, so they did not contain some of the most interesting and promising IM algorithms. We decide to test our TESA vs new IM algorithm, NoSingles, and perform a comparative analysis of the results.

Fig.1 shows testing results when TESA was compared to NoSingles [5] while processing IM for graph uk100, a medium size web graph with 100k vertices and 3M edges. Both algorithm use RIS method, with the lower bound on the hypergraph weight as defined in Theorem 4.1 in [13]. Parameters used by both algorithms were identical throughout testing: Brogs et al. number of calculated seeds varied: k = 5, 10, 25, and the probability of edge existence was taken as p = 0.1,0.01,0.001. This ensures that the IM solutions computed by TESA and NoSingles have the same approximation guarantees.

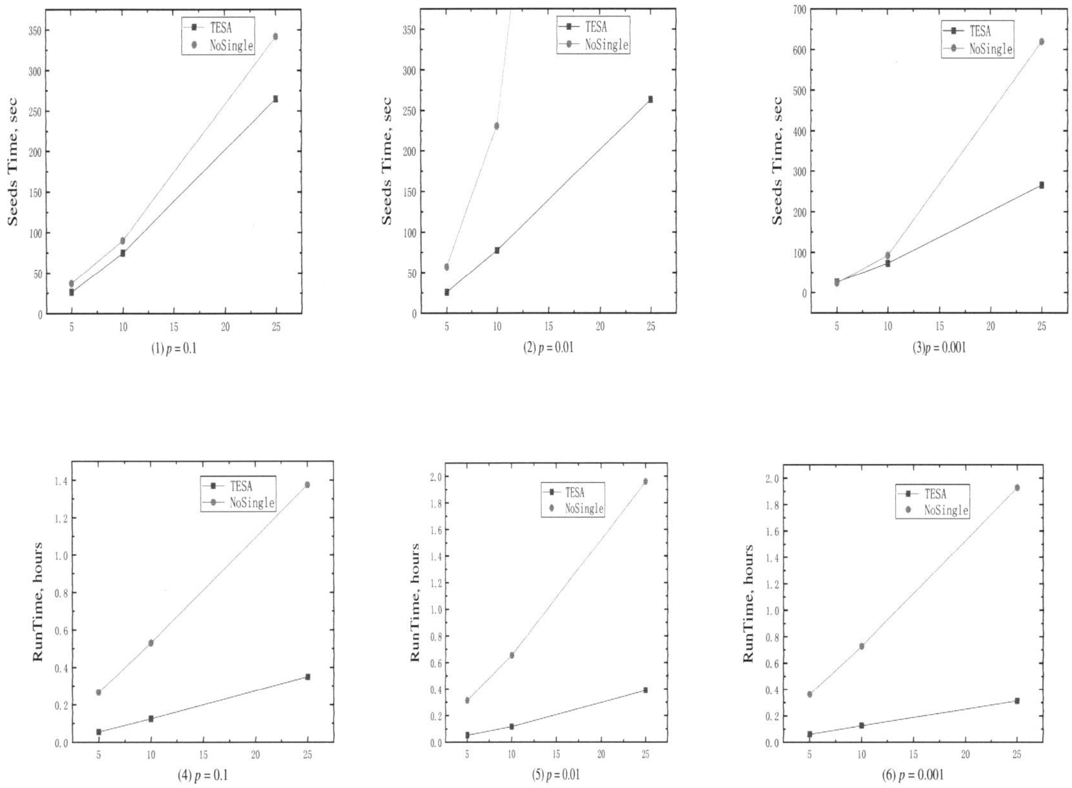

Figure 1. TESA vs NoSingles, varying k in uk100

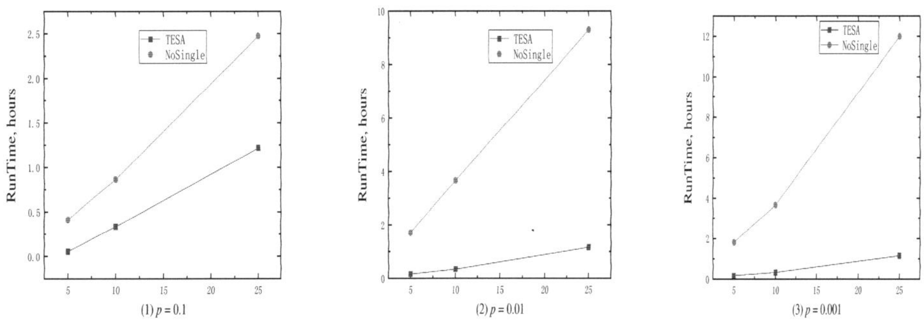

Figure 2 . TESA vs NoSingles, varying k in cnr2000Charts in Fig. 1 and Fig.2 demonstrate that:

(1) Space consumed by TESA is smaller than NoSingles, for all tested k.

IOP Conf. Series: Materials Science and Engineering 533 (2019) 012048 doi:10.1088/1757-899X/533/1/012048

(2) On two datasets, time taken by the whole processing (Total Time), is lower for TESA, for all tested *k*, and
(3) Time taken by the seed calculation (Seeds Time), is lower for TESA.

RunTime. The running tie is mainly composed of the following two parts, one is the time required to build the hypergraph, and the other is to calculate the seeds set. Fig. 1 (4), (5), and (6) shows that TESA spends overall less time for calculating IM Solutions, for all *k*. On the other hand, we have adopted a new method to calculate the hypergraph, because in the actual social network, the probability of information propagation is relatively low. When the information is transmitted multiple times, the probability of being accepted is very low [7], so we only compute the hypergraph obtained by the node in the finite number of propagation. TESA reduces the number of nodes that need to be calculated and also reduce memory.

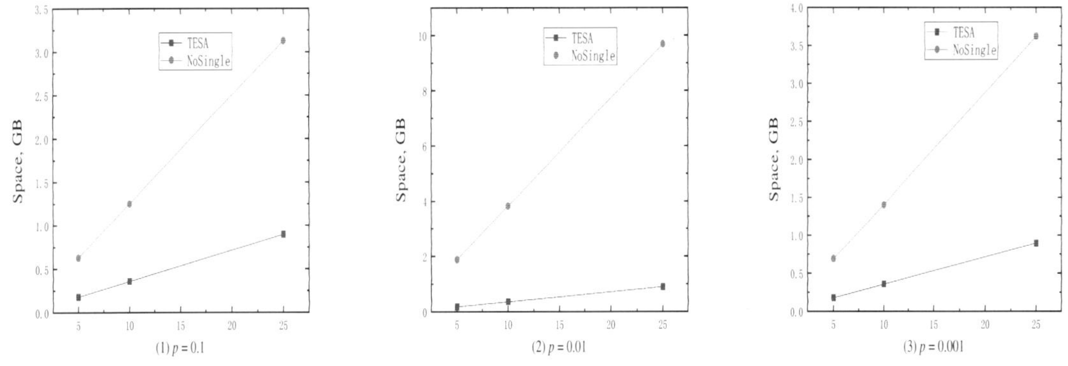

Figure 3. TESA vs NoSingles, varying k in uk100

Seeds Time. Fig. 1 (1), (2), and (3) shows that the time for computing seeds is much lower for TESA than for NoSingles. The main reason is that the number of hypergraph nodes generated by the TESA is relatively small, and have a good influence. Seeds time consists of two parts, one is to calculate the most influential node, the calculation time is the same, the second part is to update other nodes connected with the most influential node in the social network. There are fewer nodes in the hypergraph generated by TESA, so it takes less time to update the hypergraph, which makes the seeds time of TESA less than NoSingles.

Space. Space consumption is shown in Fig. 3 (1), (2). In our testing, we assigned the same p for all graph edg-es. From the Fig. 3 we can see that the TESA uses less memory than the NoSingles algorithm. The main reason is that we only calculate the finite neighbors to form the subgraph. Reduce memory while ensuring a large sprea-d range.

5. Conclusions And Future Research

We have implemented a new algorithm, which is faster and takes less memory than the existing RIS algorithm and the improved algorithm based on RIS. We tested th-e performance of the new alrgorithm on a larger graph and compared the test results. Without compromising t-heoretical guarantees, we significantly reduced the run-ning time an required memory, enabling millions of edge graphics to run on regular desktop computers. The main research in the future is to speed up the algorithm while greatly reducing the memory footprint of the data.

References

[1] D. Kempe, J. Kleinberg, and É. Tardos. Maximizing the spread of influence through a social network. In KDD, pages 137–146, 2003.

[2] W. Chen, Y. Wang, and S. Yang. Efcient influence maximization in social networks. In KDD, pages 199–208, 2009.
[3] U. Feige. A threshold of ln n for approximating set cover. Journal of the ACM (JACM), 45(4):634–652, 1998.
[4] J. Goldenberg, B. Libai, and E. Muller. Talk of the network: A complex systems look at the underlying process of word-of-mouth. Marketing Letters, 12(3):211–223, 2001.
[5] Popova D, Ohsaka N, Kawarabayashi K, et al. NoSingles: a space-efficient algorithm for influence maximization[C]//Proceedings of the 30th International Conference on Scientific and Statistical Database Management. ACM, 2018: 18.
[6] W. Chen, C. Wang, and Y. Wang. Scalable influence maximization for prevalent viral marketing in large-scale social networks. In Proceedings of the 16th ACM SIGKDD International Conference on Knowledge Discovery and Data Mining, pages 1029–1038, 2010.
[7] Wang X, Zhang X, Zhao C, et al. Maximizing the spread of influence via generalized degree discount[J]. PloS one, 2016, 11(10): e0164393.
[8] K. Huang, S. Wang, G. S. Bevilacqua, X. Xiao, and L. V. S. Lakshmanan. Revisiting the stop-and-stare algorithms for influence maximization. PVLDB, 10:913–924, 2017.
[9] H. T. Nguyen, M. T. Thai, and T. N. Dinh. A billion-scale approximation algorithm for maximizing beneft in viral marketing. IEEE/ACM Trans. Netw., 25(4):2419–2429, Aug. 2017.
[10] N. Ohsaka, T. Akiba, Y. Yoshida, and K.-i. Kawarabayashi. Fast and accurate influence maximization on large networks with pruned monte-carlo simulations. In AAAI, pages 138–144, 2014.
[11] Y. Tang, Y. Shi, and X. Xiao. Influence maximization in near-linear time: A martingale approach. SIGMOD '15, New York, NY, USA.
[12] Y. Tang, X. Xiao, and Y. Shi. Influence maximization: Near-optimal time complexity meets practical efciency. SIGMOD '14, New York, NY, USA.
[13] C. Borgs, M. Brautbar, J. Chayes, and B. Lucier. Maximizing social influence in nearly optimal time. In Proceedings of the 25th Annual ACM-SIAM Symposium on Discrete Algorithms, pages 946–957, 2014.
[14] Popova D, Khot A, Thomo A. Data Structures for Efficient Computation of Influence Maximization and Influence Estimation[J]. 2018.
[15] P. Boldi and S. Vigna. The webgraph framework i: compression techniques. In WWW, 2004.
[16] Borgs C, Brautbar M, Chayes J, et al. Maximizing social influence in nearly optimal time[C]//Proceedings of the twenty-fifth annual ACM-SIAM symposium on Discrete algorithms. Society for Industrial and Applied Mathematics, 2014: 946-957.
[17] P. Boldi and S. Vigna. The WebGraph framework I: Compression techniques. In Proceedings of the 13th International Conference on World Wide Web, pages 595–602, 2004.
[18] Arora, S. Galhotra, and S. Ranu. Debunking the myths of influence maximization: An in-depth benchmarking study. In Proceedings of the 43rd ACM SIGMOD International Conference on Management of Data, pages 651–666, 2017.

IOP Conf. Series: Materials Science and Engineering **533** (2019) 012049 doi:10.1088/1757-899X/533/1/012049

Design and Implementation of an Intelligent Operation and Maintenance Management Cloud Platform for Green Buildings Group

Qu Chenfei[1,2,a], Wang Donglin[1,2], Liu Wenxia[1,2],and Zhao Pengcheng[1,2]

[1]Tianjin Architecture Design Institute, 300074 No. 95, Hexi District Meteorological Station Road, Tianjin
[2]Tianjin Intelligent Building Design and Operation Management Control Integrated Engineering Technology Center, 300074 No. 95, Hexi District Meteorological Station Road, Tianjin

[a] Corresponding author: chinaqchf@163.com

Abstract. In recent years, green buildings have developed rapidly in China, but there are still many problems in its operation and maintenance management, it is difficult to meet the requirements of green buildings' operation. In view of the complexity of space, function and management mode, the problem of the operation and maintenance management of green buildings is more prominent. An intelligent operation and maintenance management cloud platform for green buildings of group-company was designed and implemented based on IOT technology in this paper. The platform can significantly reduce the difficulty of the operation and maintenance management of green building groups and reduce the number of operation and maintenance personnel and energy consumption, and reduce operating costs.

1. Introduction

As more and more attention has been paid to the problems of climate, environment and energy, green building has become the focus of international development. The number of green buildings in China has increased rapidly since the green building evaluation and identification work was carried out in April 2008. National green building logo evaluation project has reached 4515 items as the end of September 2016, and the total green building area has reached 523 million m2. Urban development is undergoing a green ecological transformation. Large scale green buildings group will greatly reduce the consumption of buildings and become a new trend of urban construction. [1]

Operation and maintenance management includes two aspects of operation management and maintenance management. Operation management refers to increasing its utilization rate and reducing its operating cost in the whole life period of a building. Maintenance management means to prolong its use as far as possible. In recent years, the connotation of architectural operation and maintenance management has developed into a variety of means for comprehensive management of buildings, aims at providing standardized and personalized services for customers. The information and data of the estate management would be classified, summarized, analysed (qualitatively and quantitatively evaluated) and predicted at the same time.

In the practice of operation and maintenance management of green buildings, the professional and scope of management are more extensive and the content of management is more complex. The

Content from this work may be used under the terms of the Creative Commons Attribution 3.0 licence. Any further distribution of this work must maintain attribution to the author(s) and the title of the work, journal citation and DOI.
Published under licence by IOP Publishing Ltd

original manual management mode or simple computer management mode has been unable to meet the requirements. There are many problems in the operation and maintenance management of green buildings in China, such as the lack of professional operation and maintenance management platform, lagging behind, extensive management, high maintenance cost and poor energy-saving effect. More than 80% of the energy efficiency of green buildings is not up to standard. This is also an important reason for the current domestic green building which is difficult to obtain the running logo. Many enterprise groups hold a number of regional and large-scale green buildings group, and their operation and management are divided into separate modes of operation and group management. This increases the difficulty of operation and maintenance management. [2]

In response to these problems, some domestic scholars, scientific research departments and enterprises have begun research and exploration in operation and maintenance management of green building and achieved certain results, bringing positive economic and social benefits to enterprises. [3] However, these explorations are more limited to the exploration of operation and maintenance management of single-building. There are no relevant research results and case for the complex situation of multi-space multi-purpose and multi-property of large-scale green buildings group.

The IOT is a highly integrated and comprehensive application of the new generation of information technology. It has the characteristics of strong penetration and rapid development of technology. We propose to operate green buildings group by internet, applying the IOT technology to the intelligent operation and maintenance control cloud platform to solve the problems of operation and maintenance management of green buildings group. The platform was designed and implemented in this paper including architecture design, functional design, and platform implementation and so on.

2. Architecture design

2.1. *Logic architecture design*

The cloud platform uses the fast developing Internet of things (IOT) technology, information technology (IT) and operation technology (OT). It can perceive, analyse and integrate the key information of the green building group, integrate and utilize all kinds of information and data resources to the maximum. [4] The logical structure of the platform is shown in Figure 1, which is composed of the perceptual layer, the transport layer, the data layer, the supporting layer and the application layer.

•Perceptual layer: The running images and data of green building (Group)can be collected through all kinds of sensors, meter and other special monitoring equipment, and then be passed through RFID, fieldbus, Zigbee, Bluetooth, infrared and other short distance transmission technology, through the network type equipment such as the field controller, gateway and so on to the network layer .

•Transport layer: In order to realize the communication and interconnection between the intelligent operation and maintenance management cloud platform and the system and equipment of the perceptual layer, the data transmission is carried out through the equipment special network, the wired network, the wireless network, the satellite network and so on. The transport layer has standardized protocol interface, non-standard protocol interface, and special protocol interface and so on. It can include real-time monitoring data interface, database interconnected data interface, video image data interface and other categories.

•Data layer: It includes basic support data, equipment operation data, operation and maintenance data, business data, etc., mainly based on structured data. It provides storage and computing services for other layers through the data engine service interface.

•Supporting layer: It provides data bus service, intelligent analysis service, assistant decision service, data visualization service and so on.

•Application layer: It provides the terminal users with the business processing module oriented to the intelligent operation and maintenance management of green buildings group, including the data overview module, the system log module, the property management module, the maintenance management module, the report management module and so on.

IOP Conf. Series: Materials Science and Engineering 533 (2019) 012049 doi:10.1088/1757-899X/533/1/012049

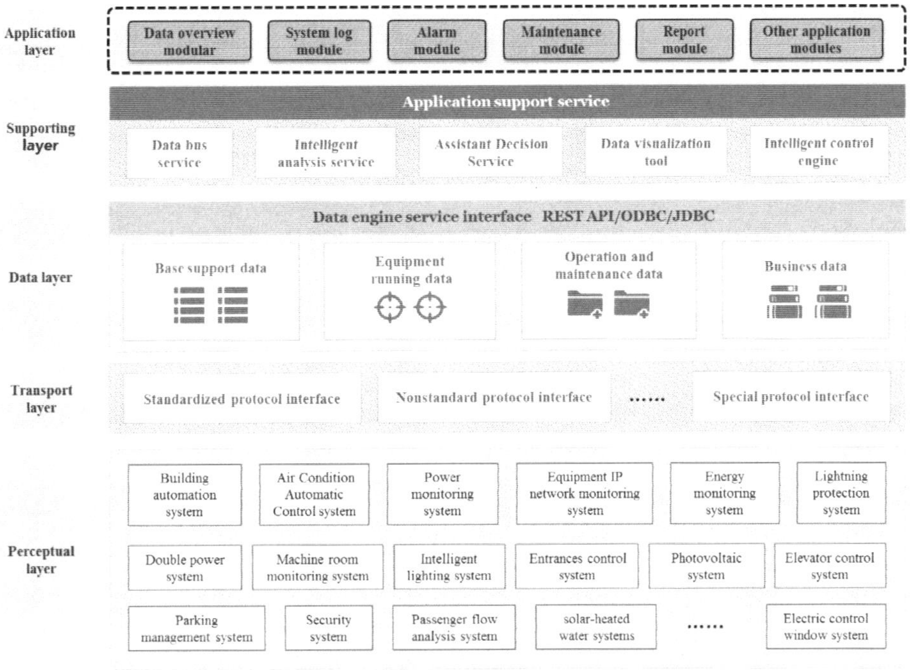

Figure 1. Logical architecture diagram of intelligent operation and maintenance management cloud platform.

2.2. *Physical architecture design*

Considering the operation and maintenance management is divided into management of a single building and unified management of the group, this platform can be implemented on two level platforms, with horizontal and vertical multidimensional management. Each project unified the data to the group intelligence operation and maintenance management cloud platform. Data storage, backup and big data analyses are carried out by group cloud platform, so that all projects can be managed uniformly, as shown in Figure 2.

The intelligent operation and maintenance management platform of each project adopts local deployment. The system and equipment of the project are connected to the platform through real-time monitoring data interface, database interconnection data interface, video and image data interface. Video and data are collected, analysed, transmitted and processed, and the optimal control and decision of the project is realized. The core data of project energy, security, equipment maintenance and operation can be transmitted to the group cloud platform through interface at the same time.

The group intelligent operation and maintenance management cloud platform can choose the mode of public cloud or private cloud deployment according to the actual requirements and conditions of the project, so as to realize the flexible deployment and application of the platform. We can manage massive project data relying on cloud platform's good data processing and analysis ability. We can expand the cloud service resources according to the project construction situation. [5] The group cloud platform uses modular service architecture to provide customized services for different managers in different dimensions, such as regional and vertical (such as the architectural use, management content).

IOP Conf. Series: Materials Science and Engineering **533** (2019) 012049 doi:10.1088/1757-899X/533/1/012049

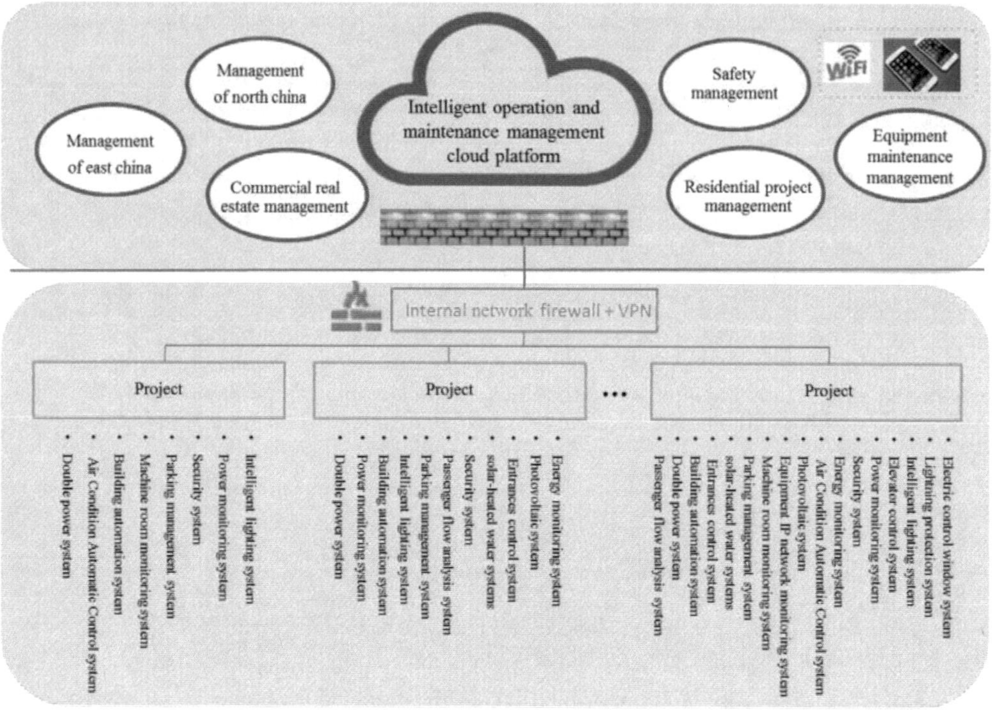

Figure 2. Physical architecture diagram of intelligent operation and maintenance management cloud platform.

3. Functional design

From the view of the service content of the intelligent transportation management of green buildings group, it is not only limited to security, cleaning, warranty and so on, but the comprehensive promotion of the traditional property management service, realizing operation and maintenance digitalization, scientific management and intelligent experience. The platform can ensure the green building mechanical and electrical equipment, power supply water supply, and public facilities and so on are in good working condition. In all circumstances, it can respond accurately, reliably and quickly to deal with and coordinate the work of each system in time, so as to save energy, save people and increase efficiency.

The intelligent operation and maintenance management (cloud) platform can monitor each subsystem, can realize energy management, alarm management, equipment maintenance management, support running log, evaluation analysis, strategy optimization, pattern management, system calendar, report query, user management, lookup management and so on. The specific functions are shown in Figure 3. The intelligent operation and maintenance management cloud platform has a complete user management scheme, which can be customized to generate user pages according to the different dimensions, such as the region and the vertical (such as the architectural use, the management content), according to the user's role and user information. The main functions of the cloud platform for intelligent operation and maintenance management are as follows:

•System monitoring: with a rich graphical function, the state of the device can be monitored in a graphical software interface. All systems and devices can be controlled by mode control or manual control.

•Energy management: energy utilization can be displayed in line with line drawings, bar charts, pie charts, energy efficiency labels, data tables, etc. And the platform can carry out trend tracking analysis of key energy consumption and primary energy consumption, and conduct abnormal energy consumption alarm.

•Alarm management: the platform can timely display the fault alarm system, equipment, location, type, level and other information on the monitoring interface when the alarm occurs, and it support the alarm definition, historical alarm query and SMS alarm.

•Maintenance information: the platform can provide the first hand maintenance data for equipment management, mainly including equipment accounts, maintenance management (for example: fault maintenance management, preventive maintenance plan management), work sheet management (for example: the implementation of the work list, the plan of the work list, the record of maintenance), spare parts management and so on.

•Running log: the platform can record the running state information of each intelligent subsystem, record the operation information of the system (such as editing user, defining authority, making mode, switching mode and so on), record equipment control instructions and feedback (such as starting and stopping equipment, modifying equipment running parameter settings, etc.).

•Evaluation and analysis: the platform can dynamically evaluate the performance of the building according to the dynamic changes of the operating performance of the building, and evaluate the performance of the indicators of the green building dynamically, and carry out the evaluation results.

•Strategy Optimization: the platform can combine the system monitoring and evaluation analysis results to realize the dynamic adjustment of the building operation performance, and finally realize the dynamic balance between the building operation cost and the building environment quality,

•Mode management: the platform has the function of global one key mode management (manual switching, automatic switching), and the function of mode adding, editing, deleting, batch uploading, batch export and so on.

•System calendar: through the platform, we can presupposition and view the schedule of each subsystem running control mode.

•Report query: the platform is rich in data analysis function, and it is presented to users in the form of various reports. For example, real time data curve, historical data curve, daily weekly monthly report, trend chart, etc.

•User management: the platform can set users according to different categories and roles and give different privileges to realize different categories, roles and users' differential operation to the system, and support class management and handover class management.

•Check Gang: users of group cloud platform can test the operation and maintenance personnel on the platform through the platform, and the operation and maintenance personnel of each project should respond in response time. It includes the functions of real-time lookup and lookup record.

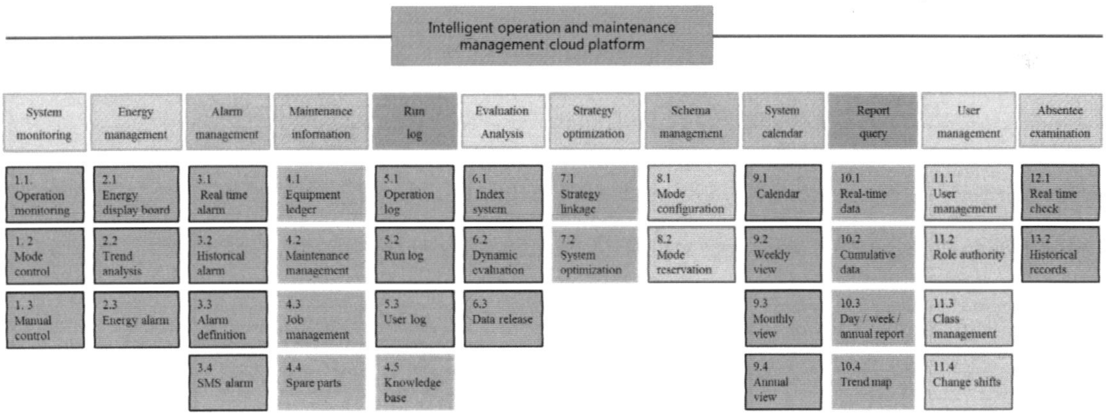

Figure 3. Function diagram of intelligent operation and maintenance management cloud platform.

4. Platform implementation

4.1. *Technical method*

JavaEE technology architecture and SQL Server database are used in this project. Internet technologies, internet of things technology, web technology are comprehensive used to develop the platform. The following technologies are adopted in the platform:

•Java language which is cross-platform and open was used to code based on the Eclipse (an integrated development environment).

•SQL Server databases was used in data storage based on structured data storage. MyBatis which is a Java-based persistence layer framework was used in the data access layer, saving a lot of work on JDBC code writing and result set retrieval. Writing SQL statements in XML files makes the system more flexible and easy to be maintained.

• The platform was implemented based on the B/S architecture. The popular SpringMVC framework was adopted which reduces the coupling between business logic and interface presentation. The same data can be displayed in different forms to meet the individual needs of different customers. As a lightweight IoC container, Spring reduces the coupling between invocation classes and implementation classes of various service, making business logic implementations more flexible and easy to be maintained, and enabling different developers to make a reasonable division of work based on their strengths. The integration architecture of SpringMVC+MyBatis is shown in Figure 4.

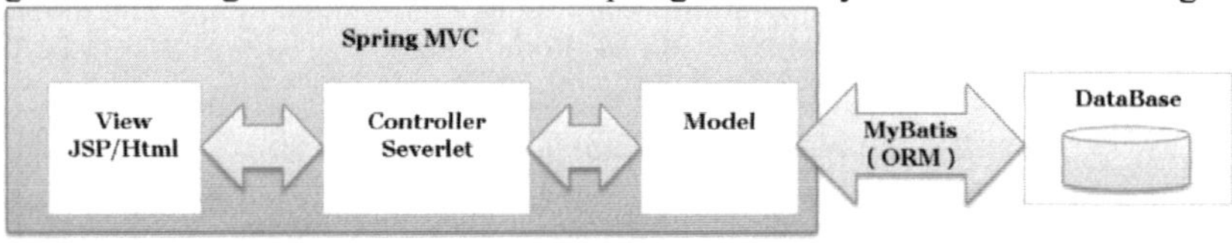

Figure 4. The integration architecture of SpringMVC+MyBatis.

•The response of the platform to the client's Http request comply with the REST style, so that the same data can be added, deleted, updated and selected by the same controller, making the web service lighter.

•In order to meet the individual needs of different customers, a large number of versatile user interface controls that can be configured and expanded, such as general lists, general forms, and general charts were used in cloud platform development. We can flexibly configure according to meet customer's needs, getting different presentation forms, making the platform more expressive.

4.2. *Case of engineering implementation*

A green buildings group consists of six green building units: science and technology museum, art museum, library, performing arts center, civic activity center, cultural corridor and underground parking lot. Each unit is operated independently and managed by a group. The control-center has the authority to manage all spaces, systems, and functions. Others such as sub-control-center, leadership management terminals, employee service terminals, merchant login terminals, display terminals, alarm front-ends and so on will be assigned different management and query modules according to different permissions, realizing horizontal and vertical multidimensional management of all green building units. The platform is shown in Figure 5.

IOP Conf. Series: Materials Science and Engineering 533 (2019) 012049 doi:10.1088/1757-899X/533/1/012049

Figure 5. The intelligent operation and maintenance management cloud platform.

The platform can realize centralized monitoring, rapid response, linkage control and global event management of intelligent system data, and can easily complete energy efficiency analysis, energy efficiency benchmark, energy consumption statistics, energy efficiency ranking and high-energy alarm. It supports equipment file, multi-level alarm management, fault management, preventive maintenance, work order management, operation log, shift management, check and other functions. Part of the platform page is shown in Figure 6-9.

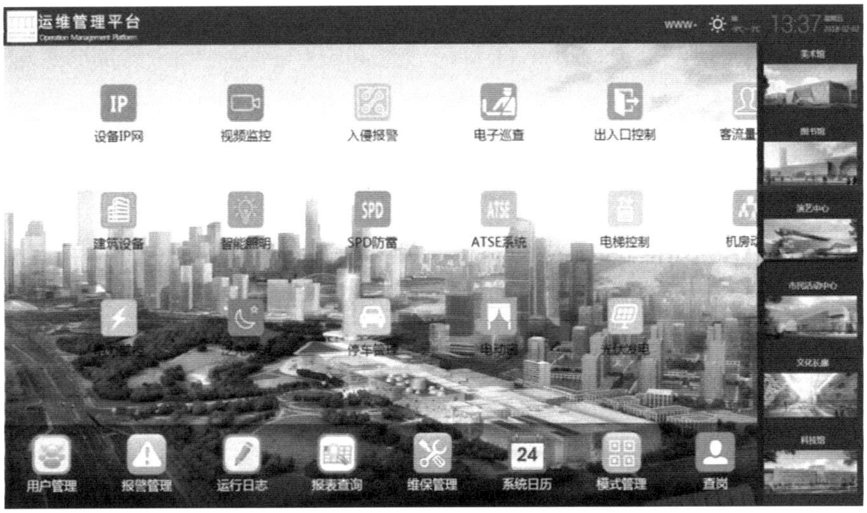

Figure 6. The homepage of the platform.

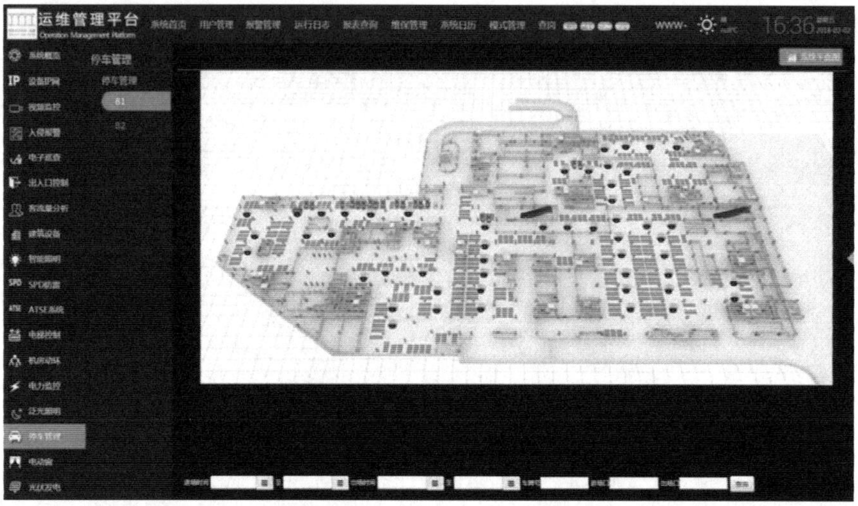

Figure 7. The system monitoring page of the platform.

Figure 8. The report query page of the platform.

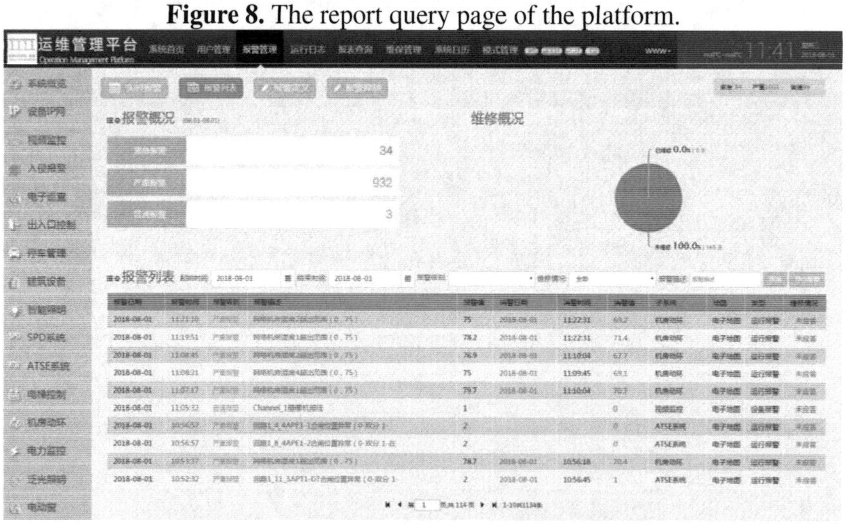

Figure 9. The alarm management page of the platform.

The platform was completed and put into use in 2017, and it has been in operation for almost a year. It can be seen from Tables 1 and 2 that the use of the project through the platform (including a total of 6 monomers) can save personnel costs of about 3.8 million yuan per year. According to the National Building Energy Standard, the project is located in severe cold and cold areas. The underground area is about 110,000 m2 (the energy consumption guiding index is about 12 kWh/m2*a), and the above ground area is about 200,000 m2 (the energy consumption guiding index is about 120 kWh/m2*a). The annual energy consumption is estimated to be 25.32 million kWh, and the annual energy consumption during the actual operation is about 18.9 million kWh, which can save energy costs of about 6.42 million yuan per year. It can be seen that the equipment maintenance cost reduced by the platform is still not counted, and the annual operation and maintenance cost of the project can be saved by about 10.22 million yuan.

Table1. Regular Property Watch Staff Table (taking one building block as an example).

	Duty room	Duty time	Duty people	Number of people
1	Fire control room	24h	Three shifts, 2 people/shift	6
2	Security control room	24h	Monday to Friday 8:30~:17:30	6
3	Substation	24h	Three shifts, 2 people/shift	6
4	Plumbing maintenance personnel	12h	7:00~19:00, One person on duty, two people for repair	3
5	Electrical maintenance staff	12h	7:00~19:00，One person on duty, two people for repair (Maintenance personnel share with the cooling room)	3
6	Other auxiliary maintenance personnel	8h		2
7		Total people		26

Table2. Property watch staff table after set up a unified operation and maintenance management platform (taking one building block as an example).

	Duty room	Duty time	Duty people	Number of people
1	Security control room	24h	Three shifts, 2 people/shift	6
2	Plumbing maintenance personnel	12h	7:00~19:00，One person on duty, two people for repair	3
3	Electrical maintenance staff	12h	7:00~19:00，One person on duty, two people for repair (Maintenance personnel share with the cooling room)	3
4	Other auxiliary maintenance personnel	8h		2
5		Total people		14

5. Platform implementation

Green buildings have been widely accepted. Large-scale green buildings which can significantly reduce the consumption of resources become a new trend in urban construction. Traditional operation and maintenance management mode has been unable to meet the requirements of current green building operation and maintenance and there are many problems. The IOT is a highly integrated and comprehensive application of the new generation of information technology. It has the characteristics of strong penetration and rapid development of technology. We propose to operate green buildings by internet. Aiming at the grouping and unified management in green buildings, the two-level platform construction is proposed, and the platform development ideas of horizontal and vertical multidimensional management are carried out, and platform development and project application are carried out. Practice has shown that the intelligent operation and maintenance control cloud platform of green building groups through IOT technology can reduce operation and maintenance personnel, reduce building energy consumption, extend equipment service life, and reduce construction operation expenses.

References

[1] X.M. Liu, Doc Dis,Chongqing University(2014).
[2] C.F. Qu, W.J. Wu, D.L. Wang , Intelligent building 6, 78(2016).
[3] S. Katipamula, K. Gowri, G. Hernandez, Hvac & R Research 23(4), 546(2016).
[4] J.R. Lin, Doc Dis, Tsinghua University(2016).
[5] X.T. Zhang, C.F. Qu, B.T. Huang, International Conference on Control Science and Systems Engineering. IEEE(2016).

IOP Conf. Series: Materials Science and Engineering 533 (2019) 012050 doi:10.1088/1757-899X/533/1/012050

Multilevel Risk Prediction of Cardiovascular Disease based on Adaboost+RF Ensemble Learning

Runchuan Li[1,2] and Shengya Shen[3] and Gang Chen[1,2,4a] and Tiantian Xie[1] and Shasha Ji[1] and Bing Zhou[1,2] and Zongmin Wang[1,2,a]

[1]Industrial Technology Research Institute, Zhengzhou University, Zhengzhou Henan, 450000, China.
[2]Cooperative Innovation Center of Internet Healthcare, Zhengzhou University, Zhengzhou Henan, 450000, China.
[3]School of Foreign Languages, Zhengzhou University, Zhengzhou Henan, 450000, China.
[4]School of Distance Learning, Zhengzhou University, Zhengzhou Henan, 450000, China.

gchen@ha.edu.cn or zmwang@ha.edu.cn

Abstract. Background: In the field of diagnostic CVD, the predecessors used a large amount of data with no missing two-category data, and obtained good results. However, in the process of electronic input of historical data, a large number of data attribute values are missing, and there are multiple levels of disease risk. Goal: On the data set of imbalance and a large number of missing values, this paper focuses on the five levels of cardiovascular disease. Methods: A new prediction model of Adaboost+RF is constructed by using the information gain ratio to analyze the feature contribution degree of the data set. The performance of this model is evaluated with Precision, Recall, F-measure and ROC Area values. Results: The results show that the four key indicators of the Adaboost+RF model on five-categories unbalanced datasets in Precision, Recall, F1 and AUC values, which are 40.9%, 49.3%, 41.4% and 71.6%. Conclusion: The experiment results demonstrate that the four key indicators of the Adaboost+RF model on five-category unbalanced missing datasets are better than other machine learning algorithms.

1 Introduction

With the improvement of individual's living standards, changes in living habits, the deepening of the aging society and the risk factors for cardiovascular disease (CVD) are becoming increasingly. At present, cardiovascular disease (CVD) has become one of the ruthless killers in individual's daily life. According to the data released by China Cardiovascular Disease Report 2016 in June 2017, cardiovascular death is the main cause, 45.01% in rural areas and 42.61% in urban areas, and the mortality of CVD is increasing [1-2]. It is estimated that by 2030, due to the increase of hypertension, obesity, diabetes and an aging population, the number of deaths caused by cardiovascular diseases will increase to 23.6 million. Early diagnosis and early intervention of cardiovascular disease is recognized as the most effective way to reduce cardiovascular morbidity and mortality [3-4]. The American Society of Cardiology/American Heart Association (ACC/AHA) proposed concepts related to

Content from this work may be used under the terms of the Creative Commons Attribution 3.0 licence. Any further distribution of this work must maintain attribution to the author(s) and the title of the work, journal citation and DOI.
Published under licence by IOP Publishing Ltd

cardiovascular disease risk assessment and early warning, pointing out that it can be based on established risk factors to predict future CVD risks, and the importance of these risk factors for CVD has been confirmed in many CVD risk prediction tools [5]. According to the identified risk factors, the disease can be detected early. The management of the disease in stages and the corresponding preventive interventions is also crucial for the diagnosis of the disease, which can be diagnosed and treated early to reduce the incidence and disability of CVD.
Machine learning [6] is widely used in medical diagnostics. With the deepening of the application of machine learning in the medical field, more and more machine learning applications have emerged in CVD diagnostics. For example, Shameer K et al. [7] examine the potential limitations and challenges of implementing machine learning method in the context of cardiovascular medicine. El-Saadawy et al. [8] used Probabilistic Neural Network (PNN), Support Vector Machine (SVM) and SoftMax regression classifiers to diagnose heart disease. Ade M R R et al. [9] proposed a cardiac classifier and explained the application of SVM and Naive Bayes (NB) in the classification of heart diseases. Li R et al. [10-11] proposed using support vector machines and random forests to establish disease risk prediction models to improve the accuracy of traditional predictive models on processed data set. Papers [8] [9] [10] [11] all use SVM classifiers. However, the SVM classifier has no general solution to nonlinear problems, and the SVM is only good at handling the two-class problem. Paper [12] used logistic regression (LR) analysis to determine risk factors for cardiovascular disease. However, the LR classifier is easy to under-fitting, and it can only deal with linear and two classification problems. Saško Ristov et al. [13] used the international CVD risk factor analysis machine learning methods, and proposed a new method for prevention and early detection of cardiovascular diseases in sample data. Jabbar M A et al. [14] proposed a new effective associative classification algorithm combining k-NearestNeighbor (KNN) and Genetic Algorithms (GA). The programming implementation of KNN and GA algorithm is more complicated and computationally intensive. In addition, they are sensitive to outliers and they are vulnerable to unbalanced data. Dominic V et al.[15] proposed how heart disease datasets and machine learning techniques can help understand the level of risk associated with heart disease. Das R et al. [16] used SAS software to build a diagnostic method for heart disease based on neural network integration. Chen M et al. [17] proposed a new multi-model disease risk prediction algorithm based on convolutional neural network (CNN) to effectively predict the occurrence of chronic diseases. Paper [16] and [17] are both based on neural networks. Neural networks is suitable for data sets with large data volumes. The computational cost is high, and its biggest disadvantage is the "black box" nature, in other words, there is no way to explain reasoning process basis. This paper uses Adaboost+RF algorithm to classify, the classification accuracy is high, and the generalization error is low.

Most of the above prediction models used the two-category data set with high standards without missing values to continuously explore and research in the field of CVD and obtained high accuracy results. However, due to the incompleteness of data information, the possibility of data loss is often caused during the electronic entry of the history form. Therefore, in response to this problem, we combined random forest and adaboost algorithm from the perspective of a basic tree-based unit to construct a new prediction model and apply it to the CVD field. In this paper, we establish a CVD risk prediction model based on Adaboost+RF ensemble learning and analyze the feature contribution degree of the data set using the information gain ratio. The innovations in this paper are as follows:

On the one hand, this paper uses the information gain ratio to analyze the feature contribution of existing CVD sample data to mine disease risk factors and improve the accuracy of identifying CVD, instead of establishing risk warnings based on the identified possible risk factors through experiments model.

On the other hand, due to the importance of prevention of disease grading, this paper selects random forests as weak classifiers, combined with adaboost algorithm to build a new prediction model, using multiple unbalanced sample with a large number of missing values for experimental studies to improve the accuracy of multilevel prediction of CVD risk.

The outline of this paper is as follows: In the second part, the pretreatment and detailed description of the original sample data and the analysis of feature contribution using information gain ratio are introduced. The third part introduces Adaboost+RF ensemble learning in detail. The fourth part analyzes and evaluates the experiment, including the selection of evaluation index, the results of different classification of Adaboost+RF ensemble learning classifier on five-categorie unbalanced sample and the influence of different random factors on the classification accuracy of Adaboost+RF ensemble learning classifier. Finally, the fifth part summarizes this paper and discusses the future work.

2. DATA PROCESSING AND ANALYSIS

2.1 Data processing

This experiment selects the data set disclosed in the UCI Machine Learning Library: Heart Disease Datasets. The total number of Heart Disease Datasets are 920 [18-19]. It is a five-categories data set: no heart disease is represented by 0, and the severity of cardiovascular disease is represented by 1-4 different levels. We need to transform the attribute types of the data set. WEKA 3.8 is a very useful software for data analysis modeling, but sometimes the algorithm modeling used requires very strict data requirements. Numeric data is used to establish a regression model, and it is used to establish a classification model. Therefore, before the establishment of the disease prediction model, the attribute format of the data set is converted from numeric to nominal. Figure 1 shows the process of classification using the Adaboost+RF algorithm.

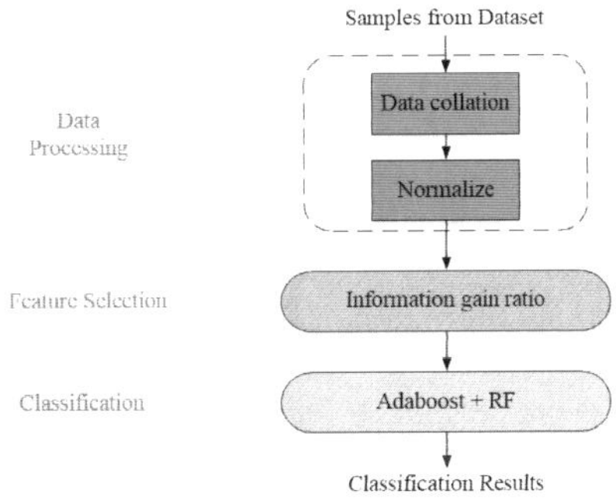

Figure 1. Adaboost+RF algorithm flowchart.

2.2 Contribution analysis of the features

For ease of description, the definition of entropy and conditional entropy is given first, and entropy is a measure of the uncertainty of a random variable. Let X be a discrete random variable taking a finite number of values with a probability distribution of

$$P(X = x_i) = p_i, \quad i = 1, 2, ..., n \tag{1}$$

Then the entropy of random variable X is defined as

$$H(X) = -\sum_{i=1}^{n} p_i \log p_i \tag{2}$$

With random variables (X, Y), the joint probability distribution is

$$P(X = x_i, Y = y_i) = p_{ij}, \quad i = 1, 2, ..., n, j = 1, 2, ..., m \tag{3}$$

Under the condition given by the random variable X, the conditional entropy of the random variable Y is defined as the mathematical expectation of the entropy of the conditional probability distribution of Y under X given conditions for X:

$$H(Y|X)=\sum_{i=1}^{n}p_i H(Y|X=x_i) \tag{4}$$

The information gain represents the degree of uncertainty of the class Y information by knowing the information of the feature X. The information gain $g(D,F)$ of the feature F to the training data set D is defined as the difference between the empirical entropy $H(D)$ of the set D and the empirical conditional entropy $H(D|F)$ of the D under the given condition of the feature F.

$$g(D,F)=H(D)-H(D|F) \tag{5}$$

Information gain ratio is based on the information gain multiplied by a penalty parameter. When the number of features is large, the penalty is small; when the number of features is small, the penalty is larger.

$$g_R(D,F)=g(D,F)/H(D) \tag{6}$$

The penalty parameter indicates the data set D with feature F as the inverse of the entropy of the random variable. The feature information gain ratio is calculated by Eq. (6), and the feature contribution degree analysis is performed according to the feature information gain ratio. The ranking results obtained are shown in Table 1. It is further illustrated whether cardiovascular disease is related to information such as exercise-induced angina, level of chest pain, serum steroid content, and gender. Table 1 shows contribution analysis of the features. At the same time, we also analyze the six top-ranked feature subsets $Heart_{section}=\{F_1, F_2,...,F_6\}$ and all the feature subsets $Heart_{all}=\{F_1, F_2,...,F_{13}\}$ on the Adaboost+RF algorithm respectively. The experimental results are shown in Table 2. As can be seen from the data in Table 2, the indexes of Adaboost+RF classifier are 39.5%, 46.4%, 41.7% and 68.6% respectively in the $Heart_{section}$ of feature subset. The three indicators in the feature subset $Heart_{all}$ are higher than the $Heart_{section}$. According to the results, it can be seen that the ranking of the feature items behind is also crucial for improving the accuracy of cardiovascular disease prediction.

Table 1. Contribution analysis of the features.

ID	Attribute	Description
F1	Exang	exercise induced angina
F2	Cp	chest pain type
F3	Chol	serum cholestoral in mg/dl
F4	Sex	sex
F5	Oldpeak	ST depression induced by exercise relative to rest
F6	Thalach	maximum heart rate achieved
F7	Trestbps	resting blood pressure
F8	Age	age in years
F9	Slope	the slope of the peak exercise ST segment
F10	Thal	normal; fixed defect; reversable defect
F11	Fbs	fasting blood sugar > 120 mg/dl
F12	Ca	number of major vessels (0-3) colored by flourosopy
F13	Restecg	resting electrocardiographic results

Table 2. The influence of different feature subsets on the results of Adaboost+RF algorithm.

Data	Precision	Recall	F1	AUC
$Heart_{sectin}$	0.395	0.464	0.417	0.686
$Heart_{all}$	0.401	0.490	0.411	0.710

3. METHODS

After the above analysis of the sample data, the existing machine learning algorithms can be used to establish five-categories CVD risk prediction models. First of all, we use WEKA 3.8 data analysis modeling software to convert the data set's attribute type, from numeric to nominal. Second, we use the information gain ratio to analyze the feature contribution degree of the data set. Finally, a 10-fold cross validation method is used to evaluate the performance of the predictive model. As an important part of disease risk prediction, the following briefly describes the principles of these two algorithms. Figure 2 shows the structure of the Adaboost+RF algorithm. The training samples of m samples are trained and weighted. Then, combined with random forest, Adaboost+RF algorithm is used for risk prediction.

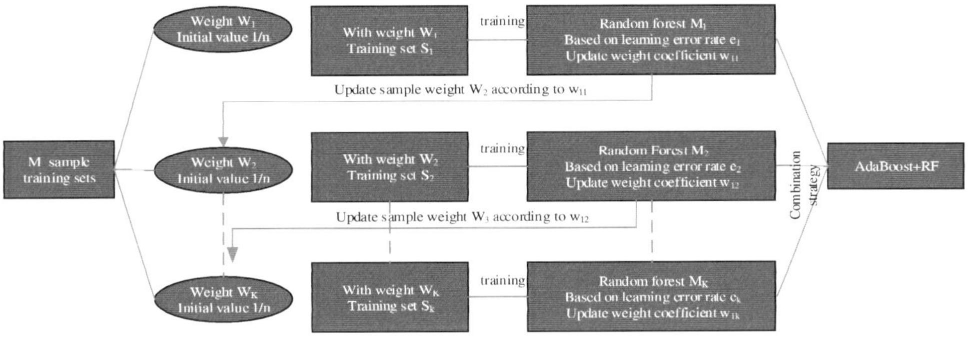

Figure 2. Adaboost+RF algorithm structure diagram.

3.1 Adaboost+RF ensemble learning

Random Forest (RF) [20] is an integrated classification algorithm proposed by L. Breiman in 2001. The algorithm uses a decision tree model as a basic classifier. It adopts bootstrap resampling method to extract multiple samples from the original sample, performs separate decision tree modeling for each bootstrap sample, and then combines the predictions of multiple decision trees to determine the final category of the sample. Then, all decision trees are allowed to vote and the final forecast results are obtained by voting. This paper uses a C4.5 decision tree to form a base classifier for a random forest.

Algorithm: Adaptive Boosting

Input: S : A set of training tuples of class markers;

K : The number of rounds (the number of rounds is the number of classifiers (classification schemes). The K_{th} round classification is equivalent to the classification of the K_{th} classifier, which is equivalent to the k-classification scheme of the training set); A basic classification algorithm;

Output: A combined classifier model: contains K classifiers, and the weight of the voting rights of the classifier;

1. Initialize the weight of each tuple in S to 1/n;
2. **for** $k = 1$ to K **do**
3. According to the weight of the tuple, the sampling is put back from S, and S_k (K_{th} training set) is obtained;
4. Deriving (training out) the classifier M_k (K_{th} classifier) according to S_k;
5. Calculate the error rate of k according to formula (8);
6. **if** e_k>0.5 **then**
7. Go to step 3;
8. **end if**
9. Calculate the weight of the voting rights of the classifier according to formula (9);
10. **for** each of S_k's correctly classified tuples **do**
11. Update the weight of the correctly classified tuple;
12. Normalize the weight of each tuple;
13. **end for**
14. **end for**

The Adaboost algorithm [21] is a lifting algorithm proposed by Freund and Schapire in 1995. It is the most representative algorithm in the lifting algorithm. Like the random forest algorithm, Adaboost algorithm is only an algorithm framework that must have a basic learning algorithm as a support. The algorithm starts from the base classification algorithm, and iteratively obtains a series of base classifiers, and then combines these base classifiers to build a final strong classifier. During each iteration, the distribution of the weights of the training samples is constantly changed so that the next iteration is more concerned with the samples that are wrongly assigned, that is, giving higher weights to the misclassified samples. According to the classification effect of each base classifier, it is given its weight, so that the base classifier with a small classification error rate has a higher weight, and the error rate has a lower weight. Assuming a training data set $D=\{(x_1,y_1),...,(x_N,y_N)\}$ for a class five classification, in which sample instance $x_i \in R^n$ and class label $y_i \in \{0,1,2,3,4\}$ are categorized using the Adaboost algorithm, the steps are as follows:

Step1: initialize training data sample weight distribution: $W_1 = (w_{1,1},...,w_{1,i},...,w_{1,N})$, where, $w_{1,i} = 1/N$, $i = 1, 2, ..., N$.

Step2: Perform k_{th} iteration, where $k = 1, 2, ..., K$

During each iteration, training is performed on the training data set with weight distribution w_k to obtain a base classifier:

$$G_k(x): x \to \{0,1,2,3,4\} \tag{7}$$

Calculate the classification error rate of this base classifier $G_k(x_i)$ in this iteration:

$$e_k = P(G_k(x_i) \neq y_i) = \sum_{i=1}^{N} w_{ki} I(G_k(x_i) \neq y_i) \tag{8}$$

From Eq. (8), we can see that the error rate e_k of $G_k(x)$ on the training dataset is the sum of the weights of $G_k(x)$ misclassified samples. Calculate the weight of the base classifier during this iteration according to Eq. (8).

$$\partial_k = \frac{1}{2}\log[(1-e_k)/e_k] \tag{9}$$

From Eq. (9), we can see that when $e_k \leq \frac{1}{2}$, $\partial_k \geq 0$, and ∂_k increase with the decrease of e_k, it means that the smaller the classification error rate, the greater the effect of the basic classifier in the final classifier. Then update the weight distribution of the training data set: $W_{k+1} = (w_{k+1,1}, w_{k+1,2},..., w_{k+1,i},..., w_{k+1,N})$

$W_{k+1} = (w_{k+1,1}, w_{k+1,2},..., w_{k+1,i},..., w_{k+1,N})$, where $G_1, G_2,..., G_k,..., G_K$, $i = 1, 2, ..., N$

Step3: After K iterations, K base classifiers $G_1, G_2, ..., G_k, ..., G_K$ can be obtained, then the K base classifiers can be linearly combined to obtain the final integrated classifier $G(x) = sign\left\{\sum_{k=1}^{K} \partial_k G_k(x)\right\}$.

4. EXPERIMENT

4.1 Experiment and result analysis

In this experiment, we select six typical classification algorithms, establish a risk prediction model in a given training set, and verify the efficiency of the Adaboost+RF algorithm on five-categories data set through analysis and comparison of experimental results. We compare Support Vector Machine (SVM), Logistic Regression (LR), Random Tree (RT), Decision Tree (DT), Random Forest (RF), and Adaboost+RF differences in the performance of the classifier over five-categories unbalanced sample.

We use the SVM algorithm in paper [11] and the LR algorithm in paper [13] to compare with the experiments in this paper. The SVM algorithm is marked as Baseline-1, and the LR algorithm is marked as Baseline-2. The experimental results with various classifiers are shown in Table 3, and the evaluation index values of different classifiers are compared as shown in Figure 3. The five-categories data set is divided into Value 0, Value 1, Value 2, Value 3, and Value 4. According to all the index values, it can be seen that the Value 0 has a higher sensitivity and the largest area. Since Value 0 has a large amount of data, there are 411 samples and 337 are correctly classified, so Value 0 has the best performance. The number of samples 1, 2, 3, and 4 are 196, 135,

135, and 43, respectively. Due to the large number of missing values in the sample data, their correct classification rate is lower than the value 0.

Table 3. Experimental result values of different classifiers.

Classifier	Precision	Recall	F1	AUC
Baseline-1	0.286	0.449	0.281	0.503
Baseline-2	0.390	0.332	0.352	0.605
RT	0.364	0.418	0.385	0.607
DT	0.384	0.488	0.416	0.681
RF	0.389	0.482	0.401	0.706
Adaboost+RF	0.409	0.493	0.414	0.716

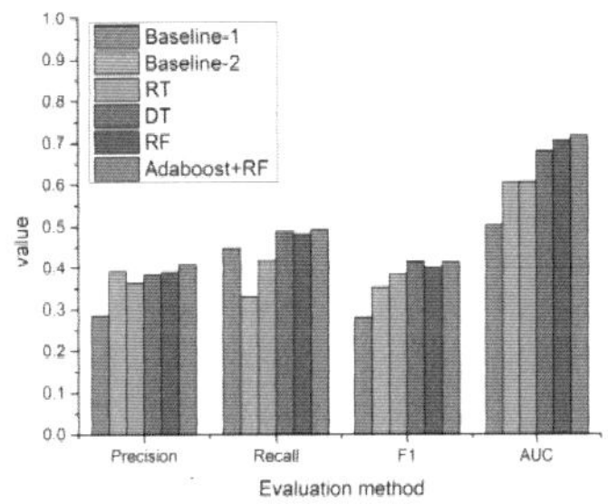

Figure 3. Evaluation index values of different classifiers.

From the data in Table 2 and Table 3, it can be seen that when dealing with a five-categories unbalanced sample with a large number of missing values, the baseline-1, baseline-2 classification results are poor compared to Adaboost+RF, especially for baseline-1, its three key indicators are lower than other machine learning algorithms. The indicators of the Adaboost+RF classifier are better than those of other classifiers, with the highest Precision value of 40.1%, Recall value of 49%, F-Measure value of 41.1%, and ROC Area value of 71%. In the Adaboost+RF algorithm, different random factors (s) also have different effects on the experimental results. From the experimental results, it can be seen that Adaboost+RF algorithm performs best when the stochastic factor is 16, and the values of Precision, Recall, F1 and AUC value of the algorithm are the highest, which are 40.9%, 49.3%, 41.4% and 71.6%. For different random factors, the highest F1 value reached 41.4%, the lowest reached 39.8%, the highest ROC Area value reached 71.6%, the lowest reached 70.5%, with little change. Therefore, it is appropriate to apply Adaboost+RF algorithm to solve the problem.
Therefore, it can be seen that the Adaboost+RF algorithm performs better, and the accuracy of CVD prediction can be further improved by using this algorithm. The reason for the analysis is as follows: In such a five-categories data set with a large number of missing values and unbalanced data, the Adaboost+RF algorithm is superior. It is a strong classifier obtained by combining Adaboost algorithm with a weak classifier random forest that is relatively simple and fast in training. It can iteratively fit even noisy data sets until the data points of each training set adapt to error-free, and it can also adapt to the respective training error rate of weak classifiers, effectively reducing the errors caused by unbalanced dataset.

5. CONCLUSION

In recent years, the mortality caused by CVD is still increasing, which has caused widespread concern in the health sector. The pathogenesis of CVD is complex and changeable, it is closely related to individual's bad behavior, such as smoking, unhealthy eating habits, long sedentary and other factors. Therefore, in the future work, we will first collect and analyze more different Occupation populations of data to extract more hidden features to more accurately assess and predict CVD risk. Second, mobile medical care is the future direction of medical development. We will develop mobile applications, provide better multilevel risk prediction services, and conduct daily health management of residents to reduce the incidence of CVD in patients.

6 COMPETING INTERESTS

The authors declare that there is no conflict of interests regarding the publication of this paper.

ACKNOWLEDGMENT

This work is supported by the following fund projects: National Key Research and Development Program of China (No.2017YFB1401200); Integration of Cloud Computing and Big Data, Innovation of Science and Education (No.2017A11017); Purcell Network Next Generation Internet Technology Innovation Project (No. NGII20161202, No. NGII20170716); Key Science and Technology Projects of Henan (No.152102210249); Key Scientific Research Projects of Colleges and Universities in Henan (No.18A520049).

REFERENCES

[1] Prochaska, J. H., et al. Herz,"Prevention of cardiovascular diseases." 43.1:87-100(2018)
[2] Chen, W. W., et al. J Geriatr Cardiol, "China cardiovascular diseases report 2015: a summary:." 14.1:1-10(2017)
[3] Hood, L, R. Balling, and C. Auffray.Biotech J,"Revolutionizing medicine in the 21st century through systems approaches. " 7.8:992-1001(2012)
[4] Hunter, Peter J., and Thomas K. Borg. NAT REV MOL CELL BIO,"Integration from proteins to organs: the Physiome Project." 4.3: 237 (2003)
[5] Weng, Stephen F., et al. PloS one , "Can machine-learning improve cardiovascular risk prediction using routine clinical data?." 12.4: e0174944 (2017)
[6] Deo, Rahul C. Circulation, "Machine learning in medicine." 132.20 : 1920-1930 (2015)
[7] Shameer, Khader, et al. Heart, "Machine learning in cardiovascular medicine: are we there yet?." : heartjnl-2017 (2018)
[8] El-Saadawy, Hadeer, et al. "Electrocardiogram (ECG) heart disease diagnosis using PNN, SVM and Softmax regression classifiers." Intelligent Computing and Information Systems (ICICIS), 2017 Eighth International Conference on. IEEE, (2017)
[9] Ade, Ms RR, Dhanashree S. Medhekar, and Mayur P. Bote. INT J ENG SCI ,"Heart disease prediction system using svm and naive bayes." 2.5 (2013)
[10] Li, Runzhi, et al. IEEE International Conference on Bioinformatics and Biomedicine IEEE,"Multi-label classification for intelligent health risk prediction." 986-993,(2017)
[11] Li, Runzhi, et al. J HEALTHC ENG ,"An Ensemble Multilabel Classification for Disease Risk Prediction." , 2017,(2017-6-15), 2,1-10,2017
[12] Kilkenny, Monique F., et al. PloS one , "Knowledge of risk factors for diabetes or cardiovascular disease (CVD) is poor among individuals with risk factors for CVD." 12.2 : e0172941 (2017)
[13] Ristov, Saško, and Aleksandar Peckov. "Machine Learning Approach for Early Detection of Cardiovascular Deceases (CVD)." (2010)
[14] abbar, M. Akhil, Bulusu Lakshmana Deekshatulu, and Priti Chandra. "Heart disease prediction system using associative classification and genetic algorithm." arXiv preprint arXiv:1303.5919 (2013)
[15] Dominic, Vinitha, D. Gupta, and S. Khare. Appl Med Inform, "An effective performance analysis of machine learning techniques for cardiovascular disease. " (2015).
[16] Das, Resul, Ibrahim Turkoglu, and Abdulkadir Sengur. EXPERT SYST APPL , "Effective diagnosis of heart disease through neural networks ensembles." 36.4: 7675-7680 (2009)
[17] Chen, Min, et al. IEEE Access,"Disease Prediction by Machine Learning over Big Data from Healthcare Communities.",PP,99:1-1, (2017)
[18] LICHMANM, "UCI machine learning repository" in http://archive.ics.uci.edu/ml, Accessing 2015.
[19] UCI. Heart Disease Data Set[EB/OL]. http://archive.ics.uci.edu/ml/datasets /Heart+Disease.
[20] Wager, Stefan. "Asymptotic theory for random forests." arXiv preprint arXiv, :1405.0352 (2014)
[21] Zhang, Yang, and Z. Zhao. International Congress on Image and Signal Processing, Biomedical Engineering and Informatics IEEE,"Fetal state assessment based on cardiotocography parameters using PCA and AdaBoost." (2018)

Word-to-word Machine Translation: Bilateral Similarity Retrieval for Mitigating Hubness

Mengting Luo[1,2], Linchao He[1,2], Mingyue Guo[1,2], Fei Han[1,2] , Long Tian[1,2], Haibo Pu[1,2,a] and Dejun Zhang[3]

[1]Lab of Agricultural Information Engineering, Sichuan Agricultural University, 0086-625014 Yaan, China
[2]Key Laboratory of Agricultural Information Engineering of Sichuan Province, 0086-625014 Yaan, China
[3]Faculty of Information Engineering, China University of Geosciences, 0086-430074 Wuhan, China

sookie0331@icloud.com

Abstract. Nearest neighbor search is playing a critical role in machine word translation, due to its ability to obtain the lingual labels of source word embeddings by searching k Nearest Neighbor (k NN) target embeddings from a shared bilingual semantic space. However, aligning two language distributions into a shared space usually requires amounts of target label, and k NN retrieval causes hubness problem in high-dimensions feature space. Although most the best- k retrievals get rid of hubs in the list of translation candidates to mitigate the hubness problem, it is flawed to eliminate hubs. Because hub also has a correct source word query corresponding to it and should not be crudely excluded. In this paper, we introduce an unsupervised machine word translation model based on Generative Adversarial Nets (GANs) with Bilingual Similarity retrieval, namely, Unsupervised-BSMWT. Our model addresses three main challenges: (1) reduce the dependence of parallel data with GANs in a fully unsupervised way. (2) Significantly decrease the training time of adversarial game. (3) Propose a novel Bilingual Similarity retrieval for mitigating hubness pollution regardless of whether it is a hub. Our model efficiently performs competitive results in 74min exceeding previous GANs-based models.

1. Introduction

Monolingual word distributed representation has great effect on majority Natural Language Process (NLP) applications, text classification [1], sentimental analysis [2], machine translation [3], short text similarity analysis [4] and so on. In addition, recent years witness an upsurge of research in bilingual embeddings learning which embedding multiple but independent monolingual word vectors in a shared semantic space. Bilingual embeddings learning is able to implement bilingual corpora semantic mining and similarity calculation. It extends the applications of monolingual embeddings to cross-semantic text similarity computation and multi-lingual word machine translation.

A common application of bilingual embeddings learning is the word-to-word machine translation. It usually implemented by two steps: the first step is linear transformation to align bilingual embeddings distribution; the second step is retrieving the k Nearest Neighbors (k NN) as source queries lingual

Content from this work may be used under the terms of the Creative Commons Attribution 3.0 licence. Any further distribution of this work must maintain attribution to the author(s) and the title of the work, journal citation and DOI.
Published under licence by IOP Publishing Ltd

labels to build bilingual dictionary or predict existing dictionary. At present, there are still two problems: (i) the lack of parallel data. The parallel corpora and the labels of target are difficult to obtain in minority languages and dialects. Furthermore, existing dictionaries cannot fully contain vocabularies with gradually changing meaning in this constantly changing world, especially cyber language, therefore, it is necessary to extract potential information from the original monolingual data. (ii) The hubness problem. The k NN retrieval in unified feature space after linear transformation is seriously affected by pollution of 'hubs'. Hub elements appears in the k NN list more than any other points but seem to be less meaningful and more disturbing.

In order to solve the above existing problems, we propose an unsupervised GANs-based machine word translation model with bilateral similarity, namely, Unsupervised-BSMWT, for mitigating hubness pollution in mapped semantic space. The goal of GANs is to learn bilingual embeddings mapping in an unsupervised manner. The generator products a transformation matrix W to gradually narrow the distance between the source vector space and the target vector space. The discriminator detects whether random inputs are from target space or from transformed source space. Furthermore, we first consider the Bilateral Similarity (BS), covering both bilingual and monolingual. For bilingualism, the BS considers the distance of source query to target candidate embeddings. For monolingualism, the BS considers the distance of source query to back-translation embeddings conversely. The BS can mitigate hubness problem effectively and make hubs become target candidates under the same probability instead of being rudely excluded. The structure of this paper is as follow: firstly, the relevant literature is briefly reported in section 2. Then, the proposed Unsupervised-BSMWT model is described in detail in section 3. We elaborate the overall word translation model in section 4 and discuss the experiment results in section 5. The paper closes with the main conclusions and future work in section 6.

2. Related work

With the focus on the unsupervised bilingual embeddings learning model for word-to-word machine translation, this section reports related research in several involving areas such as GANs for NLP [3], [5–7], the curse of dimensionality [8] and the best-k retrieval [9, 10].

2.1. GANs for NLP

Generative Adversarial Nets (GANs) proposed by Goodfellow et al. [11], is used to generate synthetic image. The GANs has been extensively exploited into a widely range of deep learning research. In speech recognition, GANs have capacity of generating ideal audio for data-hungry speech recognition system. In video representation learning, GANs is introduced to learn future-frame predictions replacing the conventional method of inherent pixel flows. Although the GANs shows competitive performance for computer vision, it is difficult to apply GANs for the field of NLP [12].

Recent years have witnessed a steady growth of the research in GANs for NLP. Li et al. [13] extends GANs to Markovian GANs for efficient real-time texture synthesis. Instead of generated natural language from a single corpus, Wang et al. [14] improve document embedding generation with multiple text corpora. In terms of Neural Machine Translation (NMT), Wu et al. [15] employ GANs to discriminate the results of NMT and human-translated sentences, which minimizes the gap between resulting sequence and expert translation. Yang et al. [16] improve NMT via conditional GANs, which could enable to regulate certain arbitrary external data. For the machine word translation, Zhang et al. [3] apply GANs for bilingual embeddings representation learning and alleviate the difficult in insufficient parallel data resources among minority language. Inspired by previous work, we apply GANs for machine word translation by transforming the source distribution to target in a unified semantic spacetext.

2.2. The curse of dimensionality

The k NN retrieval is an irreplaceable step to match transformed source query with appropriate translation in word-to-word machine translation. The best-k retrieval is affected by the intrinsic property in high dimensionality, also called the curse of dimensionality [8]. In the past few years, the factor of dimensionality curse is studied thoroughly in early research (e.g. [17, 18]). For one thing, the Euclidean distance leads to a concentrated trend as the dimension increases. The phenomenon of concentrate shows that the Euclidean distance between all pairs of points is narrowed, making it difficult to measure the discrepancy. For another thing, the curse is relative to the k NN. The k NN [9], the simplest approach, selects target translations by ranking the best-k cosine similarity value of source query. The larger the value of cosine similarity between source query and target translation, the closer the meaning of the words is. But it causes many data points are 'popular' than other points [18] and may not even be qualitatively meaningful, resulting in the emergence of hubs [17]. Hubs have high occurrences rate in k NN list of all word vectors, which arises problems in a variety of high dimensional, data mining applications, in particular, indexing, information retrieval and classification clustering [17].

2.3. *The best-k retrieval*

To our best knowledge, in word-to-word translation, the best- translation candidate retrievals still relies on the Euclidean distance. The successful proposed approaches [7, 10, 19] all try to break the curse by one of the aspects that is reliving the hubness. The latest advanced approaches are as follows: Inverted Softmax (ISF) [10] applies softmax to normalize the probability of candidates and reverses search space to source space for detecting hubs and reduces their probability. Cross-domain Similarity Local Scaling (CSLS) [7], a bi-partite neighborhood graph approach, measures cosine similarity and the average cosine similarity of NN surrounding opposite language of source query and target candidates. That is to say, CSLS is to avoid candidates from dense areas that are heavily polluted by hubs. However, the apparent defect of above retrieval is getting rid of hubs from the list of candidates. Because hubs also have corresponding correct translations. In this paper, we mainly focus on the perspective of mitigating the hubness problem too. As will be described in section 4.3 we propose Bilateral Similarity retrieval form mitigating hubness pollution without excluding hubs.

3. Model overview

The Unsupervised-BSMWT is comprised of two main components in a fully unsupervised learning scheme: the GANs and the Bilateral Similarity (BS) retrieval. Firstly, the pre-trained source and target vectors are embedded into an identical feature space. Secondly, GANs generates a transformation matrix for bilingual embeddings learning. Finally, the BS retrieval extracts bilingual dictionaries from mapped embeddings.

Generative Adversarial Nets. On the first step, we introduce GANs for unsupervised bilingual distributions alignment. The GANs contains two sub-nets the generator nets and the discriminator nets. The work of generator is to produce a transformation matrix that allows mapped source embeddings gradually similar to target embeddings. While the work of discriminator is to detect random inputs which are from source or target semantic space. In every round of cat-chase-mouse competitive game, the generator runs once while the discriminator runs c times alternatively.

Bilateral Similarity retrieval. On the second step, we apply proposed BS retrieval for the best-k translation candidate retrievals in the mapped semantic space. The BS first select the k nearest neighbors of source query as target candidates, then the candidates are translated back as back-translation, finally BS measures more much appropriate translation candidates via combining distance of source query to target candidate and source query to back-translation.

Our contributions can be summarized as follows:

1) Innovation. We proposed a novel Bilateral Similarity for the best-k target candidate retrievals to alleviate hubness pollution. Particularly, BS measuring the probability of candidates via bilingual and monolingual distance that allows model hubs to not be excluded rudely. Moreover, BS is able to retrieve more appropriate target translation candidates especially in distance language

2) Efficiency. Our Unsupervised-BSMWT is **12.9×~19.4×** faster than previous GANs-based word translation model in training phase, which is computationally efficient.
3) High Performance. Our Unsupervised-BSMWT obtains competitive performance on various tasks in a fully unsupervised way. It reaches precision@1 79.87% translating English to Spanish without parallel sign and refinement.

4. Unsupervised-BSMWT

In this paper, we focus on developing GANs-based word machine translation with BS for bilingual distribution learning and word translation. The Unsupervised-BSMWT model is divided into three parts: (1) GANs for linear transformation by the generated transformation matrix W. (2) The nearest orthogonal restriction for W. (3) The Bilateral Similarity retrieval for bilingual lexicon induction.

4.1. GANs for linear transformation bilingual embeddings

We defined $X_s = \{x^{(i)}\}_{i=1,2,\ldots,n}$ and $Y_t = \{y^{(j)}\}_{j=1,2,\ldots,n}$ where $x^{(i)}$ and $y^{(j)}$ represent the lexical item from two languages. X_s and Y_t are two contacted matrices of every word embedding in their vocabularies. Where the subscript s and t represent the row of X and Y respectively come from source and target semantic space. On the base of early research [9], the similar structures of the distribution, representation of different language allows cross-lingual distribution learning via linear transformation

Generator layer: On the one hand, the goal of generator, $G(x_s : \theta_g)$, is to generate a transformation matrix W. The X_s is linearly transformed by W mapped into WX_s which is approximate to target semantic space Y_t.

Discriminator layer: On the other hand, the goal of discriminator, $D(y_t : \theta_d)$, is to find out lexical items that are not from target semantic space Y_t Under the guidance of discriminator, the generator can product more deceptive W to narrow the distance between WX_s and Y_t.

Lost function: Our GANs contain two sub-nets the generator and discriminator networks. The objective function of the generator and discriminator is cross entropy loss, and two networks alternatively iterate with interval of c steps.

Lost function of discriminator as

$$\nabla\theta_d = -\frac{1}{m}\sum_{j=1}^{m}\log D(z(y_t^{(j)})) - \frac{1}{n}\sum_{i=1}^{n}\log D(1 - z(Wx_s^{(i)})). \quad (1)$$

Lost function of generator as

$$\nabla\theta_g = -\frac{1}{n}\sum_{i=1}^{n}\log D(z(Wx_s^{(i)})) - \frac{1}{m}\sum_{j=1}^{m}\log D(1 - z(y_t^{(j)})). \quad (2)$$

Where $z(\cdot)$ represents $N \sim (1, \sigma^2)$ Gaussian noise injection of random inputs [3]. The generated matrix W transform the source embeddings $x_s^{(i)}$ similar to $y_t^{(j)}$, and most source embeddings approximate to the corresponding target embeddings.

4.2. Orthogonal Constraint

After the adversarial game generating transformation matrix W, we recommend constraining W to orthogonal, for the following reasons.

IOP Conf. Series: Materials Science and Engineering 533 (2019) 012051 doi:10.1088/1757-899X/533/1/012051

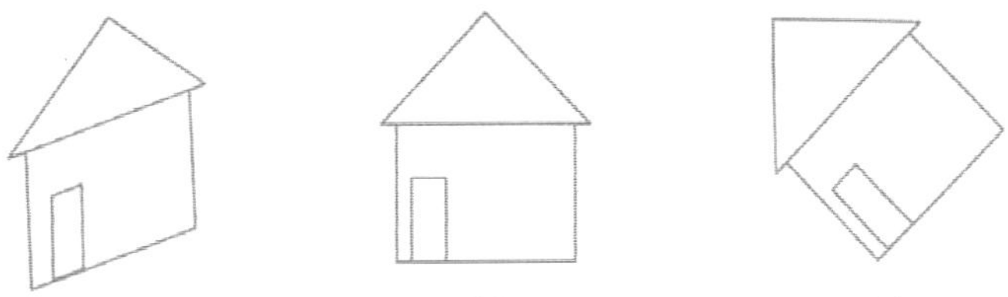

(a) (b) (c)

$\mathbf{W} = \begin{bmatrix} 0.7 & 0.3 \\ 0.3 & 0.6 \end{bmatrix}$ $\mathbf{W} = \begin{bmatrix} \cos 45° & -\sin 45° \\ \sin 45° & \cos 45° \end{bmatrix}$

Figure 1. Bilateral Similarity retrieval. As small houses are depicted in this picture, we vividly treat all the edge of the house as word vectors and the house as matrix of contacting word embeddings. Both (a) and (c) are linear transformation forms of (b). (c) is the results of a orthogonal linear transformation, while (a) is non-linear.

4.2.1. Invariance

The orthogonal constraint preserves the invariance of matrix and the semantics of original data. As Fig.1 shown, the edge of little house can be thought of word embeddings. Like Fig.1(a), implemented with nonorthogonal linear transformation, the construction information of the Fig.1(a) is destroyed and the length and the angle of embeddings have changed larger or smaller. While Fig.1(c), implemented with orthogonal linear transformation, keeps the structure unchanged.

4.2.2. Self-consistent

An excellent translation model is required to have capacity of back-translation. Therefore, in order to translate back to, should be an identity matrix, and must be orthogonal. As shown in Fig.1(c), which is implemented with orthogonal linear transformation, can convert back to Fig.1(b) by generated, but Fig.(a) cannot be revert by.

4.2.3. Speed up training

Adopting orthogonal constraint effectively narrows down the search space, and therefore the constraint accelerates speed of convergence. Because the transformation matrix will never make the mapped source embeddings with non-orthogonal mapping weight like small Fig.1(a).

The calculation of orthogonal matrix is adopted [20] as

$$W^{r+1} = (1+\beta)W^r - \beta(W^r W^{rT})W^r . \tag{3}$$

The updated is not strict orthogonal matrix, but an approximate orthogonal matrix. Where superscript r represents r-th state of in adversarial training, and we commend hyper-parameter for better performance.

4.3. Bilateral similarity for building dictionary

4.3.1. Target label retrieval in mapped space .As we discuss above, NN takes the nearest neighbors of mapped source, which causes the hubs contamination [21]. However, proposed approaches attempt to solve hubs problem by reducing the probability of hubs [10, 22] or by searching candidates in sparse space to avoid hubness contamination [7, 22]. Different from previous work, in order to address the challenge of relieving hubness pollution in high-dimensions and dense space without precluding target hubs, we proposed a novel Bilateral Similarity (BS) retrieval that considers both bilingual similarity and monolingual similarity.

Motivated by the game of passing on a message, BS treats the retrieval of searching the best-k translations as the messaging game among three agents 'A', 'B', 'A_B', as shown in Fig.2, the same icon represents same native language user. 'A' passes a message to 'B', and 'B' sends the message back to 'A_B'. Then the two identical local language users ('A' and 'A_B') compare the information they received. The same game among 'A', 'C', 'A_C'. 'B' is the nearest neighbor in k NN list of 'A', but the back-translation 'A_B' is more similar with source query 'A'. While 'C' is inferior than 'B', but the back-translation 'A_C' is closer to 'A'. As Fig.2 illustrated, a nuance between 'A' and 'A_C' proves the candidate 'C' is highly reliable. On the contrary, farther distance between 'A' and 'A_B' proves that 'B' is highly disturbing. In particular, BS additionally considers the cosine similarity of source query to back-translation embeddings for the first time, which allows BS to distinguish between true interference candidates when hubs are not excluded.

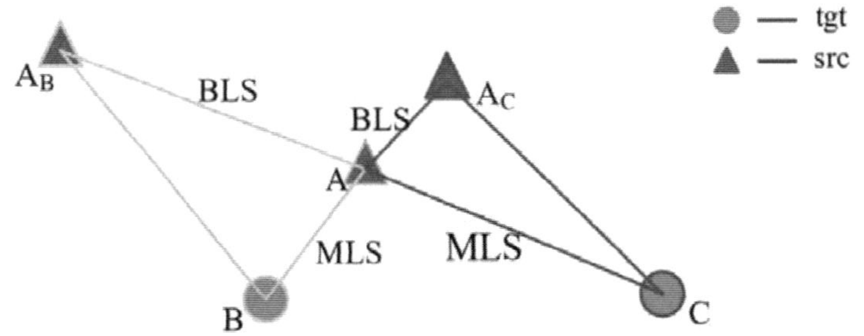

Figure 2. Bilateral Similarity retrieval. In the unified semantic space, the BS retrieves appropriate target candidates as semantic label of source query 'A'. 'A': source query. 'B': the top-1 NN. 'C': the top-2 NN. 'AB': the backtranslation of 'A' from 'B'. 'AC': the back-translation of 'A' from 'C'. BLS: BiLingual Similarity. MLS: MonoLingual Similarity. Suppose 'C' often appears in k NN lists of other points, regarded as a hub.

4.3.2. Formulation. The BiLingual Similarity (abbreviated as BLS to distinguish it from BS) is defined as the cosine similarity of source query embedding to target embedding as

$$\mathrm{BLS}(Wx_s^{(i)}, y_t^{(j)}) = \cos(Wx_s^{(i)}, y_t^{(j)}) = \frac{Wx_s^{(i)} y_t^{(j)}}{\left|Wx_s^{(i)}\right|\left|y_t^{(j)}\right|}. \tag{4}$$

Similar to BLS, the MonoLingual Similarity (MLS) is defined as the cosine similarity of query source embedding $Wx_s^{(i)}$ to $Wx_{y_t^{(j)}}$, where $Wx_{y_t^{(j)}}$ is the back translation of $Wx_s^{(i)}$ from $y_t^{(j)}$:

$$\mathrm{MLS}(Wx_s^{(i)}, Wx_{y_t^{(j)}}) = \cos(Wx_s^{(i)}, Wx_{y_t^{(j)}}) = \frac{Wx_s^{(i)} Wx_{y_t^{(j)}}}{\left|Wx_s^{(i)}\right|\left|Wx_{y_t^{(j)}}\right|}. \tag{5}$$

The weighted average of BLS and MLS between mapped source and target embeddings is defined as BS(.,.), which is represented as

$$\mathrm{BS}(Wx_s^{(i)}, y_t^{(j)}) = (1-\rho)\mathrm{BLS}(Wx_s^{(i)}, y_t^{(j)}) + \rho\mathrm{MLS}(Wx_s^{(i)}, Wx_{y_t^{(j)}}). \tag{6}$$

Where ρ is a hyper-parameter that is a continuous value between 0 and 1 balancing the two terms. After Bilateral Similarity retrieval in mapped space, the high-quality 1-to-k dictionaries are output without any given target label. The one-to-many dictionaries enhance the polysemy of semantic, and every target translation of source word vector $x_s^{(i)}$ can be calculated with identical transformation matrix W even if $x_s^{(i)}$ is not covered be resulting dictionary.

5. Experiment

We set two groups of experiments to show the performance of Unsupervised-BSMWT: (1) whether the Unsupervised-BSMWT learns cross lingual semantic without any parallel data and (2) evaluate the Unsupervised-BSMWT efficiency from both training time and translation accuracy. We report the results and discussion in each task.

5.1. *Experiments Configuration*

Word representations. Each monolingual embeddings of our model are trained using fastText [23] on the most 200,000 frequent words in Wikipedia with 300-dim, released by Facebook AI Research. We normalize the word embeddings to unit length before they are fed into GANs.

Training details. For generator, the size of W is obviously set for 300×300 to retain the dimension of embeddings after linear transformation. For discriminator, we construct two hidden layer of 2048 units with dropout rate of 0.1 and Leaky-Relu for non-linear activation. For discriminator regularization, we implement multiplicative Gaussian noise $N \sim (1, \sigma^2)$, where $\sigma = 0.5$. Then we alternately update the generator and discriminator using stochastic gradient descent with a batch size of 32 and the number of steps c is recommended $c = 5$, for hyper-parameters of BS, $\rho = 0.23$ usually works well.

5.2. *Cross-lingual Semantic Similarity*

The cross-lingual semantic similarity evaluates the relevance of the exported dictionaries by Unsupervised-BSMWT. High correlation proves that our model learns better bilingual embeddings. We apply the harmonic mean of Pearson correlation [24] for this task. The Pearson correlation refers to SemEval2017 [25] manually scored similarity scale by trained native or fluent speaker of target language user. And in this experiment, we refer NASARI [26], the official baseline of SemEval2017. As Tab.1 shown, our model achieves more than 0.7 Pearson correlation on all datasets without using parallel data. In comparison with NASARI, Unsupervised-BSMWT exceeds 0.11 score on English to German dataset, which learns more relevant bilingual semantic embeddings.

Table 1. Average pairwise Pearson correlation results of three datasets (EN to ES, DE and IT).

	EN-ES	EN-DE	EN-IT
NASARI [26]	0.64	0.60	0.65
Our	**0.71**	**0.71**	**0.71**

5.3. *Bilingual lexicon induction*

Bilingual Lexicon Induction is the most common application of word-to-word translation, and we report the training time and translate precision @k of our model on English to Spanish test set. In this task, the experiment is designed for testing the efficiency of speed and accurate, compared to previous GANs-based machine translate model. All experiments run on a single CPU.

Table 2. Word translation retrieval precision@1 for k=1. The results with symbol use tiny data.

	EN-ES		ES-EN	
	time (min)	Precision @1	time (min)	precision @1
Zhang et al. (2017) [3]	1425	00.00	1425	00.00
Zhang et al. (2017) [3]	78	68.76	78	72.78
Lamlpe et al. (2018) [7]	906	**81.70**	906	**83.30**
Our	**74**	79.87	**74**	79.20

It is noticed in Tab.2 that Zhang et al. [3] costs 1425min to learn the cross-lingual distribution embeddings of English to Spanish, but it does not seem to work at all. This method constructs a single hidden layer of discriminator with 500 size, fed tiny-scale datasets (5k most frequent word embeddings with 50-dim). While the discriminator of our network is set as two hidden layers with 2048 size, fed large-scale datasets (200k large-scope embeddings with 300-dim). We infer that the poor results are caused by small amounts of neural units that lacks the ability to fit large-scale datasets. As expected, Zhang's model works in tiny-scale datasets and obtains precision@1 68.76% costing 78min. The results are symbol with in the Tab.2 taking almost equal time, our model achieves 8.77%

higher precision. In comparison with Lample's method, the precision of it is averagely exceeds ours 2.96%, but it costs **12.9×** time than our model. Overall, our model achieves competitive performance spending minimal time

6. Conclusion and future work

This paper develops a model of unsupervised bilingual embeddings learning for word-to-word translation including the transformation of source embeddings and the retrieval of source query. For the distribution transformation, we apply GANs to generate a transformation matrix in a fully unsupervised form. Then the matrix is restricted to be orthogonal for three factors, invariance, self-consistent and speed. Finally, we propose a novel Bilateral Similarity retrieval for searching the best-transformation. For the hubness problem, considering the bilateral similarity of bilingualism and monolingualism allows BS to detect real interference and select more appropriate candidates without excluding hubs. The results of experiments show that Unsupervised-BSMWT achieve competitive performance and significantly 19.4× faster with the high efficiency, compared with previous GANs-based model. For future work, we will focus on further efficiency improvement and various refinement approaches in word translation.

Acknowledgment

This work was supported by the National Natural Science Foundation of China under Grant 61702350, and the Sichuan Provence Department of Education (NO.17ZA0297).

References

[1] X. Guo, H. Sun, T. Zhou, L. Wang, Z. Qu and J. Zang, *SAW Classification Algorithm for Chinese Text Classification,* 7, 2338–2352 (2015).

[2] A. L. Maas, R. E. Daly, P.T. Pham, D. Huang, A.Y. Ng and C. Potts, *Learning Word Vectors for Sentiment Analysis,* 142–150 (2011).

[3] M. Zhang, Y. Liu, H. Luan and M. Sun, *Adversarial Training for Unsupervised Bilingual Lexicon Induction,* 1, 1959–1970 (2017).

[4] H. Zhang, G. Zhong, *Improving short text classification by learning vector representations of both words and hidden topics,* Knowl. Based Syst., 102, 76–86 (2016).

[5] V. Premachandran, A. L. Yuille, *Unsupervised learning using generative adversarial training and clustering,* (2016).

[6] A. Radford, L. Metz, S. Chintala, *Unsupervised Representation Learning with Deep Convolutional Generative Adversarial Networks,* Int. Conf. Learn. Represent., (2016).

[7] G. Lample, A. Conneau, M. Ranzato, L. Denoyer, H. Jegou, *Word translation without parallel data,* Int. Conf. Learn. Represent., (2018).

[8] C. M. Bishop, *Pattern Recognition and Machine Learning (Information Science and Statistics)*, Springer-Verlag, (2016)

[9] T. Mikolov, Q. V. Le, I. Sutskever, *Exploiting Similarities among Languages for Machine Translation,* Comput. Sci., (2013).

[10] S. L. Smith, D.H.P. Turban, S. Hamblin, N.Y. Hammerla, *Offline bilingual word vectors, orthogonal transformations and the inverted softmax,* Int. Conf. Learn. Represent., (2017).

[11] I. J. Goodfellow, J. Pougetabadie, M. Mirza, B. Xu, D. Wardefarley, S. Ozair, A. C. Courville, Y. Bengio, *Generative adversarial nets,* NLPS, 2672-2680 (2014).

[12] L. Yu, W. Zhang, J. Wang, Y.Yu, *SeqGAN: Sequence Generative Adversarial Nets with Policy Gradient,* AAAI, 2852-2858, (2017).

[13] C. Li, W. Michael, *Precomputed Real-Time Texture Synthesis with Markovian Generative Adversarial Networks,* Eur. Conf. Comput. Vis., 702–716 (2016).

[14] B. Wang, D. Klabjan, *Generative Adversarial Nets for Multiple Text Corpora,* arXiv Comput, Lang, (2017).

[15] Y. Wu, M. Schuster, Z. Chen, Q. V. Le, M. Norouzi, W. Macherey, M. Krikun, Y. Cao, Q. Gao, K. Macherey, J. Klingner, A. Shah, M. Johnson, X. Liu, L. Kaiser, S. Gouws, Y. Kato, T. Kudo, H. Kazawa, K. Stevens, G. Kurian, N. Patil, W. Wang, C. Young, J. Smith, J. Riesa, A. Rudnick. O.

Vinyals, G. Corrado, M. Hughes, and J.Dean, *Google's Neural Machine Translation System: Bridging the Gap between Human and Machine Translation,* abs/1609.0, (2016).
[16] Z. Yang, W. Chen, F. Wang, B. Xu, *improving neural machine translation with conditional sequence generative adversarial nets,* north Am. chapter Assoc. Comput. Linguist., 1, 1346–1355 (2018).
[17] M. Radovanovic, A. Nanopoulos, M. Ivanovic, On the existence of obstinate results in vector space models, 186–193 (2010).
[18] M. Radovanovic, A. Nanopoulos, M. Ivanovic, *Hubs in Space: Popular Nearest Neighbors in High-Dimensional Data,* J. Mach. Learn. Res., 11, 2487–2531 (2010).
[19] G. Dinu, M. Baroni, Improving zero-shot learning by mitigating the hubness problem, arXiv Comput. Lang., (2014).
[20] M. Cisse, P. Bojanowski, E. Grave, Y.N. Dauphin, N. Usunier, *Parseval Networks: Improving Robustness to Adversarial Examples,* Int. Conf. Mach. Learn., 854–863 (2017).
[21] E. L. Denton, S. Chintala, A. Szlam, R. Fergus, *Deep generative image models using a Laplacian pyramid of adversarial networks,* neural Inf. Process. Syst., 1486–1494 (2015).
[22] M. Artetxe, , G. Labaka, E. Agirre, Generalizing and Improving Bilingual Word Embedding Mappings with a Multi-Step Framework of Linear Transformations, (2018).
[23] P. Bojanowski, E. Grave, A. Joulin, T. Mikolov: , *Enriching Word Vectors with Subword Information,* TACL, 5, 135-146 (2017).
[24] D. Jurgens, M.T. Pilehvar, R. Navigli, Semeval-2014 task 3: Cross-level semantic similarity, In: Proceedings of the 8th international workshop on semantic evaluation (SemEval 2014), 17-26 (2014).
[25] J. Camacho-Collados, M.T. Pilehvar, N. Collier, R. Navigli, Semeval-2017 task 2: Multilingual and cross-lingual semantic word similarity, In: Proceedings of the 11th International Workshop on Semantic Evaluation (SemEval-2017) (2017).
[26] J. Camacho-Collados, M.T. Pilehvar, R. Navigli, *Nasari : Integrating explicit knowledge and corpus statistics for a multilingual representation of concepts and entities,* Artif. Intell., 240, 36–64 (2016).

IOP Conf. Series: Materials Science and Engineering **533** (2019) 012052 doi:10.1088/1757-899X/533/1/012052

Detection of Android Applications Homology Based on the Authors' Programming Style

Jiayu Xu
Beijing haidian district west TuCheng Road 10, Beijing university of posts and telecommunications.

isxujiayu@foxmail.com

Abstract. As the mobile internet playing an increasingly important role in people's daily life, the copyright issue of Android applications attaches more and more attentions. Because Android applications are vulnerable to reverse engineering, there are many criminals illegally modify and re-sign the official android for their own interests, which has seriously damaged the user's rights and disturbed the order of the mobile internet. In this paper, a novel detects method of Android applications similarity based on the authors' programming style is described. In the proposed method, we firstly extract a series of basic features that can represent the programming style of author. Considering that the features selected manually are highly subjective, we tried ICA, PCA and LDA to optimize the features. Ultimately, an Extreme Learning Machine (ELM) model is trained with sequences of feature vectors. The results of experiments demonstrate that our proposed method has better performance than string comparison and grammar tree analysis when fighting against variable substitution, insert independent statement and data stream confusion.

1. Introduction

Nowadays as smartphones and other mobile devices spread, more and more entertainment and work that originally required computers to be migrated to mobile devices. According to China Internet Development Report 2018, as of the end of 2017, the number of mobile internet users in China reached 753 million, and the proportion of mobile internet users reached 97.5%. Mobile phones have become the most important mobile internet devices [1]. Android applications provide great convenience for people's lives, people can use mobile devices to shop, play games and more. However, unlike iOS applications, Android applications are easily decompiled, resulting from most Android applications are written in JAVA. As the "2015 Android mobile phone application piracy research report" shows, in the sample of 10305 genuine applications, every genuine android application corresponds to 92 pirated copies [2]. In this case, software homology detection is extremely important.

There are some methods for software homology detection that have been proposed. These methods can be roughly classified into four categories: string-based software homology detection, token-based software homology detection, graph-based software homology detection, and syntax-tree-based software homology detection.

However, the plagiarism technology is also constantly evolving. Edward L.Jones [3] raised the 10-difficulty level of plagiarism. At the level II, it just simply modifies the comment statements. Nevertheless, when it comes to the level X, it uses the equivalent control structure to replace the original structure, in which case it is very difficult to detect plagiarism of the android application. The

Content from this work may be used under the terms of the Creative Commons Attribution 3.0 licence. Any further distribution of this work must maintain attribution to the author(s) and the title of the work, journal citation and DOI.
Published under licence by IOP Publishing Ltd

original methods to software homology detection is no longer applicable in some cases. In order to fight against new plagiarism technology, we proposed a novel detect method of Android applications' similarity based on the author's programming style.

The contributions of this paper are as follows.

(1) A scheme for extracting a series of basic features that can represent the programming style of author.

(2) In order to improve the efficiency of the operation and eliminate the overlap between the features, we tried ICA, PCA and LDA to reduce the extracted basic features. The result shows that features are more effective after dimensionality reduction.

(3) An ELM neural network for software homology detection is introduced; it achieved an accuracy of 88.9% in our experiments.

The rest of this paper is organized as follows. Section 2 reviews the related work for software homology detection. In Section 3, we introduce the proposed method. In Section 4, we introduce the Extreme Learning Machine. Section 5 presents the result of experiments. Moreover, Section 6 concludes the paper.

2. Related work

Software homology detection can be divided into two categories: static methods and dynamic methods. Static methods extract features from source code or assembly code. Static methods can achieve high accuracy, but its accuracy is easily affected by obfuscation. The dynamic method monitors the behavior of applications, which can effectively avoid the interference of confusion, but its temporal and spatial overhead are too high.

So far, many techniques of software homology detection have been proposed. There are four main types: string-based software homology detection, token-based software homology detection, graph-based software homology detection, and syntax-tree-based software homology detection.

The software homology detection that bases on string alignment was firstly proposed by Ottenstein in 1976 [4]. Because it is a comparison of the similarity of the program at the text layer, it can be applied to almost any programming language. And because the analysis of the code structure is omitted, the implementation of the method is simple and the efficiency is high. However, the disadvantages are also obvious. As long as the source code is slightly modified, such as adding some white space, can lead to a sharp decline of the similarity of the source code. There are some mature software based on this method, such as UltraCompare, Beyond Compare and so on.

The principle of token-based software homology detection is that the language program is composed of token sequences at the lexical level. If the two programs have similar token sequences, then the two programs can be considered similar. Among software that base on this method, CP-Miner [5] and CCFinder that developed by Toshihiro Kamiya et al. [6] are relatively matured. The CCFinder firstly performs a certain level of pre-processing, and then compares the intermediate data obtained by these processes. Compared to string-based software homology comparisons, token-based detection can better detect plagiarism that changes the format of origin program.

The graph-based software homology comparison mainly constructs the Program Dependence Graph (PDG) by analysing the syntactic structure, the function call relationship, the control dependency and the data flow of the source code. The similarity of the code is then determined by matching the nodes in the PDG. However, some static analysis tools, which may be time consuming, need to be used in the construction process of PDG, causing this method less efficient.

The syntax-tree-based software homology detection builds an abstract syntax tree (AST) by analyzing the source code and calculates the similarity between these ASTs. The syntax tree-based method includes the structure information of program, so it is very effective to the plagiarism methods that do not change the program's structure. But its efficiency is influenced greatly by software scale.

3. Proposed method

3.1. Overall architecture

The method we proposed is aim to detect whether the application plagiarizes other applications. The structure of our method is shown in Figure1, and the core steps of our method are as follows:

(1) Collecting Android application packages (APKs) with the author tag as the sample set.

(2) Decompiling APK and extracting basic features from configuration files, resource files and source code files.

(3) Compressing and optimizing the extracted basic features by using ICA, PCA or LDA.

(4) Training the model using ELM with sequences of features vectors and testing the model with reserved data sets.

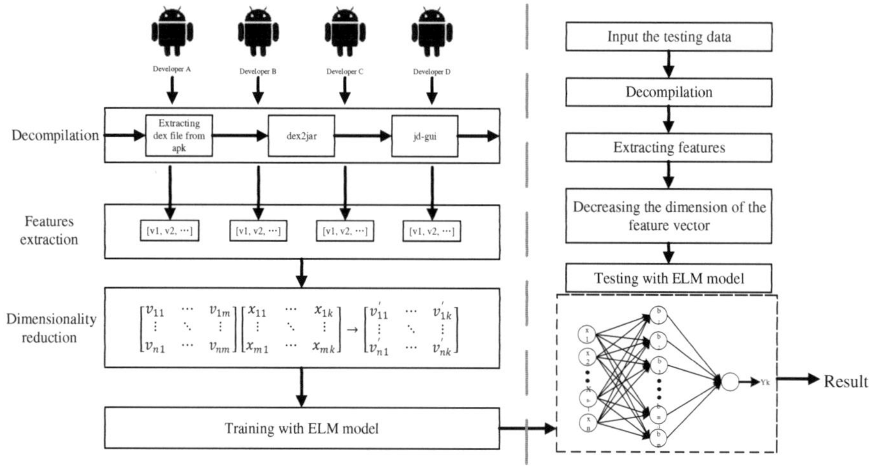

Figure 1. The overall architecture of our proposed method

3.2. Extraction of basic features

In this paper, we extract features that reflect the authors' programming style. The authors' programming style is the same as the authors' writing style. It reflects the authors' programming habits, not the functionality and architecture of the software.

To more fully describe the authors' programming style, we not only extract features from the source code, but also from the APK file. So the extracted features are divided into two parts. The first part is the information from APK file, including configuration information, path information and so on. Another part includes features that are extracted from the Android source code. We detail some of the most important features in the Section 3.2.1 and Section 3.2.2. Moreover, we list all the extracted features in the table 2.

3.2.1. Features extracted from APK

When an android program is released, all the data and resource files will be packaged into an APK file, which is a compressed file with the suffix .apk. APK file contains all the contents of an Android application, which is the file used by the Android platform to install the application. The APK file is a zip archive. After extracting this package, we can see the structure that is shown in Figure 2.

We can find many useful information about the Android application from its APK file. The following are couple of examples.

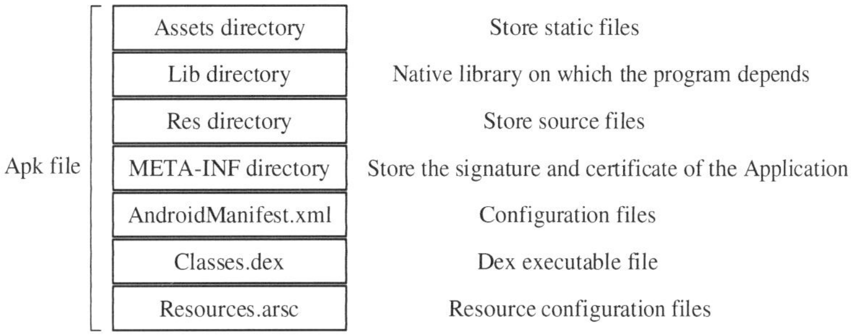

Figure 2. The structure of the APK file

- Package path

We can get the package name of an application from the AndroidManifest.xml file. For example, the package name of qqlite is "com.tencent.qqlite". We use PATH1 to record the maximum depth of package path and PATH2 to tag whether the package path make sense. Therefore, the package name above make PATH1 equals to three and PATH2 equals to one.

- External references

In an Android application, lots of constants are used, such as constants that represent colors, string constants, and so on. There are two main methods for using constants. One is to use constants directly. The other is to define constants in the resource file in advance, and then refer to them by @ or # symbol when needed. For example, we can use "@string/app_name" to refer a string constant from the "string" folder. We will use EX to represent the proportion of external references through @ or # symbol to all constants.

3.2.2. Features extracted from source code

- Naming habits

During the development of an application, there are many identifiers to be named, such as function name, class name, Activity name and so on. The naming habits in programming are similar to the handwriting habits. Every developer has its own naming habits while different people have different handwriting habits. We conclude author's naming habits and find the following pattern: double hump, capitalize the first letter, including underline, using the first letter of each word, meaningless single letter, containing uppercase letters, all uppercase, containing digitals, all letters, meaningful, the total length is more than eight. The specific description is showed in the following table (Table 1):

Table1. Naming habits

Style	*Example*	Description
DH	BinSh	Double hump
CA	Binsh	Capitalize the first letter
UL	bin_sh	Underline
SC	bs	The first letter of each word
SL	a	Meaningless single letter
CU	binSh	Containing uppercase letters
AU	BINSH	All uppercase
CD	bin1	Containing digitals
AL	abc	All letters
M	bind	Meaningful
L	binshaaaa	The total length is more than eight

For an identifier, its naming habit is defined as

$$Code1 = (DH, CA, UL, SC, SL, CU.AU, CD, AL, M, L) \quad (1)$$

- Optional functions

To implement the same functionality, a programmer can use different statements. For example, when implementing a loop function, we can use a "while" statement, a "for" statement, or a "do-while" statement. Similar statements include "if-else" and "switch-case". Let Code2 be the proportion of the while statement in all loop statements, and Code3 as the proportion of the "if" statement in all conditional statements.

Table2. Basic features

Features	Description
IMG1	Naming style of picture.
IMG2	Image format.
PATH1	Maximum depth of package path.
PATH2	Identifies whether the package path makes sense.
STR	The proportion of people names, place names and religious language in the APK file.
EX	The proportion of external references through @ or # symbol to all constants.
Per	Permissions that the Android application applies (such as camera, SMS, phone and so on).
Code1	Naming style of identifier in source code.
Code2	The proportion of the while statement to all loops.
Code3	The proportion of the "if" statement to all loops.
Code4	The ratio of the number of comment lines to the number of lines of code.
Code5	The ratio of the number of line comment lines to the number of all comment lines.
Code6	The ratio of the number of comment lines that immediately follow the code to the number of all comment lines.
Code7	The average length of comment lines.
Code8	The proportion of the left curly braces in a new line to all left curly braces.
Code9	Maximum nesting depth of parentheses (including '(', ')', '[', ']', '{', '}').
Code10	The ratio of the number of functions to the number of lines of code.
Code11	The proportion of the constant first style in statement to determine whether is equal to a constant.
Code12	The proportion of the private classes to all classes.
Code13	The proportion of the functions that have never been used to all functions.
Code14	The proportion of the global variables to all variables.
Code15	The proportion of the static variables to all variables.

3.3. Feature extraction and dimensionality reduction

Feature extraction converts data patterns to features, which are condensed representations of patterns and contain only salient information [7]. The converted features should retain the maximum amount of information and get the best classification.

Principal Components Analysis (PCA) is a technique that finds underlying variables (known as principal components) that best differentiate your data points. It is a linear mapping, which uses the eigenvectors with the largest eigenvalues. However, PCA has some limitations: maximizing spread, interpreting components, orthogonal components. The orthogonal components mean that the principal components generated by PAC must not to overlap in space. To resolve this, we can use an alternative method called Independent Component Analysis (ICA).

ICA originated from blind signal separation (BSS) [8]. Compared with Principal components analysis (PCA), ICA utilizes high-order statistical information, which is more conducive to the decomposition of observed signals. Furthermore, ICA allows its components to overlap in space, so they do not need to be orthogonal. Instead, ICA prohibits its components from overlapping in the information it contains to reduce mutual information shared between components. Therefore, ICA may be a backup

technique to use if we suspect that components need to be derived based on information beyond data spread, or that components may not be orthogonal [9].
Linear Discriminant Analysis (LDA) is a supervised linear mapping based on feature vectors, and its classification performance is usually better than PCA.

4. Extreme learning machine

The Extreme Learning Machine (ELM) is an easy-to-use, efficient single-layer feedforward neural network (SLFNS) learning algorithm. In 2004, Associate Professor Huang Guangbin of Nanyang Technological University proposed it. Traditional neural network learning algorithms (such as BP algorithm) need to set manually a large number of network training parameters, and it is easy to generate local optimal solutions. The extreme learning machine only needs to set the number of hidden layer nodes of the network. During the execution of the algorithm, it is not necessary to adjust the input weight of the network and the offset of the hidden nodes, and generate a unique optimal solution, so the learning speed is fast and general. The structure of the ELM is given in Figure 3 [10].

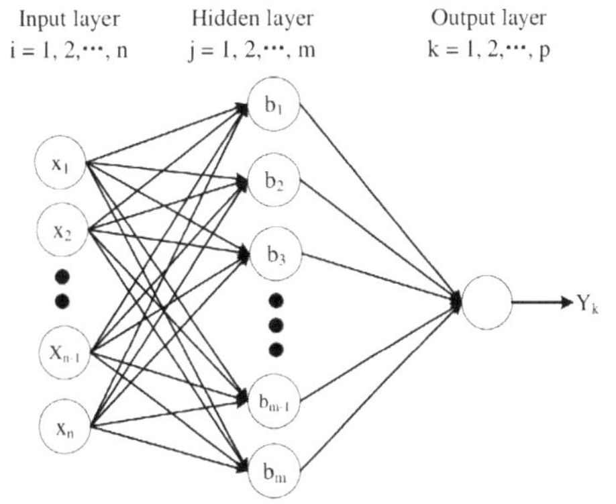

Figure 3. Architecture of Single Layer Feed-forward Network in Extreme Learning Machine Model

For a single hidden layer neural network, suppose there are N arbitrary samples (X_i, t_i), where $X_i = [x_{i1}, x_{i2}, \dots, x_{in}]^T \in R^n, t_i = [x_{i1}, x_{i2}, \dots, x_{im}]^T \in R^m$.

A single hidden layer neural network with L hidden layer nodes can be expressed as

$$\sum_{i=1}^{L} \beta_i g(W_i \cdot X_j + b_i) = o_j, j = 1, \dots, N \tag{2}$$

$g(x)$ is the activation function, $W_i = [w_{i,1}, w_{i,2}, \dots, w_{i,n}]^T$ is the input weight, β_i is the output weight, b_i is the offset of the i-th hidden layer unit. $W_i \cdot X_j$ represents the inner product of W_i and X_j.

The goal of single hidden layer neural network learning is to minimize the error of the output, which can be expressed as

$$\sum_{i=1}^{L} \| o_j - t_j \| = 0 \tag{3}$$

That is, to find the appropriate value of β_i, W_i and b_i that satisfies the following equation.

$$\sum_{i=1}^{L} \beta_i g(W_i \cdot X_j + b_i) = t_j, j = 1, \dots, N \tag{4}$$

Expressed as a matrix,

$$H\beta = T \tag{5}$$

H is the output of hidden layer nodes, β is the output weight, T is the expect output.

$$\beta = H^{+}T \tag{6}$$

IOP Conf. Series: Materials Science and Engineering **533** (2019) 012052 doi:10.1088/1757-899X/533/1/012052

5. Experiment evaluation and result

We conduct our experiment with the game works of six companies crawled from the Internet as the data sets.

5.1. Experimental steps

Our experiment is mainly divided into two parts. The first part is about comparing with other methods of software homology detection. In the second part, we measure the effect of dimensional reduction.

In the first part, we use two methods to confuse the code: variable substitution plus insert independent statement and data stream confusion. And we use totally three methods, including the string alignment, syntax tree analysis and method proposed in this paper, to detect the Android application that has been confused.

In the second part, PCA, ICA and LDA are respectively used to reduce and optimize the basic features. We gradually increase the number of dimension reduction and measure the effect of different descending dimension method.

5.2. Experimental results

Figure 4 shows the accuracy rate of string comparison, grammar tree analysis and our proposed method when detecting the homology of Android applications which are confused by means of variable substitution and insert independent statement. The horizontal axis represents the number of confusing lines of code. Our method achieved the best results, the method of grammar tree analysis is slightly worse and the method of string comparison is the worst.

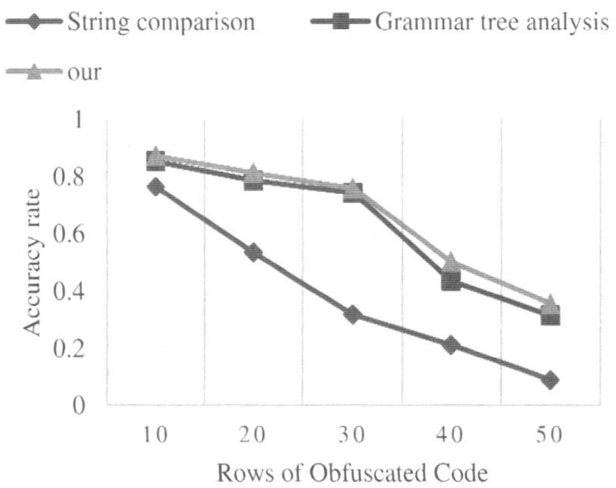

Figure 4. Variable Substitution and Insert Independent Statement

In Figure 5, we confuse Android applications with data stream obfuscation. It can be seen from the results that the proposed method is significantly better than the Grammar tree analysis and String comparison methods. When the number of obfuscated lines increases, the advantage is more obvious. For example, when the number of lines of obfuscated code reaches 30 lines, our method still has an accuracy of 75.7%, while Grammar tree analysis and String comparison have only 54.2% and 41.7% accuracy respectively.

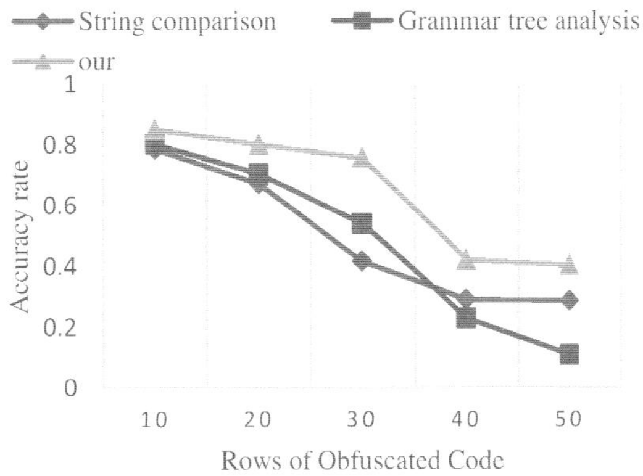

Figure 5. Data Stream Confusion

Figure 6 shows the effect of different dimensionality reduction methods on detection accuracy. The accuracy of PCA is almost not reduced when the feature is reduced by 10 dimensions because it has removed redundancy. The performance of ICA is not good. We speculate that the orthogonal components of ICA are not good feature vectors. As the number of dimensionality reductions become larger, the accuracy of both ICA and PCA is significantly reduced, resulting from some useful features are removed. At this case, the advantage of supervised learning is obvious. The accuracy of LDA decreases slowly.

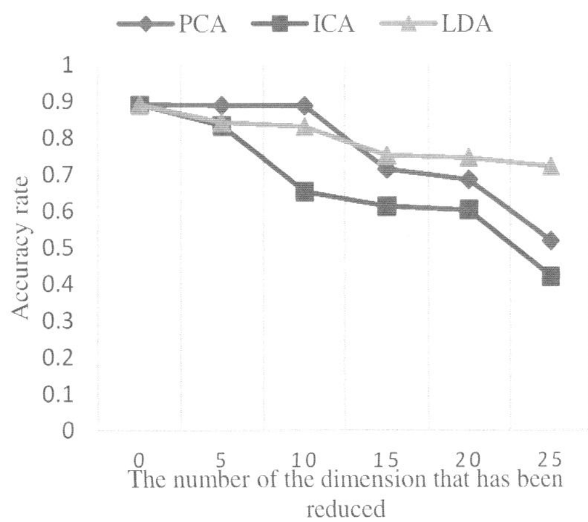

Figure 6. Feature Dimension Reduction

6. Conclusions

In this paper, we propose a novel detect method of Android applications similarity based on the authors' programming style. We extract a series of basic features that can represent the programming style of author from Android applications. In order to optimize features and improve computational efficiency, we have tried three feature compression techniques. We also build an ELM model with lots of Android applications collected from the Internet. The result of the experiment shows the proposed method performs well when detecting the applications that have been confused by means of variable substitution, insert independent statement and data stream confusion. Especially in terms of resisting data stream confusion, our approach has obvious advantages.

Reference

[1] China Internet Association. China Internet Development Report 2018[R].Beijing, 2018.
[2] 360.2015 Android mobile phone application piracy research report[R], 2016
[3] Rinewalt J D, Elizandro D W, Varnell R C, et al. 'Development and Validation of a Plagiarism Detection Model for the Large Classroom Environment[J]. Computers in Education (CoED), 1986, 6(3): 9-13.
[4] Ottenstein K J. An algorithmic approach to the detection and prevention of plagiarism[J]. Acm Sigcse Bulletin, 1976, 8(4):30-41.
[5] Li Z, Lu S, Myagmar S, et al. CP-Miner: A Tool for Finding Copy-paste and Related Bugs in Operating System Code[C]//OSdi. 2004, 4(19): 289-302.
[6] Kamiya T, Kusumoto S, Inoue K. CCFinder: a multilinguistic token-based code clone detection system for large scale source code[J]. IEEE Transactions on Software Engineering, 2002, 28(7): 654-670.
[7] Chen V C. Evaluation of Bayes, ICA, PCA and SVM methods for classification[C]//RTO SET symposium on target identification and recognition using RF systems. 2004, 37: 1-12.
[8] Taillard É. Parallel iterative search methods for vehicle routing problems[J]. Networks, 1993, 23(8): 661-673.
[9] Algobeans. (2016). Principal Component Analysis Tutorial. [online] Available at: https://algobeans.com/2016/06/15/principal-component-analysis-tutorial/ [Accessed 22 Nov. 2018].
[10] Huang G B, Zhu Q Y, Siew C K. Extreme learning machine: theory and applications[J]. Neurocomputing, 2006, 70(1-3): 489-50

DA-U-Net: Densely Connected Convolutional Networks and Decoder with Attention Gate for Retinal Vessel Segmentation

Cong Wu, Yixuan Zou, Jinhao Zhan

Hubei University of Techonology, No.28, Nanli Road, Hong-shan District, Wuchang, Wuhan, Hubei, China

Abstract. The segmentation of retinal vessels is greatly significant for doctors to diagnose the fundus diseases. However, existing methods have various problems in the segmentation of the retinal vessels, such as insufficient segmentation of retinal vessels, weak anti-noise interference ability. Aiming to the shortcomings of existed methods, this paper proposes an improved model based on the U-Net networks, which contains densely-connected convolutional network and a novel attention gate (AG) model, referred as Densely-Attention-U-Net (DA-U-Net), to automatically segment the retinal blood vessels. The method can alleviate the vanishing-gradient problem, strengthen feature propagation, substantially reduce the number of parameters, and automatically learn to focus on target structures without additional supervision. By verifying the method on the DRIVE datasets, the segmentation accuracy rate is 96.09%, higher than that of U-Net and R2U-Net.

1. Introduction

Medical image segmentation is a special computer vision area that is very important for many real-life applications. The fundus contains a large number of blood vessels, which is a deep microvascular system that can be directly observed in the human body without damage. It can provide doctors with a wealth of information about eye conditions and general system status. Ophthalmologists can detect early increases in the systemic vascular load caused by hypertension and diabetes, as well as early signs of vascular structural abnormalities such as retinal vein occlusion and retinal artery occlusion, the diseases caused by blood vessels and vascular systems can cause blindness. With the development of technology, it has been widely studied to explore an automatic method to segment retinal vessels, which can help doctors diagnose and analyze the patient's conditions and make early prevention and treatment of the fundus disease, so as to effectively avoid the visual loss caused by the degenerative disease.

In many cases of biomedical applications, a small number of objects is to be found, on the other hand, only small datasets can be acquired, class imbalance is present, and very high recognition quality and robustness is required[1]. Convolutional neural networks (CNN) have already demonstrated their success in image classification, image segmentation and object detection. For almost any computer vision problems, CNN-based approaches outperform other techniques and in many cases even human experts in the corresponding field. In this work we intend to provide a new approach to the medical image segmentation tasks, which is based on well-known and highly-performing U-Net[2] convolutional neural network (CNN) of encoder-decoder style where initial convolutional feature maps are skip-connected[3] to upsampled layers from bottleneck layers. This skip-connection is crucial to segmentation tasks as the initial feature maps maintain low-level features. We refer to the neural network built as DA-U-Net.

Content from this work may be used under the terms of the Creative Commons Attribution 3.0 licence. Any further distribution of this work must maintain attribution to the author(s) and the title of the work, journal citation and DOI.
Published under licence by IOP Publishing Ltd

IOP Conf. Series: Materials Science and Engineering **533** (2019) 012053 doi:10.1088/1757-899X/533/1/012053

With the advent of convolutional neural networks (CNNs), near-radiologist level performance can be achieved in automated medical image analysis tasks. However, this approach leads to excessive and redundant use of computational resources and model parameters. For instance, similar low-level features are repeatedly extracted by all models within the cascade. To address this general problem, we propose a simple and yet effective solution, namely attention gates (AGs), and we replace the convolution layer with a Denseblock at the same time. CNN models with AGs can be trained from scratch in a standard way similar to the training of a FCN model, and AGs automatically learn to focus on target structures without additional supervision. The Denseblock can alleviate the vanishing-gradient problem, strengthen feature propagation, encourage feature reuse, and substantially reduce the number of parameters. We demonstrate the implementation of AG and the Denseblock module in a standard U-Net architecture (DA-U-Net) and evaluate DA-U-Nets on DRIVE datasets, very high segmentation quality has been achieved.

2. Related Work

The retinal vascular network is the only deep blood vessel that carl be observed directly with non—invasive means in the human body. The changes of characteristics or morphology in the microvasculature can be leaded by any pathological changes of systemic and hematological characteristics. The segmentation of retinal blood vessels is the key step in retinal image processing and analysis, it has a good research value for the early prevention and diagnosis of systemic and hematological diseases. The characteristics of retinal images are complex, the automatic segmentation of retinal blood vessels is easy to be affected by external conditions and pathological changes, and in the retinal images, the difficulty of segmentation is increased by the reason that contrast ratio of tiny vessels and its background is low, therefore, to improve the segmentation accuracy is an important topic to research.

Convolutional neural networks (CNNs) have become the dominant machine learning approach for visual object. We embrace this observation and introduce the Densely Convolutional Network (DenseNet), which connects each layer to every other layer in a feed-forward fashion. Whereas traditional convolutional networks with L layers have L connections—one between each layer and its subsequent layer—our network has L (L2+1) direct connections. For each layer, the feature-maps of all preceding layers are used as inputs, and its own feature-maps are used as inputs into all subsequent layers. Whilst following a simple connectivity rule, DenseNets naturally integrate the properties of identity mappings, deep supervision, and diversified depth. They allow feature reuse throughout the networks and can consequently learn more compact. DenseNets have several compelling advantages: they alleviate the vanishing-gradient problem, strengthen feature propagation, encourage feature reuse, and substantially reduce the number of parameters[4]. The Denseblock can be summarised as follows Figure 1:

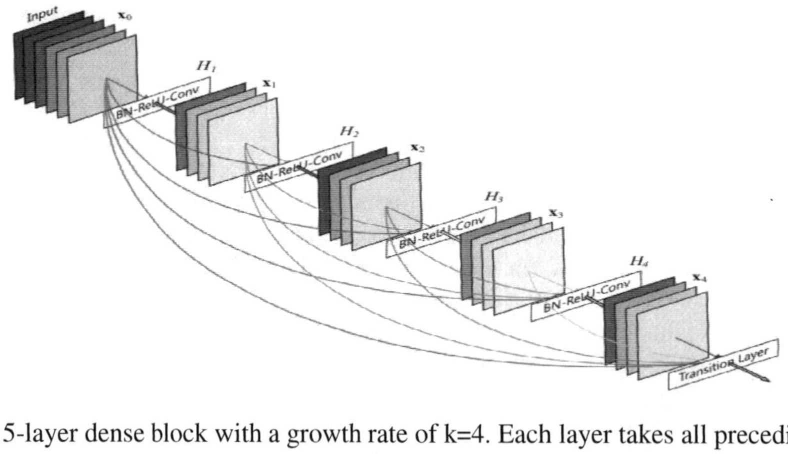

Figure 1. A 5-layer dense block with a growth rate of k=4. Each layer takes all preceding feature-maps as input.

Attention Gates: AGs are commonly used in natural image analysis, knowledge graphs, and language processing (NLP) for image captioning[5], machine translation[6], and classification[7, 8] tasks. Initial work has explored attention-maps by interpreting gradient of output class scores with respect to the input image. Trainable attention, on the other hand, is enforced by design and categorised as hard- and soft-attention. Hard attention, iterative region proposal and cropping, is often non-differentiable and relies on reinforcement learning for parameter updates, which makes model training more difficult. Recursive hard-attention is used in[9] to detect anomalies in chest X-ray scans. Contrarily, soft attention is probabilistic and utilises standard back-propagation without need for Monte Carlo sampling. For instance, additive soft attention is used in sentence-to-sentence translation[10] and more recently applied to image classification[7,8]. Channel-wise attention is used to highlight important feature dimensions, which was the top-performer in the ILSVRC 2017 image classification challenge. Self-attention techniques have been proposed to remove the dependency on external gating information. For instance, non-local self attention is used in[11] to capture long range dependencies. The Attention Gate can be summarised as follows Figure 2:

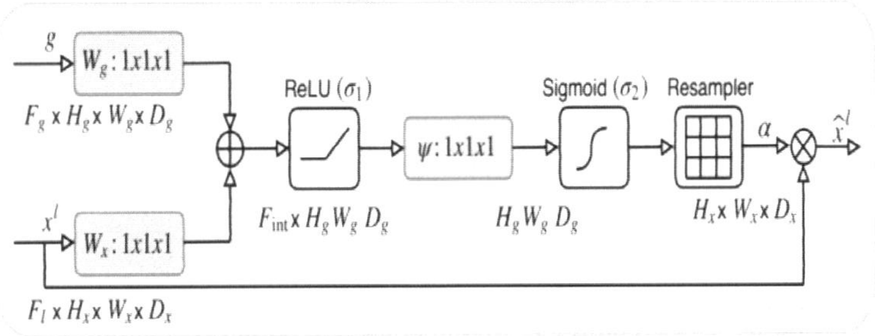

Figure 2. Schematic of the proposed additive attention gate (AG). Input features are scaled with attention coefficients (α) computed in AG. Spatial regions are selected by analysing both the activations and contextual information provided by the gating signal (g) which is collected from a coarser scale. Grid resampling of attention coefficients is done using trilinear interpolation.

3. Approach

In this experiment, we have 40 retinal vessels images, and we use 30 images to train,5 images to validation, and 5 images to test. Data augmentation is essential to teach the network the desired invariance and robustness properties, when only few training samples are available. In case of microscopical images we primarily need shift and rotation invariance as well as robustness to deformations and gray value variations. Especially random elastic deformations of the training samples seem to be the key concept to train a segmentation network with very few annotated images.
Data Augmentation refers to the method of expanding the original data through a series of random transformations to increase the amount of data. The new data set generated by this method that cannot find any two identical images. This helps to suppress overfitting and thus improves The generalization ability of the deep learning model. This paper selects translation, rotation, shearing and scaling for data augmentation.
The presented improved model, which we refer to as DA-U-Net, is depicted on Figure 4. It consists of basic blocks, and each of them follows the encoderdecoder architecture similar to U-Net[2]. We consider two kinds of basic blocks: Denseblock and Attention gate. They both feature connections (shown on the Figure 4), linking layers of the encoder and decoder, which are of very high importance. In this paper, we replace the normal convolution layers as the Bottleneck layers in our network. Although each layer only produces k output feature-maps, it typically has many more inputs. It has been noted in [12, 13] that a 1×1 convolution can be introduced as bottleneck layer before each 3×3 convolution to reduce the number of input feature-maps, and thus to improve computational efficiency. We find this design especially effective for DenseNet and we refer to our network with such a bottleneck layer to the BN-ReLU-Conv (3×3), as shown in the Figure 3. And in the stage of the

decoder, we added Attention gate mechanism, which can automatically learn to focus on target structures without additional supervision.

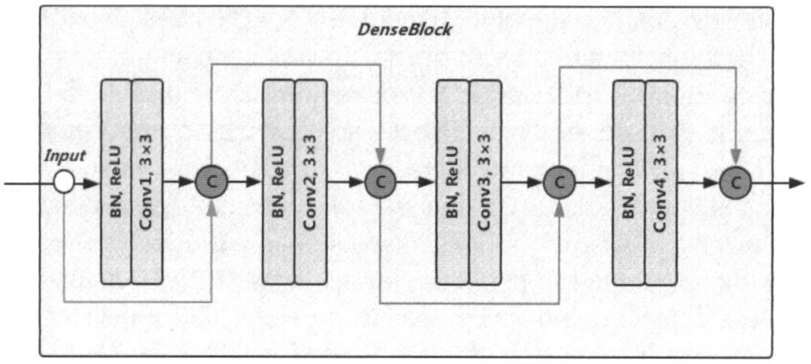

Figure 3. The DenseBlock in DA-U-Net architecture.

As a loss function, we use l(A; B):

l(A; B) = − log d(A; B);

where: $$d(A;B) = \frac{2\sum_{i,j} a_{ij} b_{ij}}{\sum_{i,j} a^2{}_{ij} + \sum_{i,j} b^2{}_{ij}}$$

where A = $(a_{ij})_{i=1\,j=1}$ is a predicted output map, containing probabilities that each pixel belongs to the foreground, and B = $(b_{ij})_{i=1\,j=1}$ is a correct binary output map.

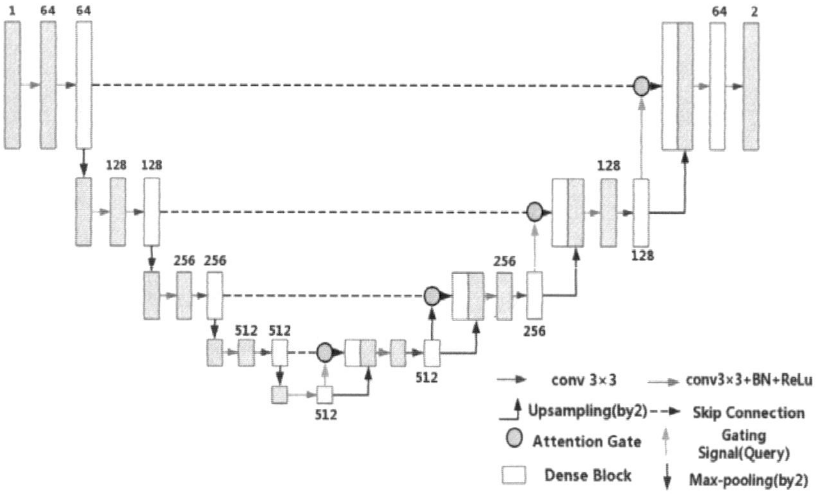

Figure 4. DA-U-Net architecture | U-Net model with DenseBlock and Attention Gate in decoder. Schematic of the AGs is shown in Figure 2. Feature selectivity in AGs is achieved by use of contextual information (gating) extracted in coarser scales.

4 Experiment

Experimental data provided by the DRIVE datasets, which is consisted of 40 color retinal images in total, in which 20 samples are used for training and remaining 20 samples are used for testing. The size of each original image is 565×584 pixels[14]. The data set is 900 sets of images generated after

adopting the data augmentation method. In the training stage, we choose the Adam as the optimization, and the learning rate is 0.0001. The experimental platform is based on Windows 7 64-bit operating system, i7-4790 CPU, 16G memory, GTX 750, using a deep learning framework for TensorFlow.
In order to further prove the effectiveness of the proposed method on retinal vascular segmentation, the comparison with the best found methods for the segmentation of retinal vessels in DRIVE datasets is presented in Table 1. Sensitivity (SE) indicates the percentage of correctly segmented blood vessel pixels as a percentage of true blood vessel pixels, and the specificity (SP) is the percentage of correctly segmented background pixels to the true background pixels. The accuracy (ACC) is the percentage of correctly segmented blood vessels and background pixels in the entire image.

Table1. Segmentation performance of retinal vessel on the DRIVE datasets.

Methods	SE	SP	ACC
Liskowsk [15]	0.7763	0.9768	0.9495
Qiaoliang Li [16]	0.7569	0.9816	0.9527
U-Net [17]	0.7537	0.9820	0.9531
Residual U-Net [17]	0.7726	0.9820	0.9553
R2U-Net [17]	0.7792	0.9813	0.9556
DA-U-Net	0.7539	0.9904	0.9609

The different prediction between the DA-U-Net and U-Net in the DRIVE can be made based on Figure 5.

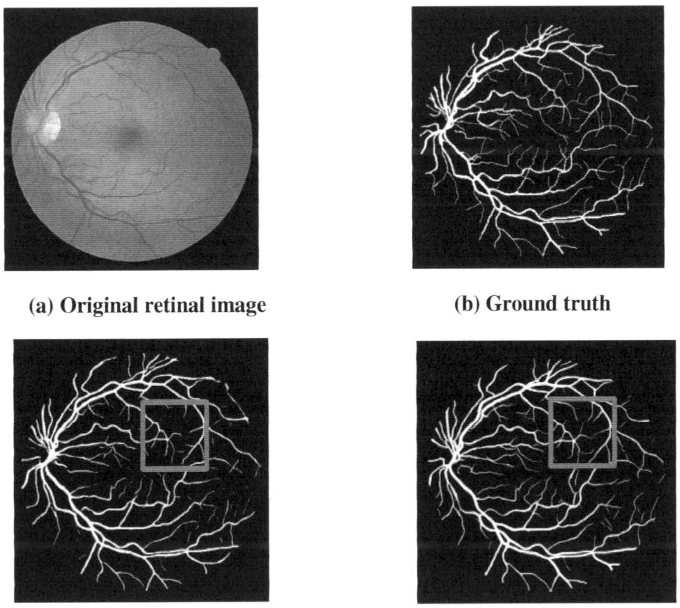

(a) Original retinal image **(b) Ground truth**

(c) Segmented result by U-Net **(d) Segmented result by DA-U-Net**

Figure 5. The figure shows the different prediction between the DA-U-Net and U-Net. From the Segmented result by U-Net, We can see there are many small blood vessel segmentation effects that are not good, the effect in (d) is obviously better than (c).

5. Conclusion and Discussion

We present the model for the medical image segmentation based on Denseblock and Attention gate used in the well-konwn U-Net models. And the Segmented result by DA-U-Net is obviously better than the U-Net. We verified the feasibility of the proposed method on the DRIVE datasets, and the accuracy rates were 96.09%. By analyzing and comparing the segmented retinal blood vessel images, the proposed method is more advantageous than other methods. It can provide doctors with a wealth of information about eye conditions to help patients receive timely treatment. Still, our results fail todetect very thin vessels that span only 1 pixel. We expect that additional prior knowledge on the vessel structures such as connectivity may leverage the performance further.

The use of artificial intelligence technology for medical image-assisted diagnosis is a rapidly developing field. With the rapid development of deep learning technology and the continuous opening of medical imaging data, artificial intelligence is gradually maturing in early stage disease screening and doctor-assisted diagnosis.

Acknowledgments

I would like to extend my sincere gratitude to my teacher, Cong Wu, for his instructive advice and useful suggestions on my thesis. I am deeply grateful of his help in the completion of this thesis.

References

[1] Milletari, F., Navab, N., Ahmadi, S.A.: V-net: Fully convolutional neural networks for volumetric medical image segmentation. In: 3D Vision (3DV), 2016 Fourth International Conference on, IEEE

[2] Ronneberger, O., Fischer, P., Brox, T.: U-net: Convolutional networks for biomedical image segmentation. In: International Conference on Medical image computing and computer-assisted intervention, Springer (2015)

[3] He, K., Zhang, X., Ren, S., Sun, J.: Deep residual learning for image recognition. In:Proceedings of the IEEE Conference on Computer Vision and Pattern Recognition.pp. 770 – 778 (2016)

[4] Huang G, Liu Z, Maaten L V D, et al. Densely Connected Convolutional Networks[C]// IEEE Conference on Computer Vision and Pattern Recognition. IEEE Computer Society, 2017:2261-2269

[5] Anderson, P., He, X., Buehler, C., Teney, D., Johnson, M., Gould, S., Zhang, L.: Bottom-up and top-down attention for image captioning and vqa. arXiv preprint arXiv:1707.07998 (2017)

[6] Vaswani, A., Shazeer, N., Parmar, N., Uszkoreit, J., Jones, L., Gomez, A.N., Kaiser,. Polosukhin, I.: Attention is all you need. In: NIPS. pp. 6000 – 6010 (2017)

[7] Jetley, S., Lord, N.A., Lee, N., Torr, P.: Learn to pay attention. In: International Conference on Learning Representations (2018)

[8] Wang, F., Jiang, M., Qian, C., Yang, S., Li, C., Zhang, H., Wang, X., Tang, X.: Residual attention network for image classification. In: IEEE CVPR. pp. 3156 – 3164 (2017)

[9] Ypsilantis, P.P., Montana, G.: Learning what to look in chest X-rays with a recurrent visual attention model. arXiv preprint arXiv:1701.06452 (2017)

[10] Shen, T., Zhou, T., Long, G., Jiang, J., Pan, S., Zhang, C.: Disan: Directional self-attentionnetwork for rnn/cnn-free language understanding. arXiv preprint arXiv:1709.04696 (2017)

[11] Wang, X., Girshick, R., Gupta, A., He, K.: Non-local neural networks. arXiv preprint arXiv:1711.07971 (2017)

[12] K. He, X. Zhang, S. Ren, and J. Sun. Deep residual learning for image recognition. In CVPR, 2016

[13] C. Szegedy, V. Vanhoucke, S. Ioffe, J. Shlens, and Z. Wojna.Rethinking the inception architecture for computer vision. In CVPR, 2016

[14] Staal, Joes, et al. "Ridge-based vessel segmentation in color images of the retina." IEEE

transactions on medical imaging23.4 (2004): 501-509
[15] Liskowski, Paweł, and Krzysztof Krawiec. "Segmenting Retinal Blood Vessels With Deep Neural Networks." IEEE transactions on medical imaging 35.11 (2016): 2369-2380
[16] Li, Qiaoliang, et al. "A cross-modality learning approach for vessel segmentation in retinal images."IEEE transactions on medical imaging 35.1 (2016):109-118
[17] Alom, Md Zahangir, et al. "Recurrent Residual Convolutional Neural Network based on U-Net(R2UNet) for Medical Image Segmentation." arXiv preprint arXiv:1802.06955 (2018)

Face Image Manipulation Detection

Lilong Wen[1], Dan Xu[1,a]
[1]School of Information Science and Engineering of Yunnan University

danxu@ynu.edu.cn

Abstract. This paper proposes a CNN-based (Convolutional Neural Network based) network to detect altered face picture, which can cover the most common face swap methods. The network uses an autoencoder which is pre-trained on the original images to reconstruct the input images. The reconstructed one and the input image are then processed by the SRM filter which can extract the noise distribution of images. We then feed the minus result of two processed results into a CNN architecture to predict whether the input image is original or tampered. The model was trained and evaluated in FaceForensics dataset and state-of-art face swap method. Experimental results demonstrate the effectiveness of our network..

1. Introduction

Images are important media that commonly used for both daily communication and crucial situation like political or commercial cases. But with the great progress achieved in computer vision and image process, changing the content of original images without detection have become much easier. With some user-friendly software like Photoshop and Meitu, everyone can do some manipulation to the original picture without authorization and post to the Internet. Distinguishing the original picture with tampered one become much more important than ever. Especially images involve face regions. There are many advanced applications can automatically swap two faces or manipulate one's facial features benefited by the face detection and recognition techniques. Plenty of pictures on the Internet that are well tampered with fake face for fun or some other reasons which can always cause misleading and confusions in some situations. And with recently rising deep learning technology, many advanced image-to-image translating methods have been proposed to change the human face based on generative model. The result can be hard to distinguish for human even after close observation.

In order to tell the difference between real and fake pictures, there has already been a great achievement in this particular field. Some traditional methods such as error level analysis (ELA) [1], local noise features [2] and camera filter array (CFA) [3] have been developed. Recently, neural network has been widely used in picture forensics and have obtained the promised results. However most of the current approaches concentrate on general image tamper techniques such as copy-move, splicing, removal et al. and cannot totally combat the well shopped fake face pictures on the Internet. Since unlike other image manipulation detection, photos with altered faces usually are well tampered and has no semantic difference with the origin one. The detection should pay more attention on the facial feature differences together with the differences between the face and the rest of the picture, which suggests more fine-grained features to learn.

Content from this work may be used under the terms of the Creative Commons Attribution 3.0 licence. Any further distribution of this work must maintain attribution to the author(s) and the title of the work, journal citation and DOI.
Published under licence by IOP Publishing Ltd

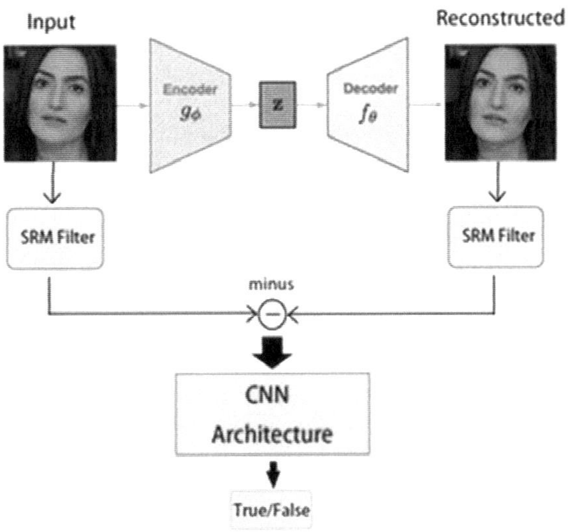

Figure 1. Architecture of Proposed Network

2. Related Work

Research on image forensics has achieved great progress in recent years and plenty of detecting methods have been proposed. Prior works include some low-level feature analysis such as error level analysis (ELA) that rely on the compression artifacts caused by the lossy compression of the JPEG image, and some improved method based on the double image compression and Discrete Cosine Transform (DTC) coefficient analysis [4,5] could achieve a better result. There are also some methods based on the fingerprints left by internal camera processers such as Camera Response Function (CRF) [6] and photo response non-uniformity (PRNU) noise pattern [7]. And the work of Yerushalmy I and HelOr H. [8] directly rely on the physical structure and characteristics of digital cameras. What's more, Color Filter Array (CFA) based methods has also been proved to be efficient such as the method proposed by Popescu AC and Farid H [9], which quantify the specific correlations introduced by CFA interpolation to automatically detect image tampering. And there are also many pixel level statistic methods such as using possibility map [10] and statistical properties of natural photographic images [11].

With recent success of deep learning approaches in computer vision and image process filed, researchers apply deep learning in photo forensic field, which have achieved a promised result. Bondi L et.al [12] use convolutional neural network (CNN) to detect different image patches under the camera model theory that assuming all the pixels from the same picture should have same character. Chen Jiansheng, et al.[13] apply convolutional neural network in median filtering image forensics, and feed the median filtering residual into the later network. Other deep learning architecture such as autoencoder and Long short-term memory (LSTM) network also perform well in image anomaly detection task except using CNN. Zhang et al. [14] use a staked autoencoder to learn complex features for each individual image patches, and then integrate the contextual information of each patch to perform the final results. Cozzolino D and Verdoliva L [15] use autoencoder to generate implicit information from the noise data to detect the anomaly. Jawadul H and Amit K. et.al employ a hybrid CNN-LSTM model to discriminate the features between altered and original areas.

Recently, local noise feature-based method like steganalysis rich model (SRM) [16] has been widely used in image forensic. Cozzolino D and Verdoliva L extract local noise features from image using SRM filters and then feed into convolutional neural network architecture. Peng Zhou et al. [17] also use a steganalysis feature extractor to pre-process images before feed into the patch triplet stream. Their later work [18] also adopt the same structure but instead of using all 30 SRM filter kernels, they only choose 3 kernels which can still achieve a decent performance with lower computational cost.

Inspired by those prior work, we designed a network that adopt both autoencoder and SRM filters as well as face classification stream to jointly detect the tampered face images.

3. Proposed Method

Inspired by the recent work and newly released dataset, this paper proposed a CNN-based architecture to detect the manipulated face images. The proposed network using a pre-trained autoencoder to reconstruct image from input ones, which works kind of like a filter that can rebuilt the input image that close to the input one when the input is untampered, but the similarity would be much less when the input is tampered. The model then combined the steganalysis rich model (SRM), which has been proved to be effective in image manipulation detection, to extract the noise distribution of two images. We use the minus result of those two processed images and then feed into the fine designed VGG16 [19] network to get the final prediction result. The full architecture is shown in figure 1.

To train and evaluate our network, we adopt a large-scale dataset for forgery detection in human face called FaceForensics [20], which is a video dataset contains 500,000 frames from 1004 videos collected from Youtube and the tampered data was created by applying state-of-the-art Face-to-Face approach [21] on the original data.

3.1. Autoencoder

An autoencoder is a type of artificial neural network used to learn the data representation through iterative forward and backward training. It includes two part, the encoder and decoder, the encoder encodes the data into low level representation and the decoder is designed to perform an opposite operation that reconstruct the data to produce a result as close as possible to input data. The encoder and decoder can be presented in (1) and (2), while the (3) shows the loss function training the network. And the (4) shows the parameters updated by gradient descent.

$$h_i = \phi(x_i) \tag{1}$$

$$\hat{x}_i = \phi^{-1}(h_i) = f(x_i) \tag{2}$$

$$J(f) = \frac{1}{n}\sum_{i=1}^{n} \| f(x_i) - x_i \|_2^2 \tag{3}$$

$$\overline{g} = \frac{dJ}{df} = \frac{1}{n}\sum_{i=1}^{n} g_i = \frac{2}{n}\sum_{i=1}^{n}(f(x_i) - x_i) \tag{4}$$

When update the parameters of autoencoder using gradient descent as (4) shows, the gradient is averaged over all training data, so an autoencoder dose not try to reduce the error of single datum but the overall error. We can measure the error reduction in single datum by using effective gradient magnitude using (5)

$$g_i^{effect} = \frac{< g_i, \overline{g} >}{| g_i |} = | \overline{g} | \cos\theta(g_i, \overline{g}) \tag{5}$$

Where $\theta(g_i, \overline{g})$ is the angle between g_i and $\overline{g}$. According to the research of Xia Y et.al [22], the angle for a positive datum is often smaller than it is for an outlier, leading to larger effective gradient magnitude. In another words, the reconstruction errors of positive data are always smaller than that of outliers.

Based on this idea, the network is first pre-trained off-line on original picture to fully learn the positive data distribution, after that, it could reconstruct the positive image with good accuracy while negative ones with large errors, which means the original images and the altered ones would get a quite

different reconstructed result through the network. And this kind of difference could be used for later distinguish work.
After pre-training the CAE, we freeze all parameters of this autoencoder and does not involve this into training. It will only be used as an image filter in the final network.

3.2. Noise Distribution of Image
Visual features are not quite enough to cover all the face tamper operations. Pictures that well altered with sophisticated methods can reduce any kinds of splicing boundary and color inconsistent cannot be detected by only using the visual features. So we adopt this noise classification stream to utilize the local noise distribution of images to get more information about the tampered face area. According to the work of [18], We use the same 3 SRM kernels to reduce the computational cost and speed up the process, which is showed in (6).

$$\frac{1}{4}\begin{vmatrix} 0 & 0 & 0 & 0 & 0 \\ 0 & -1 & 2 & -1 & 0 \\ 0 & 2 & -4 & 2 & 0 \\ 0 & -1 & 2 & -1 & 0 \\ 0 & 0 & 0 & 0 & 0 \end{vmatrix}, \frac{1}{12}\begin{vmatrix} -1 & 2 & -2 & 2 & -1 \\ 2 & -6 & 8 & -6 & 2 \\ -2 & 8 & -12 & 8 & -2 \\ 2 & -6 & 8 & -6 & 2 \\ -1 & 2 & -2 & 2 & -1 \end{vmatrix}, \frac{1}{2}\begin{vmatrix} 0 & 0 & 0 & 0 & 0 \\ 0 & 0 & 0 & 0 & 0 \\ 0 & 1 & -2 & 1 & 0 \\ 0 & 0 & 0 & 0 & 0 \\ 0 & 0 & 0 & 0 & 0 \end{vmatrix} \quad (6)$$

And to demonstrate the effectiveness of the SRM filter, we pick an original image and it's corresponding tampered one from dataset to perform minus operation before and after SRM filter, which is shown in figure 2.

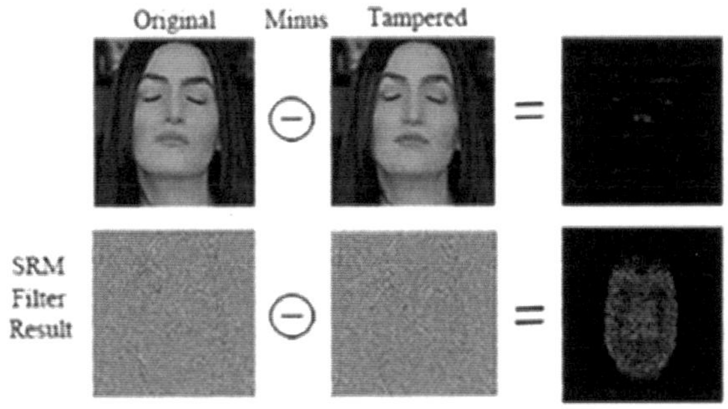

Figure 2. The Noise Feature of both Original and Altered Images Extracted by SRM Filters. The Third Column Shows the Subtraction Result of Previous Two Columns. (The Subtraction Value is Round to be Integer)

3.3. Classification
After processed by the SRM filter, we perform minus operation between two images to get an intermediate result. Figure 3 shows the minus results between original untampered input and tampered input. To speed up the training process, we adopt VGG16 and perform a transfer learning process in this part.

4. Experiments
In this section, we demonstrate our architecture on FaceForensics dataset and compared with other state-of-the-art classification method on this dataset. What's more, to prove the effectiveness beyond the dataset, we also test our model in several state-of-art open-sourced face-swap method found in github.

4.1. Setting Up Datasets
FaceForensics dataset is a large-scale dataset created by using automated version of Face2Face approach. All videos are downloaded from Youtube and cut down to short continuous clips that

contain mostly frontal faces. The dataset is divided into three parts, the training set, the validation set and the test set. All sets include number-balanced original and altered images. the training set, the test set and the validation set include 736270, 155490 and 151052 images, respectively. The figure 3 shows some examples of original and altered images from dataset.

Figure 3. Some Examples of Original and Altered Images from Faceforensics Dataset

4.2. Training Model

Firstly, we pre-train the CAE on the positive original data until the reconstruction loss is converged into an acceptable level that the reconstructed image is very close to the input one, then save the weights of the network. In addition, to prevent the over-fitting to the dataset, we also used some public available face dataset [23, 24, 25] while training the CAE.
Then, we perform an end-to-end training process using the whole dataset including the untampered original images the tampered ones.

4.3. Testing and Compared with Other Methods

We evaluated the model using the test set of the dataset which have 77745*2 pictures. And compared with traditional classification architecture VGG16 and Resnet [26]. We use the H.264 compressed data with quantization parameter 35 since the uncompressed data is too easy to classify. The test result is shown in table 1.
What's more, to prove the effectiveness of our model, we also trained three face-swap models found in github which all using state-of-art methods to produce tampered face images for testing. The face-swap models we used are PRNet [27, 28], DeepFaceLab [29] and face2face-demo [30]. All those three models are trained using video and image data from FaceForensics dataset and use the same swap paradigm that swap the same person's face in the same scenes, which make the result image looks more real and only difference would be the face region. The test result in those tampered images produced by above three model is shown in table 2, and those images are using the same compression method as well.

Table1. Compared with other classification networks

method	**accuracy**
VGG16	78.13
ResNet	82.45
Ours	90.76

Table2. Evaluate on three face swap methods

method	**accuracy**
PRNet	90.80
DeepFaceLab	94.87
face2face-demo	93.14

5. Conclusions

We propose a CNN-based architecture for tampered face detection. The experimental result shows that our approach outperforms other prior works. What's more, since this paper has proved that combined with a pre-trained autoencoder, the negative input could have much difference with the reconstructed one compared to the positive input. This method could be used in all kinds of binary classification problem and hopefully could produce a better performance compared to the previous methods.

References

[1] Barton K C. A picture's worth[J]. Social Education, 2001, 65(5): 278.
[2] Pan X, Zhang X, Lyu S. Exposing image splicing with inconsistent local noise variances[C]//Computational Photography (ICCP), 2012 IEEE International Conference on. IEEE, 2012: 1-10.
[3] Goljan M, Fridrich J. CFA-aware features for steganalysis of color images[C]//Media Watermarking, Security, and Forensics 2015. International Society for Optics and Photonics, 2015, 9409: 94090V.
[4] Bianchi T, De Rosa A, Piva A. Improved DCT coefficient analysis for forgery localization in JPEG images[C]//Acoustics, Speech and Signal Processing (ICASSP), 2011 IEEE International Conference on. IEEE, 2011: 2444-2447.
[5] Galvan F, Puglisi G, Bruna A R, et al. First quantization matrix estimation from double compressed JPEG images[J]. IEEE Transactions on Information Forensics and Security, 2014, 9(8): 1299-1310.
[6] Hsu Y F, Chang S F. Image splicing detection using camera response function consistency and automatic segmentation[C]//Multimedia and Expo, 2007 IEEE International Conference on. IEEE, 2007: 28-3.
[7] Lukáš J, Fridrich J, Goljan M. Detecting digital image forgeries using sensor pattern noise[C]//Security, Steganography, and Watermarking of Multimedia Contents VIII. International Society for Optics and Photonics, 2006, 6072: 60720Y.
[8] Yerushalmy I, Hel-Or H. Digital image forgery detection based on lens and sensor aberration[J]. International journal of computer vision, 2011, 92(1): 71-91.
[9] Popescu A C, Farid H. Exposing digital forgeries in color filter array interpolated images[J]. IEEE Transactions on Signal Processing, 2005, 53(10): 3948-3959.
[10] Li H, Luo W, Qiu X, et al. Image forgery localization via integrating tampering possibility maps[J]. IEEE Transactions on Information Forensics and Security, 2017, 12(5): 1240-1252.
[11] Lyu S. Natural Image Statistics in Digital Image Forensics[M]//Digital Image Forensics. Springer, New York, NY, 2013: 239-256.
[12] Bondi L, Lameri S, Guera D, et al. Tampering Detection and Localization Through Clustering of Camera-Based CNN Features[C]//CVPR Workshops. 2017: 1855-1864.
[13] Chen, Jiansheng, et al. "Median filtering forensics based on convolutional neural networks." IEEE Signal Processing Letters 22.11 (2015): 1849-1853.
[14] Zhang Y, Goh J, Win L L, et al. Image Region Forgery Detection: A Deep Learning Approach[C]//SG-CRC. 2016: 1-11.
[15] Cozzolino D, Verdoliva L. Single-image splicing localization through autoencoder-based anomaly detection[C]//Information Forensics and Security (WIFS), 2016 IEEE International Workshop on. IEEE, 2016: 1-6.
[16] Fridrich J, Kodovsky J. Rich models for steganalysis of digital images[J]. IEEE Transactions on Information Forensics and Security, 2012, 7(3): 868-882.
[17] Zhou P, Han X, Morariu V I, et al. Two-stream neural networks for tampered face detection[C]//2017 IEEE Conference on Computer Vision and Pattern Recognition Workshops (CVPRW). IEEE. 2017: 1831-1839.
[18] Zhou P, Han X, Morariu V I, et al. Learning Rich Features for Image Manipulation Detection[J]. arXiv preprint arXiv:1805.04953, 2018.
[19] Simonyan K, Zisserman A. Very deep convolutional networks for large-scale image recognition[J]. arXiv preprint arXiv:1409.1556, 2014.
[20] Rössler A, Cozzolino D, Verdoliva L, et al. FaceForensics: A Large-scale Video Dataset for Forgery Detection in Human Faces[J]. arXiv preprint arXiv:1803.09179, 2018.

IOP Conf. Series: Materials Science and Engineering **533** (2019) 012054 doi:10.1088/1757-899X/533/1/012054

[21] Thies J, Zollhofer M, Stamminger M, et al. Face2face: Real-time face capture and reenactment of rgb videos[C]//Proceedings of the IEEE Conference on Computer Vision and Pattern Recognition. 2016: 2387-2395.
[22] Xia Y, Cao X, Wen F, et al. Learning discriminative reconstructions for unsupervised outlier removal[C]//Proceedings of the IEEE International Conference on Computer Vision. 2015: 1511-1519.
[23] Zheng T, Deng W, Hu J. Cross-Age LFW: A Database for Studying Cross-Age Face Recognition in Unconstrained Environments[J]. 2017.
[24] Kumar N. Attribute and Simile Classifiers for Face Verification[C]// IEEE International Conference on Computer Vision. IEEE, 2010.
[25] Liu Z, Luo P, Wang X, et al. Deep Learning Face Attributes in the Wild[J]. 2014.
[26] He K, Zhang X, Ren S, et al. Deep Residual Learning for Image Recognition[J]. 2015.
[27] Feng Y, Wu F, Shao X, et al. Joint 3D Face Reconstruction and Dense Alignment with Position Map Regression Network[J]. 2018.
[28] https://github.com/YadiraF/PRNet
[29] https://github.com/iperov/DeepFaceLab
[30] https://github.com/datitran/face2face-demo

Real-time Measurement, Computing and Visualization of Plant Structure

Guo Li[1,2], Tingting Qian[3], Shenglian Lu[1,2*]

[1]College of Computer Science and Information Technology, Guangxi Normal University, 541004, Guilin, China
[2]Guangxi Key Lab of Multi-source Information Mining & Security, 541004, Guilin, China
[3]Agricultural Information Institute of Science and Technology, Shanghai Academy of Agricultural Sciences, 201403, Shanghai, China

shllu@126.com

Abstract. A method and software tool for measuring and reconstructing small plant in real time was presented. The key of this method lies in the feature points measuring with 3D digitizer and organ templates-based details 3D reconstructing. Firstly, several feature points were selected and measured by using a digitizer from the targeted plant. Then the morphological parameters of this plant could be calculated from these feature points, with its skeleton model was reconstructed in the process of measuring. For a details 3D reconstruction of the plant structure, organ's 3D template was used, which was created by using 3D point cloud data measured from real plant organs, or created with interactive designing. This method has been tested in many plants, including crops and small fruit trees. The experiment results indicated that the proposed method is suitable for real-time measurement and multi-scale structural modeling of small plants.

1. Introduction

The rapid and automatic measurement and reconstruction of plant structure has a long history both in area of agronomy research and computer science [1, 2]. Recently, with the successful applications of various measurement devices in industry, more and more people reproduce 3D plant models from measured data. The calculation of morphological parameters and reconstruction from measured data is presently considered to be the most accurate approach to quantitatively represent the 3D architecture of a plant, because the actual features of plant geometry are taken into account.
Early in 1980s electromagnetic digitizers were used to measure the spatial position and orientation of stems and leaves from plant both for accurate reconstructing structure and giving a quantitative assessment to the plant geometry [3-5]. This kind hand-held device is robust for capturing the spatial characteristics of objects as such it is still a good choice for crop reconstruction to many people [6, 7]. However, it is a tedious and time-consuming job to digitize a plant with complex branch structure and crowned leaves, or canopy of plant population with crowded organs [8], and often not enough for accurately capturing the detailed organ geometry, e.g. measuring the detail leaf surface [9]. Several species of crops (such as maize, grain sorghum, cotton and so on) has been measured by using a

Content from this work may be used under the terms of the Creative Commons Attribution 3.0 licence. Any further distribution of this work must maintain attribution to the author(s) and the title of the work, journal citation and DOI.
Published under licence by IOP Publishing Ltd

digitizer for analyzing their morphological architecture. However, the time and labor costs for measuring data still maintain a question for considering.

In recent years, with the wide applications of 3D laser scanner, measurement and reconstruction plant structure from laser scanned point clouds has been used more frequently [10-12]. Laser scanners enable us to rapid quantify the surface of an object as a dense set of points. This may be very useful for the rough measurement and reconstruction of big plant with sparse branches and leaves in their canopies. However, how to obtain the morphological parameters (such as the length, width and incidence angle of an organ) automatically from laser scanned 3D point cloud is hard to be addressed, especially for small plant. Because of limited scanning resolution and occlusion from other organs of other plants, parts of organs would be missed in the scanned point cloud. The missing information can be estimated by using existing or statistical knowledge about the morphological structure of plants [13]. However, automatic segmentation of different individual organs from 3D point cloud and then accurate calculation of morphological parameters still not an easy task [14].

Image-based method or computer vision is a more directed and low-cost solution for measuring morphological parameters. Many work have been done in this field under the rising requirement from plant phenotyping [15-17]. Multidisciplinary collaboration among the biological, engineering, and computer sciences are expected and also needed for a more automatic and accurate calculation of plant phenotyping.

This study presented a method and software tool for real-time measuring, computing and visualizing of small plant structure with low costs of time and manual labor. Our method used spatial characteristics points measured from plant with 3D digitizer as the input. These spatial characteristics points were then used to calculate morphological parameters, reconstruct skeleton model and details 3D model of the plant.

2. Data Measurement

A 3D digitizer Patriot (Polhemus Inc., Colchester, VT, USA) was used to collect data from target plant. We can describe the structure of plant with less complex structure by a small number of feature points. As such several feature points is enough for each internode on a branch of a plant. In this case, 6 points were selected from each internode, as Figure 1 shows. In which p1 is the crossing point of the stem and the petiole, p2 is the crossing point of the petiole and the leaf blade, others located on the margin of the leaf blade. Generally point p4 is the tip of the leaf blade, while p3 and p5 were digitized at the widest points of each leaf blade. While p0 means the root of the plant. The digitizing order was from root to top for each stands.

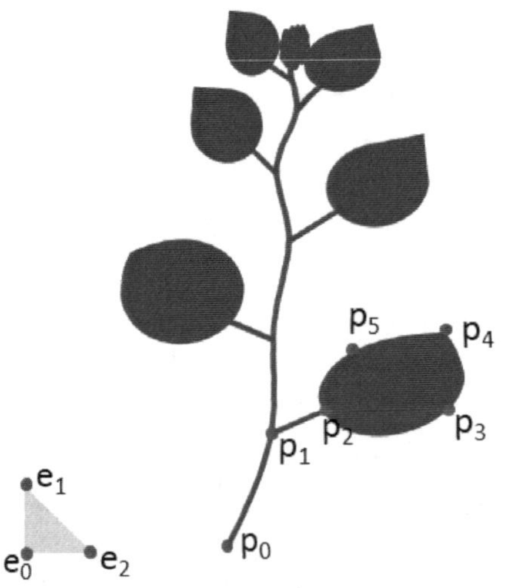

Figure 1. Diagram of the selection of Feature Points

Three extra points (from e_0 to e_2 in Figure 1) were also picked before measuring the feature points from the plant. These extra points are auxiliary and necessary for calculating azimuth angle and inclination angle of each leaf in the plant. All the extra points were selected near the root of the plant. In which $\overrightarrow{e_0e_1}$ represents the northern direction. Triangle $\Delta e_0e_1e_2$ represents the ground.

3. Real-time Visualization and Parameters Calculation

Since all the feature points mentioned at Section 2 were picked by digitizer one by one. We can show the picked points on the screen and also calculate the morphological parameters of the measured plant while digitizing.

3.1. Display the Feature Points

OpenGL, an open source 3D graphics API, was used to display the feature points measured by digitizer. Each feature point was input as a vertex. The contiguous two points from a same internode were connected by a straight line. While the stem was displayed by connecting the first point of all internodes. As Figure 2 shows, the three points on the bottom of the black area are the extra points, as such the brown triangle represents the ground. This figure shows that one and half of an internode had been measured.

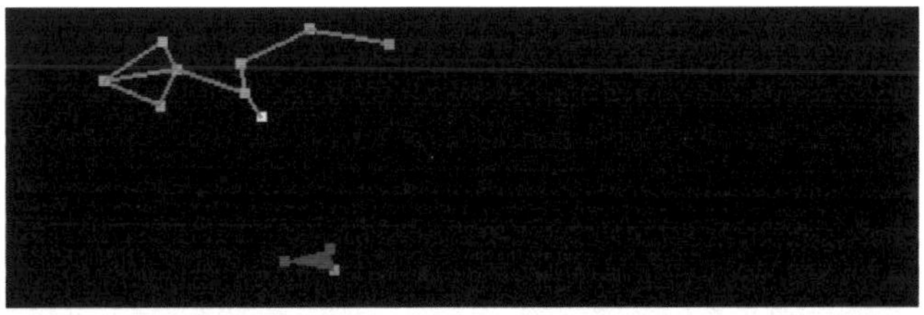

No.	len/node	len/peti	len/leaf	wid/leaf	azimuth	inclina.
1	5.95	16.40	14.85	16.44	-13.03	87.05
2	6.01	18.17				

Figure 2. Visualization of feature Points and Calculation Results

By this way, the process of digitizing feature points from a plant were visualized (see Figure 3). This would give users a more intuitive experience, and tell him how many organs had been digitized, whether the picked feature points are right.

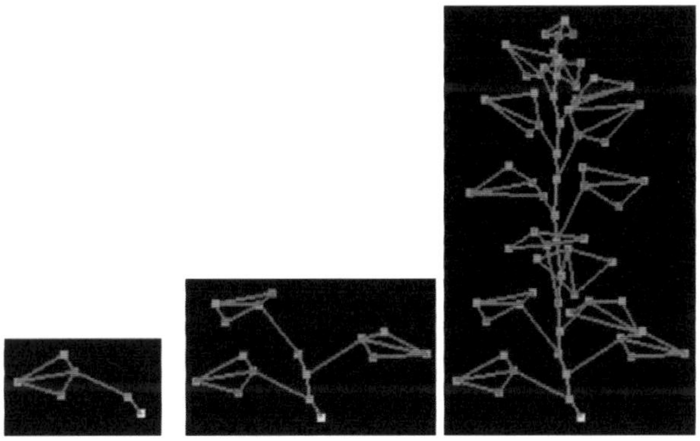

Figure 3. Visualizing the process of digitizing a plant

3.2. Morphological Parameters Calculation

Base on the digitized feature points, several morphological parameters of the plant could be calculated easily. For example, the length of leafblade is the distance between point p_2 and p_4; the width of leafblade is the distance between point p_3 and p_5. The length of internode can be calculated from point p_0 and p_4. To calculating the azimuth angle and inclination angle of each leaf in the plant, the three extra points were used as reference, in which the direction from point e_0 to e_1 represents the northern direction of the earth. And the triangle consisting of these three points represents the ground.

The calculation results were displayed immediately on the screen of our software. Once a feature point was measured, the morphological parameters were then calculated and shown, as Figure 2 shows. In other words, this calculation was real-time.

4. Structure Reconstruction

When all the selected feature points of a plant were measured, the skeleton structure of this plant was also reconstructed. However, in some case a skeleton model is too simple. For example, this skeleton model didn't produce the leaf area of the plant.

4.1. Simple Reconstruction

The simple reconstruction of a plant is to generate a 3D mesh model directly from these feature points measured from this plant. The key processes includes two tasks, one is the reconstruction of stem, another is the meshing of leafblade. The 3D model of the stem can be generated by using a sweep algorithm. The 3D model of the petiole can be also generated with the same way. We can see, from Figure 1, that four feature points were measured from a leafblade. To generate a 3D mesh from the four feature points of a leafblade, firstly B-spline curve was used to represent the edge of the leafblade, then Delaunay triangulation method was used to generation a mesh from the closed B-spline curve. As such a simple 3D model of the plant was reconstructed (see Figure 4a).

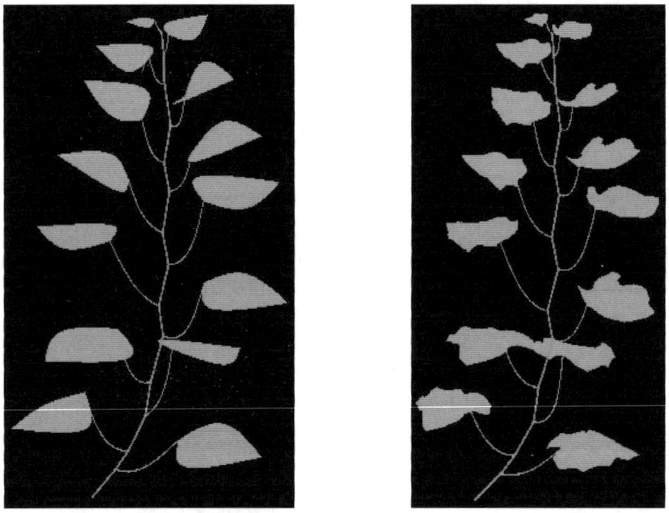

(a) simple model (b) high-precision model

Figure 4. The Reconstructed 3D Model of A Plant

4.2. High-precision Reconstruction

The surface of a leafblade in real plant is often irregularity. The leafblade mesh used in the simple model was still not very accurate, and should reach incorrect area calculation.

To give a more precise mesh model of a plant (see Figure 4b), we can use a more detail leafblade mesh to replace the mesh generated by using just four feature points. This kind detail leafblade mesh (see Figure 5) could be generated from 3D point cloud scanned from real plant leaves.

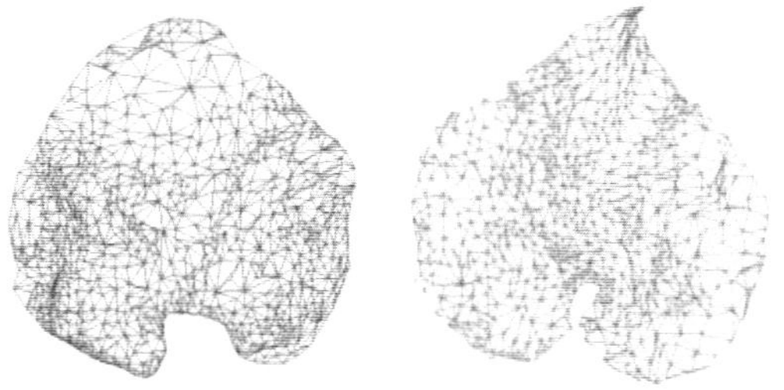

Figure 5. Two Precise Leafbalde Mesh

5. System Implementation and Results

5.1. System Implementation

With the regulation of data measurement, a software system has been implementation by using C++, and combing Polhemus Tracker SDK. This system firstly connect to Patriot 3D digitizer, then the measured feature points were detected and displayed in the interface. And the morphological parameters of the plant were calculated synchronously, as Figure 6 shows.

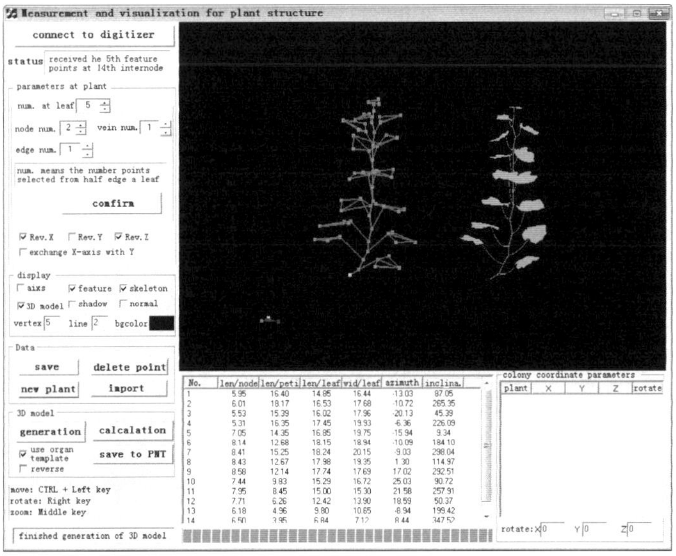

Figure 6. The Reconstructed 3D Model of A Plant

This system also provides several useful functions, including simple 3D reconstruction of plant, high-precision reconstruction, population generation, data exporting, and so on. This provides users a flexible tool for measuring and analysing plant structure.

5.2. Case of Applications

Our software has been used for morphological measuring of several plants, from seedling to maturity of the targeted plant.

Figure 7 gives an example for measuring a cucumber population. In this case 16 stands of cucumber were selected to collect spatial feature points by using Patriot 3D digitizer. It has been mentioned in section 2 that only 5 feature points were selected from each internode, as such points needed to be digitized from a plant was few. About fifteen minutes was used for the digitizing of a stand. The 3D model of each digitized plant was generated rapidly, with the morphological parameters.

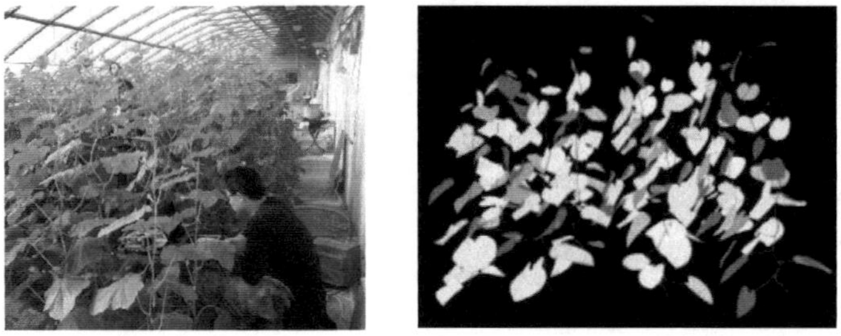

(a) work scenario (b) reconstructed population model

Figure 7. Case in Measuring A Cucumber Population

6. Conclusion

The problem of measuring the morphological parameters and reconstructing its 3D model of plant was put forwarded in this paper, with considering low time and labor consuming. A method for measuring feature points and software tool for real-time computing and visualizing was presented. This method used spatial characteristics points measured from plant by using a 3D digitizer as the input. These spatial characteristics points were then used to calculate morphological parameters, reconstruct skeleton model and details 3D model of the plant.

Currently the presented method is suitable for small plants. However, we know that this method is no enough for big plant which have more complex morphological structure. To this kind plant, computer vision-based or 3D scanned point cloud-based method should be better.

Automatic, rapid, and non-destructive measurement and 3D reconstruction of plant is of great significance to plant phenotyping. And this is presently still an open problem because of complex natural conditions in real field. The key challenge roots not only in the complication of the plant canopy, but also the measuring method. All these problems are expected to be furthered and maintain the measurement of plants canopy a challenging topic.

Acknowledgments

This work is supported by Natural Science Foundation of China (No. 61762013), basic ability improvement project for young and middle-aged teachers in universities of Guangxi province (No 2018KY0078 and 2017KY0075), Scientific Research Foundation of Guangxi Normal University (No. 2017BQ018), and Guangxi Scholarship Fund of Guangxi Education Department. The authors also thanks the support from the Foundation for University 1000 Young Key Teachers by Guangxi Province.

References

[1] T. Watanabe, J. S. Hanan, P.M. Room. Rice Morphogenesis and Plant Architecture: Measurement, Specification and the Reconstruction of Structural Development by 3D Architectural Modelling, Ann. Bot. 95, 7: 1131-1143(2005)

[2] G. Sonohat,H. Sinoquet, V. Kulandaivelu, D. Combes, F. Lescourret. Three-dimensional reconstruction of partially 3D-digitized peach tree canopies. Tree Physiol. 26,3: 337–351(2006)

[3] P. Kaitaniemi, P. Room, J. Hanan. Architecture and morphogenesis of grain sorghum, Sorghum bicolor (L.) Moench. Field Crop. Res. 61: 51-60(1999).

[4] R. P. de MoraesFrasson and W.F. Krajewski. Three-dimensional digital model of a maize plant. Arg. Forest Meteor. 150, 3:478-488(2010)

[5] J. L. Drouet, R. Bonhomme. Do variations in local leaf irradiance explain changes to leaf nitrogen within row maize canopies? Ann. Bot. 84: 61-69(1999)

[6] C. Fournier, C. Pradal. A Plastic, Dynamic and Reducible 3D Geometric Model for Simulating Gramineous Leaves. *International Symposium on Plant Growth Modeling, Simulation,*

Visualization and Applications (Shanghai, China, 2012).
[7] J. Wu, Y. Guo. An integrated method for quantifying root architecture of field-grown maize. Ann. Bot. 114: 841-851(2014)
[8] M. Rakocevic, H. Sinoquet, C. Christophe. Varlet-Grancher. Assessing the Geometric Structure of a White Clover (Trifolium repens L.) Canopy using 3-D Digitizing. Ann. Bot. 86: 519-526(2000)
[9] B. Loch, J. Belwar, J. Hanan. Application of surface fitting techniques for the representation of leaf surfaces. *International Congress on Modelling and Simulation* (Melbourne: MODSIM Press, 2005)
[10] E. Kaminuma, N. Heida, Y. Tsumoto, N. Yamamoto, N. Goto. Automatic quantification of morphological traits via three-dimensional measurement of Arabidopsis. Plant J. 38: 358–365(2004)
[11] T. Dornbusch, P. Wernecke, W. Diepenbrock. A method to extract morphological traits of plant organs from 3D point clouds as a database for an architectural plant model. Ecol. Model 200: 119-129(2007)
[12] G. Louarn, S. Carré, F. Boudon. Characterization of whole plant leaf area properties using laser scanner point clouds. *Fourth International Symposium on Plant Growth Modeling, Simulation, Visualization and Application s*(Shanghai, China, 2012)
[13] D. Sylvain, J. Christian, R. Pascal. PypeTree: A Tool for Reconstructing Tree Perennial Tissues from Point Clouds. Sensors, 14: 4271-4289(2014)
[14] A. K. Aijazi, P. Checchin, L. Malaterre and L. Trassoudaine. Automatic Detection and Parameter Estimation of Trees for Forest Inventory Applications Using 3D Terrestrial LiDAR. Remote Sens. 9, 946: 1-24(2017)
[15] D. Rousseau, H. Dee, T. Pridmore. Imaging methods for phenotyping of plant traits. Phenomics in Crop Plants: Trends, Options and Limitations. Springer, New Delhi(2015)
[16] H. Scharr, H. Dee, A.P. French. Special issue on computer vision and image analysis in plant phenotyping. Mach. Vis. Appl. 27, 607–609 (2016)
[17] S. A. Tsaftaris, M. Minervini, H. Scharr. Machine Learning for Plant Phenotyping Needs Image Processing. Trends Plant Sci. **21**,12: 989-991 (2016)

IOP Conf. Series: Materials Science and Engineering **533** (2019) 012056 doi:10.1088/1757-899X/533/1/012056

Small Aircraft Detection in Remote Sensing Images Based on YOLOv3

Kun Zhao, Xiaoxi Ren

Hunan University, Changsha, China

happyrxx@hnu.edu.cn

Abstract. Accurate and rapid detection in remote sensing images is of great significance in military navigation, environmental monitoring, and civil applications. Due to the small object detection problem of remote sensing images, higher requirements and challenges are put forward for object detection technology. In recent years, the one-stage object detector and the two-stage object detector based on convolutional neural network have made great achievements in the field of image classification and detection. The one-stage object detector is generally superior to the two-stage object detector in detection speed, but the detection accuracy is inferior to the two-stage target detector. In this paper, YOLOv3 is used as a one-stage target detector for small aircraft detection of remote sensing images. We chose appropriate anchors to cover the size distribution in our experimental data by using dimension clusters. The experimental results show that YOLOv3 not only exceeds the conventional one-stage object detector in speed, but also well match the accuracy of the two-stage object detector. The YOLOv3 achieved excellent detection accuracy and low processing time at the same time for small aircraft detection.

1. Introduction

With the development of space remote sensing technology, the resolution of remote sensing images has been greatly improved, and the information carried by remote sensing images is becoming more and more abundant. In recent years, aircraft detection in remote sensing images [1-3] has become one of the research hotspots in the fields of environmental monitoring, military and civil applications, etc. Most of the traditional aircraft detection methods use artificially extracted shallow features, such as the shape feature, or scale-invariant feature transform (SIFT) [4] and histogram of oriented gradients(HOG) [5], then select support vector machine (SVM) or Adaboost as the classifier. For example, Cheng et al. [6] used multi-scale histogram of oriented gradients (HOG) feature and trained a latent SVM for each object category. Luo et al. [7] utilized Object Proposal algorithm for airplane locating, and a SVM classifier was trained on the HOG features to detect the airplanes. However, the traditional methods extract the low-level features of remote sensing images, and can not fully exploit the high-level features. What is more, the selection and extraction of features are subjective and the extraction process is very complicated. The traditional methods are far from meeting the needs of large-scale automation applications in terms of detection speed and accuracy. For nearly 10 years, with the continuous development of big data and computing capacity, deep learning methods have been rapidly developed and widely used in various fields. Among them, convolutional neural network (CNN) proved to be a promising approach for feature extraction has caused great concern in the field of image recognition and detection. Many theoretical researches concerning CNN have been quite

Content from this work may be used under the terms of the Creative Commons Attribution 3.0 licence. Any further distribution of this work must maintain attribution to the author(s) and the title of the work, journal citation and DOI.
Published under licence by IOP Publishing Ltd

mature [8-13]. There are many detection methods based on CNN for aircraft detection have been proposed. Zhang et al. [1] proposed a weakly supervised learning framework based on coupled CNN for aircraft detection. Wu et al. [14] put forward a new aircraft detection framework based on objectiveness detection techniques and CNN. A modified Faster R-CNN [15] for the task of small object detection in optical remote sensing images was proposed in [16]. Nonetheless, superior detection speed and accuracy can hardly be achieved at the same time for small aircraft detection.
In this paper, YOLOv3 [17] is used as a object detector for small aircraft detection, and compared with two-stage object detector Faster R-CNN [15] and one-stage object detector SSD [18] in terms of performance. Experiment results show that the YOLOv3 outperforms the other two methods.

2. Methodology

2.1. Object Detection Based on CNN
Object detection method based on CNN was successfully applied to object detection as early as 1994 [19]. Due to lack of data, limitations in hardware performance, and over-fitting problem, object detection based on CNN has not made a breakthrough for a long time. Compared with the traditional object detection method at present, the object detection based on CNN has no great advantage in both detection accuracy and detection speed. Therefore, the research is gradually neglected. Until 2012, the convolutional neural network AlexNet [20] made a major breakthrough in image classification, and researchers began to refocus on how to apply CNN to object detection. Nowdays, CNN-based object detection methods occupy a dominant position in the field of object detection, which are mainly divided into two categories, one is two-stage object detector represented by R-CNN (SPP Net [21], Fast R-CNN [22], Faster R-CNN[15], Mask R-CNN [23]), and the other is one-stage object detector represented by SSD [18] and YOLO [24].

2.2. Yolov3
1) YOLOv3 has made some improvements on the basis of YOLOv2 [25], and these changes make it even better. The network structure of YOLOv3 is shown in Figure 2. The performance comparison of YOLOv3 with other detection methods in the COCO dataset [26] at 0.5 IoU(Interception- of-Union) metric is shown in Fig 1. We can observe that YOLOv3 performs very well, both accuracy and speed can be considered at the same time. These improvements are mainly reflected in the following aspects.
2) Unlike the Darknet-19 used in the YOLOv2 [25], YOLOv3 uses a 53-layer convolutional network named Darknet-53 that is superimposed by residual units as feature extraction networks. According to the author ' s experiments in [17], this model performs better than Darknet-19 in terms of classification accuracy and efficiency.
3) To achieve multilabel classification, use logistic classifier instead of softmax and use binary cross-entropy loss for the class predictions during training. YOLOv3 uses logistic regression to predict the score for each bounding box. If the bounding box prior overlaps a ground truth object by more than any bounding box prior. If the priori bounding box is not the best, but does overlap the ground truth by more than a certain threshold, then this prediction is ignored. YOLOv3 only assigns a bounding box to each ground truth. If the a priori bounding box does not match the ground truth, no coordinate or class prediction loss will occur, and only objectness.
4) Different from the previous YOLO and YOLOv2, YOLOv3 predicts tasks from three different scale feature maps, which better solve the small target detection problem.
YOLOv3 made some improvements on the basis of the original, which further improved the performance of the model.

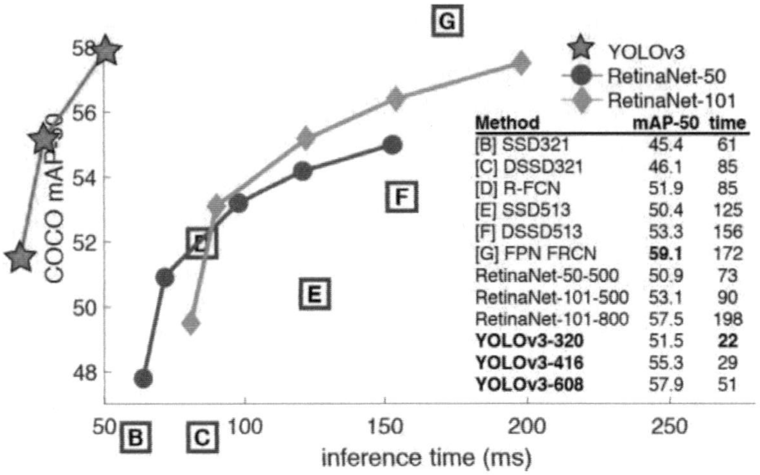

Figure 1. The Speed/Accuracy Tradeoff on the Map of the Detection Methods at 0.5 Iou Metric in the COCO Dataset.

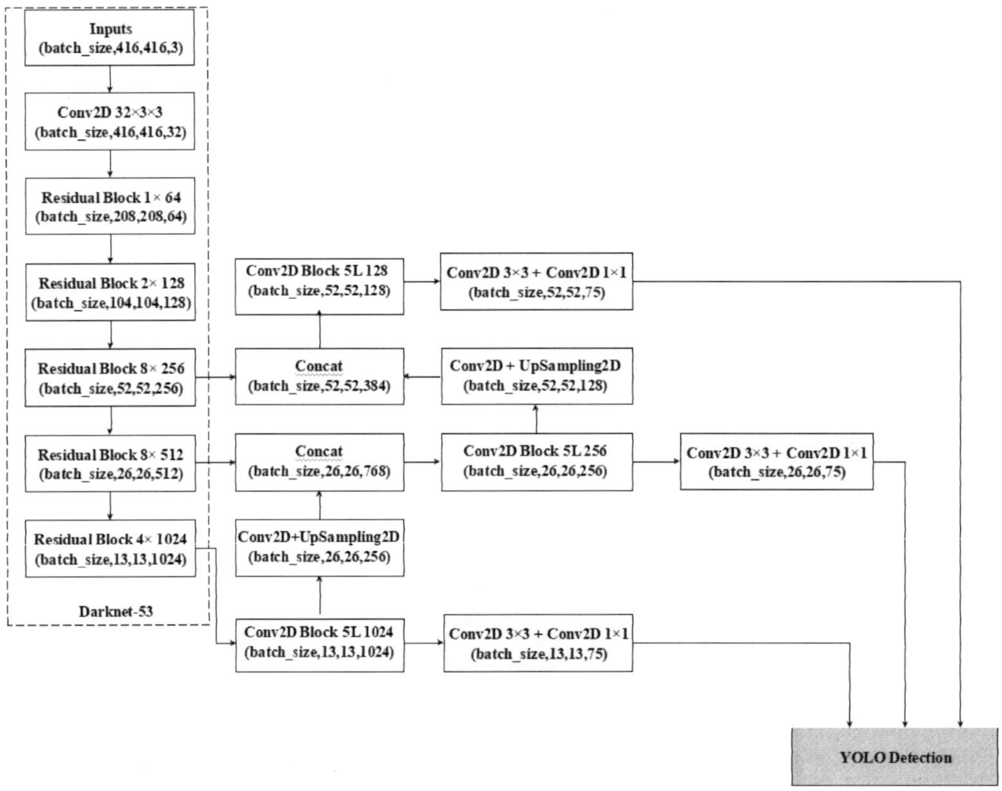

Figure 2. The Structure Diagram of YOLOv3.

3. Experiments

The experiment consists of two processes: the training phase and the detection phase. All the experiments were performed on Ubuntu with NVIDIA GTX 1080Ti GPU, and the evaluation metrics are average precision (AP) with IoU threshold as 0.5 and average detection time per image. If the overlap between the bounding box and ground truth is greater than or equal to 0.5, the bounding box is a TruePositive; otherwise, it is a FalsePositive.

3.1. Dataset Description

In our experiments, we collected 350 remote sensing images from Google Earth and DOTA dataset [27], each image is of the size in the range from about 600x600 to 1500x1500 pixels and we manually labeled aircraft targets in each image. The training validation set and the test set are randomly generated in a ratio of 8:2. The training set and the verification set are randomly generated by the training verification set according to the ratio of 8:2. The experimental data statistics are shown in Table 1. Some samples of experimental data are shown in Figure 3.

Table 1. Statistics of the Data Sets.

Class	Image Number	Aircraft Number
Training set	224	4528
Verification set	56	965
Testing set	70	1372
Total	350	6865

Figure 3. Some Samples of Experimental Data

3.2. Evaluation Indicators

In the experiment, we use the AP and the average detec-tion time of each image to evaluate the performance of the models. The indicators we used in the experiment are as follows.

1) True Positive (TP): The number of samples that are predicted to be positive and actually positive;
2) True Negative (TN): The number of samples that are predicted to be negative and practically negative;
3) False Positive (FP): The number of samples that are predicted to be positive and actually negative;
4) False Negative (FN): The number of samples that are predicted to be negative and practically positive.

The measures can be defined as following:

$$P(\text{Precision}) = \frac{TP}{TP+FP} \tag{1}$$

$$R(\text{Recall}) = \frac{TP}{TP+FN} \tag{2}$$

$$\text{AP}(\text{AveragePrecision}) = \int_0^1 \text{P}(R)dR \tag{3}$$

3.3. Experimental Results and Comparisons

We evaluated the performance of YOLOv3 object detector for small aircraft detection by comparing it with Faster R-CNN and SSD. For a fair comparison, the same arguments for training that are listed in Table 2 and the same test set are used in the experiment. Some experimental results is shown in Figs 4, 5 and 6.

Table 2. Arguments for training

Argument	Value
Learning rate	0.01(20k:0.001, 50k:0.0001)
Batch Size	32
Momentum	0.9
Weight decay	0.0005
Max iteration	80k

Figure 4. The detection results of SSD

Figure 5. The detection results of Faster R-CNN

Figure 6. The detection results of YOLOv3

The corresponding AP value and detection time are shown in Table 3.

Table 3. Detection Results in Term of the Metrics of AP and Average Detection Time

Method	AP	Average detection time
SSD	0.538	0.078s
Faster R-CNN	0.874	0.652s
YOLOv3	0.925	0.023s

Form Figs 4, 5 and 6 and Table 2, the following conclusions can be obtained: the YOLOv3 achieves much better performance than the Faster R-CNN and the SSD. The YOLOv3 outperforms the Faster R-CNN and the SSD by 0.051 and 0.387 in terms of AP, respectively. In terms of detection time, the average detection time of YOLOv3 is 0.023s. The time cost of Faster R-CNN is 27 times than that of YOLOv3. The YOLOv3 is 3.4 faster than SSD. As can be seen, a large number of mislabeling and missing labels appear in the SSD detection, followed by Faster R-CNN, and YOLOv3 can accurately detect the objects in different scenarios. To sum up, we believe that the YOLOv3 is more suitable and has obvious advantages for small aircraft detection in remote sensing images, and can meet the requirements of accurate detection and real-time in practical applications.

4. Conclusion

In this paper, we presented an effective detection method to address the small object problem in remote sensing images by using YOLOv3. We analyzed some improvements that YOLOv3 made on the previous basis, and compared the performance with other object detection algorithms. Comparisons with Faster R-CNN and SSD approaches demonstrated the effectiveness and superiority.

YOLOv3 is real-time while accurately detecting small aircraft. In the future, we will continue to work on multi-class small object detection in remote sensing images based on the current research basis.

Acknowledgments
The research was supported by the nation fund project for the study of key issues in high-availability on-board computing under Grant 61802122.

References

[1] Fan Zhang, Bo Du, Liangpei Zhang, and Miaozhong Xu. Weakly supervised learning based on coupled convolutional neural networks for aircraft detection. IEEE Transactions on Geoscience & Remote Sensing, 54(9):5553–5563, 2016.
[2] Shijin Li, Jianbin Qiu, and Yu Hui. Aircraft detection in high-resolution remote sensing imagery based on visual words selection. Journal of Data Acquisition &Processing, 2014.
[3] Yuqingyang Hou, Jicheng Quan, and Yongming Wei. Valid aircraft detection system for remote sensing images based on cognitive models. Acta Optica Sinica, 2018.
[4] Junwei Han, Dingwen Zhang, Gong Cheng, Lei Guo, and Jinchang Ren. Object detection in optical remote sensing images based on weakly supervised learning and high-level feature learning. IEEE Transactions on Geoscience & Remote Sensing, 53(6):3325–3337, 2015.
[5] Gong Cheng, Peicheng Zhou, Xiwen Yao, Chao Yao, Yanbang Zhang, and Junwei Han. Object detection in vhr optical remote sensing images via learning rotation-invariant hog feature. In International Workshop on Earth Observation and Remote Sensing Applications, pages 433–436, 2016.
[6] Gong Cheng, Junwei Han, Lei Guo, Xiaoliang Qian, Peicheng Zhou, Xiwen Yao, and Xintao Hu. Object detection in remote sensing imagery using a discriminatively trained mixture model. Isprs Journal of Photogrammetry & Remote Sensing, 85(9):32–43, 2013.
[7] Qinhan Luo and Zhenwei Shi. Airplane detection in remote sensing images based on object proposal. In Geoscience and Remote Sensing Symposium, pages 1388–1391, 2016.
[8] Ali S Razavian, Hossein Azizpour, JosephineSullivan, and Stefan Carlsson. Cnn features off-the-shelf: An astounding baseline for recognition. pages 512–519, 2014.
[9] Y Lan Boureau, Jean Ponce, and Yann Lecun. A theoretical analysis of feature pooling in visual recognition. In International Conference on Machine Learning, pages 111–118, 2010.
[10] Pulkit Agrawal, Ross Girshick, and Jitendra Malik. Analyzing the Performance of Multilayer Neural Networks for Object Recognition. Springer International Publishing, 2014.
[11] Jeffrey Dean, Greg S. Corrado, Rajat Monga, Kai Chen, Matthieu Devin, Quoc V. Le, Mark Z. Mao, Marc'Aurelio Ranzato, Andrew Senior, and Paul Tucker. Large scale distributed deep networks. In International Conference on Neural Information Processing Systems, pages 1223–1231, 2012.
[12] Christian Szegedy, Wei Liu, Yangqing Jia, Pierre Sermanet, Scott Reed, Dragomir Anguelov, Dumitru Erhan, Vincent Vanhoucke, and Andrew Rabinovich. Going deeper with convolutions. pages 1–9, 2014.
[13] Hossein Azizpour, Ali Sharif Razavian, Josephine Sullivan, Atsuto Maki, and Stefan Carlsson. Factors of transferability for a generic convnet representation. IEEE Transactions on Pattern Analysis & Machine Intelligence, 38(9):1790–1802, 2014.
[14] Hui Wu, Hui Zhang, Jinfang Zhang, and Fanjiang Xu. Fast aircraft detection in satellite images based on convolutional neural networks. In IEEE International Conference on Image Processing, pages 4210–4214, 2015.
[15] Shaoqing Ren, Kaiming He, Ross Girshick, and Jian Sun. Faster r-cnn: towards real-time object detection with region proposal networks. In International Conference on Neural Information Processing Systems, pages 91–99, 2015.
[16] Yun Ren, Changren Zhu, and Shunping Xiao. Small object detection in optical remote sensing images via modified faster r-cnn. Applied Sciences, 2018.

[17] Joseph Redmon and Ali Farhadi. Yolov3: An incremental improvement. 2018.
[18] Wei Liu, Dragomir Anguelov, Dumitru Erhan, Christian Szegedy, Scott Reed, Cheng Yang Fu, and Alexander C. Berg. Ssd: Single shot multibox detector. pages 21–37, 2015.
[19] R Vaillant, C Monrocq, and Y Le Cun. Original approach for the localisation of objects in images. Vision, Image and Signal Processing, IEE Proceedings-, 141(4):245 – 250, 1994.
[20] Alex Krizhevsky, Ilya Sutskever, and Geoffrey E. Hinton. Imagenet classification with deep convolutional neural networks. In International Conference on Neural Information Processing Systems, pages 1097–1105, 2012.
[21] Kaiming He, Xiangyu Zhang, Shaoqing Ren, and Jian Sun. Spatial pyramid pooling in deep convolutional networks for visual recognition. IEEE Trans Pattern Anal Mach Intell, 37(9):1904–1916, 2015.
[22] Ross Girshick. Fast r-cnn. Computer Science, 2015.
[23] Kaiming He, Georgia Gkioxari, Piotr Dollar, and Ross Girshick. Mask r-cnn. IEEE Transactions on Pattern Analysis & Machine Intelligence, PP(99):1–1, 2017.
[24] Joseph Redmon, Santosh Divvala, Ross Girshick, and Ali Farhadi. You only look once: Unified, real-time object detection. In IEEE Conference on Computer Vision and Pattern Recognition, pages 779–788, 2016.
[25] Joseph Redmon and Ali Farhadi. Yolo9000: Better, faster, stronger. pages 6517–6525, 2016.
[26] Tsung Yi Lin, Michael Maire, Serge Belongie, James Hays, Pietro Perona, Deva Ramanan, Piotr DollA˜a‹r, and C. Lawrence Zitnick. Microsoft coco: Common objects in context. 8693:740–755, 2014.
[27] Gui-Song Xia, Xiang Bai, Jian Ding, Zhen Zhu, Serge Belongie, Jiebo Luo, Mihai Datcu, Marcello Pelillo, and Liangpei Zhang. Dota: A large-scale dataset for object detection in aerial images. In The IEEE Conference on Computer Vision and Pattern Recognition (CVPR), June 2018.

IOP Conf. Series: Materials Science and Engineering 533 (2019) 012057 doi:10.1088/1757-899X/533/1/012057

Fast Implementation of Image Structural Similarity Algorithm

Yiwen Xu, Zhongwen Zhuang, Weimeng Li, Ying Fang[a]

College of Physics and Information Engineering, Fuzhou University, Fuzhou China

Abstract. The traditional Structural SIMilarity (SSIM) index has higher computation complexity compared with Peak-Signal-to-Noise-Ratio (PSNR). In order to solve this problem, we propose a two-stage optimization algorithm to reduce the computation complexity of SSIM. First, floating-point operations in the calculation process of SSIM are converted to fixed-point ones. Then, the repeated calculation rule of the iteration operation in SSIM is analyzed and the repeated operations are subsequently removed via modifying SSIM calculation method. Experiment results show that, in comparison with the original SSIM algorithm, the proposed algorithm obtains 24.12% of computation time reduction with negligible accuracy decrease.

1. Introduction

With the rapid development of digital image field, Image Quality Assessment (IQA), as one of the important aspects in the image processing, has attracted more and more researchers' attention and achieved remarkable results. It evaluates the performance of the algorithm by analyzing the various characteristics of the image and comparing the degree of distortion.

To our best knowledge, Peak-Signal-to-Noise-Ratio (PSNR) is one of the classic objective metrics in IQA, which is widely used in the image processing. It measures the degree of the distortion in the image by calculating the errors of the corresponding pixels between the original image and the distorted image. However, PSNR does not take into account the visual characteristics of the human eyes [1]. For example, the human eyes are more sensitive to the contrast difference of the lower frequency. In addition, the sensitivity to the contrast of the luminance is higher than that of the contrast of the chromaticity. Furthermore, the perception of the region by the human eyes is often affected by the surrounding areas. Therefore, the evaluation result of PSNR may be inconsistent with the Mean Opinion Score (MOS). Considering the characteristics of the Human Visual System (HVS), Wang *et al.* [2] proposed Structural SIMilarity (SSIM) index based on the characteristics of the human eyes that are sensitive to structural edges. It firstly extracted the brightness, contrast and structure information of the original image and the distorted image. Then, it calculated the similarities of the three features, respectively. Finally, the similarities were integrated to obtain the overall metric. SSIM ranges from 0 to 1. The larger the value is, the greater the similarity between the original image and the distorted image. Compared with the traditional PSNR, SSIM has obvious advantages in accuracy and practicability.

In practical applications, SSIM usually adds a window to calculate the local SSIM. Then it traverses the image and calculates the average value as SSIM of the entire image. As a result, this process generates a lot of repetitive operations and dramatically increases the calculation complexity of SSIM. In order to avoid unnecessary repetitive operations, in this paper, we calculate the SSIM of each image after performing block processing on the image. In addition, there are a large number of floating-point calculations in the SSIM

Content from this work may be used under the terms of the Creative Commons Attribution 3.0 licence. Any further distribution of this work must maintain attribution to the author(s) and the title of the work, journal citation and DOI.
Published under licence by IOP Publishing Ltd

calculation process, which need much more calculation time than fixed-point calculations. Thereby, we optimize the SSIM formula by converting floating-point operations to fixed-point ones without affecting the accuracy of SSIM.

2. Advantages of Structural Similarity Algorithm

In the past few decades, Video Quality Assessment (VQA) [3-5] has developed rapidly and played an important role in the video processing. VQA is a part of IQA. VQA and IQA have two main roles. The one is to be used as a test tool to select the best quality video/image from multiple reconstructed videos/images, or to select the best encoding algorithm. The other is to be used as an optimization tool to embed VQA and IQA metrics into the encoder to optimize the algorithm for optimal coding efficiency. According to the conclusion in [6-7], the performance of SSIM is significantly better than that of PSNR in terms of saturation at high rate. During the encoding process, as the Bit Rate (BR) increases, the quality of the reconstructed video or image obtained by the encoder is also increasing. However, the quality of the reconstructed video or image will be close to that of the original video or image when the BR is increased to a certain extent. It means that increasing the BR will no longer affect the visual quality. The ideal IQA or VQA model needs to accurately reflect this saturation effect. PSNR is still rising at high BRs while SSIM can reach saturation. This is because SSIM does not guide the encoder to allocate bits to the position where the spatio-temporal perceptual quality is saturated during the optimization process for video or image coding. Therefore, SSIM can more accurately reflect the degree of distortion of the video or image compared to PSNR.

In addition, SSIM takes the characteristics of HVS into consideration. Thus, the performance of SSIM is also significantly superior to PSNR in terms of perceptual quality correlation [8-10].

From the above analyses, it can be seen that SSIM is better than PSNR in terms of the overall performance. PSNR only takes into account pixel error between the original image and the reconstructed image, while SSIM considers three different properties, including brightness, contrast and structure. It indicates that the computation complexity of SSIM is significantly higher than that of PSNR, which has negative effects on its applications in realtime video/image encoding. To solve this problem, we propose a modified SSIM calculation method to reduce its calculation time.

3. The Proposed Algorithm

The proposed algorithm optimizes the SSIM from two aspects to reduce its calculation time: converting floating-point operations to fixed-point operations and reducing the number of repeated operations.

3.1. Convert Floating-Point Operations to Fixed-Point Operations

The SSIM is calculated as follows [2]:

$$SSIM(x,y) = \frac{(2\mu_x\mu_y + C_1)(2\sigma_{xy} + C_2)}{(\mu_x^2 + \mu_y^2 + C_1)(\sigma_x^2 + \sigma_y^2 + C_2)} \tag{1}$$

where μ_x and μ_y represent the average of the grayscale value of image x and image y, respectively. δ_x and δ_y represent the variance of the grayscale value of image x and image y, respectively. δ_{xy} represents the covariance of the grayscale value of image x and image y. C_1, C_2 and C_3, which aim at maintaining stability, are constants. It can be seen from equation (1) that $\mu_x, \mu_y, \delta_x, \delta_y, \delta_{xy}$ are all floating-point numbers. It is well known that, floating-point operation consumes more time than fixed-point operation. So we can reduce the calculation time of SSIM via converting floating-point operations to fixed-point ones. However, it brings heavy impact on the result of SSIM if just cast a floating-point number to an integer number. Hence, in this paper, we reduce the calculation results of floating-point type by changing the calculation formula, so as to reduce the calculation time. The specific process is as follows. Assume that the size of the window is $w \times w$, x_i and y_i represent the grayscale values of the i-th pixel of image x and image y, respectively. Then μ_x and δ_x can be represented by x_i and y_i as shown in equation (2) and equation (3).

$$\mu_x = \frac{\sum x_i}{w^2} \tag{2}$$

$$\delta_x^{\ 2} = \frac{\sum(x_i - \mu_x)^2}{w^2} = \frac{(\sum x_i - 2x_i\mu_x + \mu_i^{\ 2})}{w^2} = \\ \frac{\sum x_i^{\ 2}}{w^2} - (\frac{\sum x_i}{w^2})^2 \tag{3}$$

In a similar way, μ_y, δ_y and δ_{xy} also can be represented by x_i and y_i. Therefore, equation (1) can be described as:

$$SSIM(x, y) = \frac{2\sum x_i \sum y_i + w^4 C_1}{((\sum x_i)^2 + (\sum y_i)^2 + w^4 C_1)} \times \\ \frac{2(w^2 \sum x_i y_i - \sum x_i \sum y_i) + w^4 C_2}{w^2((\sum x_i^{\ 2}) + (\sum y_i^{\ 2})) - (\sum x_i)^2 + (\sum y_i)^2 + w^4 C_2} \tag{4}$$

After the above conversion, all the variables in equation (4) are fixed-point numbers, which means that all the floating-point operations in SSIM, except for the last process, are converted to fixed-point operations.

3.2. *Reduce the Number of Iteration Operations*

In the process of SSIM calculation, the window needs to traverse the whole image. In addition, each traversal can only move one row or one column of the current window. As a result, a large number of repeated operations are generated in the adjacent windows, which leave the gap of reducing computation complexity of SSIM.

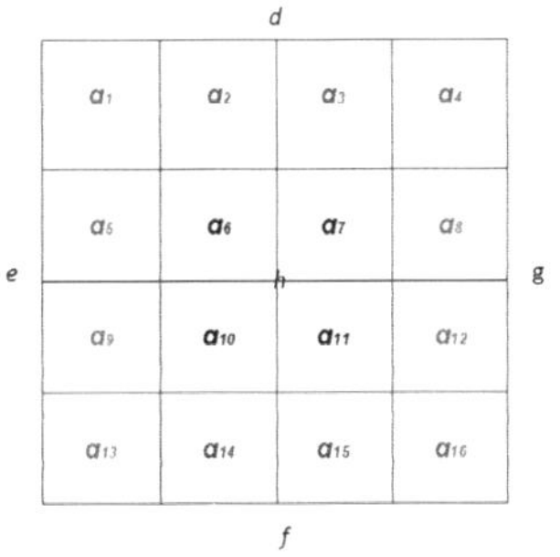

Figure 1. An example of repeated calculation in traditional SSIM

To demonstrate the derivation of computation complexity reduction, we use 3×3 size window traversing in 4×4 size image as an example. As shown in Fig.1, when the 3×3 window slide from the window centered at a_6 to the window centered at a_7, there will be 2×3 parts (that is, a_2, a_3, a_6, a_7, a_{10}, a_{11} as show in Fig.1) being repeatedly operated. Apparently, removing the repeated operations means reducing the calculation complexity of SSIM.

First, we divide the repeatedly calculated values into five parts, named as *d*, *e*, *f*, *g* and *h*, where

$$\begin{cases} d = a_2 + a_3 \\ e = a_5 + a_9 \\ f = a_{14} + a_{15} \\ g = a_8 + a_{12} \\ h = a_6 + a_7 + a_{10} + a_{11} \end{cases} \tag{5}$$

Then, we denote the calculation results of 3×3 windows centered at a_6, a_7, a_{10} and a_{11} as X_1, X_2, X_3 and X_4, respectively. Combining equation (5), it is easy to obtain the following relationships:

$$\begin{cases} X_1 = a_1 + e + d + h \\ X_2 = a_4 + d + h + g \\ X_3 = a_{13} + e + h + f \\ X_4 = a_{16} + f + g + h \end{cases} \tag{6}$$

From equation (5) and (6), it can be seen that the repeated calculations in the iteration operations of the traditional SSIM algorithm are removed.
After the above two steps, there will be no floating point calculations except for the last process in the SSIM calculation formula and the repeated calculations will be removed in the window calculation.

4. Experiments and Results

The process of converting the floating-point operations to fixed-point ones in our proposal algorithm may bring calculation errors to SSIM results. In order to verify the accuracy of the proposed algorithm, we calculate the SSIM values of typical images (*Aerial_city*, *Boston*, *Bridge*, *et al.* as shown in Fig.2) from live image database [11]. The results are shown in Table 1.
It can be seen from Table 1 that the deviation between the SSIM values calculated by the traditional algorithm and the proposed algorithm are so small that can be negligible.
In order to verify the effect of the proposed algorithm on calculation complexity, we use an 11×11 window to traverse the images with different resolutions, which are 128×128, 256×256, and 512×512. The calculation time and time decrease of the traditional algorithm and the proposed algorithm are shown in Table 2.

Table 1. Accuracy of the Proposed Algorithm

Image	**SSIM values**		**Relative error (%)**
	Traditional algorithm	**Proposed algorithm**	
Aerial_city	0.77642	0.77629	0.016
Boston	0.80971	0.80962	0.011
Bridge	0.79235	0.79231	0.005
butter_flower	0.73245	0.73113	0.180
cactus	0.81772	0.81215	0.681
child_swimming	0.83214	0.83163	0.061
couple	0.75564	0.75181	0.506
family	0.71998	0.71633	0.506
fisher	0.79877	0.79134	0.930

Table 2. Complexity Reduction of the Proposed Algorithm

Image resolution	**Traditional algorithm (s)**	**Proposed algorithm**	
		(s)	**(%)**
128x128	0.464	0.355	23.49
256x256	0.792	0.602	23.98
512x512	1.097	0.824	24.88
average	/	/	24.12

From Table 2, it can be seen that the calculation time of SSIM increases with the increase of the image resolution. The reason is that the number of iteration operations in SSIM calculation is increase with the image resolution. The more times the window is traversed, the longer the calculation time is required. The results in Table 2 also show that, compared with the traditional algorithm, the calculation time of our proposal algorithm is reduced by 24.12% in average. The result of the table 1 and the table 2 show that, the propose method improve the computational efficiency and not affect the calculation accuracy.

In summary, compared to the tradition algorithm, the proposed algorithm reduces the calculation time to a certain extent under the condition that the accuracy is basically unchanged. The proposed algorithm is effective in reducing the calculation time in the case of different image resolutions. Experimental results prove that after two steps of optimization, the proposed algorithm is better than traditional algorithm which has better practicability.

(a) (b) (c)

(d) (e) (f)

Figure 2. Live image database samples. (a)-(c) are original images. (d)-(f) are distorted images.

5. Conclusions

In this paper, we propose a modified SSIM algorithm to reduce the computation complexity of traditional SSIM algorithm. The proposed algorithm is composed of two stages, one is the conversion from floating-point operations to fixed-point ones, and the other is reduction of the number of iteration operations in SSIM calculation. The experimental results show that the proposed algorithm can effectively reduce the calculation time of SSIM without decrease of its accuracy. The proposed optimization scheme can be applied to various image quality evaluations.

This method reduces the calculation time to a certain extent. However, there is still much room for improvement. In future works, we will optimize the algorithm by further analyzing the image structure and studying on the application of SSIM in image quality evaluation to reduce more computation time.

References

[1] Z. Wang, A.C. Bovik, "Mean squared error: love it or leave it? A new look at signal fidelity measures," IEEE Signal Processing Magazine, vol.26, no. 1, pp. 98-117, Jan. 2009.

[2] Z. Wang, A.C. Bovik, H.R. Sheikh, and E.P. Simoncelli, "Image quality assessment: From error visibility to structural similarity," IEEE Transactions on Image Processing, vol. 13, no. 4, pp. 600-612, Apr. 2004.

[3] Z. Wang, A. C. Bovik, "A universal image quality index," IEEE Signal Processing Letters, vol. 9, no.3, pp. 81-84, Mar. 2002.
[4] Z. Wang, L. Lu, and A.C. Bovik, "Video quality assessment based on structural distortion measurement," Signal Processing: Image Commu-nication. Special issue on Objective Video Quality Metrics, vol. 19, no.2, pp. 121-132, Feb. 2004.
[5] Z. Wang, E. P. Simoncelli, and A. C. Bovik, "Multi-scale structural similarity for image quality assessment," Proc IEEE Asilomar Conference on Signals, vol. 2, no. 2, pp.1398-1402, Nov. 2003.
[6] T. Zhao , K. Zeng, A. Rehman, and Z. Wang. "On the use of SSIM in HEVC," Asilomar Conference on Signals, Systems and Computers, 2013, pp. 1107–1111.
[7] S. Wang, A. Rehman, Z. Wang, et al. "SSIM–motivated rate–distortion optimization for video Coding," IEEE Transactions on Circuits and Systems for Video Technology, vol. 22, no. 4, pp. 516–529, Apr. 2012.
[8] A. Rehman, "SSIM-inspired quality assessment, compression, and processing for visual communications", Ph.D. Thesis. University of Waterloo: Canada, 2013.
[9] M.-J. Chen and A.C. Bovik, "Fast structural similarity index algorithm," Journal of Real-Time Image Processing, vol. 6, no. 4, pp. 281-287, Dec. 2011.
[10] D. Brunet, E.R. Vrscay, and Z. Wang, "On the mathematical properties of the structural similarity index," IEEE Transactions on Image Processing, vol. 21, no. 4, pp. 1488-1499, Apr. 2012.
[11] R. L. DeValois and K. K. DeValois, Spatial Vision, Oxford University Press, Oxford, UK, 1990.

Ocular Rectus Muscle Segmentation Based on Improved U-net

Cong Wu, Jinhao Zhan, Yixuan Zou, Fagang Jiang, Junjie Yang

Hubei University of techonology, No.28, Nanli Road, Hong-shan District, Wuchang, Wuhan, Hubei China

zhanjinhao5@gmail.com

Abstract. In recent years, deep learning has made great progress in computer vision, and shown a good application prospect in reading medical images. TAO (thyroid-associated ophthalmopathy) is one of the most common orbital diseases in adults is autoimmune disease, the exact pathogenesis is not clear. In hospitals, CT images are usually used for disease analysis and the distortion of ocular rectus muscles is one of the main causes of TAO, thus the separation of the ocular rectus muscle is of great practical significance. However, the artificial segmentation of the ocular rectus muscle is a time-consuming task and relies heavily on the experience of the operato. In this paper, we improved the traditional U-net with GoogLeNet inception module and apply it on the segmentation of ocular rectus muscles. The experimental results show that the segmentation results of our method are more accurate and efficient than the traditional algorithm, and it is helpful for doctors to diagnose patients' diseases.

1. Introduction

TAO (thyroid-associated ophthalmopathy) is one of the most common orbital diseases in adults is autoimmune disease, the exact pathogenesis is not clear. [1] The severity of TAO is divided into three grades: mild, moderate and severe. In mild grade of TAO, It had little effect on the quality of life of the patients. Although mild TAO did not significantly threaten the vision of the patients, it had a great impact on their life and work. When it comes to severe TAO, the patient developed thyroid disease-related optic neuropathy and/or corneal lesions, and should take immediate intervention measures such as glucocorticoid intravenous shock [2]. The disease of TAO will cause swelling of the extraocular muscles and stimulate the differentiation of some pre-adipocytes into adipocytes [3].There are three types of eye CT scanning in hospitals: transverse position, sagittal position and coronary position. Among them, the CT images of the transverse position scan shows the stripe structure of the left and right rectus muscle of the patient's eyes. The contour of the ocular rectus muscle was most clearly observed and was diagnosed by its morphology, such as the degree of obesity of ocular rectus muscle. See figure 1 for details.

Content from this work may be used under the terms of the Creative Commons Attribution 3.0 licence. Any further distribution of this work must maintain attribution to the author(s) and the title of the work, journal citation and DOI.
Published under licence by IOP Publishing Ltd

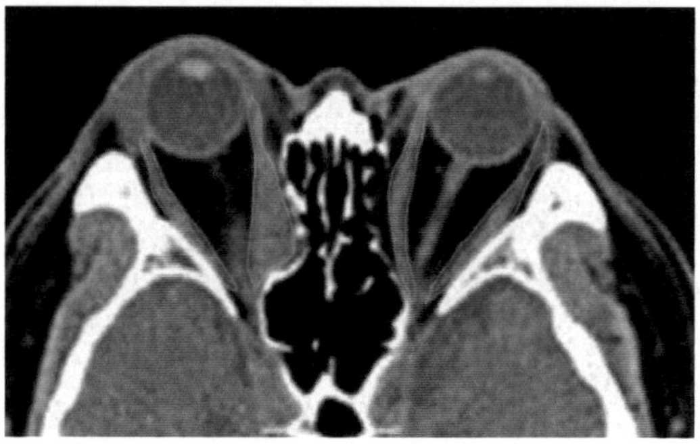

Figure 1. As shown in the figure, this is the image of the transverse position of the human eye CT image. The green contour delineates the left and right rectus muscle of the eye, which is the target area to be segmented.

2. Related Work

At present, in the field of computer vision, convolution neural network is the most effective depth model of deep models. The ImageNet Large Scale Visual Recognition Challenge (ILSVRC) witnessed the development of CNN model in the field of computer vision. From AlexNet[4] in 2012, VGGNet[5] and GoogLeNet[6] in 2014 to the Residual Network in 2015, the size and depth of CNN increased sharply. At the same time, the error rate of objects recognition task is also decreasing rapidly. At present, in 1 000 categories of object recognition task, CNN model achieved less error rate than human do. In addition, CNN has been successfully applied in a lot of general recognition tasks, such as object detection, semantic segmentation and contour detection.

Our approach draws on recent successes of deep nets for image segmentation. The most straightforward way of improving the performance of deep neural networks is by increasing their size. This includes both increasing the depth, the number of levels of the network and its width: the number of units at each level[6].The progress of Image segmentation give the credit to the development of image classification. Image segmentation changes the final full connection layer into the full convolution layer so as to achieve the purpose of segmentation. FCN [7] proposed the full convolution network and achieved remarkable results in different deep learning tasks.

On the respect of medical images segmentation, U-net [8] has been unanimously recognized, and has been greatly ahead in the IBIS cell segmentation task with only 30 training images. And that, the related scholar in The field of deep learning put forward a few variants of U-net, literature [9] discusses The utilize of resnet in U-net, It deepen the U-net without causing vanishing gradient problem and enhance the feature extraction capacity of the network. Literature [10] introduced dense block, which can deliver low-level features in the previous layer, and work parallel with higher-level features in higher layers so as to realize the efficient reuse of features. This paper explores the performance improvement of inception module in U-net.

3. Network Achitecture Design

Inspired by NIN [11] and GoogLeNet V3[12], we found that 1×1 convolution layer can fuse information of network layers in different channels, in other words, enlarging the width of the network helps to extract image features. In GoogLeNet v3, 3 inception module_a, 5 inception module_b and 2 inception module_c are inserted into the network structure after several convolution layers.

It has been proved that dimensional reduction and parallel structures of the Inception modules which allows for mitigating the impact of structural changes on nearby components[12] and the utilise of inception modules can reduce parameters of the whole network: GoogleNet employed only 5 million parameters, which represented a 12× reduction with respect to its predecessor AlexNet, which used 60 million parameters. Furthermore,VGGNet employed about 3× more parameters than AlexNet[12].

The structure of the three inception modules are shown in figure 2.

IOP Conf. Series: Materials Science and Engineering 533 (2019) 012058 doi:10.1088/1757-899X/533/1/012058

The fact that different inception module with different amount being placed in different parts of the network illustrates their different functions: module_a can do feature extraction in a relatively shallow network depth, accordingly, inception module_b and inception module_c can perform better in middle depth and a deeper depth in network.

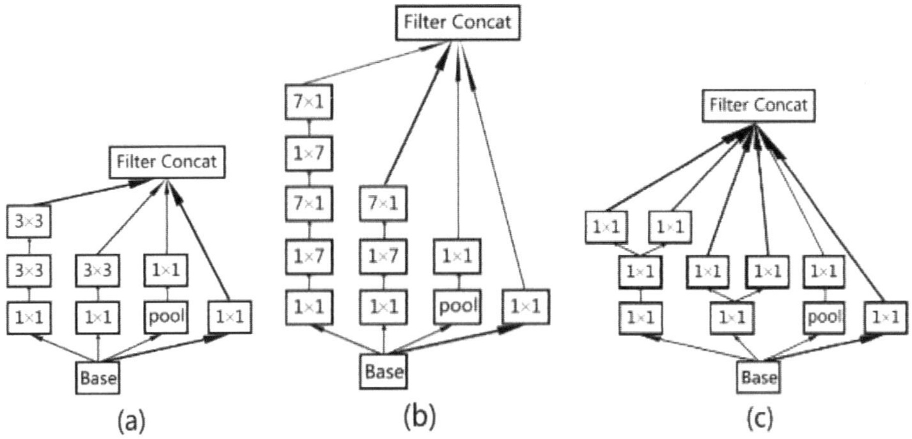

Figure 2. GoogLeNet inception module. (a).inception module_a appears after a few convolutional layers in the network,(b).inception module_b appears after inception module_a,(c).inception module_c appears after inception module_b.

Among varies image segmentation algorithm, U-net can train without large amount of training samples and perform well in segmentation tasks. In ISBI cell tracking challenge 2014 and 2015 ,U-net achieve an average IOU (intersection over union) of 92%,which is signicantly better than the second best algorithm with 83%[8]. So It's a segmentation algorithm for medical image segmentaion.

In this paper, we design several network based on U-net to explore how to insert inception module into U-net to get ideal network architecture thus to increase segmentation accuracy. U-net network belongs to encoder-decoder frame structure, in the encoding phase, the network extract image feature through several convolutional layers, after getting the feature map, the network do a few transpose convolution and meanwhile merge layers in encoder phase in same shape to get segmentation result. U-net is the most effective segmentation network in the field of medical image segmentation. The Several network structures designed in this paper are as shown in table1:

Table 1. The configuration of seven improved U-net.

net	A			B			C			D			E			F			G		
Smallest image size	8			8			8			8			8			32			32		
Insert when picture size is	128	64	32	64	32	16	32	16	8	32	16	8	32	16	8	128	64	32	128	64	32
Insert phase	Encoding phase			Encoding phase			Encoding phase			Encoding phase			Encoding phase			Encoding-decoding phase			Encoding phase		
Module	a	b	c	a	b	c	a	b	c	a	b	c	a	b	c	a	b	c	a	b	c
amounts	1	1	1	1	1	1	1	1	1	2	3	1	3	5	2	1	1	1	1	1	1

Through the comparison between A and G networks, we can know the influence of network depth on network training. Through comparative experiments of A, B and C networks, it can be seen that inception module is best placed in that location. Comparison experiment of network F and G can tell us whether it is better to insert inception module in encoding-decoding stage or not. Through the comparison experiment of C, D and E networks, we can know the influence of different amount of inception module on the network, and finally obtain the optimal network in current task.

4. Experiments

The experimental environment is as follows:
Hardware: Inter(R) Core(TM) i7-6700, 12G, NVIDIA GeForce GTX TITIAN X
Software: Windows 64bit, tensorflow1.7.0, keras2.0.8, CUDA8.0, cunn5.1

4.1. Dataset

Since there is no way to get the label of the test set of IBSI data set, 30 training sets are divided into 20 training sets and 10 test sets. The data expansion method is shown in table 2:

Table 2. Configuration of data augmentation.

Data expansion configuration	
Rotation	0.20
Width adjustment	0.05
Height adjustment	0.05
Shear	0.05
Scale	0.05

The TAO data set comes from the partner hospital. There are 89 CT image data sets of the transecting position of TAO. The distribution of data sets is training set: test set =1:1.All images in the dataset have been desensitized to patient information. To preprocess data sets, we first segment the ROI area which contains the ocular rectus muscle, and then resize the images to 512×512. After that the labels of the dataset are to be made. The black and white mask tags are made with the image labeling tool – labelme.

4.2. Experiment Result

We designed seven network based on U-net and inception module, aiming to explore ideal network architecture of improved U-net under different network configuration. Through experiment, all experiment result in IBSI task are as shown in table 3:

Table 3. Configuration of data augmentation.

net	**Training accuracy**	**Training loss**	**Test PA**	**Test SN**	**Test SP**
A	0.9637	0.0833	0.9106	0.9446	0.6858
B	0.9577	0.1826	0.9068	0.9487	0.6693
C	0.9586	0.1816	0.9214	0.9307	0.7312
D	0.9623	0.5041	0.8851	0.9624	0.5768
E	0.9603	0.5067	0.8785	0.9651	0.5481
F	0.9486	0.1948	0.8937	0.9547	0.6154
G	0.9621	0.5096	0.8830	0.9588	0.5695

The definitions of evaluation indexes PA, SN and SP are as shown in table 4:

Table 4. Defination of Segmentation Indexes

Index	PA	SN	SP
defination	$\frac{S_{pred} \cap S_{label}}{S_{pred}}$	$\frac{S_{pred} \cap S_{label}}{S_{label}}$	$\frac{S_{pred\_bg} \cap S_{label\_bg}}{S_{label\_bg}}$

S_{pred} represents the target area predicted result, S_{label} represents the target area of label image, S_{pre_bg} represents the background area of the predicted result, S_{label_bg} represents background area of the label image.

According to the data in the table, it can be seen that: 1) when the smallest image size in the network is 8, the network performs better.2) from the perspective of insertion location of inception module, insert module_a when image size is 128×128 and module_b and module_c when image size is 64×64 and 32×32 respectively can make the network perform better;3) To insert 3 module_a, 5 module_b and 2 module_c is the better;4) It's better to inert inception module in encoding-decoding phase. In conclusion, in the current task, the ideal network structure is to insert 3 module_a when image size is 128×128, 3 module_b when image size is 64×64 and 2 module_c when image size is 32×32 in encoding-decoding phase of the network symmetrically. With this ideal improved U-net, the segmentation accuracy is significantly improved.

The comparison between loss and accuracy in the training process of original u-net and optimal improved u-net is shown in figure 3:

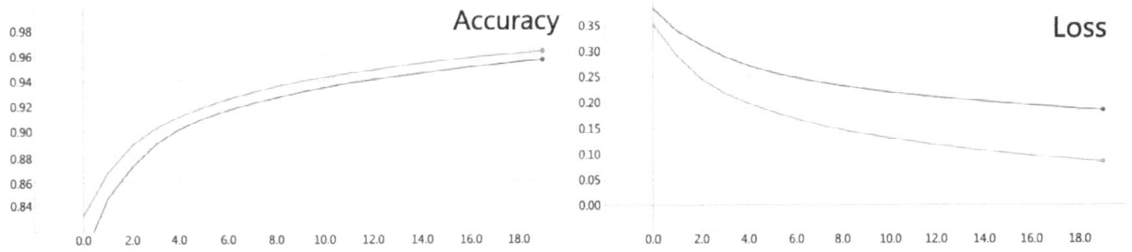

Figure 3. The upper figure shows comparison of accuracy convergence curve and the figure below shows the loss reduction curves of original U-net and improved U-net on the same task

As can be seen from figure 3, the improved U-net has a faster convergence speed, a higher segmentation accuracy and a faster loss reduction speed than the original U-net training. The segmentation results of the improved U-net on the ocular rectus muscle in the transverse position are shown in figure 4:

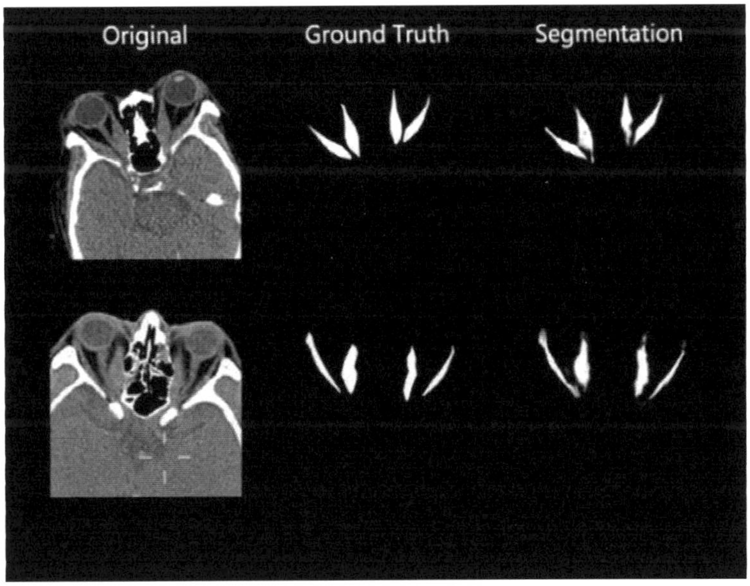

Figure 4. segmentation result of TAO ocular rectus muscle

Compared with original U-net, the accuracy of the improved u-net in the segmentation task of the ocular rectus muscle is shown in table 5:

Table 5. Configuration of data augmentation

	PA	SN	SP
U-net	0.8799	0.8017	0.9955
Our method	0.9537	0.8388	0.9984

5. Conclusion

This paper presents an Ocular rectus muscle segmentation algorithm based on U - Net, the CT images of the ROI area is extracted firstly, then after preprocessing the image as input of network, we trained the ideal network do segmentation of the CT image and finally obtain the ocular rectus muscle. The results of finalization and quantitative experiments show that compared with the traditional segmentation algorithm, the segmentation effect of our method is more accurate and efficient, and it is an auxiliary for doctors to do segmentation of ocular rectus muscle.

Acknowledgments

Thank you for the valuable data provided by Wuhan Union Medical College Hospital, Hubei province, China and relevant technical and knowledge guidance provided by professor Fa-Gang Jiang and doctor Jun-Jie Yang.

References

[1] Guoqiang li. The application of CT and MRI in thyroid - related eye diseases diagnosis [J] . Journal of hubei university of traditional Chinese medicine, 2013, 15 (3) : 103-104.
[2] Fuping liu. Evaluation and treatment strategy of thyroid - related eye diseases [J]. Chinese journal of clinical physicians: electronic edition, 2010, 4(1):4-7.
[3] Yanjiao Zhang. Current status of thyroid - related immunosuppressive therapy for eye diseases [J]. Medical review, 2004, 10(9):553-555.
[4] Krizhevsky A, Sutskever I, Hinton G E. ImageNet classification with deep convolutional neural networks[C]// International Conference on Neural Information Processing Systems. Curran Associates Inc. 2012:1097-1105.
[5] Simonyan K, Zisserman A. Very Deep Convolutional Networks for Large-Scale Image Recognition[J]. Computer Science, 2014.
[6] Szegedy C, Liu W, Jia Y, et al. Going deeper with convolutions[C]// IEEE Conference on Computer Vision and Pattern Recognition. IEEE, 2015:1-9.
[7] Long J, Shelhamer E, Darrell T. Fully convolutional networks for semantic segmentation[C]// Computer Vision and Pattern Recognition. IEEE, 2015:3431-3440.
[8] Ronneberger O, Fischer P, Brox T. U-Net: Convolutional Networks for Biomedical Image Segmentation[J]. 2015, 9351:234-241.
[9] Drozdzal M, Vorontsov E, Chartrand G, et al. The Importance of Skip Connections in Biomedical Image Segmentation[J]. 2016:179-187.
[10] Jégou S, Drozdzal M, Vazquez D, et al. The One Hundred Layers Tiramisu: Fully Convolutional DenseNets for Semantic Segmentation[C]// Computer Vision and Pattern Recognition Workshops. IEEE, 2017:1175-1183.
[11] Lin M, Chen Q, Yan S. Network In Network[J]. Computer Science, 2013.
[12] Szegedy C, Vanhoucke V, Ioffe S, et al. Rethinking the Inception Architecture for Computer Vision[C]// IEEE Conference on Computer Vision and Pattern Recognition. IEEE Computer Society, 2016:2818-2826

IOP Conf. Series: Materials Science and Engineering **533** (2019) 012059 doi:10.1088/1757-899X/533/1/012059

Target Pedestrian Tracking Algorithm Based on Occlusion Scene

Guocai Du

Department of Computer Science and Engineering, Yunnan University, 650000 Kunming, China

495139052@qq.com

Abstract. The traditional incremental learning tracking algorithm tends to drift or lose track when the pedestrian is occluded and the background is similar to the pedestrian target. This paper proposes a pedestrian tracking algorithm based on improved incremental learning. According to the difference between the central position of the current observation model and that of the previous one, the reliability of the observation model can be judged, and then the forgetting factor can be adjusted, so that the observation model can adapt to the changes of pedestrian targets. The experimental results show that the improved algorithm can track the pedestrian target accurately and effectively in the occlusion scene.

1. Introduction

Pedestrian tracking is one hot research topic in computer vision field, its main task is to track the target in image sequence feature detection, to further achieve semantic layer analysis (action recognition, scene recognition, etc). At present, pedestrian tracking can be divided into three categories according to the design of the models. The tracking methods based on the generative model[1-4], the tracking methods based on the discriminant model[5-6], and the tracking methods based on the hybrid generative-discriminant model[7]. Generative model-based tracking methods focus on target self-modeling and learning. When the illumination intensity changes and the target moves quickly, these will seriously affect the appearance and background of the tracking target, and eventually lead to the failure or drift of the tracking target. Discriminant model-based tracking method emphasizes the separation of target and background, but when complex background appears, due to the similarity and ambiguity of background, it will eventually lead to the failure of tracking target. Literature[8] presents a method based on incremental learning, which can adapt to the changes of some tracking targets and the surrounding environment without drastic changes. However, the disadvantage of this algorithm is that it can not adapt to these complex environmental changes when the change of illumination intensity is the largest and the fast movement of the final target is encountered. Finally, the tracking target failed. In this paper, the incremental learning algorithm is used to improve the forgetting factors, and the stability and robustness of the method are verified by three video tests. The structure of this paper is as follows: the first section is the introduction, the second section is the detailed introduction of the algorithm, the third section is the experimental results of the algorithm, and the last section is the summary of the paper.

Content from this work may be used under the terms of the Creative Commons Attribution 3.0 licence. Any further distribution of this work must maintain attribution to the author(s) and the title of the work, journal citation and DOI.
Published under licence by IOP Publishing Ltd

2. Our Algorithm

2.1. Singular Value Decomposition Based on Incremental Learning

The eigenvalues and eigenvectors of the training set can be obtained by the singular value decomposition of the training set image. The feature space constituted by feature vector contains the information of image in training set. In the process of target tracking, training set is constantly increasing. Therefore, when faced with the ever-increasing training set, it should be able to make some changes to the training set, so as to be able to learn the knowledge hidden in the new training set. The idea of incremental learning is to add the latest training set on the basis of the previous training set, and then calculate the mean value vector and singular value decomposition again to obtain the new eigenvalue and eigenvectors.

n frame image sequences $A=[\mathrm{I}_1,\ldots,\mathrm{I}_n]$, m frame new image sequences $\mathrm{B}=[\mathrm{I}_{n+1},\mathrm{I}_{n+2}\ldots,\mathrm{I}_{n+m}]$, let $\mathrm{C}=[A \quad B]$, $\bar{\mathrm{I}}_\mathrm{A}, \bar{\mathrm{I}}_\mathrm{B}$ represent the mean of A and B, which $\bar{\mathrm{I}}_\mathrm{A}=\frac{1}{\mathrm{n}}\sum_{i=1}^{n}\mathrm{I}_i, \bar{\mathrm{I}}_B=\frac{1}{m}\sum_{i=n+1}^{n+m}\mathrm{I}_i$. First, image sequence A is centralized $A=[I_1-I_A,\cdots,I_n-\bar{I}_A]$, and then singular value decomposition is performed to obtain the matrix U_A and diagonal matrix Σ_A. Then, the solving steps of matrix C and diagonal matrix are as follows:

1. Calculate the mean value of the new matrix $\bar{\mathrm{I}}_C=\frac{fn}{fn+m}\bar{\mathrm{I}}_A+\frac{m}{fn+m}\bar{\mathrm{I}}_B$, where f is the forgetting factor.

2. Structure the matrix: $\hat{B}=\left[(\mathrm{I}_{n+1}-\bar{\mathrm{I}}_B),\ldots(\mathrm{I}_{n+m}-\bar{\mathrm{I}}_B),\sqrt{\frac{nm}{n+m}}(\bar{\mathrm{I}}_B-\bar{\mathrm{I}}_A)\right]$.

3. The QR decomposition is applied to the matrix $(\hat{B}-U_\mathrm{A}U_\mathrm{A}{}^T\hat{B})$ to obtain the orthogonal matrix Q and $R=\begin{bmatrix} f\Sigma & U_A^T\hat{B} \\ 0 & Q(\hat{B}-U_AU_A^T\hat{B}) \end{bmatrix}$.

4. Compute the singular value decomposition of R : $\mathrm{R}\overset{SVD}{=}\tilde{U}\tilde{\Sigma}\tilde{\mathrm{V}}^\mathrm{T}\approx\tilde{U}_k\tilde{\Sigma}_k\tilde{V}_k^T$. Retain the first k singular values and update the mean $\mathrm{I}_A=\mathrm{I}_C$.

5. Finally $U_C=[U_A \quad Q]\tilde{U}_k$ and $\Sigma_\mathrm{C}=\tilde{\Sigma}_k$.

2.2. Target Tracking Based on Incremental Learning

The problem with video tracing can be thought of as reasoning in a markov model with hidden state variables, the state variables X_t describe the target affine motion parameters and position at time t. Given a series of observation images $\mathrm{I}_t=\{\mathrm{I}_1,\ldots,\mathrm{I}_n\}$, the purpose is to estimate the value of hidden state variables X_t. According to Bias theorem and Markoff model[9], the formula can be obtained as follows:

$$p(X_t|\mathrm{I}_t)\propto p(\mathrm{I}_t|X_t)\int p(X_t|X_{t-1})p(X_{t-1}|\mathrm{I}_{t-1})d\mathrm{X}_{t-1} \tag{1}$$

The motion of a tracking target in a continuous frame can be represented by an affine change. When at time t, the target state is composed of six affine parameters $\mathrm{X}_t=(x_t,y_t,\theta_t,s_t,\alpha_t,\phi_t)$. Where respectively represent x, y translation, rotation Angle, proportion, width and height ratio of coordinates, and tilt direction at the time t. Each parameter in the equation follows the gaussian distribution and is independent from each other. The model is established by Brown motion, and the system state transition model[10] is as follows:

$$p\left(X_t \middle| X_{t-1}\right)=\mathrm{N}\left(X_t;X_{t-1},\Psi\right) \tag{2}$$

Where Ψ is the covariance matrix with corresponding changes of affine parameters, the diagonal elements of this matrix are the variance of each affine parameter.

Given the image block I_t, it is predicated by X_t, the hypothesis I_t is generated by the feature subspace of the target object, and the probability of sample generation is inversely proportional to the reference point μ of the subspace, and inversely proportional to the distance d between the sample points. The distance between the reference point of the feature subspace and the sample is d , and d can be decomposed into two parts, namely the distance d_t from the sample to the feature subspace, and the distance d_w from the observation sample to the feature subspace. According to the gaussian distribution[11], the probability of generating the sample of the subspace is as follows:

$$p_{d_t}\left(\mathrm{I}_t \middle| \mathrm{X}_t\right)=\mathrm{N}\left(\mathrm{I}_t;\mu,UU^T+\varepsilon I\right) \tag{3}$$

where I is the identity matrix, μ is the mean of the image's subspace, εI is the Gauss noise added to the observation process, U represents the eigenvector of the subspace after the t time is added to the new image.

According to the markov distance[12], the probability of affine sample is expressed as follows:

$$p_{d_w}\left(\mathrm{I}_t \middle| \mathrm{X}_t\right)=N\left(\mathrm{I}_t;\mu,\ U\Sigma^{-2}U^T\right) \tag{4}$$

Where Σ is the singular value matrix corresponding to U .

According to the above, the similar function of observation image and feature space can be deduced[13]. The formula is as follows:

$$p\left(\mathrm{I}_t \middle| \mathrm{X}_t\right)=p_{d_t}\left(\mathrm{I}_t \middle| \mathrm{X}_t\right)p_{d_w}\left(\mathrm{I}_t \middle| \mathrm{X}_t\right)=N\left(\mathrm{I}_t;\mu,UU^T+\varepsilon\mathrm{I}\right)N\left(\mathrm{I}_t;\mu,U\Sigma^{-2}U^T\right) \tag{5}$$

2.3. *Adaptive Adjusted Forgetting Factor*

Incremental learning algorithm is with the addition of new data, will continue to update the eigenvalue and eigenvector, used to express the appearance of the pedestrian targets, finally can adapt to the target part of the change of appearance, but if the pedestrian target or background under the condition of relatively complex, the incremental learning algorithm can't deal with these problems, eventually leading to the pedestrian tracking failure. There are some improvements to incremental learning algorithm, such as target tracking by sparse representation, which only uses a small number of samples to describe the shape of the tracking object. If there are very similar pedestrian targets and backgrounds, the tracking results will also fail. In this paper, a new idea is proposed. According to the difference between the central position of the current observation model and the previous frame, the magnitude of the forgetting factor is finally adjusted. It is necessary to adjust the proportion of the observation value, increase the proportion of the new observation value and reduce the proportion of the former observation value, so that the past samples can be forgotten to adapt to the current environmental changes, so the forgetting factor can be increased. The difference of the center position is higher than that of the novel observation model. The proportion of the former observation value and the new observation value is maintained. This can improve the accuracy of the observation model in tracking the target and reduce the value of the forgetting factor.

The central position of the target $C_t^n=\left(x_t^n,y_t^n\right)$ is calculated according to the current observation model, the observation model of the previous frame calculates the central position of the target $C_t^o=\left(x_t^o,y_t^o\right)$, the maximum difference d_t can be calculated according to formula (6). and the effect of tracking target can be judged according to d_t . Finally, the size of the forgetting factor can be adjusted according to formula (7), where E_m and E_M represent the thresholds of the accepted and rejected subspace models.

It was found that the forgetting factor was the most robust between 0.5 and 0.95, so parameters were set as follows: $E_M = 30$, $Em = 10$, $FF_M = 0.95$, $FF_m = 0.5$.

$$d_t = \max\left(\left|x_t^n - x_t^o\right|, \left|y_t^n - y_t^0\right|\right) \tag{6}$$

$$f = \begin{cases} FF_M, & d_t < Em \\ FF_M - \dfrac{(FF_M - FF_m)}{(E_M - E_m)} d_t, & E_m < d_t < E_M \\ FF_m, & d_t > E_M \end{cases} \tag{7}$$

2.4. *Summary of the Algorithm*
The main steps of the algorithm described above are as follows:

1. Initialize the relevant parameters and the size of the tracking window, and calculate the U and Σ in the tracking image subspace.
2. According to formula (2), the particle set generated by the state transfer function is $x_t^i \sim P(X_t | X_{t-1}^i)$.
3. According to the particle set generated in step 2 and formula (5), the similarity function is calculated and the particle weight is calculated by the weight formula of particle filter.
4. The weight is normalized, and if the effective value of the particle is less than the threshold value, the resampling is performed.
5. Locate the location of the pedestrian target to be tracked according to the particle weight calculated in steps 3 and 4.
6. The increment singular value decomposition proposed in 2.1 and the dynamically adjusted forgetting factor proposed in 2.3 updated U and Σ.
7. Read the next frame and return to step 2, otherwise, until the end of the track.

3. Experiment Results
In order to verify the reliability of the proposed algorithm, several experiments were carried out on pedestrian video. The CPU is Intel (R) Core (TM) i5-7500 CPU@3.40GHz, and the memory is 8GB. The algorithm in this paper was verified based on Matlab R2016a, and three groups of pedestrian video were used for experiment. The total frame number of video was 330, 256 and 208, including simple background, complex background and target occlusion.

3.1. *Qualitative Comparison*
Fig.1, the target pedestrian (marked in red boxes) has a zebra crossing at about 330 frames, but the algorithm in this paper is still capable of real-time tracking. Fig.2 shows that the algorithm in this paper can effectively track the target pedestrian when multiple people are side by side and the pedestrian turns from left to right. At 166 frames and 252 frames, there are pedestrian, motorcycle, zebra crossing and tree on the right side as background interference, which can still effectively track the target. Fig.3, target pedestrians appeared in the process of tracking other pedestrians sheltered, the algorithm still can effectively track, also the 80th frame, appear on the left side of the pedestrian walking from left to right, completely hides are tracking target pedestrian, pedestrian and target the right there are some side by side to interference, eventually to be able to effectively track pedestrians.

3.2. *Quantitative Comparison*
This section is a quantitative analysis of the tracking results of the target pedestrians. The average pixel error Average Pixel Error (APE) and the average overlap rate Average Overlap Rate (AOR)[14] are used to analyze the results. It is measured by the area. Table 1 and Table 2 compare the tracking results of different algorithms in different videos. It can be seen that the proposed algorithm has better

tracking accuracy and robustness than other algorithms. According to the above experimental results, it has a small error on the whole and better robustness than the other two algorithms.

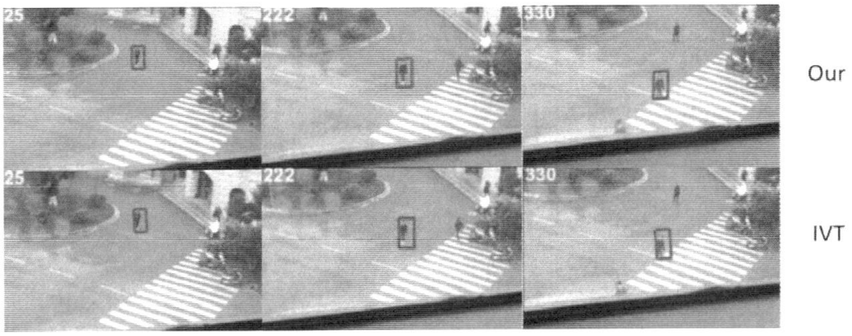

Figure 1. Tracking Results of A Target Pedestrian Against A Simple Background

Figure 2. Tracking Results of Target Pedestrians in Complex Background

Figure 3. Tracking Results of A Target Pedestrian in Occlusion

Table 1. Our Algorithm is Based on the APE Of 3 Test Videos.

Test video	Video 1	Video 2	Video 3
Our algorithm	4.6	11.5	8.0
IVT	10.0	20.2	25.1
LOT	11.0	18.6	16.7

Table 2. Our Algorithm is Based on the AOR in 3 Test Videos.

Test video	Video 1	Video 2	Video 3
Our algorithm	0.88	0.75	0.80
IVT	0.72	0.56	0.62
LOT	0.59	0.65	0.70

4. Conclusion

Traditional incremental learning tracking algorithms focus on the similarity between image samples, and do not take into account the past samples sometimes negligible, but sometimes must take into account the past samples, so the forgetting factor is set to a fixed value. Therefore, this paper improves the size of the forgetting factor by frame difference method, balances the proportion between new observation data and old observation data, and improves the accuracy of the subspace model in describing the appearance of the target at the current moment. Therefore, when the subspace model deviates from the current target appearance, the tracking algorithm in this paper can still maintain the tracking accuracy and adapt to the changes of the target appearance and illumination intensity. By adaptively adjusting the value of forgetting factor, the accuracy of the observation model in describing the appearance is enhanced. The algorithm can still maintain stability and robustness in the case of similar background or other pedestrian occlusion. The qualitative and quantitative results of experiments also verify the effectiveness of the proposed algorithm.

References

[1] Tkachenko, Maksim, Lauw, IEEE Computer Society, **29(4)**, 771-783 (2017).
[2] Feng Ping, Xu Chunyan, Elsevier B.V, **308**, 245-254 (2018).
[3] Kawamoto Kazuhiko, Yonekawa Tatsuya, *IEEE Computer Society*, 711-714 (2012).
[4] Bao C L, WuY, Linh H Bet, Proceeding soft the IEEE Conferenceon Computer Vision and Pattern Recognition, Rhode Island, 1830-1837 (2012).
[5] Yang M, Zhang C X, Wu Y W, Electronic Proceedings of the 2013 IEEE International Conference on Multimedia and Expo Workshops, San Jose, 1-4 (2013).
[6] Babenko B, Yang M H, Belongie S, IEEE Transactions on Pattern Analysis and Machine Intelligence, **33(8)**, 1619-1632 (2011).
[7] Zhong Wei, Lu Hu-chuan, IEEE Transactions on Image Processing, **23(5)**, 2356-2368 (2014).
[8] Ross D A, Lim J W, Lin R S, International Journal of Computer Vision, **77(1-3)**, 125-141 (2008).
[9] MIHAYLOVA L, CARMIAY, Digital signalprocessing, **25(1)**, 1-16 (2013).
[10] MUTHUSWAMY K, IET computer vision, **9(3)**, 428—438 (2015).
[11] Zhao Jing, Li Zhiyuan. Expert Systems with Applications, **37(12)**, 8910-8914 (2010).
[12] Aughenbaugh, Jason Matthew, IEEE Transactions on Aerospace and Electronic Systems, **47(1)**, 503-523 (2011).
[13] Xiao Jingjing, Stolkin Rustam, IEEE Sensors Journal, **16(8)**, 2639-2649 (2016).
[14] Gao Shan, Ye Qixiang, IEEE Transactions on Image Processing, **18(6)**, 5575-5589 (2017).

A Gait Patterns Recognition Approach Based on Surface Electromyography and Three-axis Acceleration Signals

Jinghan Hao, Peng Yang, Lingling Chen and Yanli Geng

School of Artificial Intelligence, Hebei University of Technology, No.8 Guangrong Road, Hongqiao District, Tianjin, China

Email:aq87778@163.com

Abstract. In this paper, an approach based on combining surface electromyography(sEMG) and three-axis acceleration(ACC) signal was proposed to recognize 5 different kinds basis daily gait patterns, including walking on the ground, going up stairs, going down stairs, going up slope and going down slope. Firstly, the gait related sEMG signal and three-axis ACC signal were collected from the lower limbs of subjects. Secondly, the de-noising of sEMG signal was finished and the segmentation of the fusion signal was done. Thirdly, the features of fusion signal were extracted. Finally, a classifier based on 2-stream hidden Markov model (HMM) was built to recognize 5 kinds of basis daily gait patterns. The experiment obtained an average recognition accuracy of 94.32%, which is 4.15% higher than the accuracy by adopting sEMG signal only (Average 90.17%), and 9.60% higher than the accuracy by adopting ACC signal only (Average 84.72%). The result demonstrated that it can improve the recognition accuracy of gait patterns effectively to combine sEMG signal and three-axis ACC signal.

1. Introduction

In recent years, due to the frequent occurrence of accidental injuries such as traffic accidents, work-related injuries and natural disasters, as well as the continuous spread of such chronic diseases as cerebrovascular diseases, diabetes and osteoarthrosis, a considerable number of people have lost the ability to walk, bringing heavy burdens to families and society. Improving the quality of life of such patients and gradually restoring their walking ability have become the focus of social concern and the subject of medical rehabilitation [1].

The electromyography signal (EMG) is the source of the electrical signal that generate muscle force, reflecting the connection between neuromuscular activity and functional state. The electromyography signal is a weak bioelectrical signal from the neuromuscular activity in the process of human movement. It occurs after the intention of the brain movement is generated, and before the muscle actually contracts. It is a signal that is closer to the original intention of human movement [2]. Compared with only gathering dynamic information or image information such as posture and angle, gathering the myoelectric signal to recognize movement intention has obvious advantages. As a signal source, it can guarantee the real-time control [3,4].

Accelerometer(ACC) is a kind of motion sensor, which is wearable. Its analysis algorithm is relatively simple. And the real-time performance of accelerometer is good. The acceleration signal cannot only reflect the velocity information generated by daily behavioral movements over time and the motion

Content from this work may be used under the terms of the Creative Commons Attribution 3.0 licence. Any further distribution of this work must maintain attribution to the author(s) and the title of the work, journal citation and DOI.
Published under licence by IOP Publishing Ltd

IOP Conf. Series: Materials Science and Engineering **533** (2019) 012060 doi:10.1088/1757-899X/533/1/012060

track information in space, but also provide the angular inclination information of daily behavioral movements relative to the direction of gravity acceleration. Therefore, it is widely used in daily behavioral movement detection and fall detection. Time domain features of acceleration signal are often extracted, such as simple threshold value, kurtosis, signal magnitude area (SMA) [5] and instantaneous change value of acceleration. Irene S et al. [6] used acceleration to recognize sitting, standing and walking, and obtained 99% accuracy.

As both sEMG electrodes and ACC sensors have the advantages of low cost, miniaturization and portability, the research results of many scholars[7-9] showed that SEMG and ACC signal fusion technology was superior to single sensor in human movement intention recognition, which benefited from the advantages of sEMG signal in detecting fine movements and ACC signal in detecting large-scale movements and behaviors. Roy et al.[10] used 8 accelerometers and 8-channel surface myoelectric signal to classify 11 recognition tasks and 10 non-recognition tasks, and obtained an average recognition rate of 90%. Zhang et al.[11] and Kosmidou et al.[12] applied SEMG and ACC signal fusion technology to continuous sign language gesture research, and the experimental results showed that the classification accuracy of two types of sensors was significantly improved than that of one type of sensor alone.

However, most of the identified targets of the above research were the gestures recognition based on sEMG signal and ACC signal. There are few researches on human gait recognition.

In this paper, the sEMG signal which can reflect the original intention of the human body and the ACC signal which can reflect the motion track information in space were combined. And a classifier based on 2-stream hidden Markov model (HMM) was built to recognize 5 kinds of human body basis daily gaits, including walking on the ground, going up stairs, going down stairs, going up slope and going down slope.

2. EXPERIMENTAL SETUP

2.1. Experimental Acquisition Equipment

In this experiment, DELSYS Trigno™ Wireless EMG System (Delsys Inc., 20-450Hz band pass filter) was used as data acquisition system, as shown in Figure 1. The device is composed of Trigno charging base station and EMG smart sensors. Every sensor can simultaneously collect sEMG signal and three-axis acceleration signal.

5 typical lower limb gait patterns in human daily movement were selected, including walking on the ground, going up stairs, going down stairs, going up slope and going down slope.

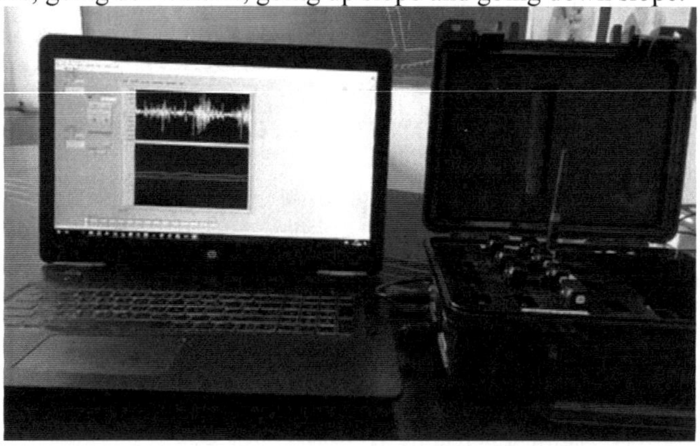

Figure 1. Acquisition of the sEMG singals and ACC signal.

2.2. Experimental Platform

The experimental platform consists of staircase, slope and flat panels, as shown in Figure 2. The staircase has 6 steps. The height of the steps is 150mm, which is widely used. The slope is 15°, which

IOP Conf. Series: Materials Science and Engineering **533** (2019) 012060 doi:10.1088/1757-899X/533/1/012060

is the barrier-free access angle. The number of the gait cycle that subjects can complete on the experimental platform relate to personal walking habits, which is generally 3-4.

Figure 2. Experimental Platform.

2.3. Experimental Procedure
5 subjects (3 males and 2 females) who are with an age range from 20 to 27 years participated in this experiment. As Figure 3 shown, five muscle groups were selected to collect signal:
(1) rectus femoris(sensor1);
(2) vastus medialis(sensor2);
(3) vastus lateralis(sensor3) ;
(4) medial gastrocnemius(sensor4) ;
(5) lateral gastrocnemius(sensor5).

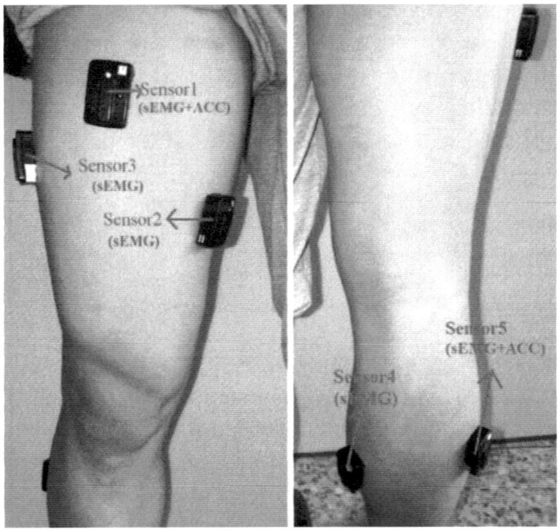

Figure 3. Positions of sensors.

3. Data Processing Method

3.1. SEMG Signal De-noising
In the process of signal collection, the sEMG signal was affected by noise, environment and other factors. Analyzing the signal directly will lead to many errors and reduce the recognition rate of gait patterns. So it is necessary to reduce sEMG signal noise.

In this paper, wavelet packet de-noising method was adopted. Sym8 wavelet was selected to decompose the sEMG signal in 4 layers. Then the threshold value of each layer was obtained. Soft threshold method was used to reduce noise of the signal. Figure 4 shows the comparation between the original sEMG signal and the de-noising sEMG signal of vastus medialis when subject was going upstairs.

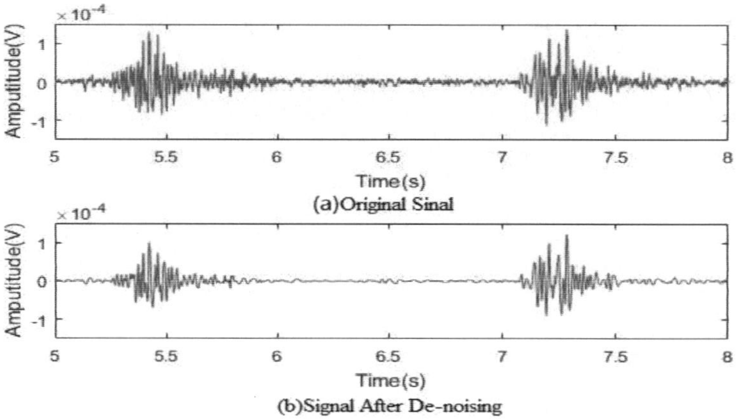

Figure 4. Comparison of SEMG Signal Before and After the Signal De-noising

3.2. Segmentation

This article defined that the movement from the measuring side foot just touching land to the next time measuring side foot just touching land as an active segment. According to the previous research, the general time of an active segment is 1.3s-1.6s. When the fusion signal is used to recognize various gaits directly, sending the fusion signal to the recognition system is too long, which will cause a perceptible delay to the user. Therefore, the active segment was segmented by sliding window method. It is generally believed that the user will not feel the delay if the time interval between the muscle contraction of the lower limb and the corresponding gait by the recognition system is less than 300ms [13]. In the experiment, the sliding window length was set as 128ms. And the window increment was set as 64ms.

3.3. Feature Extraction

Each windowed data was extracted.

3.3.1. sEMG Feature Extraction

5 kinds features were calculated, including the mean absolute value (MAV), standard deviation (STD), zero crossings (ZC), waveform length (WL) and 4-order AR coefficients of sEMG signal of each window for every channel as the features. Then, all the features of 5 channels will combine to form the sEMG feature vector. These features are capable of representing the waveform amplitude, frequency and duration [13-14]. Finally, a 40-dimension sEMG feature vector(named sEMG) was obtained.
The mean value (MEAN) represents the average intensity of a segment of sEMG signal. The calculation formula is as follows:

$$MAV = \frac{1}{N}\sum_{i=1}^{N}|x(i)| \tag{1}$$

The standard deviation (STD) reflects the intensity of signal deviating from its average value, and reflects the change range and intensity of sEMG signal during movement. The calculation formula of standard deviation is as follows:

$$STD = \sqrt{\frac{1}{N-1}\sum_{i=1}^{N}(S(i)-\bar{S})^2} \tag{2}$$

where the calculation formula of $\bar{S}$ is as follows:

$$\bar{S} = \sqrt{\frac{1}{N}\sum_{i=1}^{N} S^2(i)} \tag{3}$$

The zero crossings (ZC) reflects the change of the intensity of the sEMG signal. The calculation formula is as follows:

$$ZC = \sum_{i=1}^{N-1} sgn(-x(i)x(i+1)) \tag{4}$$

$$sgn(x) = \begin{cases} 1 & x > 0 \\ 0 & else \end{cases} \tag{5}$$

The waveform length (WL) is the cumulative value of the waveform over a period of time. The calculation formula is as follows:

$$WL = \sum_{i=1}^{N-1}|x(i+1) - x(i)| \tag{6}$$

Auto-Regressive(AR) model is widely used for its outstanding performance in time series analysis, which is suitable for short data analysis. The sEMG meets the requirements of this parameter model. In the AR model, sEMG can be seen as the product of a linear system motivated by a white noise $e(i)$ with zero mean [15-16]. Therefore, the feature parameters of AR model can be extracted as the features of gait pattern recognition. The AR model was established as follows:

$$x(i) = \sum_{k=1}^{\mathrm{p}} a_k \times x(i-k) + e(i) \tag{7}$$

where $x(i)$ is the sEMG of ith sample point, a_k is the ith coefficient of AR model, p is the order number of AR model, $e(i)$ is white noise residual, which obeys normal distribution. In this experiment, 4 order AR model empirically, i.e. p=4。

3.3.2. Acceleration Signal Feature Extraction

For a three-axis acceleration sensor, the data collected each time can be represented as a vector $\boldsymbol{a} = (a_x, a_y, a_z)$. Firstly the mean value (MEAN) and the standard deviation (STD) of each window for each axis were caculated. Then a 12-dimension feature vector from acceleration data of both sensors(named ACCa) was extracted. Secondly, the modulus of vector $\boldsymbol{a}$ was calculated:

$$|\boldsymbol{a}| = \sqrt{a_x^2 + a_y^2 + a_z^2} \tag{8}$$

then, unitize $\boldsymbol{a}$:

$$\boldsymbol{a_e} = \frac{\boldsymbol{a}}{|\boldsymbol{a}|} \tag{9}$$

The three-axis data of three-axis accelerometer are related to the attitude of the sensors, i.e. the attitude of thighs and calves. Even if the attitude of body cannot be calculated accurately by the ACC signal when the lower limbs is moving, the accelerometer sensor data can be used to get some useful information that can represent the attitude in a certain extent. For an acceleration sensor, an assumption can be made that kinematic acceleration will not significantly influence the direction of resultant acceleration when subjects are moving, which means the magnitude of the acceleration of gravity dominates [17].Then a estimation is made that the direction of gravity is the same as $\boldsymbol{a}$ under five movement patterns[18]. So the modulus of kinematic acceleration is:

$$a_m = |\boldsymbol{a} - g \times \boldsymbol{a}_0| \tag{10}$$

where g is the modulus of the gravitational acceleration.
Although this is not an accurately estimate of motion posture, they can represent posture information in a certain extent. Finally, the MEAN and STD of $\boldsymbol{a}_e$, $|\boldsymbol{a}|$ and a_m in each window of both acceleration sensors was calculated to obtain a 20-dimensional feature vector(named ACCb).
So, by combining ACCa and ACCb to get a 32-dimensional acceleration feature vector (named ACC).

4. Classifier

A 2-stream hidden Markov model (HMM) classifier was selected to recognize the gait patterns. HMM is a statistical analysis model, which was established in the 1970s. It was spread and developed in the 1980s and became an important direction of signal processing. HMM is a dual-random process. On the one hand, it is a kind of Markov chain. Its state cannot be directly observed, but can be reflected through the observation sequence. On the other hand, each observation vector is represented as each state by some probability density distribution, so each observation vector is generated by a state sequence with a specific probability density distribution. HMM can be described by five elements:

$$\lambda=(A, B, \pi, N, M) \tag{11}$$

Where N is the number of hidden states, and M is the number of observations that can certain probability distribution for each state. A , B and π are three different probability matrices. A is the hidden state probability matrix. B is the observation sample probability matrix. And π is the initial probability distribution matrix, which is usually assigned the same value during initialization.
The main steps of training and classification of gait patterns adopting 2-stream HMM[19], are as follows: set sEMG and ACC signal feature sequence as $O = \{O^E, O^A\}$. O^E is sEMG signal observation sequence, O^A is ACC signal observation sequence. For training samples of known gait patterns, $O = \{O^E, O^A\}$, the Baum-Welch algorithm was used to train the classifier. Then a 2-stream HMM was gotten of each gait pattern, i.e. $\lambda_c = \{\lambda_c^E, \lambda_c^A\}$, where c represents a kind of unknown gaits patterns. In this paper, the hidden state number N of HMM was empirically set to 5. And the probability distribution of each observation value was modeled by a 3-order Gaussian mixture model [20].
In the testing phase, for unknown gait patterns, the likelihood probability of the observation sequence was calculated in all the gait pattern HMMs, i.e. $\mathrm{P}(\mathrm{O}|\lambda_c^E), \mathrm{P}(\mathrm{O}|\lambda_c^A), 1 \le c < \mathrm{C}$, where C is the number of all the movement patterns. This paper will recognize 5 kinds gait patterns so that C=5. Then the gait pattern was identified as the gait pattern of the 2-stream HMM which produced the maximum sum of likelihood probability, i.e.:

$$c^* = \arg\left(\max_{1 \le c < C} \mathrm{P}(\mathrm{O}^E|\lambda_c^E) + \mathrm{P}(\mathrm{O}^A|\lambda_c^A)\right) \tag{12}$$

5 Result and Discussion

The size of window length was set as 128ms meanwhile the increment size was set as 64ms. Half of the feature vector dataset was used as training sample to train classifier. And the other half was used as testing dataset to test the performance of classifier. For comparing the recognition accuracies of differient data sources, the experiment was performed with different datasets from:
(1) sEMG;
(2) ACC;
(3) sEMG+ACC.
The recognition accuracy for 5 kinds of basis daily gait for 3 kinds of datasets was calculated, as shown in Table 1. And Figure 5 is the bar chart form of Table 1.

Table 1. Recognition accuracy of 3 kinds of datasets.

	SEMG ONLY	ACC ONLY	SEMG+ACC
WALK	91.18%	87.27%	94.63%
UP STAIR	89.34%	81.06%	93.36%
DOWN STAIR	89.33%	86.16%	95.33%
UP SLOPE	89.36%	83.16%	94.65%
DOWN SLOPE	91.65%	85.96%	93.65%
AVERGE	90.17%	84.72%	94.32%

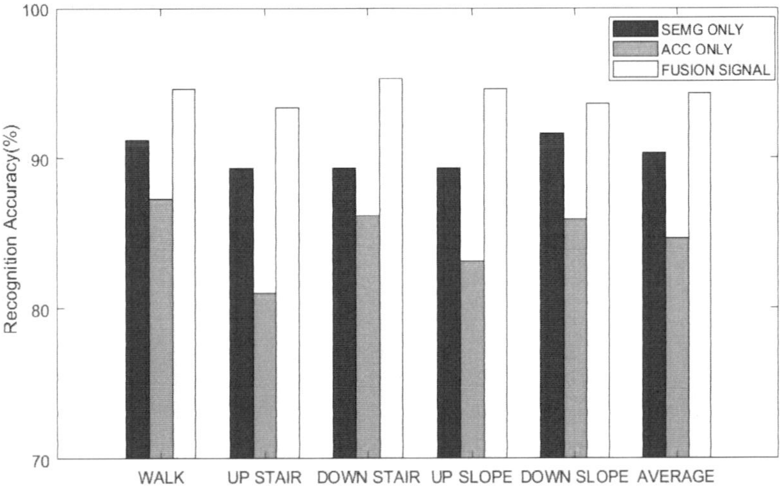

Figure 5. Recognition accuracy bar chart of 3 kinds of datasets.

Figure 5 and Table 1 show that the recognition accuracy by adopting the fusion of sEMG and ACC (Average: 94.32%), is 4.15% higher than that by adopting sEMG only (Average: 90.17%), and 9.60% higher than that by adopting ACC only (Average: 84.72%). This indicates that combining sEMG signal and ACC signal can improve the recognition accuracy compared to only adopting sEMG signal or ACC signal.

Combining sEMG signal and ACC signal is a good choice for lower-limb exoskeleton to identify human gait patterns. But for people with disabilities who need prostheses because of amputation, it is impossible to collect the sEMG signal and ACC signal from calf. Therefore, the signal from thigh only was selected to identify the gait pattern. The recognition accuracy of five motion patterns was obtained, which adopted sEMG and ACC signal of thigh. The result is shown in Table 2. And the results are compared with the results adopting all the signal, as shown in Figure 6.

Table 2. Recognition accuracy of adopting thigh dataset only.

	WALK	UP STAIR	DOWN STAIR	UP SLOPE	DOWN SLOPE
Accuracy	91.23%	89.27%	87.65%	88.99%	88.29%
Average					89.09%

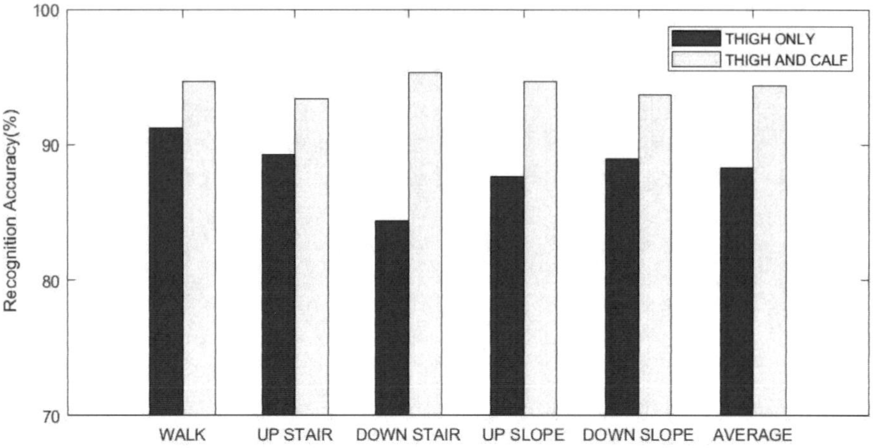

Figure 6. Recognition accuracy bar chart of adopting thigh signal only and adopting all signal.

Only through adopting sEMG signal and ACC signal of thigh, the recognition accuracy obtained is low. This is because the data provided by the three sEMG signal collection points at the thigh and an acceleration sensor at the front of the thigh cannot accurately distinguish the five basic gait patterns. This cannot meet the requirements of active prosthesis for identifying the motion intention of prosthesis wearers. In the future, research will be continued to study how to obtain higher lower-limb intention recognition rate without collecting data information of the calf. The gait patterns recognition rate of prosthetic wearers will be improved by increasing the sEMG signal collection points at the thigh and placing an acceleration sensor at the upper body, etc.

6. Conclusions

Gait patterns recognition has been studied and achieved good recognition results in other previous study. But the selection of sensor signal in those studies usually focused on the sEMG signal. The experiment in this paper showed that more useful movement information could be obtained by combining sEMG signal and ACC signal, which help improve the recognition accuracy.

This paper proposes methods to improve the recognition accuracy of gait pattern by combining sEMG signal and ACC signal. Experimental result shows that this method can successfully improve the recognition accuracy in a statistically significant way. The recognition accuracy by adopting the fusion of sEMG signal and ACC signal (Average: 94.32%), is 4.15% higher than that by adopting sEMG signal only (Average: 90.17%), and 9.60% higher than that by adopting ACC only (Average: 84.72%). This indicates that combining sEMG signal and ACC signal can improve the recognition accuracy compared to only adopting sEMG signal or ACC signal.

Acknowledgments

This work was supported by the National Natural Science Foundation of China under Grants 61703135, the National Natural Science Foundation of China under Grants 61803143 and the Hebei Natural Science Foundation of China under grant F2017202119. The authors would like to acknowledge all the participants of the signals acquisition experiment, and thank the contribution and guidance of the members in our lab.

Reference

[1] P. Langhorne, J. Bernhardt, G. Kwakkel. “Stroke rehabilitation”. The Lancet, Vol.377(9778): 1693-1702 (2011)

[2] C. Lingling, “Research of Electromyography Pattern Recognition Based on Support Vector Machine”. Tianjin: Hebei University of Technology (2006)

[3] Daniel Zlatni, Beatrice, “Gerhard Schweitze. Finite-state control of a trans-femoral (TF) prosthesis” . IEEE Transactions On control system technology, Vol.10(3):408-420 (2002)

[4] A. Kyoung, N. Huynh, “Intelligent switching control of a pneumatic muscle robot arm adopting learning vector quantization neural network” . Mechatronics, Vol.17(4-5):902-905 (2007)

[5] M. Karantonis, R. Narayanan, M. Mathie, et al. "Implementation of a real-time human movement classifier adopting a triaxial accelerometer for ambulatory monitoring". IEEE Transactions on Information Technology in Biomedicine, Vol.10(1): 156-167 (2006)
[6] S. Irene, Chennai C-DAC, M. Shwetha, et al. "Improving classification of sit, stand, and lie in a smartphone human activity recognition system". IEEE International Conference on Computer and Information Technology. Liverpool. 1460-1466 (2015)
[7] Chen, X. Zhang, X. Zhao, et al. "Hand gesture recognition research based on surface EMG sensors and 2D-accelerometers". Eleventh IEEE International Symposium on Wearable Computers, Proceedings: 11-14 (2007)
[8] J. Kim, J. Wagner, M. Rehm, et al. "Bi-channel Sensor Fusion for Automatic Sign Language Recognition". 2008 8th IEEE International Conference on Automatic Face & Gesture Recognition, Vols.1-2: 647-652 (2008)
[9] Y. Li, X. Chen, X. Zhang, et al. "A Sign-Component-Based Framework for Chinese Sign Language Recognition Adopting Accelerometer and sEMG Data". IEEE Transactions on Biomedical Engineering, Vol.59(10): 2695-2704 (2012)
[10] Roy, S.H., Cheng, M.S., Chang, S.S., et al. "A Combined sEMG and Accelerometer System for Monitoring Functional Activity in Stroke." IEEE Transactions on Neural Systems and Rehabilitation Engineering, Vol.17(6): 585-594 (2009)
[11] X. Zhang, X. Chen, Y. Li, et al. "A Framework for Hand Gesture Recognition Based on Accelerometer and EMG Sensors." IEEE Transactions on Systems Man and Cybernetics Parta-Systems and Humans, Vol.41(6): 1064-1076 (2011)
[12] Kosmidou, V. E., et al. "Sign language recognition adopting intrinsic-mode sample entropy on sEMG and accelerometer data." IEEE Transactions on Biomedical Engineering, Vol.56(12): 2879-2890 (2009)
[13] K. Englehart, B. Hudgins, "A robust real-time control scheme for multifunction myoelectric control". IEEE Transactions on Biomedical Engineering, Vol.50, no. 7, pp. 848 (2003)
[14] B. Hudgins, P. Parker, R. N. Scott. "A new strategy for multifunction myoelectric control". IEEE transactions on bio-medical engineering, Vol. 40, no. 1, 82-94 (1993)
[15] L. Hong-Tao, C. Yu-Zhen, X. Xiao-Bo, H. Yong. "Estimation of muscle fatigue degree adopting time-varying autoregressive model parameter estimation of surface electromyography." Chinese Journal of Biomedical Engineering, Vol.26(4): 493-497 (2007)
[16] Z. Pei-Lin, L. Sheng. "Fault feature selection method based on wavelet and GA-PLS algorithm". Journal of Vibration, Measurement & Diagnosis, Vol.34(2): 385-391 (2014)
[17] P. Batista, C. Silvestre, P. Oliveira, B. Cardeira, "Accelerometer Calibration and Dynamic Bias and Gravity Estimation: Analysis Design and Experimental Evaluation". IEEE Transactions on Control Systems Technology, Vol. 19, no. 5, 1128-1137 (2011)
[18] Z. Yongjie, L. Bo, S. Xinjun, Z. Xiangyang. "Towards Chinese sign language recognition adopting surface electromyography and accelerometers". 2017 24th International Conference on Mechatronics and Machine Vision in Practice (M2VIP), 1-5 (2017)
[19] S. Zhong, J. Ghosh, "HMMs and coupled HMMs for multi-channel EEG classification." Proceeding of the 2002 International Joint Conference on Neural Networks, Vols.1-3:1154-1159 (2002)
[20] R. Allen, E. Ambikairajah, H. Lovell, G. Celler "An Adapted Gaussian Mixture Model Approach to accelerometry-based movement classification adopting time-domain features". Engineering in Medicine and Biology Society, 28th Annual International Conference of the IEEE. 3600-3603 (2006)

IOP Conf. Series: Materials Science and Engineering **533** (2019) 012061 doi:10.1088/1757-899X/533/1/012061

Femur segmentation in X-ray image based on improved U-Net

Fan Lianghui[1], Han JunGang[1], Jia Yang[1], Yang Bin[2]

1. School of Computer, Xi'an University of Posts and Telecommunications, Xian Shanxi 710121, China
2. Xi`an Honghui Hospital, Xi`an 710054, China

592556555@qq.com, School of Computer, Xi'an University of Posts and Telecommunications

Abstract. Segmentation of Femur bone from X-ray images is an indispensable step in computer aided analysis of medical images and orthopaedic examinations. It is more complex than segmentation from CT and MR images, due to some associated less dense tissues that are hard to distinguish from the femur bone in X-ray images. This paper presents an improved method based on U-Net to automatically extract the femurs from hip X-ray images. This method changes the structure of the U-Net network, which can effectively map the non-linear relationship between hip image and femur image, and accurately segment femur image. The paper also added the absolute deviation loss function to improve the segmentation effect. Experimental results show that this method is accurate, robust, and achieves an average dice similarity coefficient of 0.966. The segmentation results are satisfactory.

1. Introduction

Analysis of proximal femur images plays an important role in Musculoskeletal Disorders (MSDs) diagnosis, preoperative planning and surgical planning of orthopedic conditions. Femur segmentation from X-ray image of whole hip bone is the prerequisite condition for bone morphologic feature analysis.

Because of the high noise, low contrast, tissue overlapping with hip bones and the narrow gap between femur and acetabulum[1], femur segmentation is a challenging task. In the previous works, segmentation error always occurs in joint regions [2].

Classical natural image segmentation algorithms apply homogeneous features[2, 3]to segment object regions. Unfortunately, such features are appropriate to natural pictures but not satisfied by X-ray images. For example, Figure 1 is a sample X-ray image of hip joint. Region 1 and 2, region 3 and 4 are with high homogeneity, while 2 and 4 belong to femur, 1 and 3 belong to hipbone. In this circumstance, color segmentation with threshold is ineffective for segmenting femur from the image. The arrows show some blurring regions on the edge of proximal femur. To determine the accurate edge in blurring region is also challengeable.

Most proposed segmentation methods are performed on CT and MRI images [4-11], rather than on X-ray images. Chen et al.[12] developed a model-based approach for automatically extracting femur contours from standard hip X-ray images. The method begins with detecting the positions of the femoral shaft, head and turning points, then a model femur contour is registered piecewise to the X-ray image according to the detected features. Finally, active contours with curvature constraints are applied to accurately identify the femur contours. 81.4% femur contours were successfully segmented from 172 X-ray images. Unfortunately, the proposed model-based method is not fit in fracture or odd

Content from this work may be used under the terms of the Creative Commons Attribution 3.0 licence. Any further distribution of this work must maintain attribution to the author(s) and the title of the work, journal citation and DOI.
Published under licence by IOP Publishing Ltd

femurs segmentation and is time consuming. Feng et.al. [13] presented an atlas-based approach for automatic femur segmentation in X-ray images, which deforms the aligned atlas to extract detailed femur contours. Test results show that the proposed algorithm is robust and accurate in segmenting femur contours. However, it is extremely difficult to segment femurs with major fractures, i.e., not only the shapes but also the topologies of femur contours can be changed. Behiels et.al. [14] proposes an improved search procedure for Active Shape Model (ASM) to segment femur from X-ray Images. During the shape searching, shape model needs to change smoothly. But the method is sensitive to the morphological parameters (such as translation, rotation and scaling). Dong and Zheng [15] use a 3D statistical model for the femur bone contour extraction, but in their model the existence of occlusion would lead to a failure in extraction the contour around the femur neck area. Chav et.al. [16] proposed a prior shape segmentation method to extracted femur boundary. The method applies subspace to achieve segmentation of rectangular image, but the result of segmentation is not satisfactory. Ouertani et.al. [1] improved hip joint segmentation by simultaneously solving a dual edge-detection problem, but the results differed between the recovered contours and the manual segmentations.

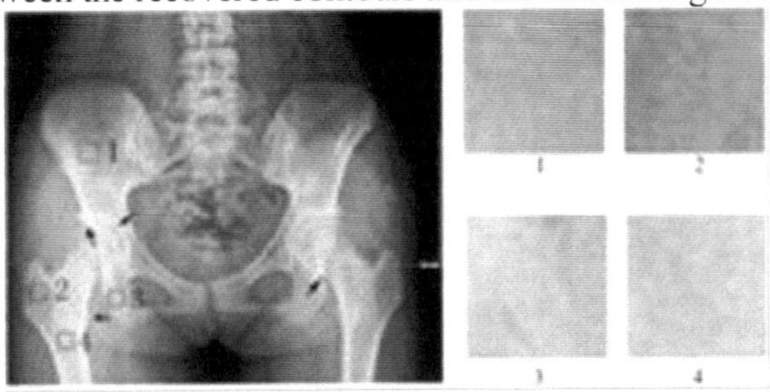

Figure 1. Sample X-ray Image of hip joint.

In recent years, deep learning has been widely used and often outperforms traditional methods in the field of medical image analysis because of its great modeling capacity, and ability of extraction features. In this paper, we present a method to automatically segment the X-ray image of hip joint based on the improved u-net network， and the method is evaluated on the real X - ray data set. The experimental results show that this method can accurately segment the X-ray image of femur, which is better than the existing method.

2. The Network

CNN has achieved superior performance on visual object recognition and image classification [17]， and the resulting U-Net network has made outstanding achievements in image segmentation [18]. In view of this, we improved the U-Net network model based on practical problems. The network improvements are described in detail below.

2.1. The Structure of Network

In the structure of U-Net network, there are two stages of convolution and deconvolution. In the convolution stage, the pooling layer is following two successive convolution layers. In this stage, the network retains the feature of the pooling layer. In the deconvolution phase, the feature of deconvolution layer is connected the feature of the corresponding pooled layer that previously retained for deconvolution.

The improvement of the network is removing the pooling layer and adds the batch normalized layer. Instead of cutting out the features extracted from the convolution layer, we merge the feature that have same dimension from convolution. The pooling layer is used to reduce the dimension of the output results by down sampling the input features. As reduces the output feature dimension, it discards the image's global information about the location that is importance for segmentation task. Therefor, we removed the pooling layer from the network for retaining the global information of the image. After removing the pooling layer in the network, the network training speed is slowed down because of the increase of data dimension. We add batch normalization layer to improve the convergence speed of the network, and also avoid the influence of abnormal values in the image. In the improved network, the features of the convolution layer and the corresponding deconvolution layer have the same dimension. So, we can directly merge those features that have same dimension without cutting operation. The network structure is shown in figure 2.

In the shallow convolution layer, the global position information of pixel points is clear. In the deep convolution layer, the classification of pixels is accurate. These two kinds of information are both critical for medical image segmentation. In the U-Net network，skip to fuse can make use of both information to improve the accuracy of image segmentation. We continued this technique. The network proposed in this paper has 8 convolution layers and 8 deconvolution layers. The number of convolution layer filters is 64, 128, 256, 512, 512, 512, 512, and 512 respectively. The size of filter is fixed at 4×4 and set the strides value as 2. And the activation function of those filter is LReLu. The deconvolution layer is symmetric with the convolution layer. And the number of filters is 512, 512, 512, 512, 512, 256, 128, and 64 respectively. The parameter setting is the same as the convolution layer. Except for the final output layer, the activation function of filter in deconvolution use ReLu function. The Tanh function is used as the activation function in the final output of the network. The batch normalization layer is added after each layer of convolution and deconvolution. In order to increase the generalization capability of the network, we add dropout[19] layer after the first three layers of deconvolution and set dropout keep a probability of 0.5.

Another difference in our network is that what the network outputs is not a binary image of the femur, but the original femur image. As shown in the figure 2.

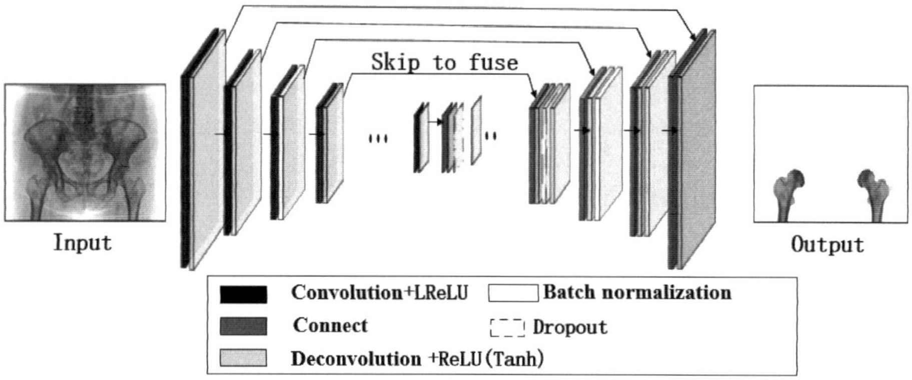

Figure 2. The framework of network

2.2. Training

In the original U-Net network, a new weighted cross entropy loss function was defined according to the problem of cell segmentation. In our femur segmentation task, we use the original cross entropy loss function because the femur has a relatively fixed form.

$$L_H(x,z) = -\sum_{k=1}^{d} x_k \log z_k + (1-x_k)\log(1-z_k) \quad (1)$$

Where x and z stand for the input hip image and the corresponding ground-truth femur image respectively. Their vector length is d.

We also add the absolute loss error function, which is used as the loss function of the network together with the cross-entropy loss function. The loss function of the network is shown in eq (2).

$$L_U = \lambda_1 L_H(x,z) + \lambda_2 \sum_{k=1}^{d} |x_k - z_k| \quad (2)$$

Where λ_1 and λ_2 are used to balance the loss.

IOP Conf. Series: Materials Science and Engineering 533 (2019) 012061 doi:10.1088/1757-899X/533/1/012061

In practical, we set the minibatch size, the value of λ_1 and the value of λ_2 are set as 1, 1 and 100, respectively. The learning rate of the Adam optimizer is fixed in 2e-4, and parameter β1 and β2 are set as 0.5 and 0.999.

3. Experiment

3.1. Datasets and data pre-processing

The hip image dataset includes 50 DICOM X-ray images collected from the Xi'an Honghui Hospital. Data Masking is done to remove the sensitive information about the patient. The DICOM images are with a size of 2,688 × 2,208 (width × height) and a display window of [0 4,096]. In order to balance the network computation and retain sufficient image information, the DICOM images were converted into PNG format and the image size was changed accordingly. Each image is transformed into a square with 512 × 512 pixels by cutting and stitching operations. Taking the femur segmentation image as an example, in Fig. 3, image (b) is a part of image (a), while image (c) is the combination of images (a) and (b).

An example of the hip and femur image is given in Figure 4. Each sample consists of a hip image and a femur image labeled manually.

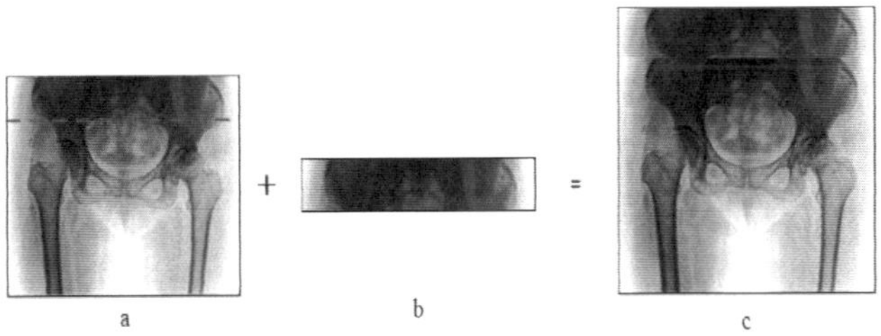

Figure 3. Femur image pre-processing

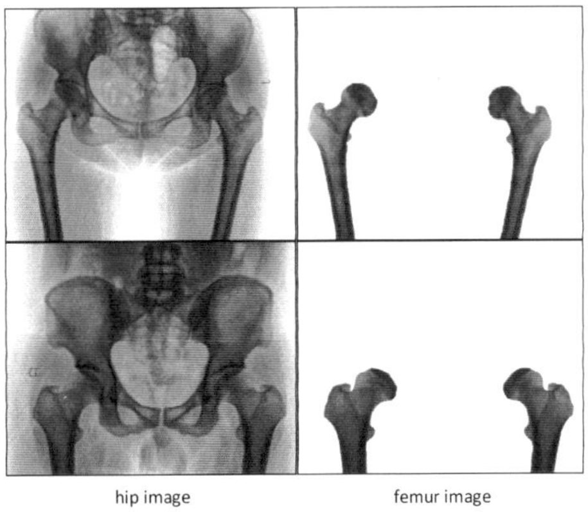

Figure 4. hip image (left) and its manually segmented femur (right)

To augment the amount of data, we use the left-right flip for each image. 83 images are used for training and tested with the rest 17 images.

3.2. Qualitative result

The results of femur segmentation of improved U-Net and original U-Net are showed in Figure 5. It can be clearly seen from Figure 5 that the improved U-Net can accurately segment the femur from the hip joint, while the original U-Net network segmentation results are not clear, and the contour of the femur is incomplete.

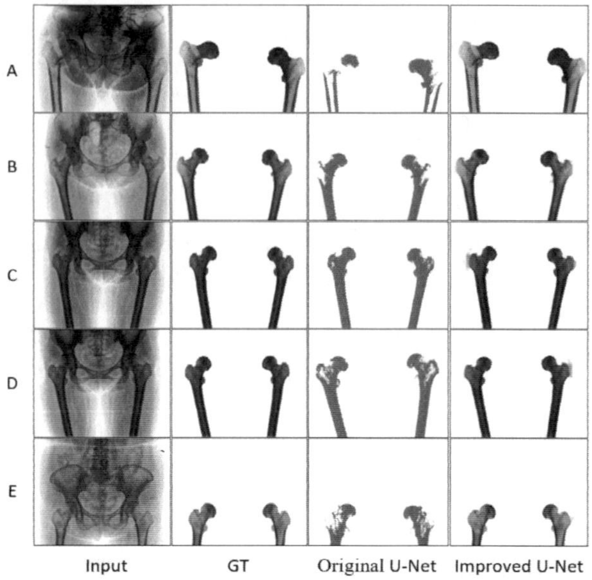

Figure 5. Comparison of original U-Net and improved U-Net segmentation results

3.3. *Qualitative result*

Table 1 summarizes Dice coefficient, Hausdorff distance, recall, specificity and precision measured in pixel for 17 different subjects. Dice coefficient is a similarity measure related to the Jaccard index. The formula is defined as follows:

$$s = \frac{2|X \cap Y|}{|X|+|Y|} \tag{3}$$

where X stands for the femur pixel set of segmentation result from the output of G, and the Y indicates the femur pixel set of ground truth image. The larger the Dice coefficient is, the more similar the two images are.

The Hausdorff distance measures the dissimilarity of two shapes. As shown in eq (4)

$$d_H(X,Y) = \max\{sup_{x\in X} inf_{y\in Y} d(x,y), sup_{y\in Y} inf_{x\in X} d(x,y)\} \tag{4}$$

where sup and inf stands for the supremum and the infimum respectively, X and Y denote the contours set of the segmentation result and the ground truth respectively. The smaller the distance, the closer the match of two images, the better the segmentation results.

We also calculate the three evaluation criteria of image segmentation in this paper. the three criteria are calculated on the pixel level based on the segmented image and the ground truth image. TP stands for the pixel number of to correctly segmented femur. FN stands for the pixel number of that the femur is incorrectly segmented as background. FP stands for the pixel number of the background incorrectly segmented as femur, and TN stands for the pixel number of correctly segmented background. The three definitions are as follows:

$$Recall = \frac{TP}{TP+FN} \tag{5}$$

$$Specificity = \frac{TN}{FP+TN} \tag{6}$$

$$Precision = \frac{TP}{TP+FP} \tag{7}$$

Recall represents the ratio of the number of pixels belong to correctly segmented femur to the number of pixels in ground truth. Specificity represents the ratio of the number of pixels belong to correctly segmented background to the number of pixels belong to the whole segmented background. Precision represents the ratio of the number of pixels belong to correctly segmented femur to the number of pixels in the whole segmented femur.

Table 1. Quantitative segmentation results

ID	Dice	Hausdorff	Recall	Specificity	Precise
1	0.9684	9.00	0.9675	0.9595	0.9249
2	0.9751	39.02	0.9355	0.9399	0.8981
3	0.9850	36.02	0.9576	0.9414	0.9074
4	0.9787	8.25	0.9213	0.9390	0.9365
5	0.9760	13.00	0.9373	0.9598	0.9483
6	0.9661	9.07	0.9183	0.9198	0.9281
7	0.9722	45.64	0.9491	0.9287	0.9222
8	0.9592	48.05	0.8536	0.9386	0.9304
9	0.9686	29.08	0.9291	0.9596	0.9362
10	0.9776	40.81	0.9247	0.9192	0.9331
11	0.9586	5.00	0.9419	0.9294	0.9445
12	0.9483	20.21	0.9388	0.9493	0.9139
13	0.9682	2.83	0.9500	0.9390	0.9497
14	0.9483	9.49	0.9293	0.9393	0.9339
15	0.9677	7.00	0.9417	0.9287	0.9124
16	0.9567	7.00	0.9579	0.9484	0.9214
17	0.9481	2.83	0.9497	0.9193	0.9147
average	0.9660	19.55	0.94	0.9388	0.9268

The average Dice coefficient reached 0.966， and the average Hausdorff distance is 19.55， indicating a high similarity between femur segmentation results and ground true. The value of three indexes: Recall, Specificity and Precision are very high, proving that the method is robust and accurate. From the data in the table 1, we can conclude that our method has an impressive effect on the segmentation of the femur.

4. Conclusion

In this paper, we present an improved U-Net for segmentation femur. By comparing the segmentation results, we can conclude that the improved method in this paper is effective. In many adverse conditions, such as extraneous straight lines caused by the analogue imaging process, blurry boundary, femoral deformities, the proposed method can accurately segment femur image. The experimental results show that this method can effectively segment the femur in the original hip image. The next step is to apply this method to the segmentation of other X-ray bone images.

References

[1] Ouertani, F. Simultaneous extraction of two adjacent bony structures in x-ray images: Application to hip joint segmentation. in International Conference on Image Processing. 2015.
[2] Bischof, Seeded Region Growing. IEEE Transactions on Pattern Analysis & Machine Intelligence, 2002. 16(6): p. 641-647.
[3] Otsu, N., A Threshold Selection Method from Gray-Level Histograms. IEEE Transactions on Systems Man & Cybernetics, 2007. 9(1): p. 62-66.
[4] Chu, C., et al., MASCG: Multi-Atlas Segmentation Constrained Graph method for accurate segmentation of hip CT images. Medical Image Analysis, 2015. 26(1): p. 173-184.
[5] Cheng, Y., et al., Automatic segmentation technique for acetabulum and femoral head in CT images. Pattern Recognition, 2013. 46(11): p. 2969-2984.
[6] Seim, H., et al. Automatic segmentation of the pelvic bones from CT data based on a statistical

shape model. in Eurographics Conference on Visual Computing for Biomedicine. 2008.
[7] Chandra, S.S., et al., Focused shape models for hip joint segmentation in 3D magnetic resonance images. Medical Image Analysis, 2014. 18(3): p. 567-578.
[8] Xia, Y., Automated segmentation and quantitative analysis of the hip joint from magnetic resonance images. Bone, 2015.
[9] Xia, Y., et al. Automated MR Hip Bone Segmentation. in International Conference on Digital Image Computing Techniques and Applications. 2012.
[10] Schmid, J., J. Kim, and N. Magnenat-Thalmann, Robust statistical shape models for MRI bone segmentation in presence of small field of view. Medical Image Analysis, 2011. 15(1): p. 155-168.
[11] Li, W., et al., Human Hip Joint Cartilage: MRI Quantitative Thickness and Volume Measurements Discriminating Acetabulum and Femoral Head. IEEE Trans Biomed Eng, 2008. 55(12): p. 2731-2740.
[12] Chen, Y., et al., Automatic Extraction of Femur Contours from Hip X-Ray Images. 2005: Springer Berlin Heidelberg. 200-209.
[13] Feng, D., W.K. Leow, and T.S. Howe. Automatic segmentation of femur bones in anterior-posterior pelvis x-ray images. in International Conference on Computer Analysis of Images and Patterns. 2008.
[14] Behiels, G., et al., Active Shape Model-Based Segmentation of Digital X-ray Images. Lecture Notes in Computer Science, 1999. 1679: p. 128-137.
[15] Xiao, D., M.A.G. Ballester, and G. Zheng, Automatic Extraction of Femur Contours from Calibrated X-Ray Images using Statistical Information. Journal of Multimedia, 2007. 2(5): p. 55-55.
[16] Chav, R. and C. Kauffmann, Method for fast and accurate segmentation processing from prior shape: application to femoral head segmentation on x-ray images. 2009. 72594Y-72594Y-8.
[17] J. Gu, Z. Wang, J. Kuen, L. Ma, A. Shahroudy, B. Shuai, T. Liu, X. Wang, L. Wang, G. Wang, J. Cai, T. Chen, Recent Advances in Convolutional Neural Networks, Computer Vision and Pattern Recognition (2015).
[18] Ronneberger, O., P. Fischer, and T. Brox. U-Net: Convolutional Networks for Biomedical Image Segmentation. in International Conference on Medical Image Computing and Computer-Assisted Intervention. 2015.
[19] Hinton, G.E., et al., Improving neural networks by preventing co-adaptation of feature detectors. Computer Science, 2012. 3(4): p. págs. 212-223.

IOP Conf. Series: Materials Science and Engineering **533** (2019) 012062 doi:10.1088/1757-899X/533/1/012062

Efficient Small Object Detection with an Improved Region Proposal Networks

DongWen Ma[1,2], XiaoJun Wu[1,2,a] and Honghong Yang[1,2]

[1]*Key Laboratory of Modern Teaching Technology, Ministry of Education (Shaanxi Normal University), 710062 Xi'an, China*
[2]*School of Computer Science (Shaanxi Normal University), Xi'an, China*

[a]Corresponding author:xjwu@snnu.edu.cn

Abstract. Although the state-of-the-art object detection methods, which depend on region proposal algorithms to hypothesize object locations, have achieved high detection accuracy, it still struggles in small-size object detection. In this paper, we present a novel method with a multi-scale and multi-tasking region proposal method to effectively detect small object. In the proposed method, multi-scale features and high-level features are employed to locate object position and identify object category, respectively. The main contributions of the proposed approach are two-fold: (1) A simpler way is used to improve the accuracy performance of small object detection, instead of the complex image pyramids and the complex combination framework, and make the object detection task more flexible. (2) Based on multi-scale and multi-tasking approaches, object location information in low layers and object semantic information in deep layers are made fully advantage respectively. The experimental results on the PASCAL VOC dataset show that the proposed method achieves the state-of-the art object detection accuracy.

1. Introduction

The task of object detection is to find the correct location of the objects and determine the corresponding object category of these objects. It is a challenging task in computer vision field and widely used in our daily lives. Recently, advances in object detection are driven by deep learning methods. Although the object detection method based on Convolutional Neural Network (CNN) achieves satisfied accuracy by using very deep network, it stills a difficult task for the small-size objects detection.

Traditional object detection methods via merging multiple segments or scoring candidate windows that probably contain objects to generate region proposals. These methods unusually adopt cues such as super-pixels, edges [1, 2], saliency [3] and shapes [4] as features. Since the traditional methods always with poor accuracy, several methods based on CNN are proposed [5, 6] to improve the detection performance due to the deep learning show great success in object recognition [7].

The object detection algorithms based on deep learning can be roughly categorized into a couple classes, the regional proposal methods, such as the R-CNN [8], SPPNet [5], Fast R- CNN [6], Faster R-CNN [9], R-FCN [10], FPN [11], Mask R-CNN [12], and the regression methods such as You Only Look Once (YOLO) [13,14,15], Single Shot Detector (SSD) [16].

In [8], Ross B. Girshick et al. proposed an object detection network depend on regional proposal method with CNN features (R-CNN). In their work, they combine candidate regions with

Content from this work may be used under the terms of the Creative Commons Attribution 3.0 licence. Any further distribution of this work must maintain attribution to the author(s) and the title of the work, journal citation and DOI.
Published under licence by IOP Publishing Ltd

convolutional neural networks. This is fulfilled by extracting ~2k region proposals by Selective Search [17] and classifies them with a pre-trained CNN model (VGG-16).

Ross Girshick proposed Fast R-CNN to improve the performance of R-CNN in [6]. In their work, Fast R-CNN employs multiple regions of interest (RoIs) and bounding box regression to improve the speed of training and testing and also improve the detection accuracy. In RoIs, the max pooling is used to convert the features inside any valid region of interest into a small feature map. Moreover, the bounding box regression is used as the multi-tasking loss function in the convolutional neural network for training. Despite Fast R-CNN achieves superior performance in object detection, it still relies on Selective Search, which is the major drawback for Fast R-CNN since the process to generate the region proposal is very time consumes. It nearly cost half time of all training.

As the Faster R-CNN has a good detection performance, it became a fundamental algorithm in object detection field. Faster R-CNN introduced a Region Proposal Network (RPN) that utilizes convolutional neural networks to generate candidate regions and combines region proposal generation and detection into a unified network. Usually, RPN is employed to detect regions of interest in the convolutional neural network and many State-of-the-art regional proposed methods also followed this idea, such as literature [11] and [18]. However, the poor localization performance exists in the deep layer, making it still difficult to locate small-size objects. The deep layers of the network can find the object of interest with high recall but poor localization performance. While the low layers of the network can better locate the object but with a low recall. So finding features that includes both strong semantics information and high resolution information is the key to solving the problem of small object detection.

Intuitively, an effective way to solve above problem is combining the features from deep and low layers. Several algorithms are proposed to solve this issue such as [11] [18] and [19]. However, the existed methods tend to find a feature with rich semantic information and precise location information. It is a difficult task to build a feature with both those two kinds information at a same time.

In our paper, we propose a multi-scale and multi-tasking method for object detection. Firstly, the image is fed a pretrained convolutional neural network to obtain the feature maps of each layer. Then achieve the deep feature map through the last convolutional layer and obtain the multi-scale features via fusing feature maps from different convolutional layer. Moreover, we use slight region proposal generative network, which includes two branches–box-regression (*reg*) and box-classification (*cls*), to estimate existence of an object and generate the proposals, respectively. The box-classification is used to estimate the existence of an object by calculating the probability of it belonging to the foreground or background. This is based on that whether the current pixel is in the ground-truth box or not. The box-regression is used to predict the approximate position of each object instance. For each image, the neural networks generate 2000 proposals through non-maximum suppression (NMS) [8]. Finally, these proposals are classified and adjusted via the detection module. The pipeline of this work is shown in Figure.1.

The main contributions of this paper are as follows.

(1) A more concise and effective feature fusion method with less parameters and simpler architecture is proposed.

(2) We use deep convolutional layers for classification and multi-scale features from different layers for positioning since the low-level features are more suitable for the region of interest (RoI), and high-level semantic features have a great effect on classification. Our experimental results show that the proposed method has a good detection performance, especially for small objects detection.

2. Related Works

The goal of object detection is to localize and recognize object instances in the image with a bounding box [20,21,22]. The traditional approaches generate regional proposals based on ranking the object scores of a bunch of candidate windows. With the emergence and development of deep learning, the neural networks are used to generate regional proposals. Compared with the traditional methods, the neural networks based on region proposal methods are less computational complexity and less time-consuming.

S. Bell et al. [19] proposed inside-Outside Net, an object detector that exploits information both inside and outside of the interest region. The inside information is a feature that uses a skip pooling to connect feature maps from different convolutional layers, while the outside information is contextual information from the outside of interest region, which is extracted by using spatial recurrent neural networks.

In [18], T. Kong et al. proposed a network based on Hyper Feature. The Hyper Feature is a hierarchical feature that combines the fine information and coarse information from deep and shallow level, respectively. However, it makes the features more abundant.

In [11], T.-Y. Lin et al. proposed a clever feature pyramid method that uses top-down structure and lateral connections to build high-level semantic feature maps from all scales. In their work, the top-down structure is combined by high-level features and low level features. The lateral connection method combines the upsampled result with the same size feature map from the bottom up.

However, these methods tend to find features with rich semantic information and precise location information. As known, high semantic information and high-resolution information are inherently contradictory. Therefore, building features with both types of information is a difficult task.

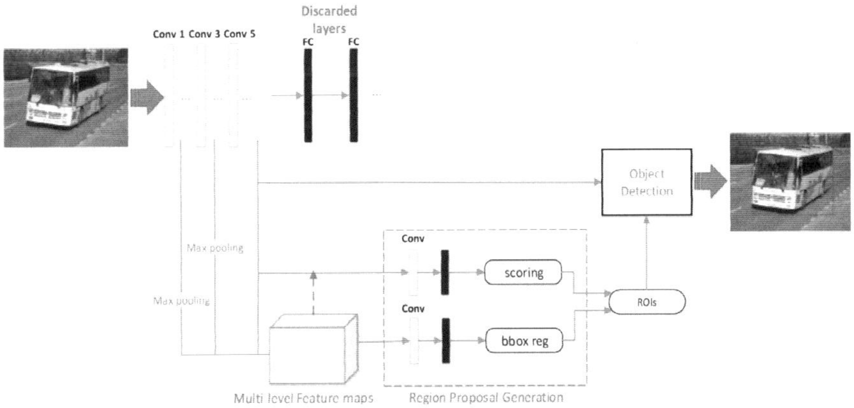

Figure 1. Our Detection Architecture.

3. Multi-scale and Multi-Tasking Implementation for Object Detection

3.1. Multi-Tasking

The proposed multi-tasking approach in our work is consisted by two main parts: (1) multi scale features from different layers are merged together as input of the regression tasks. (2) The features from the deep layer are adopted to represent the high-level semantic for classification.

Different from the region proposal network based on the single scale feature, the region proposal network in our work is build based on the features from different layers. However, keep the same dimension of the features is important for object detection since the classification and regression task of object detection are performed on a same object instance. This means that we need to keep the feature map of positioning and the corresponding feature map for classification in same dimension. It indicates that a candidate box only correspond to one category.

As shown in Figure.1, the proposed method is implemented with two 3×3 convolution layers, followed by two sibling 1×1 convolution layers (*reg* and *cls*, respectively). It consists by two parallel steps running through the region proposal network. However, it is not appropriate to use the same multi-scale features in regression and classification tasks. The first reason is that for a candidate region, the multi-scale features generated from different layers may contain noise, which will affect the classification performance. The second one is that adding more location information to category detection also affects the classification performance. Thus, we can add the semantic information on the original feature maps to improve localization performance.

3.2. Multi-scale Features
There are several ways to combine the multi-scale maps together with subsampling and pooling operations in CNN. In [18], a multi-scale scheme called Hyper Feature is proposed (HyperNet). For different scales of different layers, a max pooling layer is added on the lower level for subsampling. For higher layers, a deconvolutional operation is added to carry out upsampling. Then, the Hyper Feature is generated by performing local response normalization (LRN) [20] on multiple feature maps and concatenates them to one single output cube. By this method, the author claims that more semantic information is achieved. In [11], the proposed FPN uses the idea of a feature pyramid network, which is modified the original network directly. Specifically, each feature map is merged with the adjacent higher-resolution feature maps (which undergo a 1×1 convolutional layer to reduce channel dimensions) by element-wise addition. By this way, the feature map (for prediction) of each layer can combine the resolution information of different layers and the semantic information of different features map. Then, the combined feature maps of different resolutions are used to detect objects with corresponding resolutions.

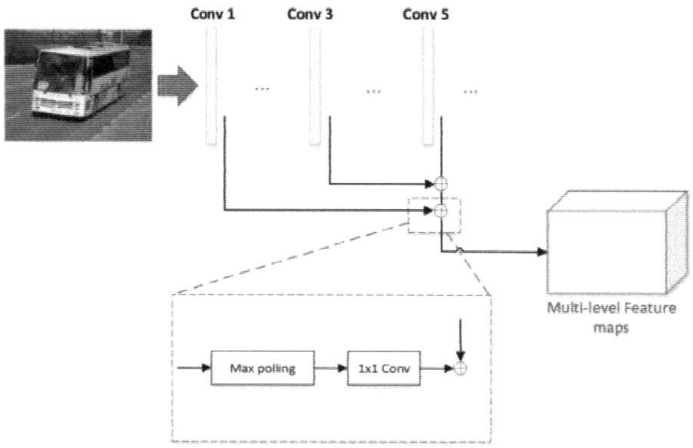

Figure 2. Multi-Scale Feature Map in Our Method.

As shown in Figure.2, the combination method used in this paper is different from the above two. In our work, we add max pooling layer to the low-level features for subsampling, and change the channel size to 512 by a 1x1 convolution operation, then we perform element-wise addition for the high-level features. In terms of connection, contrast to [18], this uses a direct connection method to handle the features of different layers. We use element-wise addition to features. The proposed approach requires fewer parameters. As shown in Table 1, assuming the size of input image is 1000×600, the channel of multi- scale features in our work will be 512, while the channel of HyperNet will be 768.

Table 1. HyperNet *vs.* Our multi-scale features (Assume that the size of the input image is 1000 600)

feature map	Scale	Channel
HyperNet [18]	215×150	768
Ours	62×32	512

3.3. End-to-end Training
As known, YOLO is a typical end-to-end training method. In the traditional method, CNN is only used to classify. The end-to-end method can achieve the object location and identification at the same time via the CNN. In our work, we use the end-to-end training method, the input is the original image and the output is the object instances. Compared with HyperNet, our training process does not require

separate training of object proposal and detection networks, and many fine-tuning steps for region proposals. Hence, in our work, we only need to concern the input and output.
We minimize the multi-tasking loss function:

$$L(k,k^*,t,t^*) = L_{reg}(k,k^*) + L_{reg}(t,t^*) \quad (1)$$

The Eq.1 is consists by L_{cls} and L_{reg}. Where L_{cls} denotes the classification loss, which is log loss over two classes (object and non object). k^* and k are the true and predicted label respectively. The second task L_{reg} is the loss that learns the weight of the network by reducing the difference between the ground truth box and the predicted anchor box. The t is ground truth box and vector t^* is the predicted anchor box. In our work, all the parameters for L_{cls} are obtained by single scale features, while the parameters for L_{reg} are obtained by the multi-scale feature.
In addition, in the proposed object detection method, we have made some minor changes for training. Firstly, we use the *crop_and_resize* [23] operator to crop and resize feature maps, which can make the feature information more complete. Secondly, we cancel the step of deleting small proposal (at the original height or width less than 16 pixels), since it would reduce the detection performance, especially for small objects. Thirdly, we exclude ground-truth bounding boxes in the RoIs for training, since they are not accessible for testing and they bias the input distribution for region classification.

4. Experiments

4.1. PASCAL VOC Results

For the performance evaluation, we experiment the proposed method on two challenging publicly available datasets: PASCAL VOC dataset and Coco dataset. The PASCAL VOC dataset is consisted by about 5k trainval images and 5k test images over 20 object categories. The proposed end-to-end training is performed on an NVIDIA Tesla K40 GPU and the public VGG-16 model. In training stage, a mini-batch, involves 1 image and 256 anchors per image, is displayed on the TensorBoard as shown in Figure 3. While the learning rate for training is starts with 0.001 and is reduced after 50k iterations. It can be seen that the loss is basically stable after 120k iterations. For evaluation, we use the standard evaluation metric mean Average Precision (mAP) for quantitatively evaluating the detection performance.

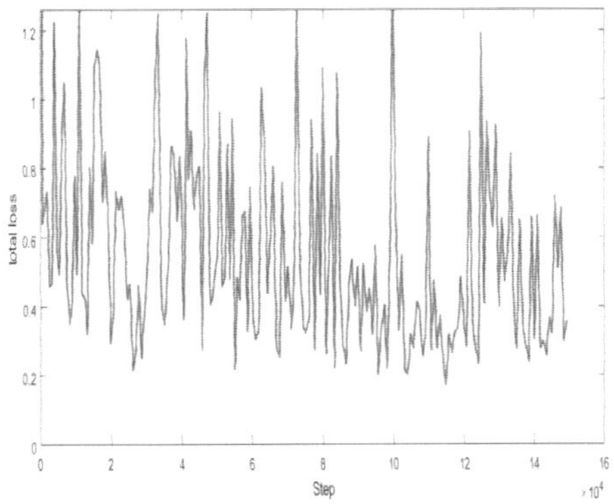

Figure 3. The results of our training step displayed on the TensorBoard (with IoU =0.7).

We compare the proposed approach with Fast R-CNN and Faster R-CNN for generic object detection on PASCAL VOC 2007. Table 2 shows the detection results with IoU=0.7. As shown in Table 2, the average precision of the proposed method is 70.9%, which is higher than the Faster R-CNN and the Fast R-CNN. The precision for Faster R-CNN and the Fast R-CNN are 69.9% and 66.9%, respectively.

IOP Conf. Series: Materials Science and Engineering **533** (2019) 012062 doi:10.1088/1757-899X/533/1/012062

Table 2. VOC 2007 test objection average precision (%). The pre-trained CNN model is VGG-16. Training data: VOC 2007 trainval.

Approach	mAP	aero	bike	bird	boat	bottle	bus	car	cat	chair	cow	table	dog	horse	mbike	person	plant	sheep	sofa	train	tv
Fast R-CNN[6]	66.9	74.5	78.3	69.2	53.2	36.6	77.3	78.2	82.0	40.7	72.7	67.9	79.6	79.2	73.0	69.0	30.1	65.4	70.2	75.8	65.8
Faster R-CNN[9]	69.9	70.0	80.6	70.1	57.3	49.9	78.2	80.4	82.0	52.2	75.3	67.2	80.3	79.8	75.0	76.3	39.1	68.3	67.3	81.1	67.6
Ours	70.9	69.8	78.6	69.9	57.6	57.2	77.2	81.2	82.4	52.3	75.4	67.3	80.2	82.2	75.9	78.0	43.2	71.9	66.0	76.1	74.7

Table 3. VOC 2007 test objection average precision (%). The pre-trained CNN model is VGG-16. Training data: union set of VOC 2007 trainval and VOC 2012 trainval.

Approach	mAP	aero	bike	bird	boat	bottle	bus	car	cat	chair	cow	table	dog	horse	mbike	person	plant	sheep	sofa	train	tv
Fast R-CNN[6]	70.0	77.0	78.1	69.3	59.4	38.3	81.6	78.6	86.7	42.8	78.8	68.9	84.7	82.0	76.6	69.9	31.8	70.1	74.8	80.4	70.4
Faster R-CNN[9]	73.2	76.5	79.0	70.9	65.5	52.1	83.1	84.7	86.4	52.0	81.9	65.7	84.8	84.6	77.5	76.7	38.8	73.6	73.9	83.0	72.6
ION[19]	75.6	79.2	83.1	77.6	65.6	54.9	85.4	85.1	87.0	54.4	80.6	73.8	85.3	82.2	82.2	74.4	47.1	75.8	72.7	84.2	80.4
HyperNet [18]	76.3	77.4	83.3	75.0	69.1	62.4	83.1	87.4	87.4	57.1	79.8	71.4	85.1	85.1	80.0	79.1	51.2	79.1	75.7	80.9	76.5
Ours	76.6	78.8	82.4	75.6	67.2	64.9	85.3	88.2	87.3	59.8	83.2	73.6	85.2	86.3	77.9	79.3	48.9	76.0	72.8	83.6	75.4

In addition, we train the proposed approach on the union set of PASCAL VOC 2007 trainval and 2012 trainval. The experimental results for Faster R-CNN, HyperNet and the proposed method are shown in Table 3. We can see the mAP for Faster R-CNN and HyperNet are 73.2% and 76.3%, respectively. The mAP for our method is 76.6%, which is higher 3.4% than Faster R-CNN and is slightly higher than the HyperNet. For detail, the proposed method has higher precision than Faster R-CNN and HyperNet on small sized object detection, such as bottle, chair and cow. The precision of bottle detection for Faster R-CNN, HyperNet and our approach are 54.9%, 62.4% and 64.9%, respectively. The precision of the proposed method is higher 12.8% and 2.5% than Faster R-CNN and HyperNet, respectively. Therefore, above experimental results demonstrate that the multi-scale and multi-tasking object detection method proposed in our work has a good detection performance, proved the effectiveness of our detection method.

4.2. Analysis for Detection Performance with Different Layers

Due to the deep layer contains higher-level information than low layer, many methods adopt the last CNN layer for region proposal generation or detection [9, 6]. However, recent research [19, 18, 11] shows that combine the multi-scale information can achieve better performance than a single layer.

Table 4. Detection performance with different layer combination strategies. All networks are trained with the same configuration.

Layers	Detection mAP
3	68.4%
5	73.2%
3+5	76.6%
3+4+5	76.0%
1+3+5	75.6%

In the experiment, we trained several different models based on VGG-16. Then, we fuse different layers, such as layer 3 and 5, layer 1, 3 and 5. In this section, all models are trained with the same configuration. The experimental results are shown in Table 4. The combining layer 3 and 5 get the best detection mAP, which proves that the detection performance can be improved by fuse different layers. In addition, since the first layer and final layer are far each other, the information from those two layers is very different. Hence, the merged feature, achieved by element-wise operation the

IOP Conf. Series: Materials Science and Engineering **533** (2019) 012062 doi:10.1088/1757-899X/533/1/012062

features of layer 1 and layer 5, has lot of redundant information. Furthermore, as shown in Figure 4, we explore the impaction for mAP with different number of proposals for different layers. This result indicates that the multi-scale combination works better than single layer, both for proposal and detection.

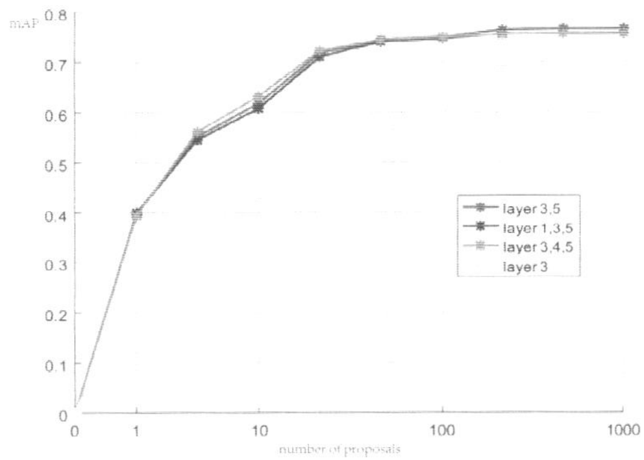

Figure 4. The mAP versus number of proposals for different layer combinations (with IoU =0.7).

4.3. *COCO Results*

In this section, we present more experimental results on the Microsoft COCO object detection dataset [24], which contains involves 80 object categories. We use the COCO trainval35k split (union of 80k images from *train* and a random 35k subset of images from the 40k image *val* split). Then, we evaluate the mAP averaged for IoU $\in$[0.5:0.05:0.95] (COCO's standard metric, simply denoted as mAP@[.5, .95]) and mAP@0.5 (PASCAL VOC's metric).

The model in our work is trained on a GPU, the learning rate starts with 0.001 and is reduced after 350k iterations. Training finishes at 900k iterations. We fix one image per batch, using *crop_and_resize* pooling, and keep the small region proposals, which gives better average recall than original method. The rest of the implement details are the same as on PASCAL VOC.

Table 5. Object detection results (%) on the MS COCO dataset. The model is VGG-16.

Method	data	Avg. Precision, IoU:			Avg. Precision, Area:			Avg. Recall, #Dets:			Avg. Recall, Area:		
		0.5:0.95	0.5	0.75	S	M	L	1	10	100	S	M	L
Faster R-CNN[9]	trainval	21.9	42.7	-	-	-	-	-	-	-	-	-	-
ION[19]	train	23.6	43.2	23.6	6.4	24.1	38.3	23.2	32.7	33.5	10.1	37.7	53.6
R-FCN[10]	trainval	29.9	51.9	-	10.8	32.8	45.0	-	-	-	-	-	-
SSD300*[16]	trainval35k	25.1	43.1	25.8	6.6	25.9	41.4	23.7	35.1	37.2	11.2	40.4	58.4
SSD512	trainval35k	26.8	46.5	27.8	9.0	28.9	41.9	24.8	37.5	39.8	14.0	43.5	59.0
YOLOv2[14]	trainval35k	21.6	44.0	19.2	5.0	22.4	35.5	20.7	31.6	33.3	9.8	36.5	54.4
Ours	trainval35k	25.3	44.5	25.9	10.5	28.8	37.0	24.3	35.2	35.6	16.9	40.0	53.4

We evaluate the proposed approach on the COCO dataset and compare test-dev results with the state-of-the-art methods, including both one-stage and two-stage models. The experiment results are shown in Table 5. Compared with the two-stage methods, our approach has higher precision performance and as shown in Table 5, our multi-scale and multi-tasking method achieves better performance than one-stage methods, especially in the detection of small-size objects. For size S objects, our approach achieves a 2.9 point AP (Average Precision) gap (10.5 *vs.* 9.0) and 2.9 point AR (Average Recall) gap (16.9 *vs.* 14.0) with the SSD512 [16]. The good performance of the proposed method is achieved owning to the proposed method has a good multi-scale feature representation, and the multi-tasking implementation can gives consistent boost on small-size objects. Figure 5 shows some results of the proposed method and the original Faster R-CNN on the PASCAL VOC 2007 dataset.

IOP Conf. Series: Materials Science and Engineering **533** (2019) 012062 doi:10.1088/1757-899X/533/1/012062

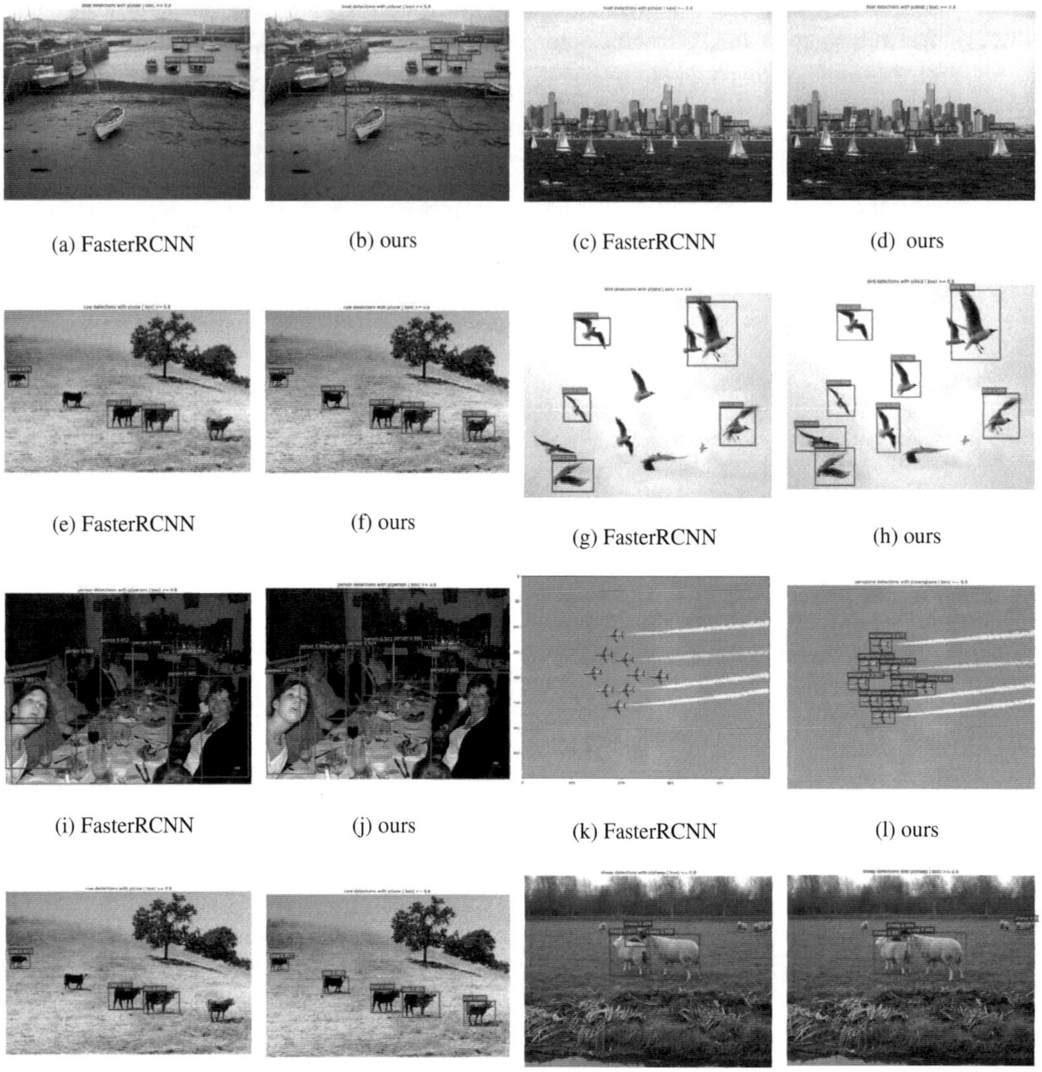

(a) FasterRCNN (b) ours (c) FasterRCNN (d) ours

(e) FasterRCNN (f) ours (g) FasterRCNN (h) ours

(i) FasterRCNN (j) ours (k) FasterRCNN (l) ours

(m) FasterRCNN (n) ours (o) FasterRCNN (p) ours

Figure 5. Some results compare the original Faster R-CNN

5. Conclusions

In this paper, we have proposed a clean and simple framework for object detection, which is which is use the information from different convolutional layers. Compared with the original Faster R-CNN and the state-art-of-the method, our method shows obvious improvement on the object detection, especially in small-size object detection. In our future work, we will consider to accelerate the training speed of the proposed method and apply it to video object detection.

Acknowledgments

This work was partially supported by the National Key Research and Development Program of China under Grant 2017YFB1402102, the National Natural Science Foundation of China under Grant 11872036, Grant 11502133, Grant 11772178, Grant 11372167, the 111 Project under Grant B18032, and the Fundamental Research Funds for the Central Universities under Grant GK201703082, Grant GK201801004.

References

[1] Zitnick, C. Lawrence, and P. Dollár. *Edge Boxes: Locating Object Proposals from Edges. Computer Vision – ECCV 2014*. Springer International Publishing, (2014):391-405.

[2] Cheng, Ming Ming , et al. "BING: Binarized Normed Gradients for Objectness Estimation at 300fps." *Computer Vision & Pattern Recognition* IEEE,(2014).

[3] Alexe, B, T. Deselaers, and V. Ferrari. "What is an object?." *Computer Vision and Pattern Recognition* IEEE, (2010):73-80.

[4] Lee, Tom , S. Fidler , and S. Dickinson . "Learning to Combine Mid-Level Cues for Object Proposal Generation." *IEEE International Conference on Computer Vision* IEEE, (2016).

[5] Zhang, Yuting, et al. "Improving object detection with deep convolutional networks via Bayesian optimization and structured prediction." *IEEE Conference on Computer Vision and Pattern Recognition* IEEE Computer Society, (2015):249-258.

[6] Girshick, Ross. "Fast R-CNN." *Computer Science* (2015).

[7] Krizhevsky, Alex, I. Sutskever, and G. E. Hinton. "ImageNet classification with deep convolutional neural networks." *International Conference on Neural Information Processing Systems* Curran Associates Inc. (2012):1097-1105.

[8] Girshick, Ross, et al. "Rich Feature Hierarchies for Accurate Object Detection and Semantic Segmentation." *IEEE Conference on Computer Vision and Pattern Recognition* IEEE Computer Society, (2014):580-587.

[9] Ren, S., et al. "Faster R-CNN: Towards Real-Time Object Detection with Region Proposal Networks." *IEEE Transactions on Pattern Analysis & Machine Intelligence* **39.6**(2017):1137-1149.

[10] Dai, Jifeng, et al. "R-FCN: Object Detection via Region-based Fully Convolutional Networks." (2016).

[11] Lin, Tsung Yi, et al. "Feature Pyramid Networks for Object Detection." *IEEE Conference on Computer Vision and Pattern Recognition* IEEE, (2017):936-944.

[12] He, Kaiming, et al. "Mask R-CNN." *IEEE International Conference on Computer Vision* IEEE Computer Society, (2017):2980-2988.

[13] Redmon, Joseph, et al. "You Only Look Once: Unified, Real-Time Object Detection." *Computer Vision and Pattern Recognition* IEEE, (2016):779-788.

[14] Redmon, Joseph, and A. Farhadi. "YOLO9000: Better, Faster, Stronger." *IEEE Conference on Computer Vision and Pattern Recognition* IEEE, (2017):6517-6525.

[15] Redmon, Joseph , and A. Farhadi . "YOLOv3: An Incremental Improvement." (2018).

[16] Liu, Wei, et al. *SSD: Single Shot MultiBox Detector. Computer Vision – ECCV 2016*. Springer International Publishing, (2016):21-37.

[17] Van, de Sande, Koen E. A, et al. "Segmentation as selective search for object recognition. " **104.2**(2011):1879-1886.

[18] Kong, Tao, et al. "HyperNet: Towards Accurate Region Proposal Generation and Joint Object Detection." *Computer Vision and Pattern Recognition* IEEE, (2016):845-853.

[19] Bell, Sean, et al. "Inside-Outside Net: Detecting Objects in Context with Skip Pooling and Recurrent Neural Networks." *IEEE Conference on Computer Vision and Pattern Recognition* IEEE, (2016):2874-2883.

[20] Everingham, Mark, et al. "The Pascal, Visual Object Classes Challenge: A Retrospective." *International Journal of Computer Vision* **111.1**(2015):98-136.

[21] Russakovsky, Olga, et al. "ImageNet Large Scale Visual Recognition Challenge." *International Journal of Computer Vision* **115.3**(2015):211-252.

[22] Yang H, Qu S, Chen C, et al. Multiple objects tracking with improved sparse representation and rank based dynamic estimation[J]. IEEE Access, (2018), PP(6):42264 – 42278.

[23] Chen, Xinlei, and A. Gupta. "An Implementation of Faster RCNN with Study for Region Sampling." (2017).

[24] Lin, Tsung Yi, et al. "Microsoft COCO: Common Objects in Context." **8693**(2014):740-755.

Institute of Physics
Dirac House, Temple Back
Bristol BS1 6BE UK

ISSN: 1757-8981
ISBN 978-1-5108-8737-4

CURRAN ASSOCIATES INC.
proceedings.com